The Future of Post-Human Probability

The Future of Post-Human Probability

Towards a New Theory of Objectivity and Subjectivity

By

Peter Baofu

Nova Science Publishers, Inc.
New York

The Future of Post-Human Probability: Towards a New Theory of Objectivity and Subjectivity, by Peter Baofu

For permission to use material from this book please contact us:
Telephone 631-231-7269; Fax 631-231-8175
Web Site: http://www.novapublishers.com

Library of Congress Cataloging-in-Publication Data

ISBN 978-1-62948-671-0

Published by Nova Science Publishers, Inc. † New York

To Those in the Future World beyond Objectivity and Subjectivity

Books also by Peter Baofu

Category I: Social Sciences and Related Fields

- 59. *The Future of Post-Human Waste* (2012)
- 58. *The Future of Post-Human Migration* (2012)
- 57. *The Future of Post-Human Criminality* (2012)
- 56. *The Future of Post-Human Geography* (2011)
- 55. *The Future of Post-Human Gambling* (2011)
- 54. *The Future of Post-Human Education* (2011)
- 53. *The Future of Post-Human War and Peace* (2010)
- 52. *The Future of Post-Human Law* (2010)
- 51. *The Future of Post-Human Organization* (2009)
- 50. *The Future of Post-Human Mass Media* (2009)
- 49. *The Rise of Authoritarian Liberal Democracy* (2007)
- 48. *Beyond the World of Titans, and the Remaking of World Order* (2007)
- 47. *Beyond Capitalism to Post-Capitalism* (2005)
- 46. Volume 1: *Beyond Democracy to Post-Democracy* (2004)
- 45. Volume 2: *Beyond Democracy to Post-Democracy* (2004)
- 44. *The Future of Capitalism and Democracy* (2002)

Category II: Natural Sciences and Related Fields

- 43. *The Future of Post-Human Health Care* (2013)
- 42. *Beyond Natural Resources to Post-Human Resources* (2013)
- 41. *The Future of Post-Human Transportation* (2012)
- 40. *The Future of Post-Human Architecture* (2012)
- 39. *The Future of Post-Human Chemistry* (2011)
- 38. *The Future of Post-Human Acoustics* (2011)
- 37. *The Future of Post-Human Geology* (2010)
- 36. *Beyond Cosmology to Post-Cosmology* (2010)
- 35. *The Future of Post-Human Personality* (2010)
- 34. *The Future of Post-Human Sexuality* (2010)
- 33. *The Future of Post-Human Urban Planning* (2009)
- 32. *The Future of Post-Human Engineering* (2009)
- 31. *The Future of Post-Human Unconsciousness* (2008)
- 30. *The Future of Complexity* (2007)
- 29. *Beyond Nature and Nurture* (2006)
- 28. *The Future of Post-Human Space-Time* (2006)
- 27. *The Future of Post-Human Consciousness* (2004)

Category III: The Humanities and Related Fields

- 26. *The Future of Post-Human Thanatology* (2013)
- 25. *The Future of Post-Human Sports* (2013)
- 24. *The Future of Post-Human Culinary Art* (2012)
- 23. *The Future of Post-Human History* (2012)
- 22. *The Future of Post-Human Performing Arts* (2012)
- 21. *The Future of Post-Human Literature* (2011)
- 20. *The Future of Post-Human Humor* (2011)
- 19. *Beyond Ethics to Post-Ethics* (2011)
- 18. *The Future of Post-Human Religion* (2010)
- 17. *The Future of Post-Human Chess* (2010)
- 16. *The Future of Post-Human Martial Arts* (2009)
- 15. *The Future of Post-Human Creative Thinking* (2009)
- 14. *The Future of Post-Human Knowledge* (2008)
- 13. *The Future of Aesthetic Experience* (2007)
- 12. *Beyond Civilization to Post-Civilization* (2006)
- 11. Volume 1: *The Future of Human Civilization* (2000)
- 10. Volume 2: *The Future of Human Civilization* (2000)

Category IV: Formal Sciences and Related Fields

- 09. *The Future of Post-Human Morphology* (2013)
- 08. *The Future of Post-Human Semantics* (2012)
- 07. *The Future of Post-Human Data Analysis* (2011)
- 06. *The Future of Post-Human Computing* (2011)
- 05. *The Future of Post-Human Formal Science* (2010)
- 04. *The Future of Post-Human Language* (2009)
- 03. *The Future of Post-Human Geometry* (2009)
- 02. *The Future of Post-Human Mathematical Logic* (2008)
- 01. *The Future of Information Architecture* (2008)

CONTENTS

Part One: Introduction

Part Two: Objectivity

Part Three: Subjectivity

Part Four: Conclusion

TABLES

Category III: Visions on the Mind

Category IV: Visions on History

ACKNOWLEDGMENTS

This book, like all other previous books of mine, is written to challenge conventional wisdom or, more precisely, in the current context, the one on probability—and to replace it with an original way of thinking about its future fate (in relation to objectivity and subjectivity).

Thus, this book receives no external funding nor help from any formal organization or institution, because of its political incorrectness—as this is something that I often stressed (and repeated) in all my previous books.

My only reward is the joy of conceiving something original in the intellectual history of probability.

In any event, I bear the sole responsibility for the ideas presented in this book.

ABBREVIATIONS

ALD = Peter Baofu. 2007. *The Rise of Authoritarian Liberal Democracy: A Preface to a New Theory of Comparative Political Systems*. Cambridge, England: Cambridge Scholars Publishing, Ltd.

BCIV = Peter Baofu. 2006. *Beyond Civilization to Post-Civilization: Conceiving a Better Model of Life Settlement to Supersede Civilization*. NY: Peter Lang Publishing, Inc.

BCPC = Peter Baofu. 2005. *Beyond Capitalism to Post-Capitalism: Conceiving a Better Model of Wealth Acquisition to Supersede Capitalism*. NY: The Edwin Mellen Press.

BCOS = Peter Baofu. 2010. *Beyond Cosmology to Post-Cosmology: A Preface to a New Theory of Different Worlds*. Cambridge, England: Cambridge International Science Publishing, Ltd.

BDPD1 = Peter Baofu. 2004. Volume 1. *Beyond Democracy to Post-Democracy: Conceiving a Better Model of Governance to Supersede Democracy*. NY: The Edwin Mellen Press.

BDPD2 = Peter Baofu. 2004. Volume 2. *Beyond Democracy to Post-Democracy: Conceiving a Better Model of Governance to Supersede Democracy*. NY: The Edwin Mellen Press.

BEPE = Peter Baofu. 2011. *Beyond Ethics to Post-Ethics: A Preface to a New Theory of Morality and Immorality.* Charlotte, NC: Infomration Age Publishing.

BNN = Peter Baofu. 2006. *Beyond Nature and Nurture: Conceivng a Better Way to Understand Genes and Memes.* Cambridge, England: Cambridge Scholars Publishing, Ltd.

BNR = Peter Baofu. 2013. *Beyond Natural Resources to Post-Human Resources: Towards a New Theory of Diversity and Discontinuity.* Cambridge, England: Cambridge Scholars Publishing, Ltd.

BWT = Peter Baofu. 2007. *Beyond the World of Titans, and the Renaking of World Order: A Preface to a New Logic of Empire-Building.* Cambridge, England: Cambridge Scholars Publishing, Ltd.

FAE = Peter Baofu. 2007. *The Future of Aesthetic Experience: Conceiving a Better Way to Understand Beauty, Ugliness and the Rest.* Cambridge, England: Cambridge Scholars Publishing, Ltd.

FC = Peter Baofu. 2007. *The Future of Complexity: Conceiving a Better Way to Understand Order and Chaos.* London, United Kingdom: World Scientific Publishing Co.

FCD = Peter Baofu. 2002. *The Future of Capitalism and Democracy.* MD: The University Press of America.

FHC1 = Peter Baofu. 2000. Volume 1. *The Future of Human Civilization.* NY: The Edwin Mellen Press.

FHC2 = Peter Baofu. 2000. Volume 2. *The Future of Human Civilization.* NY: The Edwin Mellen Press.

FIA = Peter Baofu. 2008. *The Future of Information Architecture: Conceiving a Better Way to Understand Taxonomy, Network, and Intelligence.* Oxford, England: Chandos Publishing (Oxford) Limited.

FPHA = Peter Baofu. 2011. *The Future of Post-Human Acoustics: A Preface to a New Theory of Sound and Silence.* Cambridge, England: Cambridge International Science Publishing, Ltd.

FPHARCH = Peter Baofu. 2012. *The Future of Post-Human Architecture: A Preface to a New Theory of Form and Function.* Cambridge, England: Cambridge International Science Publishing, Ltd.

FPHC = Peter Baofu. 2004. *The Future of Post-Human Consciousness.* NY: The Edwin Mellen Press.

FPHCA = Peter Baofu. 2012. *The Future of Post-Human Culinary Art: Towards a New Theory of Ingredients and Techniques.* England: Cambridge Scholars Publishing, Ltd.

FPHCHEM = Peter Baofu. 2011. *The Future of Post-Human Chemistry: A Preface to a New Theory of Substances and their Changes.* England: Cambridge Scholars Publishing, Ltd.

FPHCHESS = Peter Baofu. 2010. *The Future of Post-Human Chess: A Preface to a New Theory of Tactics and Strategy.* Cambridge, England: Cambridge International Science Publishing, Ltd.

FPHCOM = Peter Baofu. 2011. *The Future of Post-Human Computing: A Preface to a New Theory of Hardware and Software.* Cambridge, England: Cambridge International Science Publishing, Ltd.

FPHCRIM = Peter Baofu. 2012. *The Future of Post-Human Criminality: A Preface to a New Theory of Heroes and Villains.* Cambridge, England: Cambridge International Science Publishing, Ltd.

FPHCT = Peter Baofu. 2009. *The Future of Post-Human Creative Thinking: A Preface to a New Theory of Invention and Innovation.* Cambridge, England: Cambridge Scholars Publishing, Ltd.

FPHDA = Peter Baofu. 2011. *The Future of Post-Human Data Analysis: A Preface to a New Theory of Quantitative and Qualitative Research.* Cambridge, England: Cambridge International Science Publishing, Ltd.

FPHE = Peter Baofu. 2009. *The Future of Post-Human Engineering: A Preface to a New Theory of Technology.* Cambridge, England: Cambridge Scholars Publishing, Ltd.

FPHEDU = Peter Baofu. 2011. *The Future of Post-Human Education: A Preface to a New Theory of Teaching and Learning.* Cambridge, England: Cambridge International Science Publishing, Ltd.

FPHFS = Peter Baofu. 2010. *The Future of Post-Human Formal Science: A Preface to a New Theory of Abstraction and Application.* Cambridge, England: Cambridge Scholars Publishing, Ltd.

FPHG = Peter Baofu. 2009. *The Future of Post-Human Geometry: A Preface to a New Theory of Infinity, Symmetry, and Dimensionality.* Cambridge, England: Cambridge Scholars Publishing, Ltd.

FPHGAM = Peter Baofu. 2011. *The Future of Post-Human Gambling: A Preface to a New Theory of Risk and Caution.* Cambridge, England: Cambridge International Science Publishing, Ltd.

FPHGEOG = Peter Baofu. 2011. *The Future of Post-Human Geography: A Preface to a New Theory of Environments and their Interactions.* Cambridge, England: Cambridge International Science Publishing, Ltd.

FPHGEOL = Peter Baofu. 2010. *The Future of Post-Human Geology: A Preface to a New Theory of Statics and Dynamics.* Cambridge, England: Cambridge International Science Publishing, Ltd.

FPHH = Peter Baofu. 2011. *The Future of Post-Human Humor: A Preface to a New Theory of Joking and Laughing.* Cambridge, England: Cambridge International Science Publishing, Ltd.

FPHHC = Peter Baofu. 2013. *The Future of Post-Human Health Care: Towards a New Theory of Mind and Body.* Hauppauge, NY: Nova Science Publishers, Inc.

FPHHIST = Peter Baofu. 2012. *The Future of Post-Human History: A Preface to a New Theory of Universality and Relativity.* Cambridge, England: Cambridge Scholars Publishing, Ltd.

FPHK = Peter Baofu. 2008. *The Future of Post-Human Knowledge: A*

Preface to a New Theory of Methodology and Ontology. Oxford, England: Chandos Publishing (Oxford) Limited.

FPHL = Peter Baofu. 2009. *The Future of Post-Human Language: A Preface to a New Theory of Structure, Context, and Learning*. Cambridge, England: Cambridge Scholars Publishing, Ltd.

FPHLAW = Peter Baofu. 2010. *The Future of Post-Human Law: A Preface to a New Theory of Necessity, Contingency, and Justice*. Cambridge, England: Cambridge Scholars Publishing, Ltd.

FPHLIT = Peter Baofu. 2011. *The Future of Post-Human Literature: A Preface to a New Theory of Fiction and Non-Fiction*. Cambridge, England: Cambridge International Science Publishing, Ltd.

FPHMA = Peter Baofu. 2009. *The Future of Post-Human Martial Arts: A Preface to a New Theory of the Body and Spirit of Warriors*. Cambridge, England: Cambridge Scholars Publishing, Ltd.

FPHMIG = Peter Baofu. 2012. *The Future of Post-Human Migration: A Preface to a New Theory of Sameness, Otherness, and Identity*. Cambridge, England: Cambridge Scholars Publishing, Ltd.

FPHML = Peter Baofu. 2008. *The Future of Post-Human Mathematical Logic: A Preface to a New Theory of Rationality*. Cambridge, England: Cambridge Scholars Publishing, Ltd.

FPHMM = Peter Baofu. 2009. *The Future of Post-Human Mass Media: A Preface to a New Theory of Technology*. Cambridge, England: Cambridge Scholars Publishing, Ltd.

FPHMORP = Peter Baofu. 2013. *The Future of Post-Human Morphology: Towards a New Theory of Typologies and Rules*. Hauppauge, NY: Nova Science Publishers, Inc.

FPHO = Peter Baofu. 2009. *The Future of Post-Human Organization: A Preface to a New Theory of Communication, Decision-Making, and Leadership.*.Cambridge, England: Cambridge Scholars Publishing, Ltd.

FPHP = Peter Baofu. 2010. *The Future of Post-Human Personality: A Preface to a New Theory of Normality and Abnormality*. Cambridge, England: Cambridge International Science Publishing, Ltd.

FPHPHON = Peter Baofu. 2013. *The Future of Post-Human Phonology: Towards a New Theory of Generality and Specifcity*.

FPHPA = Peter Baofu. 2012. *The Future of Post-Human Performing Arts: A Preface to a New Theory of the Body and its Presence*. Cambridge, England: Cambridge Scholars Publishing, Ltd.

FPHPROB = Peter Baofu. 2013. *The Future of Post-Human Probability: Towards a New Theory of Objectivity and Subjectivity.*Hauppauge, NY: Nova Science Publishers, Inc.

FPHR = Peter Baofu. 2010. *The Future of Post-Human Religion: A Preface to a New Theory of Spirituality.* Cambridge, England: Cambridge International Science Publishing, Ltd.

FPHS = Peter Baofu. 2010. *The Future of Post-Human Sexuality: A Preface to a New Theory of the Body and Spirit of Love-Makers.* Cambridge, England: Cambridge Scholars Publishing, Ltd.

FPHSEM = Peter Baofu. 2012. *The Future of Post-Human Semantics: A Preface to a New Theory of Internality and Externality.* Cambridge, England: Cambridge Scholars Publishing, Ltd.

FPHSPORT = Peter Baofu. 2013. *The Future of Post-Human Sports: Towards a New Theory of Training and Winning.* Cambridge, England: Cambridge Scholars Publishing, Ltd.

FPHST = Peter Baofu. 2006. *The Future of Post-Human Space-Time: Conceivng a Better Way to Understand Space and Time.* New York: Peter Lang Publishing, Inc.

FPHTHAN = Peter Baofu. 2013. *The Future of Post-Human Thanatology: Towards a New Theory of Death and After-Death.* Berlin, Germany: Logos Verlag Berlin.

FPHTRAN = Peter Baofu. 2012. *The Future of Post-Human Transportation: A Preface to a New Theory of Networks and Operations.* Cambridge, England: Cambridge Scholars Publishing, Ltd.

FPHU = Peter Baofu. 2008. *The Future of Post-Human Unconsciousness: A Preface to a New Theory of Anomalous Experience.* Cambridge, England: Cambridge Scholars Publishing, Ltd.

FPHUP = Peter Baofu. 2009. *The Future of Post-Human Urban Planning: A Preface to a New Theory of Density, Void, and Sustainability.* Cambridge, England: Cambridge Scholars Publishing, Ltd.

FPHW = Peter Baofu. 2012. *The Future of Post-Human Waste: Towards a New Theory of Uselessness and Usefulness.* Cambridge, England: Cambridge Scholars Publishing, Ltd.

FPHWP = Peter Baofu. 2010. *The Future of Post-Human War and Peace: A Preface to a New Theory of Aggression and Pacificity.* Cambridge, England: Cambridge Scholars Publishing, Ltd.

• Part One •

Introduction

CHAPTER 1
INTRODUCTION—THE USE OF PROBABILITY

[Probability is] the logic of partial belief.
—Frank Ramsey (WK 2012)

A Questioning of Probability

Is the degree of probability that an individual holds when betting on a particular outcome really so subjective that, as Frank Ramsey once argued, "objective logical relations" do not exist and that "probability is 'the logic of partial belief'"? (WK 2012a)

This subjective interpretation of probability can be contrasted with an objective view by John Keynes, who argued instead that "logical probabilities are conceived…to be objective, logical relations between propositions (or sentences), and hence not to depend in any way upon belief." (WK 2012)

Contrary to these opposing interpretations (and other ones as will be discussed in the book), probability (in relation to both objectivity and subjectivity) are neither possible (or impossible) nor desirable (or undesirable) to the extent that the respective ideologues (on different sides) would like us to believe.

Surely, this questioning of the opposing interpretations on probability does not entail that probability is useless, or that those fields (related to probability) like "statistics, finance, gambling, science, artificial intelligence/machine learning and philosophy" are not worth studying. Needless to say, neither of these extreme views is reasonable.

Instead, this book offers an alternative (better) way to understand the future of probability, especially in the dialectic context of objectivity and subjectivity—while learning from different approaches in the literature but without favoring any one of them (nor integrating them, since they are not necessarily compatible with each other).

More specifically, this book offers a new theory (that is, *the interpretivist theory of probability*) to go beyond the existing approaches in a novel way.

If successful, this seminal project is to fundamentally change the way that we think about probability in relation to objectivity and subjectivity from the combined perspectives of the mind, nature, society, and culture, with enormous implications for the human future and what I originally called its "post-human" fate.

Definition and History of Probability

A good way to start this inquiry is to provide a working definition of the term "probability" and its brief history.

Definition

The term "probability" derives from "the Latin *probabilitas*, which can also mean *probity*, a measure of the authority of a witness in a legal case in Europe, and often correlated with the witness's nobility. In a sense, this differs much from the modern meaning of probability, which, in contrast, is a measure of the weight of empirical evidence, and is arrived at from inductive reasoning and statistical inference." (WK 2012b; I. Hacking 2006)

This modern meaning of "probability" refers to "a measure of the expectation that an event will occur or a statement is true. Probabilities are given a value between 0 (will not occur) and 1 (will occur). The higher the probability of an event, the more certain we are that the event will occur." (WK 2012b; W. Feller 1968)

In probability theory, there are three central objects surrounding the concept of probability: (WK 2012c)

- "Event"—is "a set of outcomes (a subset of the sample space) to which a probability is assigned." (WK 2012g; A. Leon-Garcia 2008)
- "Random Variable"—is "a variable whose value is subject to variations due to chance (i.e. randomness, in a mathematical sense). As opposed to other mathematical variables, a random variable conceptually does not have a single, fixed value (even if unknown); rather, it can take on a set of possible different values, each with an associated probability." (WK 2012d) Thus, a variable is different from a "constant" (which remains unchanged).

- "Stochastic (or Random) Process"—refers to "the evolution of some random value, or system, over time. This is the probabilistic counterpart to a deterministic process (or deterministic system)." (WK 2012e) In "science, mathematics and statistics," randomness in the stochastic process "recognizes a lack of predictability when referring to randomness, but admits regularities in the occurrences of events" whose evolution over time can be calculated. (WK 2012f)

For illustration, if an individual rolls a dice, its outcome is not predictable, but, "if repeated many times the sequence of random events will exhibit certain patterns, which can be studied and predicted"—like its "probability distribution," which "assigns a probability to each of the possible outcomes" over time. (WK 2012c & 2012h)

History

With this definition in mind—historically, "the scientific study of probability is a modern development. Gambling shows that there has been an interest in quantifying the ideas of probability for millennia, but exact mathematical descriptions arose much later. There are reasons of course, for the slow development of the mathematics of probability. Whereas games of chance provided the impetus for the mathematical study of probability, fundamental issues are still obscured by the superstitions of gamblers." (WK 2012b; J. Freund 1993)

Later, "aside from elementary work by Girolamo Cardano in the 16th century, the doctrine of probabilities dates to the correspondence of Pierre de Fermat and Blaise Pascal (1654). Christiaan Huygens (1657) gave the earliest known scientific treatment of the subject. Jakob Bernoulli's *Ars Conjectandi* (posthumous, 1713) and Abraham de Moivre's *Doctrine of Chances* (1718) treated the subject as a branch of mathematics." (WK 2012b; W. Abrams 2003; V. Ivancevic 2008)

Then, "Adrien-Marie Legendre (1805) developed the method of least squares, and introduced it in his *Nouvelles méthodes pour la détermination des orbites des comètes* (*New Methods for Determining the Orbits of Comets*)." (WK 2012b)

In the nineteenth century, "authors on the general theory included Pierre-Simon Laplace, Sylvestre Lacroix (1816), Littrow (1833), Adolphe Quetelet (1853), Richard Dedekind (1860), Friedrich Robert Helmert (1872), Hermann Laurent (1873), Liagre, Didion, and Karl Pearson. Augustus De Morgan and George Boole improved the exposition of the theory." (WK 2012b)

And "Andrey Markov introduced the notion of Markov chains (1906), which played an important role in stochastic processes theory and its applications. The modern theory of probability based on the measure theory was developed by Andrey Kolmogorov (1931)," among others.

Objectivity and Subjectivity in Probability

With this definition and brief history of probability in mind—the study of probability seems straightforward enough "in a purely theoretical setting (like tossing a fair coin)," where "probabilities describe the statistical number of outcomes considered divided by the number of all outcomes (tossing a fair coin twice will yield HH with probability 1/4, because the four outcomes HH, HT, TH and TT are possible)." (WK 2012b)

However, problems occur "when it comes to practical application," because "the word probability does not have a singular direct definition. In fact, there are two major categories of probability interpretations, whose adherents possess conflicting views about the fundamental nature of probability" in practice, as shown below: (WK 2012b)

- The Objectivist Interpretation—refers to the interpretation, in which "objectivists assign numbers to describe some objective or physical state of affairs. The most popular version of objective probability is frequentist probability, which claims that the probability of a random event denotes the relative frequency of occurrence of an experiment's outcome, when repeating the experiment. This interpretation considers probability to be the relative frequency 'in the long run' of outcomes. A modification of this is propensity probability, which interprets probability as the tendency of some experiment to yield a certain outcome, even if it is performed only once." (WK 2012b; I. Hacking 1965; B. Finetti 1970)
- The Subjectivist Interpretation—refers to the interpretation, in which "subjectivists assign numbers per subjective probability, i.e., as a degree of belief. The most popular version of subjective probability is Bayesian probability, which includes expert knowledge as well as experimental data to produce probabilities. The expert knowledge is represented by some (subjective) prior probability distribution. The data is incorporated in a likelihood function. The product of the prior and the likelihood, normalized, results in a posterior probability distribution that incorporates all the information known to date. Starting from arbitrary, subjective probabilities for a group of agents,

> some Bayesians claim that all agents will eventually have sufficiently similar assessments of probabilities, given enough evidence." (WK 2012b; B. Finetti 1970; R. Hogg 2004)

This debate in the literature then leads this book to focus on the dialectic relationship between *objectivity* and *subjectivity*, as explained below (and summarized in *Table 1.1*).

Objectivity

In accordance to the Merriam-Webster dictionary, the word "objectivity" refers to "an object, phenomenon, or condition in the realm of sensible experience independent of individual thought and perceptible by all observers: having reality independent of the mind." (MWD 2012)

But the concept of "objectivity" can be understood in different ways, and, for illustration, consider three main examples below, namely, (a) measurement, (b) mind independence, and (c) logical relation.

Measurement

The first way to understand objectivity has to do with the issue of "measurement" (in relation to "testability" and "reproducibility" in science). (WK 2012i)

In science, "objectivity...is a value that informs how science is practiced and how scientific truths are created. It is the idea that scientists, in attempting to uncover truths about the natural world, must aspire to eliminate personal biases, a priori commitments, emotional involvement, etc. Objectivity is often attributed to the property of scientific measurement, as the accuracy of a measurement can be tested independent from the individual scientist who first reports it. It is thus intimately related to the aim of testability and reproducibility. To be properly considered objective, the results of measurement must be communicated from person to person, and then demonstrated for third parties, as an advance in understanding of the objective world. Such demonstrable knowledge would ordinarily confer demonstrable powers of prediction or technological construction." (WK 2012i; L. Daston 2007)

Mind Independence

The second way to understand objectivity has to do with the issue of "mind independence" (in philosophy). (WK 2012j)

In philosophy, "a proposition is generally considered to be objectively true when its truth conditions are met and are 'mind-independent'—that is, existing freely or independently from the thoughts of a conscious entity or subject. In a simpler form, objectivity is the ability to judge fairly." (WK 2012j)

In this sense, "objectivism" is the stand that "describes a branch of philosophy that originated in the early nineteenth century. Gottlob Frege was the first to apply it, when he expounded an epistemological and metaphysical theory contrary to that of Immanuel Kant. Kant's rationalism attempted to reconcile the failures he perceived in Philosophical realism. Objectivism, in this context, is an alternative name for philosophical realism, the view that there is a reality, or ontological realm of objects and facts, that exists independent of the mind." (WK 2012j)

Logical Relation

And the third way to understand objectivity has to do with the issue of "logical relation" (in probability theory). (WK 2012)

In probability theory, "physical probabilities, which are also called objective or frequency probabilities, are associated with random physical systems such as roulette wheels, rolling dice and radioactive atoms. In such systems, a given type of event (such as the dice yielding a six) tends to occur at a persistent rate, or 'relative frequency,' in a long run of trials. Physical probabilities either explain, or are invoked to explain, these stable frequencies." (WK 2012)

Thus, these physical probabilities tell us about some "logical relations" (or "logical probabilities") which "are conceived (for example in Keynes' *Treatise on Probability*) to be objective, logical relations between propositions (or sentences), and hence not to depend in any way upon belief. They are degrees of (partial) entailment, or degrees of logical consequence, not degrees of belief….They do, nevertheless, dictate proper degrees of belief…." (WK 2012)

Subjectivity

And the opposite of objectivity is subjectivity. In accordance to the Merriam-Webster dictionary, the word "subjectivity" refers to the quality of being "characteristic of or belonging to reality as perceived rather than as independent of mind" or of being "modified or affected by personal views, experience, or background." (MWD 2012a)

Like its counterpart (objectivity)—the concept of "subjectivity" can be understood in different ways, and, for illustration, consider three main examples below, namely, (a) individual perspective, (b) thought (mind), and (c) personal belief, as explained below.

Individual Perspective

The first way to understand subjectivity has to do with the issue of "individiual perspective" (in science).

In science, "subjectivity refers to the subject and his or her perspective, feelings, beliefs, and desires….Subjectivity refers to individual interpretations of experiences consisting of emotional, intellectual, and spiritual perceptions and misperceptions." (WK 2012k)

This individual perspective therefore rejects the objectivivist view which describes "humans as 'seeing' the universe exactly for what it is from a standpoint free from human perception and its influences, human cultural interventions, past experience and expectation of the result." (WK 2012k; R. Solomon 2005)

Thought (Mind)

The second way to understand subjectivity has to do with the issue of "thought" (in philosophy). (WK 2012l)

In philosophy, "a subject is a being that has subjective experiences, subjective consciousness or a relationship with another entity (or 'object'). A *subject* is an observer and an *object* is a thing observed. This concept is especially important in continental philosophy, where 'the Subject' is a central term in debates over human autonomy and the nature of the self." (WK 2012l)

The history of this "sharp distinction between subject and object" can be traced back to "the distinction, in the philosophy of René Descartes, between thought and extension. Descartes believed that thought (subjectivity) was the essence of the mind, and that extension (the occupation of space) was the essence of matter. In the modern continental tradition, which may plausibly be said to date from Descartes, debates over the nature of the Subject play a role comparable to debates over personhood within the distinct Anglo-American tradition of analytical philosophy." (WK 2012l)

Personal Belief

And the third way to understand subjectivity has to do with the issue of "personal belief" (in probability theory). (WK 2012)

In probability theory, there is a school of probability known as "Bayesian probability," which states that "the state of knowledge corresponds to personal belief." (WK 2012m)

For the advocates, "the Bayesian interpretation of probability can be seen as an extension of logic that enables reasoning with propositions whose truth or falsity is uncertain. To evaluate the probability of a hypothesis, the Bayesian probabilist specifies some prior probability, which is then updated in the light of new, relevant data." (WK 2012n; J. Paulos 2011)

But "there are two views on Bayesian probability that interpret the probability concept in different ways. According to the objectivist view, the rules of Bayesian statistics can be justified by requirements of rationality and consistency and interpreted as an extension of logic. According to the subjectivist view, probability quantifies a 'personal belief.'" (WK 2012n; E. Jaynes 1986; R. Cox 2001)

Diverse Aspects of Probability

This introduction to "objectivity" and "subjectivity" in probability thus shows their more complicated nature.

To illustrate this more complicated nature of probability further, consider the diverse aspects of probability below (as summarized in *Table 1.1*).

Rational vs. Logical

The first aspect of probability to be illustrated here in regard to its more complicated nature concerns the distinction between "rational" and "logical" categories.

On the one hand, "reason is the capacity for consciously making sense of things, for establishing and verifying facts, and changing or justifying practices, institutions, and beliefs based on new or existing information….Reason or 'reasoning' is associated with thinking, cognition, and intellect. Reason, like habit or intuition, is one of the ways by which thinking comes from one idea to a related idea. For example, it is the means by which rational beings understand themselves to think about cause and effect, truth and falsehood, and what is good or bad." (WK 2012o; N. Kompridis 2010) Of course, there is also a slight distinction between "reason" and "rationality," in that, as Jesús Mosterín (2008) once suggested, "reason and rationality are not the same: reason would be a psychological faculty, whereas rationality is an optimizing strategy." (WK 2012v)

On the other hand, according to Douglas Hofstadter (1979) in *Gödel, Escher, Bach*, "logic is done inside a system while reason is done outside the system by such methods as skipping steps, working backward, drawing diagrams, looking at examples, or seeing what happens if you change the rules of the system....Reason is a type of thought, and the word 'logic' involves the attempt to describe rules or norms by which reasoning operates, so that orderly reasoning [i.e., logic] can be taught. The oldest surviving writing to explicitly consider the rules by which reason operates are the works of the Greek philosopher Aristotle, especially *Prior Analysis and Posterior Analysis*. Although the Ancient Greeks had no separate word for logic as distinct from language and reason, Aristotle's newly coined word 'syllogism' (*syllogismos*) identified logic clearly for the first time as a distinct field of study. When Aristotle referred to 'the logical' (hē *logikē*), he was referring more broadly to rational thought." (WK 2012o)

Independent vs. Conditional

The second aspect of probability to be illustrated here in regard to its more complicated nature concerns the distinction between "independent" and "conditional" probability.

On the one hand, there is "independent probability," which means that, "if two events A and B are independent," then they do not affect each other, so "the joint probability" is shown below: (WK 2012)

$$P(A \text{ and } B) = P(A \cap B) = P(A)\ P(B)$$

In other words, "if two coins are flipped, the chance of both being heads is ½ + ½ = ¼." (WK 2012)

On the other hand, there is "conditional probability," which means that, if some event A is conditional on some event B (that is, some event A is conditional on the occurrence of some other event B), then "the probability of A, given B" is defined below: (WK 2012)

$$P(A / B) = [\ P(A \cap B)] / P(B)$$

Of course, there are other probabilities too, so the ones here are illustrative, not exhaustive.

Continuous vs. Discrete

The third aspect of probability to be illustrated here in regard to its more complicated nature concerns the distinction between "discrete" and "continuous" probabilty distributions.

On the one hand, there are "discrete" probability distributions, which are about "events that occur in countable sample spaces," and good examples are "throwing dice, experiments with decks of cards, and random walk." (WK 2012c)

On the other hand, there are "continuous" probability distributions, which "intuitively" "can take many different possible values"; for example, "if one measures the width of an oak leaf, the result of 3½ cm is possible," but "there are uncountably many other potential values even between 3 cm and 4 cm....Formally, each value has an infinitesimally small probability, which statistically is equivalent to zero." (WK 2012h) Of course, there appears to be a paradox here.

Abstract vs. Applied

And the fourth aspect of probability to be illustrated here in regard to its more complicated nature concerns the distinction between "abstract" and "applied" levels.

On the one hand, at the theoretical level, that is, "in a purely theoretical setting (like tossing a fair coin),...probabilities describe the statistical number of outcomes considered divided by the number of all outcomes (tossing a fair coin twice will yield HH with probability 1/4, because the four outcomes HH, HT, TH and TT are possible)." (WK 2012b)

On the other hand, at the applied level, problems occur "when it comes to practical application," because "the word probability does not have a singular direct definition. In fact, there are two major categories of probability interpretations, whose adherents possess conflicting views about the fundamental nature of probability" in practice, namely, the objectivist and subjectivist interpretations (as described above): (WK 2012b)

The Theoretical Debate

These diverse aspects of probability, together with the introduction to "objectivity" and "subjectivity," then lead us to a more important question concerning the theoretical debate on the very nature of probability. Indeed, this theoretical debate can also say much about the study of probability as a discipline over time.

Although there can be many different theoretical approaches in the debate—there are five main ones to be summarized hereafter, and they can be called, in the absence of better words, namely, (a) the *mental* argument, (b) the *natural* argument, (c) the *societal* argument, (d) the *cultural* argument, and (e) the *interpretivist* argument (which is my original contribution to the debate), to be elabroated hereafter, respectively.

Lest misunderstanding occurs, three clarifications should be made here, as this is something that I regularly emphasize in my previous books, whenever a theoretical debate is summarized for a topic, *almost verbatim*.

Firstly, there can be other arguments besides the first four here, that is, the "mental," the "natural," the "societal," and the "cultural" arguments (as indicated above). The advantage to select these four here has to do with their illustrative representation of the diverse theories in the literature (which are deemed sufficient for the purpose at hand in this book).

Secondly, the examples (theories) given to illustrate each argument are also supposed to be illustrative, not exhaustive. After all, the arguments will be addressed again in the rest of the book, with other examples (theories) to be introduced.

And thirdly, the four arguments are not necessarily mutually exclusive. For instance, those who advocate the "mental" argument can also consider the "cultural" argument, although they do not focus on the latter.

And the reverse is also true, in that those who make the "cultural" argument can also consider the "mental" argument, although they do not focus on the latter.

In other words, their disagreement is often one in degree, not in kind.

The Mental Argument

With these three clarifications in mind—the first major theoretical approach is called, in the absence of better words, the *mental* argument, which focuses, relatively speaking of course, more on the biological and psychological aspects of probability.

Thus, the word "mental" in the title of the argument is suggestive, because it refers to the focus on the physiological, biological and psychological aspects of probability so understood.

But the mental argument has different issues of concern, and different defenders of the argument offer their own distinctive stands on the issues, depending on their own interests.

For illustration, consider two major versions of the mental argument (with the first on objectivity and the second on subjectivity), namely, (a) reasoning and mental models, and (b) Bayesian probability and personal belief—to be addressed below, in that order.

Reasoning and Mental Models

The first version of the mental argument has to do with reasoning and mental models (in the context of objectivity) in probability study.

An argument in the literature is that, as Kenneth Craik suggested in 1943,…the mind constructs 'small-scale models' of reality that it uses to anticipate events." (WK 2012p)

In psychology, "the term 'mental models' is sometimes used to refer to mental representations or mental simulation generally….People infer that a conclusion is valid if it holds in all the possibilities….They reject a conclusion if they find a counterexample, i.e., a possibility in which the premises hold, but the conclusion does not." (WK 2012p) (W. Schroyens 2003; N. Verschueren 2005)

However, for the critics, this argument on reasoning and mental models has some major problems. Consider, for illustration, a few examples below.

Firstly, one main criticism of the argument on reasoning and mental models is that there are alternative explanations of human reasoning.

For instance, for the critics, there are alternative views on human reasoning, which are not based on mental models, and good examples are those based on "formal rules of inference (e.g., O'Brien, 2009), domain-specific rules of inference (e.g., Cheng & Holyoak, 2008; Cosmides, 2005), or probabilities (e.g., Oaksford and Chater, 2007)." (WK 2012p)

And secondly, another main criticism of the argument on reasoning and mental models is that there is no general consensus of whether mental models are better than other models in explaining human reasoning.

For example, for the critics, "scientific debate continues about whether human reasoning is based on mental models….Many empirical comparisons of the different theories have been carried out." (WK 2012p; K. Oberauer 2006)

Bayesian Probability and Personal Belief

And the second version of the mental argument has to do with Bayesian probability and personal belief (in the context of subjectivity) in probability study.

An argument in the literature is that, as Thomas Bayes once argued, "the Bayesian interpretation of probability can be seen as an extension of logic that enables reasoning with propositions whose truth or falsity is uncertain. To evaluate the probability of a hypothesis, the Bayesian probabilist specifies some prior probability, which is then updated in the light of new, relevant data." (WK 2012n; J. Paulos 2011)

According to the subjectivist interpretation, in this way, "probability quantifies a 'personal belief.'" (WK 2012n; B. de Finetti 1974)

However, for the critics, the argument on Bayesian probability and personal belief has some major problems. Consider, for illustration, a few examples below.

Firstly, one main criticism of the argument on Bayesian probability and personal belief is that there is an alternative interpretation of Bayesian probability.

For instance, for the critics, "according to the *objectivist* view, the rules of Bayesian statistics can be justified by requirements of rationality and consistency and interpreted as an extension of logic." (WK 2012n; E. Jaynes 1986; R. Cox 2001)

And secondly, another main criticism of the argument on Bayesian probability and personal belief is that the probability of a random event can be quantified without the need of Bayesian statistics.

For instance, the proponents of "frequentist probability…claim…that the probability of a random event denotes the *relative frequency of occurrence* of an experiment's outcome, when repeating the experiment. This interpretation considers probability to be the relative frequency 'in the long run' of outcomes." (WK 2012n; I. Hacking 1965)

The Natural Argument

The second theoretical approach is called, in the absence of better words, the *natural* argument, which focuses, relatively speaking of course, more on the natural aspects of probability, especially in relation to the state of nature.

Therefore, the word "natural" in the title of the argument is suggestive, because it refers to the focus on the natural aspects of probability so understood.

Like the mental argument—the natural argument has different issues of concern, and different defenders of the argument offer their own distinctive stands on the issues, depending on their own interests.

For illustration, consider two major versions of the natural argument (with the first on objectivity and the second on subjectivity), namely, (a) discreteness and the physical world, and (b) quantum mechanics and uncertainty—to be summarized below, in that order.

Discreteness and the Physical World

The first version of the natural argument has to do with discreteness and the physical world (in the context of objectivity) in probability study.

An argument in the literature is that discreteness in the physical world can be predicted in a precise way, and a good example is the game of rolling a dice (or more than one dice). In other words, "for a single roll of a fair *s*-sided die, the probability of rolling each value is exactly $1/s$; this is an example of a discrete uniform distribution." (WK 2012q)

However, for the critics, this argument on discreteness and the physical world has some major problems. Consider, for illustration, a few examples below.

Firstly, one main criticism of the argument on discreteness and the physical world is that what is true in mathematics may not be true in the real world.

For instance, for the critics, a dice in the real world is not perfectly manufactured; moreover, "due to circumstances or character skill" (like "movements in the thrower's hand," the texture and shape of the table, and so on), the probability of rolling each value is not $1/s$ in practice. (WK 2012q)

And secondly, another main criticism of the argument on discreteness and the physical world is that there are alternative probability distributions which are not discrete.

For instance, for the critics, a non-discrete probability distribution is known as "a probability distribution that has a probability density function. Mathematicians also call such a distribution absolutely continuous....There are many examples of continuous probability distributions: normal, uniform, chi-squared, and others." (WK 2012h)

Quantum Mechanics and Uncertainty

And the second version of the natural argument has to do with quantum mechanics and uncertainty (in the context of subjectivity) in probability study.

An argument in the literature is that probability theory contributes to the understanding of uncertainty in the quantum mechanical world.

For instance, "in quantum mechanics, the uncertainty principle is any of a variety of mathematical inequalities asserting a fundamental limit to the precision with which certain pairs of physical properties of a particle, such as position *x* and momentum *p*, can be known simultaneously. The more precisely the position of some particle is determined, the less precisely its momentum can be known, and vice versa"—and "according to the Copenhagen interpretation of quantum mechanics, there is no fundamental reality that the quantum state describes, just a prescription for calculating experimental results. There is no way to say what the state of a system fundamentally is, only what the result of observations might be....Niels Bohr believed that the probability distributions...depend on which measurements we choose to perform." (WK 2012r)

However, for the critics, this argument on quantum mechanics and uncertainty has some major problems. Consider, for illustration, a few examples below.

Firstly, one main criticism of the argument on quantum mechanics and uncertainty is that the loss of certainty in quantum mechanical measurements can be due to our ignorance of some hidden variables in the real world.

For instance, "Albert Einstein believed that randomness is a reflection of our ignorance of some fundamental property of reality" and "disliked this loss of determinism" in quantum mechanics, so he held that there must be some "hidden variables" to be understood and, "consequently, that the present theory [of quantum mechanics] was incomplete." (WK 2012r & 2012s)

And secondly, another main criticism of the argument on quantum mechanics and uncertainty is that quantum mechanics itself has gone through different competing formulations over time.

For instance, "the earliest versions of quantum mechanics were formulated in the first decade of the 20th century....[Then] quantum theory was significantly reformulated in the mid-1920s by Werner Heisenberg, Max Born and Pascual Jordan, who created matrix mechanics; Louis de Broglie and Erwin Schrodinger (Wave Mechanics); and Wolfgang Pauli and Satyendra Nath Bose (statistics of subatomic particles). And the Copenhagen interpretation of Niels Bohr became widely accepted. By 1930, quantum mechanics had been further...formalized by the work of David Hilbert, Paul Dirac and John von Neumann, with a greater emphasis placed on measurement in quantum mechanics, the statistical nature of our knowledge of reality, and philosophical speculation about the role of the observer. Quantum mechanics has since branched out into almost every aspect of 20th century physics and other disciplines, such as quantum chemistry, quantum electronics, quantum optics, and quantum information science." (WK 2012s) And this changing history led "the renowned physicist Richard

Feynman" to once say, "I think I can safely say that nobody understands quantum mechanics." (WK 2012s; W. Greiner 1994)

The Societal Argument

The third theoretical approach is called, in the absence of better words, the *societal* argument, which focuses, relatively speaking of course, more on the societal aspects of probability (e.g., organizational, institutional, structural, and technological issues).

In this sense, the word "societal" in the title of the argument is suggestive, because it refers to the societal aspects of probability so understood.

Like the mental and natural arguments—the societal argument has different issues of concern, and different defenders of the argument offer their own distinctive stands on the issues, depending on their own interests.

For illustration, consider two major versions of the societal argument (with the first on objectivity and the second on subjectivity), namely, (a) optimality and decision making, and (b) feminism and rational constructs for social action—to be summarized in what follows, in that order.

Optimality and Decision Making

The first version of the societal argument has to do with optimality and decision making (in the context of objectivity) in probability study.

An argument in the literature is that optimality can be achieved in decision making, and an optimal decision is "a decision such that no other available decision options will lead to a better outcome"; "in order to compare the different decision outcomes, one commonly assigns a relative utility to each of them. If there is uncertainty in what the outcome will be, the optimal decision maximizes the expected (average) utility. Sometimes, the equivalent problem of minimizing loss is considered, particularly in financial situations, where the utility is defined as economic gain….For example, it may well be the optimal decision for someone to buy a sports car rather than a station wagon, if the outcome in terms of another criterion (e.g., effect on personal image) is more desirable, even given the higher cost and lack of versatility of the sports car." (WK 2012t)

However, for the critics, this argument on optimality and decision making has some major problems. Consider, for illustration, a few examples below.

Firstly, one main criticism of the argument on optimality and decision making is that optimization is seldom achieved in real life, as decision making can be more intuitive and less formal.

For instance, for the critics, "in practice, few people verify that their decisions are optimal, but instead use more intuitive approaches to make decisions that are 'good enough.'" (WK 2012t)

And secondly, another main criticism of the argument on optimality and decision making is that optimal and non-optimal decision-making strategies are relative to different personality traits.

For instance, for the critics, "some research has suggested that satisficing/maximizing and other decision-making strategies, like personality traits, have a strong genetic component and endure over time. This genetic influence on decision-making behaviors has been found through classical twin studies, in which decision-making tendencies are self-reported by pairs of twins and then compared between monozygotic and dizygotic twins. This implies that people can be categorized into 'maximizers' and 'satisficers,' with some people landing in between." (WK 2012u; I. Simonson 2011)

Feminism and Ratonal Constructs for Social Action

And the second version of the societal argument has to do with feminism and rational constructs for social action (in the context of subjectivity) in probability study.

An argument in the literature is that the conventional rational constructs for social action are "sexist." For instance, when Max Weber proposed different types of rationality (e.g., "instrumental," "intrinsic," and "traditional"), such that "rational people should derive conclusions in a consistent way given the information at disposal" (in that "it refers to the conformity of one's beliefs with one's reasons to believe, or...with one's reasons for action")—many femnists were quick to reject the formulations as "imbued with masculine values and oriented toward the maintenance of male power," because these constructs do not give sufficient attention to feminine values (like equality, love, and the like). (WK 2012v; M. Eagleton 2003)

However, for the critics, this argument on feminism and the rationality for social action has some major problems. Consider, for illustration, a few examples below.

Firstly, one main criticism of this argument on feminism and the rationality for social action is that it dismisses natural differences between men and women.

For instance, for the critics, "some people have opposed feminism on the grounds that they believe it is contrary to traditional values or religious beliefs. These anti-feminists argue, for example, that social acceptance of divorce and non-married women is wrong and harmful, and that men and women are fundamentally different and thus their different traditional roles in society should be maintained." (WK 2012w; C. Lukas 2006; M. Kassian 2005; P. Schlafly 1977)

Secondly, another main criticism of this argument on feminism and the rationality for social action is that feminism is oppressive.

For instance, for the critics, "writers such as Camille Paglia, Christina Hoff Sommers, Jean Bethke Elshtain, Elizabeth Fox-Genovese and Daphne Patai oppose some forms of feminism, though they identify as feminists. They argue, for example, that feminism often promotes misandry and the elevation of women's interests above men's, and criticize radical feminist positions as harmful to both men and women. Daphne Patai and Noretta Koertge argue that the term 'anti-feminist' is used to silence academic debate about feminism." (WK 2012w; C. Sommers 1995; D. Patai 2003)

The Cultural Argument

And the fourth theoretical approach is called, in the absence of better words, the *cultural* argument, which focuses, relatively speaking of course, more on the cultural aspects of probability (e.g., issues in relation to values and beliefs).

Therefore, the word "cultural" in the title of the argument is suggestive, because it refers to the focus on the cultural aspects of probability so understood.

Like the mental, natural, and societal arguments—the cultural argument has different issues of concern, and different defenders of the argument offer their own distinctive stands on the issues, depending on their own interests.

For illustration, consider two major versions of the cultural argument (with the first on objectivity and the second on subjectivity), namely, (a) statistics and common cause, and (b) individual competence and cultural consensus—to be summarized in what follows, in that order.

Statistics and Common Cause

The first version of the cultural argument has to do with statistics and common cause (in the context of objectivity) in probabiliy study.

An argument in the literature is that there are certain historical causes which can be quantitied and predicted, and these causes are known as "common causes," which are "the usual, historical, quantifiable variation in a system..., being recognised and discussed as early as 1703 by Gottfried Leibniz." (WK 2012x) This common-cause variation is "predictable, at least approximately in frequency. This variation is evident from the experience base," and a good illustration is "the outcomes of a perfectly balanced roulette wheel....Common-cause variation is the noise within the system." (WK 2012x)

However, for the critics, the argument on common causes and statistics has some major problems. Consider, for illustration, a few examples below.

Firstly, one main criticism of the argument on common causes and statistics is that there are "special causes" outside the historical experience base which are not predictable.

For instance, for the critics, "new, unanticipated, emergent or previously neglected phenomena (e.g. 'new diseases') result in variation outside the historical experience base. [Walter] Shewhart and [Edwards] Deming argued that such special-cause variation is fundamentally unpredictable in frequency of occurrence or in severity." (WK 2012x)

And secondly, another main criticism of the argument on common causes and statistics is that there are causes which can be within the historical experience base but are subjective in orientation.

For instance, for the critics, one excellent example is "the subjective Bayesian probability," which lies outside both common cause and special cause. (WK 2012x)

Individual Competence and Cultural Consensus

And the second version of the cultural argument has to do with individual competence and cultural consensus (in the context of subjectivity) in probability study.

An argument in the literature is that there exists "a framework for the measurement and evaluation of beliefs as cultural" and "shared to some extent by a group of individuals," as shown by "cultural consensus models" which "guide the aggregation of responses from individuals to estimate (a) the culturally appropriate answers to a series of related questions (when the answers are unknown) and (b) individual competence (cultural competence) in answering those questions." (WK 2012y)

However, for the critics, this argument on individual competence and cultural consensus has some major problems. Consider, for illustration, a few examples below.

Firstly, one main criticism of this argument on individual competence and cultural consensus is that cultural consensus theory does not explain the causes of consensus.

For instance, for the critics, "cultural consensus models do not create consensus or explain why consensus exists; they simply facilitate the discovery and description of possible consensus." (WK 2012y)

And secondly, another main criticism of this argument on individual competence and cultural consensus is that cultural consensus theory also does not tell us which answers are "true."

For instance, for the critics, "these methods do not provide a best estimate of the 'true' answers nor do they estimate competence of the raters." (WK 2012y)

The Interpretivist Argument

And fifthly, unlike the previous four arguments (viz., the mental, the natural, the societal, and the cultural)—the fifth one to be introduced here can be called, in the absence of better words again, the *interpretivist* argument, which learns from all other theoretical approaches (which include the previous four arguments and also other examples as will be introduced in the rest of this book), without, however, favoring any of them, nor trying to integrate them (as they are not necessarily compatible with each other, as the introduction of the theoretical debate in the previous sub-sections has shown).

The Interpretivist Theory of Probability

My *interpretivist* argument can be more precisely called *the interpretivist theory of probability*, since it is to propose a new theory of objectivity and subjectivity in probability study.

My *interpretivist theory of probability* has four major distinct features.

Firstly, my theory makes good use of all theoretical approaches to probability study in relation to objectivity and subjectivity, be they about the mental argument, the natural argument, the societal argument, or the cultural argument—especially from the most comprehensive combined perspectives of the mind, nature, society, and culture (as will be dealt with in Chapter Two and Chapter Three).

Secondly, just like many other theories of mine in my previous books, my theory here does not heavily favor any specific theory over others in the literature, nor trying to integrate them (as they are not necessarily compatible with each other).

Thirdly, my theory treats the issue of probability in the distinctive dialectic context of no objectivity without subjectivity, and vice versa.

And fourthly, it contains thirty-one major theses, namely, (a) the first thesis: the formalness-informalness principle, (b) the second thesis: the absoluteness-relativeness principle, (c) the third thesis: the partiality-totality principle, (d) the fourth thesis: the predictability-unpredictability principle, (e) the fifth thesis: the explicability-inexplicability principle, (f) the sixth thesis: the fiction-reality principle, (g) the seventh thesis: the cognitiveness-noncognitiveness principle, (h) the eighth thesis: the finiteness-transfiniteness principle, (i) the ninth thesis: the preciseness-vagueness principle, (j) the tenth thesis: the simpleness-complicatedness principle, (k) the eleventh thesis: the openness-hiddenness principle, (l) the twelfth thesis: the denseness-emptiness principle, (m) the thirteenth thesis: the change-constancy principle, (n) the fourteenth thesis: the order-chaos principle, (o) the fifteenth thesis: the slowness-quickness principle, (p) the sixteenth thesis: the expansion-contraction principle, (q) the seventeenth thesis: the theory-praxis principle, (r) the eighteenth thesis: the convention-novelty principle, (s) the nineteenth thesis: the evolution-transformation principle, (t) the twentieth thesis: the symmetry-asymmetry principle, (u) the twenty-first thesis: the softness-hardness principle, (v) the twenty-second thesis: the seriousness-playfulness principle, (w) the twenty-third thesis: the activeness-inactiveness principle, (x) the twenty-fourth thesis: the selfness-otherness principle, (y) the twenty-fifth thesis: the regression-progression principle, (z) the twenty-sixth thesis: the sameness-difference principle, (aa) the twenty-seventh thesis: the stability-reaction principle, (bb) the twenty-eighth thesis: the functionality-nonfunctionality principle, (cc) the twenty-ninth thesis: the intentionality-nonintentionality principle, (dd) the thirtieth thesis: the survivability-non-survivability principle, and (ee) the thirty-first thesis: the post-human interpretiveness—to be elaborated in the rest of the book, with a summary in the concluding chapter.

Because of the last thesis on "the post-human interpretiveness," the word "interpretivist" in the title of my theory is suggestive, because it refers to the different interpretations of probability in the face of changing circumstances over time (like the opposing objectivist and subjectivist interpretations)—which then require a multi-faceted inquiry of the dialectic relations between objectivity and subjectivity, as well as an understanding of the different views regarding these dialectic relations in probability study (together with other debates as will be discussed in the rest of the book)—and in the end it calls for a need to transcend them by future advanced humans and later post-humans in the context of the great transformations of probability in the future world (as will be discussed in the concluding chapter, Chapter Four).

With this in mind—the seminal project here, if successful, will fundamentally change the way that we think about objectivity and subjectivity in probability study (together with other debates as will be discussed in the rest of the book), from the combined perspectives of the mind, nature, society, and culture, with enormous implications for the human future and what I originally called its "post-human" fate.

Theory and Meta-Theory

The *interpretivist* argument, or *the interpretivist theory of probability*, presupposes some kind of meta-theory, in special relation to methodology and ontology—just as all theories, either explicitly or implicitly, require their own versions of meta-theory.

Meta-Theory

As so often clarified in my previous books—theory and meta-theory, broadly speaking, can be further classified into different categories, as shown below:

- On Theory
  - —Ex: in relation to the mind
  - —Ex: in relation to nature
  - —Ex: in relation to society
  - —Ex: in relation to culture
- On Meta-Theory
  - —Ex: in relation to methodology
  - —Ex: in relation to ontology

Each category, in its turn, can then be further sub-divided into different levels of analysis—which will be elaborated in the next two sections (and summarized in *Table 4.43*).

With this reminder in mind, I already proposed (in my previous books) my distinctive contribution to meta-theory, namely, methodology (like "theory of methodological holism," "complex theory of data analysis," and "critical-dialectic theory of formal science," for instance) and ontology (like "theory of existential dialectics" and "contrastive theory of rationality," for instance).

The next two sections are to introduce my meta-theory—especially, existential dialectics and sophisticated methodological holism, respectively—to be repeated, *often verbatim* (with only a few updated revisions), from my previous works, as this is something that I often do in each new book of mine as an introduction to some background information for the convenience of the readers.

In the process, I shall also introduce my numerous new theories (or to be exact, 65 of them in 57 books so far, as of December 13, 2012) in relation to the four categories of the mind, nature, society, and culture in the context of theory (as also summarized in *Table 4.43*).

A Unified Theory of Everything

In the end, my distinctive contribution to both meta-theory and theory serves as a foundation to unify all domains of knowledge (i.e., the social sciences, the formal sciences, the natural sciences, and the humanities) for an overarching understanding of all forms of reality (by way of some ontological principles at the meta-theoretical level and the comprehensive perspectives of the mind, nature, society, and culture at the theoretical level—as will be explained in the next two sections).

This is so, without commiting the sins of reductionism and reverse-reductionism (as often seen in many holistic approaches, with the fad of systems approach as a most recent notorious example, as already critically explained in *The Future of Post-Human Formal Science*, or *FPHFS*), since my distinctive contribution makes good use of different schools of thought without heavily favoring any of them nor trying to integrate them (as they are not necessarily compatible with each other), so as to include (or adjust for) heterogeneity, conflict, subjectivity, and complexity, for example.

In this sense, the word "unified" does not have to mean an integrative approach, in a narrow sense, nor a systems approach, in a broad one, both of which I have critically rebuked (in my previous books), while learning from them.

This overarching project (consisting of numerous theories of mine, as will be summarized in the next section), thus, constitutes my grand unifed theory of everything in the world and beyond.

The Logic of Existential Dialectics

To start, the ontology of existential dialectics can be summarized in this section—*often verbatim* (with only a few updated revisions), from my previous works, as this is something that I regularly do in each new book of mine as an introduction to some background information for the convenience of the readers.

After all, this book, like all others of mine, is in conversation with all previous ones of mine, for the final aim to converge all of them into a larger grand project on the future of intelligent life, both here on earth and there in deep space unto multiverses.

The summary can be organized in relation to (a) the conception of existential dialectics (or its ontics), (b) the syntax of existential dialectics (or its ontomethodology and ontologic), (c) the semantics of existential dialectics (or its ontosemantics), and (d) the pragmatics of existential dialectics (or its ontopragmatics), respectively hereafter—again *often verbatim*.

This means that, for those readers who had read some of my previous books, the summary below is really not much new (*almost verbatim*, with only some updated revisions).

But for those who have never read my previous books, it is a good review of them.

The Conception of Existential Dialectics (or Its Ontics)

This ontology was originally designated as "existential dialectics" in *Beyond Capitalism to Post-Capitalism* (herein abbreviated as *BCPC*), although it was already analyzed in other books of mine like the 2 volumes of *The Future of Human Civilization* (*FHC*), *The Future of Capitalism and Democracy* (*FCD*), *The Future of Post-Human Consciousness* (*FPHC*), and the 2 volumes of *Beyond Democracy to Post-Democracy* (*BDPD*).

Lest any confusion occurs, it is important to stress at the outset that the word 'existential' in "existential dialectics" has nothing to do with Existentialism, which I rebuked in *FHC*, *FCD*, and also *FPHC*. Instead, the word here merely refers to the existence of intelligent life (both primitive and advanced) in a broad sense.

Some Basic Concepts

The conception of existential dialectics (or its ontics) makes use of different concepts (like "sets," "elements," "relations," "operations," "functions," "truth values," "axioms," "postulates," and "principles"—as shown in *Table 4.3*), which are important for the understanding of any logic of ontology.

That clarified—existential dialectics, as a language of ontology, can be analyzed in different ways that a language, as an analogy, is often analyzed, namely, in relation to phonology ("the study of patterns of a language's sounds"), phonetics ("the study of the physical aspects of sounds of human language"), morphonology ("the study of the internal structure of words"), syntax ("the study of how words combine to form grammatical sentences"), semantics ("the study of the meaning of words [lexical semantics] and fixed word combinations [phraseology]"), and pragmatics ("the study of how utterances are used…literally, figuratively, or otherwise…in communicative acts"). (WK 2007)

But since my theory of existential dialectics makes use of the English language for communication here, it does not propose a new way to make the patterns of a language's sounds (as in phonology), to study the physical aspects of sounds (as in phonetics), or even to strive for a new organization of the internal structure of words for the English language (as in morphonology).

For this reason, the language of existential dialectics to be analyzed hereafter is not concerned with the phonological, phonetic, and morphonological aspects of the English language. Instead, the analysis will explore the syntax, semantics, and pragmatics of existential dialectics as a language of ontology to improve our understanding of reality.

The Ontic Dispute

But first, there is an ontic dispute to be addressed, which is related to what I called "the dilemma of ontology" as introduced earlier in *Sec. 1.4* of *The Future of Post-Human Knowledge* (*FPHK*).

A good point of departure concerns what I want to call *the ontic dispute*, on how to understand the very nature of ontology.

Traditionally, scholars in the field used to study the conception of ontology in terms of its entities (e.g., material objects, abstract numbers), properties (e.g., duration, plurality), and relationships (e.g., causation).

This is a good starting point, but it suffers from different problems. Consider three examples of bias below, which are quite common in the literature, and the recent version (of treating concepts and principles in ontology on the basis of "plurality," "dynamism," "duration," "interaction," "life," "consciousness," and "volition") by Reginald Firehammer (2005) only constitutes a latest addition to this old (impoverishing) tradition.

Firstly, the classification of entities, properties, and relationships is too material-centric, in often focusing more on the existence of material entities than on non-material ones.

Secondly, the classification is too anthropo-centric, in often favoring the factor of human agency in depicting reality.

And lastly, the classification is too confused among the three categories relating to each other, in often lumping them together without adequate consideration of their intricated interrelationships (both in relation to each and in relation to all).

A good alternative is precisely what I want to offer in this debate, that is, a new classification of ontological entities, properties, and their relationships in terms of *method*, *structure*, *process*, *agency*, and *outcome* instead. Let me explicate what these words mean below.

In relation to "method"—it concerns how ontology is to be studied. In this sense, the word 'method' here should not be confused with the different usage of the same term 'method' as a technique of investigation in the context of methodology (as already introduced in *Sec. 1.2* of *FPHK*).

In relation to "structure"—it concerns what the general patterns of entities, properties, and relationships are.

In relation to "process"—it concerns how these general patterns change over time.

In relation to "agency"—it concerns how agency affects any change of these general patterns over time (without putting, however, too much emphasis on the role of human agency).

And in relation to "outcome"—it concerns what the outputs of this change over time are.

The Syntax of Existential Dialectics (or Its Ontomethodology and Ontologic)

The syntax of existential dialectics, analogously speaking, refers here to the structure of an ontology which can be studied by way of combining ideas into complex relationships like ontological principles to depict reality in the world—and can be analyzed in seven sub-sections, namely, (a) the selection criteria, (b) further clarifications on classification, (c) ontomethodology, (d) ontologic, (e) the principles as short cuts, (f) the principles as family resemblances, and (g) the dialectic constraints imposed by the principles, respectively hereafter.

The Selection Criteria

To start, not any pair of relationship can be chosen as an ontological principle in existential dialectics; otherwise, there could be billions of them under the sky.

Six illustrations below suffice to clarify this point.

(a) Firstly, an ontological principle cannot *overlap* with other principles, in that it is too closely related with other ones.

Of course, one cannot totally rule out any relationship between two entities, especially in a complex world of everything being linked to everything else—but the comparison here is relative, not absolute.

In this sense, the flexibility-inflexibility pair cannot be an ontological principle, since it has something closely in common with other ones (like the simpleness-complicatedness principle, for instance).

As an illustration, something which is flexible in interpretation is more likely to allow multiple viewpoints and thus makes the overall picture more complicated than otherwise (simple).

(b) Secondly, an ontological principle cannot be *redundant* in relation to other principles, in that it somehow duplicates other principles.

In this sense, for instance, the directness-indirectness pair cannot be an ontological principle, because it somehow duplicates other ontological principles (like the preciseness-vagueness principle, for instance).

As an illustration, something which is direct means to be right to the point and does not need to go to an unnecessarily long loop—so it is less vague, in being clearer about the thing in question.

(c) Thirdly, an ontological principle cannot be *derived* from other ones—as if it is a child-parent relationship.

In this sense, the convenience-inconvenience pair cannot be an ontological principle, because it can be derived from other ones (like the slowness-quickness principle, for instance).

For instance, something which is convenient already implies that it can be obtained quickly (rather than slowly)—all things considered.

(d) Fourthly, an ontological principle cannot be *trivial* (or *parochial*)—as if it is only one component competing with other ones for inclusion in a set.

In this sense, the consistency-inconsistency pair cannot be an ontological principle, because its validity is limited (or parochial), as it can be easily challenged and replaced by opposing major (not minor) theories (e.g., the correspondence theory of truth, the pragmatic theory of truth, the semantic theory of truth, the disclosedness theory of truth, and whatnot), not just in accordance to the consistency (or coherence) theory of truth (for consistency).

Thus, the consistency argument is only one among different opposing arguments in relation to the logic of what constitutes "truth."

(e) Fifthly, an ontological principle cannot be *one-sided*, that is, part of one side (category) of a dialectic pair.

In this sense, the extroversion-introversion pair cannot be an ontological principle, since it is part of the side (category) of "normality" in the normality-abnormaltiy pair (because both extroversion and introversion are two different "normal" personality traits in psychology).

In addition, the normality-abnormality pair is already related to the symmetry-asymmetry principle, the density-emptiness principle, the convention-novelty principle, and the same-difference principle (as will be explained in the section on "the principles as family resemblances" later).

(f) And sixthly, an ontological principle cannot be *misleading*, in that the two apparent opposites are not really opposites at all, upon closer examination.

In this sense, the east-west pair cannot be an ontological principle, since east and west are not really opposites, as there are also north, south, and center to form a set of five opposing elements.

By the same logic, the left-right pair cannot be an ontological principle, since left and right (width) are not really opposites, as there are also top and bottom (height), back and front (depth), and center to form a set of seven elements in four categories. In the field of visual design, this set of seven elements is known as "form" (as opposed to "function"), as addressed in *FPHARCH*. (WK 2011nn)

Although these six examples (criteria) are not exhaustive, they illustrative the selectiveness of any ontological principle to be existentially dialectic.

Further Clarifications on Classification

Lest any misunderstanding carelessly occurs, some further clarifications on classification should be made here (and summarized in *Table 4.8*).

(a) Firstly, the total number of ontological principles is unknown, to be discovered later, as our knowledge of the world becomes more advanced.

Consequently, the principles as introduced in my books are not exhaustive, with new ones being added, whenever more of them are discovered in later research.

At least, future generations can pick up where I leave off and continue the discovery.

In this sense, my work should be treated as a pioneering effort for the development of a systematic, comprehensive analysis of a new general ontology for the future of knowledge.

(b) Secondly, the principles are not rigidly classified, as they can be reclassified in a different way.

For instance, the preciseness-vagueness principle is classified under the category of "structure" but can be reclassified under the category of "method," although in so doing, it has a different meaning in the context of method.

The same logic applies to the same-difference principle under the category of "outcome," which can be reclassified under the category of "structure," although in so doing, once more, it has a different meaning in the context of structure.

(c) Thirdly, the principles are to be understood together, not that each prinicple is to be treated independently of others. After all, the principles are mutually constraining, in that they work together as a whole.

(d) Fourthly, the principles are relevant to all subject matters, but some principles are, relatively speaking of course, more useful to some subject matters than others under certain conditions—as implied in the symmetry-asymmetry principle.

(e) And fifthly, the principles constitute only two levels of analysis, this time, at the ontological and methodological levels—while other levels of analysis (from the perspectives of the mind, nature, society, and culture) are also needed, in order to understand reality in its totality (as explained in the section on "sophisticated methodological holism," so as to avoid the dual dangers of reductionism and reverse-reductionism).

Ontomethodology

With these selection criteria and clarifications in mind—the syntax of existential dialectics can be divided into two main smaller sections for analysis, since they are related to each other in the context of some fundamental principles, that is, (a) ontomethodology and (b) ontologic, to be addressed hereafter, respectively.

In the context of ontomethodology, some ontological principles are proposed here in relation to the nature of methodology, which however is tied up with the general issue of ontology in the context of knowledge.

In this sense, ontomethodology (in the context of these ontological principles in the category of method) are related with ontologic in general (in the context of ontological principles in general).

Some good instances of ontological principles involving ontomethodology include the formalness-informalness principle, the absoluteness-relativeness principle, the partiality-totality principle, the predictability-unpredictability principle, the explicability-inexplicability principle, the fiction-reality principle, and the cognitiveness-noncognitiveness principle, which will be introduced in conjunction with ontological principles in the ontologic of existential dialectics in general hereafter.

Ontologic

With this clarification of the relationship between ontomethodology and ontologic in mind (which is comparable to the relationship between methodology and ontology as described in *Sec. 1.3* of *FPHK*)—the first three principles in existential dialectics were introduced in *BCPC*. And more principles were later added to the logic in my subsequent books (like *FC*, *FAE*, and others), as explained below.

More specifically, in *BCPC*, I proposed three principles (based on previous works of mine, not just something out of the blue) for the ontological logic, namely, (a) the regression-progression principle on the direction of history, (b) the symmetry-asymmetry principle on the relationships among existents, and (c) the change-constancy principle on the alteration of things—or in a more elegant term, the dynamics of space-time (in *FC*).

In *The Future of Complexity* (*FC*), I added three more principles, on top of the three aforementioned, namely, (d) the partiality-totality principle on the relationships between parts and whole, (e) the order-chaos principle on the pattern of things, and (f) the predictability-unpredictability principle on the occurrence of things—as part of the ontological logic of existential dialectics.

In *The Future of Aesthetic Experience* (or simply *FAE*), another principle is added, namely, (g) the evolution-transformation principle (on the multiple kinds of agency).

Also in *FAE*, I then reclassified the 7 principles into four categories, namely, (i) in relation to method, for the partiality-totality principle and the predictability-unpredictability principle, (ii) in relation to process, for the change-constancy principle and the order-chaos principle, (iii) in relation to agency, for the symmetry-asymmetry principle and the evolution-transformation principle, and (iv) in relation to outcome, for the regression-progression principle.

In *The Rise of Authoritarian Liberal Democracy* (*ALD*), I added two more principles, namely, (h) the softness-hardness principle on the force of agency (which is to be classified under the category about agency) and (i) the same-difference principle on the metamorphosis of change (which is to be classified under the category about outcome).

In *The Future of Information Architecture* (*FIA*), I introduce three additional principles, that is, (j) the simpleness-complicadness principle on the interconnection among things, (k) the preciseness-vagueness principle on the refinement of things, and (l) the slowness-quickness principle on the speed of change—with the first two in relation to structure and the third in relation to process.

In *The Future of Post-Human Unconsciousness* (*FPHU*), I added three more principles, that is, (m) the openness-hidden principle on the detection of things, in relation to structure, (n) the explicability-inexplicability principle on the underlying mechanisms of things, in relation to method, and (o) the expansion-contraction principle on the growth of things, in relation to process.

In *The Future of Post-Human Knowledge* (*FPHK*), I added one more principle, that is, (p) the absoluteness-relativeness principle on the multiplicity of entities, in relation to method.

In *The Future of Post-Human Mathematical Logic* (*FPHML*), I added one more principle, that is, (q) the formalness-informalness principle on the formal requirements of systems, in relation to method.

In *The Future of Post-Human Engineering* (*FPHE*), I added one more principle, that is, (r) the theory-praxis principle on the duality of knowledge, in relation to agency.

In *The Future of Post-Human Creative Thinking* (*FPHCT*), I added one more principle, that is, (s) the convention-novelty principle on the nature of creative thinking, in relation to agency.

In *The Future of Post-Human Geometry* (*FPHG*) here, I added one more principle, that is, (t) the finiteness-transfiniteness principle on the nature of numbers, in relation to structure.

In *The Future of Post-Human Urban Planning* (*FPHUP*), I added one more principle, that is, (u) the denseness-emptiness principle on the distribution of entities in space, in relation to structure.

In *The Future of Post-Human Humor* (*FPHH*), I added one more principle, that is, (v) the seriousness-playfulness principle on the extent of seriousness, in relation to agency.

In *The Future of Post-Human Literature* (*FPHLIT*), I added one more principle, that is, (w) the fiction-reality principle on the different faces of reality, in relation to method.

In *The Future of Post-Human Chemistry* (*FPHCHEM*), I added one more principle, that is, (x) the stability-reaction principle on the feedback mechanism for further changes, in relation to outcome.

In *The Future of Post-Human Transportation* (*FPHTRAN*), I added one more principle, that is, (y) the activeness-inactiveness principle on the extent of activeness, in relation to agency.

In *The Future of Post-Human Culinary Art* (*FPHCA*), I added two more principles, that is, (z) the selfness-otherness principle on the magnitude of self-centeredness in relation to agency and (aa) the functionality-nonfunctionality principle on the presence of function in relation to outcome.

In *The Future of Post-Human Waste* (*FPHW*), I added one more principle, that is, (bb) the intentionality-nonintentionality principle on the planning for outcomes in relation to outcome.

In *Beyond Natural Resources to Post-Human Resources* (*BNR*), I added one more principle, that is, (cc) the cognitiveness-noncognitiveness principle on the understanding of meaning in relation to method.

And in *The Future of Post-Human Probability* (*FPHPROB*),), I add one more principle, that is, (dd) the survivability-nonsurvivability principle on the survival of things in relation to outcome.

With this update, there are thirty principles so far in existential dialectics, that is:

- seven principles in the category of "method"—viz., the formalness-informalness principle, the absoluteness-relativeness principle, the partiality-totality principle, the predictability-unpredictability principle, the explicability-inexplicability principle, the fiction-reality principle, and the cognitiveness-noncognitiveness principle
- five principles in the category of "structure"—viz., the finiteness-transfiniteness principle, the simpleness-complicatedness principle, the preciseness-vagueness principle, the openness-hiddenness principle, and the denseness-emptiness principle
- four principles in the category of "process"—viz., the change-constancy principle, the slowness-quickness principle, the order-chaos principle, and the expansion-contraction principle
- eight principles in the category of "agency"—viz., the theory-praxis principle, the convention-novelty principle, the symmetry-asymmetry principle, the evolution-transformation principle, the softness-hardness principle, the seriousness-playfulness principle, the activeness-inactiveness principle, and the selfness-otherness principle
- six principles in the category of "outcome"—viz., the regression-progression principle, the same-difference principle, the stability-reaction principle, the functionality-nonfunctionality principle, the intentionality-nonintentionality principle, and the survivability-nonsurvivability principle

There can be more than thirty principles in existential dialectics, but they are what I have discovered so far (as summarized in *Table 4.4*).

The Principles as Short Cuts

Yet, these principles should be treated with caution, lest misunderstanding occurs, since they do not constitute rigid dualities (or dichotomies).

The reason is that each pair in an ontological principle consists of two opposites, which are, however, merely short cuts both for multiple variations and degrees, as well as for different interactions with multiple other entities unlike them.

In this light, each pair can end up having hundred (or even thousand, if not more) different versions, which interact with hundred (or even thousand, if not more) other entities—as summarized in *Table 4.5*.

There are two clarifications here.

Firstly, this conception of shortcuts is not mutually exclusive nor absolute, in that the opposites can come in all shapes and sizes, with different degrees. For instance, by analogy, just as there are different degrees of the two colors "white" and "black"—there are likewise different degrees of the opposites in each ontological pair, to the extent that there can be multiple entities (not only two) interacting with each other in each pair.

And secondly, this conception of shortcuts do not ignore other possible entities in interacting with the two opposites (with their different versions) in each pair. For instance, by analogy, there are not only the two colors "white" and "black" as opposites, since there are other colors too besides them like "yellow," "green," "purple," or else. The same logic can be applied to each ontological pair, in that they also interact with other entities, not with only two of them (with their different versions).

Therefore, with these two clarifications in mind—each pair in an ontological principle serves only as an abbreviation for something more complicated and, therefore, although it contains two opposites, it should not be confused as a duality (dualism).

Existential dialectics rejects any dualism (or dichotomy) as too rigid and instead allows the multiplicity of entities, to the extent that between the two opposites in each pair exist many other alternatives to choose from. They are named in that short form for aesthetic elegance only, instead of listing all possible entities between the two opposites in the title (which would be too cumbersome and sound extremely awkward).

To be dialectic is to go beyond any rigid dichotomy and transcend into something different altogether in the long haul. One may be tempted to call the dialectic logic here with a different name like existential "multilectics" (instead of "dialectics"), but this naming is incorrect (or even misleading), for the two reasons aforestated.

After all, all the possible entities between any two opposites in an ontological pair are still about the two opposites, although each ontological pair come in all shapes and sizes (with different degrees) and do not exist by themselves but also interact with other entities unlike them. The virtue of revealing an ontological pair is to show how they relate within themselves (in multiple versions) and also interact with others unlike them (also in multiple versions).

This will be clear in a later sub-section titled "the dialectic constraints imposed by the principles."

The Principles as Family Resemblances

In addition, each ontological principle is generic, with some other comparable ontological pairs to be put in the same family (like a *family resemblance*)—as summarized in *Table 4.6*.

Ludwig Wittgenstein (1953) once suggested the idea of "family resemblance" in explaining different games classified under the same family called *games*.

Why should, for instance, playing football and chess as playing "games," when it is well understood that football is not the same as chess? The answer is that, although each game is different and has different rules, many of them (though not all) share, more or less, some commonalites (e.g., scoring as necessary for winning).

And this is so, even though some games share more than some others in any given selection of critieria, and no two games are exactly identical. So, Wittgenstein's point here is that there is no essential core which is common to all games, and the best that one can look for is some characteristics which are common to many (but not all) games. (A. Biletzki 2006)

(a) Appropriate Family Resemblances

In the context of existential dialectics, I want to distinguish two types of family resemblances, and, in the absence of better words, let me call them (a) *appropriate* family resemblances and (b) *inappropriate* family resemblances.

In regard to appropriate family resemblances—good examples include some of the pairs as already explained in an earlier section on selection criteria, like the flexibility-inflexibility pair, which can be put in the family resemblance of the simpleness-complicatedness principle, although the two pairs are not exactly identical. Likewise, the directness-indirectness pair can be put in the family resemblance of the preciseness-vagueness principle, although, again, the two pairs are not exactly identical.

In this sense, which specific pair in a family should be used to designate the name of the family can be at times a bit arbitrary, but with good reason.

For illustration, in the context of method, the partiality-totality principle can take the different form like individualisticness-holisticness, just as the explicability-inexplicability principle can take the different form like underlyingness-regularness—although each two pairs are not exactly identical and have slightly different meanings and usages.

A more comprehensive listing of this family resemblance for all other principles is shown in *Table 4.6* on the syntax of existential dialectics.

In fact, this section on family resemblances overlaps with the semantics of existential dialectics, which will be introduced later in the next section (and summarized in *Table 4.11*).

(b) Inappropriate Family Resemblances

However, there are some family resemblances which are not appropriate, and let me distinugish them by calling them, in the absence of better words, (b1) *inadequate* family resemblances, (b2) *compound* family resemblances, (b3) *one-sided* family resemblances, and (b4) *vague* family resemblances (as summarized in *Table 4.10*).

(b1) In regard to "inadequate" family resemblances—there are some pairs which are not adequate, because they fail the test of selection criteria (as already pointed out in an earlier section), especially in relation to the criterion of not being *trivial* (or *parochial*)—as if it is only one component competing with other ones for inclusion in a set.

For instance, the *consistency-inconsistency* pair is inadequate, because it is only one component competing with other ones for inclusion in the larger set of the formalness-informalenss principle (like the *completeness-incompleteness* pair and the *soundness-unsoundness* pair, as discussed in my book titled *The Future of Post-Human Mathematical Logic*, or *FPHML* in short).

The same is true for the *cognition-noncognition* pair, which is related with both the completeness-incompleteness pair and the soundness-unsoundness pair in the larger set of the formalness-informalenss principle.

(b2) In regard to "compound" (or non-atomic) family resemblances—there are some pairs which are not appropriate for a different reason, because they fail the test of selection criteria (as already pointed out in an earlier section), especially in relation to the criterion of not *overlapping* with other principles, in that it is too closely related with other ones.

For instance, the *normality-abnormality* pair is compound (and thus inappropriate), because it overlaps with several other principles, like the symmetry-asymmetry principle, the density-emptiness principle, the convention-novelty principle, and the same-difference principle, for instance (as already discussed in *FPHP*).

The normality-abnormality pair overlaps with the symmetry-asymmetry principle, because being "abnormal" (say, for those with mental illnesses) is often treated not equally by others who are "normal." So, the relationship reveals the symmetry-asymmetry principle.

The normality-abnormality pair also overlaps with the density-emptiness principle, because being "abnormal" (say, for those with mental illnesses) is often not as dense in the overall (not local) distribution of a population as those who are "normal." So, the relationship reveals the density-emptiness principle.

The normality-abnormality pair also overlaps with the convention-novelty principle, because being "abnormal" (say, for those with mental illnesses) often exhibits non-conventional (non-conformist) ideas and behaviors when contrasted with those who are "normal." So, the relationship reveals the convention-novelty principle.

The normality-abnormality pair also overlaps with the same-difference principle, because being "abnormal" (say, for those with mental illnesses) shares some similarities in outcome when compared with those who are "normal" (in that both contribute to the control and regulation of human thoughts and behaviors in society, albeit in different ways). Yet, being "abnormal" also has something different from being "normal," in that abnormal thoughts and behaviors follow a different pattern when contrasted with those which are normal. So, the relationship reveals the same-difference principle.

(b2.1) There is a special case of compound family resemblances (as summarized in *Table 4.10*), which can be called, in the absence of better words, *one-sided and compound* family resemblances.

For instance, the extroversion-introversion pair is a good example of *one-sided and compound* family resemblances, because, firstly, it is "one-sided," in being part of the side (category) of "normality" in the normality-abnormality pair—and, secondly, the normality-abnormality pair is in turn "compounded," because it is related to the symmetry-asymmetry principle, the density-emptiness principle, the convention-novelty principle, and the same-difference principle, as already explained above.

(b3) In regard to "one-sided" family resemblances—there are some pairs which are not appropriate for another different reason, because they fail the test of selection criteria (as already pointed out in an earlier section), especially in relation to the criterion of not being *one-sided,* that is, part of one side (category) of a dialectic pair.

For instance, the extroversion-introversion pair cannot be an ontological principle, since it is part of the side (category) of "normality" in the normality-abnormaltiy pair (because both extroversion and introversion are two different "normal" personality traits in psychology).

By the same logic, the vision-blindness pair and the sound-silence pair cannot be ontological principles, because the pairs are part of the side (category) of "asymmetry" in the symmetry-asymmetry principle (since being able to see and hear for some individuals and being unable to see or hear for others are two contrastive abilities in the world of animals and humans, for instance).

These illustrations of indequate, compound, and one-sided family resemblances show how difficult it is to come up with an ontological principle in existential dialectics out of the blue, because many of them violate the selection criteria in one way or the other. It is all too easy to join two opposing words and simply make them into a pair, but it is all too difficult to pass the test of the selection criteria to be qualified as an ontological principle in existential dialectics.

(b4) And in regard to "vague" family resemblances—there are some pairs which are not appropriate for still another different reason, because the terms in the pairs are vague in meaning, so that they are inappropriate family resemblances if defined in some particular ways, although they can be appropriate family resemblances if defined in other particular ways.

For instance, such pairs like shortness-longness, narrowness-broadness, smallness-largeness, smallness-greatness, microness-macroness, and littleness-muchness can be understood in at least three different meanings and thus have three different possibilities. The first possibility is that they are one-sided as part of finiteness (in the finiteness-transfiniteness principle), if both opposites are defined on a finite scale. The second possibility is that they are one-sided as part of transfiniteness (in the finiteness-transfiniteness principle) if both opposites are defined on a transfinite scale. And the third possibility is that they are appropriate family resemblances in relation to the finiteness-transfiniteness principle if one opposite is defined on a finite scale and the other opposite is defined on a transfinite scale.

Other examples of inappropriate family resemblances are shown in *Table 4.10* on "The Syntax of Existential Dialectics VII: Types of Inappropriate Family Resemblances."

The Dialectic Constraints Imposed by the Principles

The principles, as they constitute the syntax of existential dialectics, are dialectic in character, such that, when they are applied, they impose dialectic constraints on how reality is to be understood.

Consider, say, the symmetry-asymmetry principle as an illustration here, in order to summarize two main characters of the dialectic constraints in question (as summarized in *Table 4.7*).

Firstly, to be dialectic here is to go beyond the narrow dichotomies (and, for that matter, any rigid multi-dimensional classificatory scheme), be they about "self" vs. "world," "freedom" vs. "unfreedom," "barbarity" vs. "civilization," "individuality" vs. "communality," and so on.

One way to do so (to go beyond) is to consider them all in terms of co-existence (without heavily favoring one over the rest). For instance, my theory of "post-civilization" (to be summarized later in the section on the pragmatics of existential dialectics) is to go beyond barbarity and civilization in terms of understanding barbarity and civilization as being co-existent.

And the same logic can be said in relation to my theories of "post-democracy," "post-capitalsm," and others (also to be introduced later in the section on the pragmatics of existential dialectics), in regard to freedom vs. unfreedom, equality vs. inequality, communality vs. individuality, spirituality vs. materiality, and so on.

But to consider them all (in the dichotomies—and, for that matter, in any rigid multi-dimensional classifactory scheme) as co-existent is not the same as to imply that the opposing categories in any classificatory scheme are all equal, since, in accordance to the symmetry-asymmetry principle (as an illustration here), if they are equal in terms of being considered as co-existent, they are asymmetric in terms of being unequal in dialectic interaction (e.g., X can be more dominant than Y in case A, or Y is more dominant than X in case B), although in some special cases, they can be relatively equal (e.g., X and Y are relatively equal in case C).

For this reason, there are different versions of "post-democracy" and "post-capitalism" in my theories. As an illustration, in version I of the theory of post-democracy, freedom is more dominant than equality, whereas in version II of the theory of post-democracy, equality is more so than freedom.

But this "X more than Y" has to be understood in the context of dialectic logic (not in symbolic logic, as conventionally understood), in that both "X" and "Y" are important in post-democracy (in the context of dialectic logic), but in an asymmetry way.

By contrast, in symbolic logic, it often favors one over the other—be it in regard to privileging freedom over equality (in Fascism), favoring freedom relatively more than equality (in Liberal Democracy), or favoring equality relatively more than freedom (in Socialist Democracy). In regard to the latter two cases (about Liberal Democracy and Socialist Democracy), the difference between dialectic logic and symbolic logic can be one in degree, not in kind—in this sense, albeit not in other senses.

The same reasoning can be said about the relatonships between individuality and communality, between spirituality and materiality, and between formal legalism and informal legalism in the different versions of my theory of "post-capitalism."

Secondly, to be dialectic is to go beyond the narrow dichotomies (and, for that matter, any rigid multi-dimensional classificatory scheme) in another way, this time, in a transcendent way, that is, in exploring other possibilities or even other issues not considered within the narrow confines of narrow dichotomies (and, for that matter, any rigid multi-dimensional classificatory scheme).

As an analogy, in this second meaning, to go beyond the narrow dichotomy of "black" vs. "white" is not just to choose both "black" and "white" (as in the first meaning) but also to explore other options (e.g., "green," "purple," "blue," etc.—and, alternatively, "shade," "line," "curve," etc.).

By the same reasoning, to go beyond "democracy" is to transcend democracy (as in version III of my theory of "post-democracy") and to explore other possibilties of lifeforms (e.g., "floating consciousness," "hyper-spatial consciousness," etc., to live beyond the narrow obsession with freedom and equality).

This dialectic character of the principles in existential dialectics has important implications for the pragmatics of existential dialectics (as will be clear shortly, in the section on the pragmatics of existential dialectics).

The Semantics of Existential Dialectics (or Its Ontosemantics)

The syntax of existential dialectics so understood in terms of ontological principles only gives us the structure of ontology in the world, in an abstract (general) sense. These principles by themselves do not tell us the specific meanings in a given context.

In order to grasp the specific meanings of the principles in a given context, it is necessary to study the semantics of existential dialectics (as summarized in *Table 4.11*).

And the analysis of the ontological principles as family resemblances in the previous section on syntax is only a starting point (and thus overlaps a bit with the semantics of existential dialectics here).

For this reason, I have often gone to great lengths in my previous books on different subjects to explain the specific meanings of the principles when applied in different contexts.

Consider two examples below for illustration, which are not exhaustive.

In *FPHST*, for instance, I used the first three principles (i.e., the change-constancy principle, the regression-progression principle, and the symmetry-asymmetry principle) to propose "the perspectival theory of space-time," for a better way to understand space and time—especially, though not exclusively, in relation to future post-human history (as summarized in *Table 1.3*, *Table 1.4*, *Table 1.5*, and *Table 1.6*).

In so doing, I had to introduce concepts and theories specific to the field of physics and other related fields (e.g., "absolute space" and "absolute time" in "classical mechanics" and "relative space-time" in "the theory of relativity").

And in *BNN*, I also exploited the three principles to propose the "transcendent" approach to the study of genes and memes as a new way to understand the interaction between nature and nurture.

In so doing, I had to explore concepts and theories in the world of evolutionary biology (e.g., "mutation," "variation," "adaptation," "selection," and "inheritance" in classical evolutionary theory) and neural biology (e.g., "chromosome," "gene," "DNA," "RNA," "protein," "neuron," "neural network," and "behavior").

Other examples, besides these two here, are given in other books of mine on different topics.

Therefore, this semantic dimension of existential dialectics is important, since, in each of my books when the ontological principles are used, each principle becomes relevant in a different meaning when contextualized within a given case study in question.

In this book, for instance, the principles are relevant in relation to a different subject matter (as summarized in the concluding chapter), and they can be shown in all shapes and sizes (since they are also to be treated as "short cuts," as indicated before), without, however, being reduced to "specific ontology" (as opposed to "general ontology"), as explained in *Table 4.9*.

In the technical jargons of semantics, these exercises perform two essential functions of translation (or interpretation), namely, (a) "intensive" interpretation (about the properties or relations of the principles in existential dialectics), and (b) "extensive" interpretation (about the classes or cases to which the properties and relations of the principles in existential dialectics are applied). (WK 2011oo)

But there are limits of translation (or interpretation), and an excellent example is that "there is a problem with the 'objectivity' of translation, as it is relative to different cultural (or sub-cultural) contexts. For instance, one of the still unresolved problems in linguistics is: 'Is there an objective gauge for the quality of translation?'" (*FPHCOM* & WK 2011pp) And the problems of "radical translation" by W. V. Quine and "radical interpretation" by Donald Davidson are another well-known examples, as already discussed in *FPHSEM*.

This is one serious problem about the logic of translation and interpretation that I already addressed in *Sec. 2.5.2* of *FPHSEM*, *Sec. 3.5.1* of *FPHCOM* (and also, for that matter, in *FHC*, in the context of hermeneutics).

The Pragmatics of Existential Dialectics (or Its Ontopragmatics)

Even the semantics of existential dialectics is not enough to understand reality, since semantics by itself does not tell us the nature of intentionality and interests in relation to the matrix of power formations and conflicting interests when existential dialectics as an ontology is applied.

Surely, one cannot completely separate the pragmatics of existential dialectics from its semantics (and for that matter, its syntax), but the point here is to give each a more elaborated analysis.

For this reason, the pragmatics of existential dialectics needs to be studied too, in special relation to the inclusion of power formation and conflicting interests in the application of the principles of existential dialectics (as summarized in *Table 4.12*).

It is thus no wonder that all my previous books have shown in different ways how and why these ontological principles and their theoretical applications can reveal the future world to come in a way that humans have never known, in a totally different battleground for competing human (and later post-human) interests by myriad groups fighting for their own versions of hegemony.

Perhaps a summary of my previous works in relation to existential dialectics and their contributions to my new theories as proposed over the years is deemed revealing here, for the further understanding of the pragmatics of existential dialectics (in close relation to the semantics of existential dialectics too).

That stressed—the following summary in four sub-sections is something that I often do in my previous books too, as a way to introduce the intellectual background of my ideas.

So, for those readers who had read my previous works, the following sub-sections serve as a reminder, as they are repeated hereafter *often verbatim* (with only some updated changes, with the rest kept intact).

These sub-sections are, namely, (a) the two-way street connecting theory with meta-theory, (b) the theoretical application of existential dialectics, (c) direct and indirect applications of existential dialectics, and (d) multiple applications of existential dialectics—to be summaried herefafter, respectively.

The Two-Way Street Connecting Theory and Meta-Theory

However, lest any misunderstanding occurs, it is important to remember that the pragmatics of existential dialectics is not a one-way street (that is, using the ontological principles for theoretical insights in praxis) but a two-way one, namely, (a) from meta-theory to theory, and (b) from theory to meta-theory.

(a) On one side of the street, the ontological principles can inspire some theoretical insights in praxis, that is, in relation to some specific fields.

(b) On the other (opposing) side of the street, however, the study of a subject matter in the specific fields in turn reveals some more hitherto unknown ontological principles to be discovered and identified.

For this reason, three new principles were added in *FC*, one in *FAE*, two in *ALD*, and three in *FIA*, on top of the original three in *BCPC*, for example (as there are more)—after some research on the specific subject matters.

Both theory and meta-theory enrich each other in all my works over the years.

The Theoretical Application of Existential Dialectics

This two-way street connecting theory and meta-theory has contributed to my original construction of many theories in numerous books of mine on numerous fields in all domains of knowledge, which can be classified in terms of specific "meta-theories" (about epistemology, methodology, logic, ontology, etc.) and specific "theories" (about the mind, nature, culture, society, and the rest—as will be further explained later in the sub-section on "the holistic organization of an inquiry" in the section about "sophisticated methodological holism").

The final aim of my numerous new specific theories (on both theory and meta-theory) is to converge all of them (as proposed in my numerous books) into a grand project about the future of intelligent life, both here on earth and in deep space.

In this sense, all these theories of mine serve as a foundation to unify all domains of knowledge for an overarching understanding of all forms of reality (by way of some ontological principles and the comprehensive perspectives of the mind, nature, society, and culture), while allowing for heterogeneity, conflict, subjectivity, and complexity, for example.

With this grand project in mind—the summary of my specific theories on meta-theory are given in this section (on "existential dialectics") and the next (on "sophisticated methodological holism"), whereas the specific theories on theory will be summarized in the following categories of (a) society, (b) culture, (c) the mind, (d) nature, and (e) the rest, as shown below, in that order.

(a) Society

The category of "society" here can be divided into four sub-categories, i.e., (a1) social organizations, (a2) social institutions, (a3) social structure, and (a4) social systems (or trends).

(a1) Social Organizations

In relation to social organizations, I proposed a new thesis on the ambivalent nature of virtual organizations in the future, known as "the theory of e-civic alienation" in *FCD* (and summarized in *Table 4.19*).

And in *FPHO*, another theory called "the combinational theory of organization" was worked out, although it was first proposed in Ch.6 of *ALD*.

(a2) Social Institutions

There are different social institutions, be they about (a2.1) politics, (a2.2) economics, (a2.3) military affairs, (a2.4) mass media, (a2.5) law, (a2.6) education, and (a.7) gambling, for instance.

(a2.1) In relation to politics, I offered different visions on (a2.1.1) political theory, (a2.1.2) comparative politics, and (a2.1.3) international relations, to cite three main ones.

(a2.1.1) In the context of political theory, my vision of a path-breaking political system in future times concerns the different forms of "post-democracy" to supersede democracy unto the post-human age.

My vision of post-democracy was called "the theory of post-democracy" (as summarized in *Table 4.22*, *Table 4.23*, and *Table 4.24*), whereas *Table 4.25* explains the distinctions among democracy, non-democracy, and post-democracy.

And *Table 4.26* stresses the multiple causes of the emergence of post-democracy, while *Table 4.27* clarifies some possible misunderstandings in regard to post-capitalism and post-democracy.

(a2.1.2) In the context of comparative politics, I also showed, in *ALD*, how and why authoritarianism can be an advanced form of liberal democracy (just as it has its own counterpart in non-liberal democracy), or what I originally called "authoritarian liberal democracy" (as summarized in *Table 4.21*), before it is to be superceded by "post-democracy."

(a2.1.3) In the context of international relations (both at the institutional and structural levels), I provided, in *BWT*, "the theory of cyclical progression of empire-building," for a better way to understand the logic of empire-building on earth and beyond—with the rise of what I originally envisioned as "hyper-empires" (like "the Chinese Union" and "the Indian Union"), "meso-empires" (like "the European Union" and "the North American Union") and "micro-empires" (like "the Middle Eastern Union" and "the Latin American Union") in the coming "post-post-Cold War era" as a visionary illustration, as summarized in *Table 4.20*.

This is also related to the issues of "social structure" in (a3), "migration" in (a4,6), "identity" in (b11), and "geography" in (d4), for instance.

In addition, in *FPHWP*, I proposed "the multifaceted theory of war and peace" to understand war and peace, in a small sense—or aggression and pacificity, in a broad sense.

(a2.2) In relation to economics, I also offered another vision, this time to understand the future forms of economic systems, that is, the different forms of "post-capitalism" to supersede capitalism unto the post-human age.

My vision of post-capitalism was referred to (in *FCD*, *BDPD* and *BCPC*) as "the theory of post-capitalism" (as summarized in *Table 4.28*, *Table 4.29*, *Table 4.30*, *Table 4.31*, and *Table 4.32*). And *Table 4.33* clarifies the differences among capitalism, non-capitalism, and post-capitalism, whereas *Table 4.34* shows multiple causes of the emergence of post-capitalism.

This vision on "post-capitalism" constitutes the third theoretical application of existential dialectics, this time, in that there is no wealth without poverty (or the wealth/poverty dialectics).

(a2.3) In relation to military affairs, I also proposed "the multifaceted theory of war and peace" (in *FCD*, although it was first so labeled in

FPHK) for a new way to understand the causes of war—and, for that matter, the rise and fall of different forms of warfare over time.

This theory was further expanded in *FPHWP*, where I suggested a new way to understand war and peace, in a small sense—or aggression and pacificity, in a broad sense.

(a2.4) In relation to mass media, I also proposed "the totalistic theory of communication" in *The Future of Post-Human Mass Media* (or in short, *FPHMM*).

(a2.5) In relation to law, I also proposed "the reconstruction theory of law" in *The Future of Post-Human Law* (or in short, *FPHLAW*). And in relation to criminality, I proposed "the reflective theory of criminality" in *FPHCRIM* for a different way to understand heroes and villains.

(a2.6) In relation to education, I proposed "the heterodox theory of education" in *The Future of Post-Human Education* (or in short, *FPHEDU*).

(a2.7) And in relation to gambling, I proposed "the detached theory of gambling" in *The Future of Post-Human Gambling* (or in short, *FPHGAM*).

(a3) Social Structure

In relation to social structure, I first proposed two additional theories, that is, "the theory of the cyclical progression of hegemony" in *FCD* (and also in *FPHC* and *BDPD*), and "the theory of the cyclical progression of empire-building" (in *BWT*, as already indicated above)—with important implications for the symmetry-asymmetry princple in existential dialectics, just to cite one example.

In addition, this is also related to the issues of "migration" in (a4,6), "identity" in (b11), and "geography" in (d4), for instance.

Then, in *BNN*, I proposed another theory, namely, "the theory of contrastive advantages" to understand the persistence of social structure by gender, race, ethnicity, or any other groupings.

(a4) Social Systems (or Trends)

In relation to social systems, there are several sub-categories for illustration, which can be about (a4.1) urbanization, (a4.2) technology, (a4.3) demography, (a4.4) the environment, (a4.5) social trends as a whole (which also include the current phenomenon of globalization), (a4.6) migration, and (a4.7) resources.

(a4.1) In relation to urbanization, I proposed "the contingent theory of urban planning" in *FPHUP* for a new way to understand density and void—which is also related to the issues of "social trends" (especially in relation to globalization) in (a4.5), "migration" in (a4.6), and "identity" in (b11).

(a4.2) In relation to technology, I worked out "the ambivalent theory of technology" in *FPHE* for a new way to understand the nature of technology.

Some other books (like *FPHMM*, *FAE*, *FCD*, and *FHC*) also deal with technology, and a good example concerns "the synthetic theory of information architecture" in *FIA* on taxonomy, network, and intelligence, in conjunction with the rise of technology.

(a4.3) In relation to demography, I proposed, first in *FHC* (and later in other books like *FCD*, *FPHC*, *FPHU*, etc.), "the theory of post-humanity," in relation to the "post-human" age at some distant point of "after-postmodernity," long after human extinction, to be eventually superseded by "post-humans" of various forms.

Some good candidates of the post-human lifeforms include, for instance, "thinking robots," "thinking machines," "cyborgs," "genetically altered superior beings," "floating consciousness," "hyper-spatial consciousness," "unfolding unconsciousness," "the hyper-sexual body," and "the hyper-martial body." This post-human vision of mine was first originally worked out in *FHC* and further elaborated in both *FCD*, *FPHC*, *FPHU*, *FPHS*, and *FPHMA*, for instance.

Lest any confusion occurs, I need to emphasize two clarifications here about the term "post-human" as a neologism in my works.

Firstly, the word "post-human" here should *not* be confused with another term which looks similar but has a totally different meaning in the literature of postmodernism, namely, "post-humanism"—which constitutes a critique of "humanism" as tradionally understood (especially, though not exclusively, in relation to the idea of progress in science and reason in the Enlightenment project). (WK 2008)

And secondly, the word "post-human" here should also *not* be confused with a similar term which is used to champion the ideology of technology for the future co-existence between humans and cyborgs in "trans-humanism." (WK 2008a)

Instead, my term "post-human" in relation to "posthuman-ism" also rejects "transhumanism" (especially, though not exclusively, in relation to the promises of technology) and refers to something else altogether, that is, the future extinction of humans and its post-human successors in deep space and beyond unto multiverses.

Later, in *BEPE*, I went further and provided a more comprehensive critique of "transhumanism," together with other issues.

With these two clarifications in mind (as summarized in *Table 1.31*)—in the end, a most fundamental question about intelligent life now has an answer, in that, if asked, "What is the future of human civilization?"—my answer in *FCD* (89) is thus: "As addressed in Ch.7 of *FHC*, a later epoch of the age of after-postmodernity (that is, at some point further away from after-postmodernity) will begin, as what I called the 'post-human' history (with the term 'post-human' originally used in my doctoral dissertation at M.I.T., which was finished in November 1995, under the title *After Postmodernity*, still available at M.I.T. library, and was later revised and published as *FHC*). The post-human history will be such that humans are nothing in the end, other than what culture, society, and nature (with some luck) have shaped them into, to be eventually superseded by post-humans (e.g., cyborgs, thinking machines, genetically altered superior beings, and others), if humans are not destroyed long before then."

(a4.4) In relation to the environment, I also addressed the issue of sustainability, especially in *FHC*, *FCD*, and *FPHUP*. For instance, in *FPHUP*, I specifically proposed a new theory, "the contingent theory of urban planning," on the need to go beyond the contemporoary obsession with sustainability, in the context of not only the environment but also other issues.

I also proposed "the interventive-reshaping theory of geography" in *FPHGEOG*—which is related to the issues of "migration" in (a4.6), "geography" in (d4), and "geology" in (d5).

In addition, I proposed "the transfigurative theory of waste" in *FPHW*.

(a4.5) In relation to social trends as a whole, I proposed in Ch.9 of *FCD* (367-8), that "civilizational history will continue into the following cyclical progression of expansion, before it is to be superseded (solely as a high probability, since humans might be destroyed sooner either by themselves or in a gigantic natural calamity) by posthumans at some distant point in after-postmodernity (which I already discussed in *FHC*)" unto multiverses (different constellations of universes):

Local → Regional → Global →Solar →
Galactic → Clustery…→ Multiversal

The current phenomeon of "globalization" lies in the third category on the line above and was extensively analyzed in *FCD*, in close connection with "globalism" at the cultural level—in the context of "postmodernity" (which will be introduced later in the section on history).

In *BDPD*, this thesis of mine was referred to as "the theory of the cyclical progression of system integration and fragmentation," at the systemic level—in close relation to another thesis analyzed in Ch.9 of

FCD, which was called, in the absence of better words, "the multifaceted theory of war and peace" in *FPHK* (as already indicated above).

In *FC*, I also proposed "the dialectic theory of complexity" to account for the factor of uncertainty, which can have a major role to play in the world.

All this is also related to the issues of "urbanization" in (a4.1), "the environment" in (a4.4), "migration" in (a4,6), "identity" in (b11), "geography" in (d4), and "geology" in (d5).

(a4.6) In relation to migration, I proposed "the resettlement theory of geology" in *FPHGEOL* and "the theory of the cyclical progression of migration" in *FPHMIG*—which are also related to the issues of "social structure" in (a3), "urbanization" in (a4.1), "the environment" in (a4.4), "social trends" in (a4.5), "identity" in (b11), "geography" in (d4), and "geology" in (d5).

(a4.7) And in relation to resources, I proposed "the resilient theory of natural resources" in *BNR* —which is also related to the issue of "natural resources" in (d10).

(b) Culture

The category of "culture" here can also be divided into different sub-categories like (b1) aesthetics, (b2) religion, (b3) popular (mass) culture, (b4) logic, (b5) civilization as a whole, (b6) ethics, (b7) epistemology (methodology), (b8) metaphysics (ontology), (b9) ideology, (b10) language, and (b11) identity.

(b1) Aesthetics

In relation to aesthetics, I proposed "the transformative theory of aesthetic experience" in *FAE* about some great transformations of aesthetic experience in future history.

In *FPHLIT*, I further proposed "the comparative-impartial theory of literature" for a new way to understand fiction and non-fiction.

In *FPHPA*, I further proposed "the transdisciplinary theory of performing arts" for a new way to understand the body and its presence—which is also related to the issues of "popular (mass) culture" in (b3) and of "the human body" in (c1.2). And the aesthetic issue also overlaps with (d8) on architecture.

And in *FPHCA*, I further proposed "the inquisitive theory of culinary art" for a new way to understand ingredients and techniques.

(b2) Religion

In relation to religion, I proposed "the comparative theory of religion" in *FHC* (and also in *FCD*, although it was first so labeled in *FPHK*).

And in *FPHR*, I provided a solid foundation of this theory with a better name, that is, "the comparative-substitutive theory of religion" for a new way to understand spirituality.

(b3) Popular (Mass) Culture

In relation to popular (mass) culture, there are some sub-categories for illustration, which can be, for example, (b3.1) mass culture in general and (b3.2) sports—both of which are also related to the issues of "aesthetics" in (b1) and of "the human body" in (c1.2).

(b3.1) In relation to mass culture in general, I proposed, in *FHC*, a new way to understand mass culture in terms of some fundamental dilemmas, which is called, in the absence of better words, "the dualistic theory of mass culture" in *FPHO*.

(b3.2) In relation to sports (like chess), I proposed, in *FPHCHESS*, "the mediative-variative theory of chess" for a new way to understand tactics and strategy (in different traditions).

And in *FPHSPORT*, I proposed "the modificative theory of sports" for a new way to understand training and winning.

(b4) Logic

In relation to logic, I also worked out, in *FPHML*, "the contrastive theory of rationality" for a new way to understand rationality and knowledge (in the context of mathematical logic).

And in *FPHPROB*, I propose "the interpretivist theory of probability" for a different way to understand objectivity and subjectivity—which is related to the issue of logic.

(b5) Civilization as a Whole

In relation to civilization as a whole, I also argued, in *FHC*, no freedom without unfreedom (as shown in *Table 4.13*) and no equality without inequality (as shown in *Table 4.14*), especially in relation to the seven dimensions of life existence first worked out in *FHC* (i.e., the technological, the everyday, the true, the holy, the sublime/beautiful, the good, and the just).

Even more interestingly, in *BDPD* and later in *BCIV*, I further revealed a theoretical refinement of existential dialectics, this time, in arguing that there is no civilization without barbarity, with human civilization to be eventually superseded by what I originally analyzed as "post-human post-civilization" (which should not be confused with "post-human civilization"), in the context of the freedom/unfreedom and equality/inequality dialectics.

In the final analysis, civilization cannot live without barbarity and has to learn to co-exist with it in ever new ways. Preposteorus as this may seem to many contemporaries—it is no more imperative to preserve civilization than necessary to destroy barbarity, and the ideal of civilization is essentially bankrupt, to be eventually superseded by "post-civilization" (as summarized in *Tables 4.35-4.43*).

This freedom/unfreedom and equality/inequality dialectics holds in pre-moderny, modernity, postmodernity, and what I orignally called "after-postmodernity" in *FHC* and *FCD*—especially, though not exclusively, in the context of the emergence of "post-humans."

There are two specific theories about "after-postmodernity" here. The first theory about "after-postmodernity" is about the trinity of pre-modernity, modernity, postmodernity, and after-postmodernity at the cultural level—and was called, in *BDPD*, "the theory of the trinity of modernity to its after-postmodern counterpart," although it was first analyzed in *FHC* and later in *FCD* and also *BCIV* (as summarized in *Table 1.34*, *Table 1.35*, *Table 1.36*, and *Table 1.37* and will be elaborated further in the later sub-section on "the rest" about future history).

The second theory about "after-postmodernity" is about pre-modernity, modernity, postmodernity, and after-postmodernity at the historical level. For instance, a conclusion in *FHC*, which is shocking indeed for many contemporaries, is none other than that "[t]he post-human history will therefore mark the end of human history as we know it and, for that matter, the end of human dominance and, practically speaking, the end of humans as well. The entire history of human civilization, from its beginning to the end, can be summarized by four words, linked by three arrows (as already discussed in *FHC*)":

Pre-Modernity → Modernity →
Post-Modernity → After-Postmodernity

In *BDPD*, this thesis of mine was specifically called "the theory of the evolution from pre-modernity to after-postmodernity," at the historical level—as shown in *Table 1.34*, *Table 1.35*, *Table 1.36*, and *Table 1.37* (and will be elaborated further in the later sub-section on "the rest" about future history).

Therefore, "[t]he end of humanity in the coming human extinction is the beginning of post-humanity. To say an untimely farewell to humanity is to foretell the future welcome of post-humanity." (P. Baofu 2002: 89) This thesis of mine was known in *BDPD* as "the theory of post-humanity," at the systemic level—as already indicated in (a4.3) above.

(b6) Ethics

In relation to ethics, I proposed, in *BEPE*, "the theory of post-ethics" for a new way to understand morality and immorality.

(b7) Epistemology (Methodology)

In relation to epistemology (methodology), I proposed "the complex theory of data analysis" in *FPHDA* for a different way to understand quantitative and qualitative research.

In addition, I already proposed "the critical-dialectic theory of formal science" in *FPHFS* and "the theory of methodological holism" in several books (like Ch.1 *FCD*, Ch.1of *FPHC*, *Sec.2.1* and *Sec.2.5* of *BCPC*, *FC*, *FPHK*, etc.) for a different way to do methodology.

And in *FPHPROB*, I propose "the interpretivist theory of probability" for a different way to understand objectivity and subjectivity.

(b8) Metaphysics (Ontology)

I also proposed "the theory of existential dialectics" in several books (like *FHC*, *FCD*, *FPHC*, *BDPD*, *FC*, *FAE*, *ALD*, *FIA*, *FPHK*, etc,) and "the contrastive theory of rationality" in *FPHML* for a different way to understand ontology.

(b9) Ideology

The issue about ideology was already discussed under the categories of "social institutions" and of "civilization as a whole," especially in relation to my theories of "post-democracy," "post-capitalism," "post-civilization," and so on.

(b10) Language

I already proposed "the multilogical theory of learning" in *FPHL* for a new way to understand the nature of language structure and context .

And I also worked out, in *FPHSEM*, "the interactive theory of semantics" for a new way to understand internality and externality (in the context of semantics).

(b11) Identity

I proposed "the theory of the cyclical progression of migration" in *FPHMIG* for a new way to understand the nature of sameness, otherness, and identity—which is also related to the issues of "social structure" in (a3), "urbanization" in (a4.1), "social trends" (especially in relation to globalization) in (a4.5), "migration" in (a4.6), religion in (b2), and ideology in (b9).

(c) The Mind

The important vision about "post-humans" (as summarized earlier) brings us to the category of "the mind" here, which, as is often understood in everyday's usage, refers to the "brain" and the "body" (together with their interactions) and has different levels of consciousness which can be both "spiritual" and non-spiritual.

Therefore, the "mind" already presupposes the "brain," "body," and "spirit" (or "soul"), so that one does not fall into the intellectual trap of the "mind-body dualism" as debunked in modern (and especially, postmodern) philosophy (which I already went to great lengths to explain in *The Future of Human Civilization*).

That said—the category of "the mind" can be divided into two sub-categories, namely, in relation to (c1) biology and (c2) psychology

(c1) Biology

In relation to biology, there are two concerns here, namely, (c1.1) genes and intelligence and (c1.2) the human body.

(c1.1) In relation to genes and intelligence, I worked out "the theory of contrastive advantages" (which was originally proposed in *FCD* and further worked out in *BNN*, as summarized in *Table 1.24*), to show the nature-nurture interactions of multiple levels in action (e.g., the biological, the psychological, the structural, the systemic, the cultural, and others) for humans and post-humans on earth and beyond, to the extent that different groups, be they on the basis of race, gender, ethnicity, class, age, or else, are not equal, on average, and have not been, nor will they be, even if there can be exceptions.

(c1.2) In relation to the human body, for instance, in *FPHMA*, I suggested "the expansive-contractive theory of martial arts," in which two great future transformations of the martial body were envisioned, that is, both "with the hyper-martial body" and "without the body" (as summarized in *Table 1.32*)—although this issue can also be classified under the category of "culture."

In *FPHS*, I proposed "the theory of virtual sexuality," in which three great future transformations of the sexual body were envisioned, that is, "virtual pleasure," "the hyper-sexual body," and "the post-sexual floating mind" (as summarized in *Table 1.33*).

And in *FPHPA*, I further proposed "the transdisciplinary theory of performing arts" for a new way to understand the body and its presence, with seven great future transformations of the body in performing arts—which is also related to the issues of "aesthetics" in (b1) and "popular (mass) culture" in (b3).

(c2) Psychology

In relation to psypchology (which also overlaps with biology too, however), I also proposed some greater transformations of the mind to come in the future, especially though not exclusively in the post-human age.

These greater transformations of the mind in psychology can be further sub-divided into different categories, namely, (c2.1) consciousness, (c2.2) unconsciousness (together with subconsciousness), (c2.3) creativity, (c2.4) learning, (c2.5) aggression and pacificity, (c2.6) personality, (c2.7) humor, and (c2.8) artificial intelligence.

(c2.1) In relation to consciousness, I analyzed, in *FPHC*, the nature of consciousness in relation to the conceptual dimensions of consciousness (as shown in *Table 1.12*), the theoretical levels of consciousness (as shown in *Table 1.13*), the thematic issues of consciousness (as shown in *Table 1.14*), and the different dimensions of human existence (viz., having-ness, belonging-ness, and being-ness) in relation to consciousness (as shown in *Table 1.15*, *Table 1.16*, *Table 1.17*, and *Table 1.18*).

In *BCPC*, the nature of consciousness was further explored in relation to cognition, emotion, and behavior, which can be in consciousness and other mental states (e.g., subconsciousness and unconsciousness), in the context of existential dialectics, as shown in *Table 1.19*, *Table 1.20*, *Table 1.21*, *Table 1.22*, and *Table 1.23*.

In fact, in *FCD*, I already proposed three different theories in relation to the limits of cognition, emotion, and behavior, namely, "the theory of cognitive partiality," "the theory of emotional non-neutrality," and "the theory of behavioral alteration."

Then, I argued that the existence of human consciousness will be superseded one day with "floating consciousness" (as first proposed in *FCD*) and "hyper-spatial consciousness" (as first proposed in *FPHC*) as a climax of evolution in consciousness, after the future extinction of human consciousness (as shown below):

Primordial consciousness → Human consciousness →
Post-human consciousness (with floating consciousness and hyper-spatial consciousness as a climax in the evolution of consciousness)

In *BDPD*, these latest theses of mine were called "the theory of floating consciousness" and "the theory of hyper-spatial consciousness," both at the cosmological and psychological levels (as summarized in *Table 1.26* and *Table 1.25*).

(c2.2) In relation to unconsciousness, for instance, in *FPHU*, the unconsious mind will undergo tremendous change, to the rise of what I originally called "unfolding unconsciousess" in relation to anomalous experience, as a climax of evolution in unconsciousness, in conjunction with the conscious mind (both human now and post-human later), especially at the psychological level, as shown below:

Primordial unconsciousness → Human unconsciousness →
Post-human unconsciousness (with unfolding unconsciousness
as a climax in the evolution of unconsciousness)

In *FPHU*, this thesis of mine is called "the theory of unfolding unconsciousness" in a small sense, or "the unfolding theory of anomalous experience," in a larger one, as shown in *Table 1.27* and *Table 1.28*.

(c2.3) In relation to creativity, for instance, in *FPHCT*, I also proposed "the comprehensive theory of creative thinking" to explore the role of creative thinking, in relation to technology and other factors, together with my original proposal of important creative techniques and traits (as shown in *Table 1.29*) and its double nature of desirability and undesirability (as shown in *Table 1.30*).

(c2.4) In relation to learning, for instance, in *FPHL*, I worked out "the multilogical theory of learning" for a new way to understand the nature of structure and context in language, in relation to learning and its future development in the context of the mind.

(c2.5) In relation to aggression (and pacificity), I proposed "the multifaceted theory of war and peace" in *FPHWP* (as mentioned above).

(c2.6) In relation to personality, I worked out "the contrarian theory of personality" in *FPHP* for a new way to understand normality and abnormality.

(c2.7) In relation to humor, I worked out "the metamorphic theory of humor" in *FPHH* for a new way to understand joking and laughing.

(c2.8) And in relation to artificial intelligence, I proposed "the supersession theory of computing" in *FPHCOM* for a new way to understand hardware, software, and the mind.

In addition, I also proposed "the complex theory of data analysis" in *FPHDA* for a different way to understand quantitative and qualitative research in the context of the evolution of new minds.

(d) Nature

The category of "nature" here refers to the state of nature, which can be divided into some sub-categories like (d1) physics, (d2) cosmology, (d3) biology, (d4) geography, (d5) geology, (d6) acoustics, (d7) chemistry, (d8) architecture, (d9) transportation, and (d10) natural resources.

But since biology in (d3) overlaps with biology in (c1)—only the rest of the categories are addressed hereafter.

(d1) Physics

In relation to physics, I also proposed "the perspectival theory of space-time" in *FPHST* (based on *FPHC*) for a new way to understand space and time, as shown in *Table 1.3*, *Table 1.4*, *Table 1.5*, and *Table 1.7*.

In *FPHG* (based on *FPHC* and *FPHST*), I proposed "the selective theory of geometry" for a new way to understand geometry, with implications for the understanding of space-time in the context of infinity, symmetry, and dimensionality for future lifeforms that our world has never known (as shown in *Table 1.6*, *Table 1.7*, *Table 1.8; Table 1.9*, *Table 1.10,* and *Table 1.11*).

And in *FC*, I proposed "the dialectic theory of complexity" for a new way to understand order and chaos in the state of nature and beyond.

(d2) Cosmology

In relation to cosmology, my theories on physics (as indicated above) already have important implications for understanding the nature and the future of the universe (which was also addressed in *FHC*, among other works aforcited).

I also proposed "the theory of hyper-spatial consciousness" in *FPHC* (and also in *FPHG*) for a new way to understand the future evolution of the mind to enter into different dimensions of space-time that our current world has never known.

In addition, in *BCOS*, I proposed "the theory of post-cosmology" for a new way to understand the beginnings and ends of the cosmos and beyond.

(d4) Geography

In relation to geography, I proposed "the theory of the geopower of nature" in *ALD* for a different way to understand the relationships between geography and other entities (like the mind, culture, and society).

I further refined this issue of geography by proposing "the interventive-reshaping theory of geography" in *FPHGEOG*—which is related to the issues of "the environment" in (a4.4), "migration" in (a4.6), and "geology" in (d5).

(d5) Geology

In relation to geology, I proposed "the resettlement theory of geology" in *FPHGEOL* for a different way to understand the relationships between statics and dynamics—which is related to the issues of "the environment" in (a4.4), "migration" in (a4.6), and "geography" in (d4).

(d6) Acoustics

In relation to acoustics, I proposed "the multilateral theory of acoustics" in *FPHA* for a different way to understand the relationships between sound and silence.

(d7) Chemistry

In relation to chemistry, I proposed "the creational theory of chemistry" in *FPHCHEM* for a different way to understand substances and their changes.

(d8) Architecture

In relation to architecture, I proposed "the transcendent theory of architecture" in *FPHARCH* for a different way to understand form and function.

(d9) Transportation

In relation to transportation, I proposed "the panoramic theory of transportation" in *FPHTRAN* for a different way to understand networks and operations.

(d10) Natural Resources

And in relation to natural resources, I proposed "the resilient theory of natural resources" in *BNR*—which is also related to the issue of "resources" in (a4.7).

(e) The Rest

The category of "the rest" here refers to what cannot be strictly classified in any of the previous categories (i.e., society, culture, the mind, and nature). A good illustration of "the rest" refers to (e1) history.

(e1) History

The category of history can be divided into some sub-categories like (e1.1) empirical history and (e.1.2) meta-history.

(e1.1) In relation to empirical history, I proposed "the theory of the evolution from pre-modernity to after-postmodernity" (as first worked out in *FHC* and later in *FCD* and *FPHC*), which overlaps—as a reminder from (b5)—with the issue concerning "civilization as a whole" in (b5).

For instance, in both *FCD* and *FPHC*, I worked out the structure of "post-human civilization" in terms of the trinity of after-postmodernity (i.e., "free-spirited after-postmodernity," "post-capitalist after-postmodernity," and "hegemonic after-postmodernity").

Both conceptually and theoretically, the trinity of after-postmodernity is a sequential extension of the trinity of modernity (i.e., "free-spirited modernity," "capitalist modernity,"and "hegemonic modernity") and the trinity of postmodernity (i.e., "free-spirited postmodernity," "capitalist postmodernity," and "hegemonic postmodernity") as first proposed in *FHC*.

And the trinity of pre-modernity (i.e., "pre-free-spirited pre-modernity," "pre-capitalist pre-modernity" and "hegemonic pre-modernity") was later conceived in *BCIV* to complete the historical set from pre-modernity to after-postmodernity.

In *BDPD*, this thesis about the trinity of pre-modernity, modernity, postmodernity, and after-postmodernity was collectively known as "the theory of the trinity of modernity to its after-postmodern counterpart," at the cultural level—as already so mentioned in (b5) earlier, and summarized in *Table 1.34*, *Table 1.35*, *Table 1.36*, and *Table 1.37*.

At the structural level, all these trinities are subject to the existential constraints (e.g., the freedom/unfreedom and equality/inequality dialectics in the context of "the cyclical progression of hegemony"), be the historical epoch in pre-modernity, modernity, postmodernity, or after-postmodernity in future times. Each of the historical epochs has its ever new ways of coming to terms with the ever new (different) mixtures of freedom/unfreedom and equality/inequality.

This is importantly so, not because, as is falsely assumed in conventional wisdom, one certain way is superior (or better) than another in terms of achieving more freedom and less unfreedom, or more equality with less inequality.

On the contrary, indeed, in each of the historical epochs, each increase of unfreedom greets each freedom achieved, just as each increase of inequality welcomes each equality achieved, albeit in ever new (different) ways. In *BDPD*, this thesis of mine was labeled as "the theory of the cyclical progression of hegemony," at the structural level, though it was first analyzed in *FCD*.

In *BDPD*, more theoretical applications of existential dialectics were further examined, in relation to five main features, in the context of the duality of oppression, namely, (a) that each freedom/equality achieved is also each unfreedom/inequality created, (b) that the subsequent oppressiveness is dualistic, both by the Same against the Others and itself and by the Others against the Same and themselves, (c) that both oppression and self-oppression can be achieved by way of downgrading differences (between the Same and the Others) and of accentuating them, (d) that the relationships are relatively asymmetric among them but relatively symmetric within them, even when the Same can be relatively asymmetric towards itself in self-oppression, and the Others can be likewise towards themselves, and (e) that symmetry and asymmetry change over time, with ever new players, new causes, and new forms, be the locality here on Earth or in deep space unto multiverses—as summarized in *Table.4.13*, *Table 4.14*, *Table 4.15*, and *Table 4.16*.

The same logic also holds both in relation to wealth and poverty (as addressed in *BCPC* and summarized in *Table 4.17* on the wealth/poverty dialectics) and in relation to civilization and barbarity (as addressed in *BCIV* and summarized in *Table 4.18* on the civilization/barbarity dialectics).

In *BDPD*, this thesis on existential dialectics was labeled as "the theory of existential dialectics."

(e.1.2) And in relation to meta-history, I proposed "the multifold theory of history" in *FPHHIST* for a different way to understand universality and relativity (together with other debates).

Direct and Indirect Applications of Existential Dialectics

A different way to appreciate the usefulness of existential dialectics is by way of the analysis of its (a) direct and (b) indirect applications.

(a) In direct applications, on the one hand, the logic of existential dialectics can shed some theoretical insights on diverse phenomena in the world, and good instances are the usage of the principles of existential dialectics for the theoretical insights on the freedom/unfreedom dialectics, the equality/inequality dialectics, and the wealth/poverty dialectics—as introduced earlier in (b5) and (e) in the previous sub-section.

For instance, my books like *FPHST*, *BNN*, *FAE*, *FC*, *FIA*, *FPHU*, *FPHK*, *FPHML*, *FPHE*, *FPHMM*, *FPHCT*, *FPHG*, *FPHUP*, *FPHL*, *FPHO*, *FPHMA*, *FPHS*, *FPHLAW*, *FPHFS*, *FPHWP*, *FPHP*, *BEPE*, *BCOS*, *FPHGEOL*, *FPHCHESS*, *FPHR*, *FPHEDU*, *FPHH*, *FPHA*, *FPHGAM*, *FPHCOM*, *FPHLIT*, *FPHCHEM*, *FPHDA*, *FPHGEOG*, *FPHPA*, *FPHSEM*, *FPHARCH*, *FPHCRIM*, *FPHHIST*, *FPHMIG*, *FPHTRAN*, *FPHCA*, *FPHW*, *BNR*, *FPHSPORT*, and *FPHPROB* use the principles to reveal some theoretical insights on the perspectives of space and time (as in *FPHST*), of nature and nurture (as in *BNN*), of beauty and ugliness (as in *FAE*), of order and chaos (as in *FC*), of taxonomy and network (as in *FIA*), of the unconscious mind (as in *FPHU*), of the nature of knowledge (as in *FPHK*), of mathematical logic (as in *FPHML*), of engineering (as in *FPHE*), of mass media (as in *FPHMM*), of creative thinking (as in *FPHCT*), of geometry (as in *FPHG*), of urban planning (as in *FPHUP*), of language (as in *FPHL*), of organization (as in *FPHO*), of martial arts (as in *FPHMA*), of sexuality (as in *FPHS*), of law (as in *FPHLAW*), of formal science (as in *FPHFS*), of aggression and pacificity (as in *FPHWP*), of normality and abnormality (as in *FPHP*). of morality and immorality (as in *BEPE*), of the beginnings and ends of the universe (as in *BOCS*), of statics and dynamics (as in *FPHGEOL*), of tactics and strategy (as in *FPHCHESS*), of secularness and sacredness (as in *FPHR*), of teaching and learning (as in *FPHEDU*), of joking and laughing (as in *FPHH*), of sound and silence (as in *FPHA*), of risk and caution (as in *FPHGAM*), of hardware and software (as in *FPHCOM*), of fiction and non-fiction (as in *FPHLIT*), of substances and their changes (as in *FPHCHEM*), of quantitative and qualitative research (as in *FPHDA*), of environments and their interactions (as in *FPHGEOG*), of the body and its presence (as in *FPHPA*), of internality and externality (as in *FPHSEM*). of form and function (as in *FPHARCH*), of heroes and villains (as in *FPHCRIM*), of universality and relativity (as in *FPHHIST*), of Sameness and Otherness (as in *FPHMIG*), of networks and operations (as in *FPHTRAN*), of ingredients and techniques (as in *FPHCA*), of uselessness and usefulness (as in *FPHW*), of diversity and discontinuity (as in *BNR*), of training and winning (as in *FPHSPORT*), and of objectivity and subjectivity (as in *FPHPROB*).

(b) In indirect applications, on the other hand, the theoretical insights can further be used to reveal other phenomena directly from them (viz., the theoretical insights) and therefore indirectly from the principles themselves.

A good illustration is of course the use of the theoretical insights on the freedom/unfreedom and equality/inequality dialectics for the understanding of the civilization/barbarity dialectics.

This distinction between direct and indirect applications may sound a bit academic, since even in indirect applications, the phenomena under study can still be directly related back to the principles themselves.

In the previous example, as an illustration, the civilization/barbarity dialectics can be directly related to the principles of existential dialectics without the intermediate role of the freedom/unfreedom and equality/inequality dialectics.

Multiple Levels of Application

There is another issue to be clarified, however. In other words, the theoretical insights can be applied to multiple levels of analysis—even though, in a given example, it may refer to one level or a few only.

For instance, in the example concerning the freedom/unfreedom dialectics, it can be used at the structural level (e.g., in relation to the theory of the cyclical progression of hegemony), but it can also be exploited for other levels (e.g., the theory of post-capitalism at the institutional level).

All these levels of application should not be misleadingly construed, as I stress this before, as a one-way street (that is, to use the ontological principles for theoretical application at multiple levels of specific analysis) but a two-way one, in which theoretical insights in praxis, when studied in more specific contexts, can in turn refine the nature of existential dialectics (for example, with the addition of new principles).

Sophisticated Methodological Holism

The summary of my original meta-theory on methodology is provided below, again, to be repeated *often verbatim* from my previous works—as this is something I regularly do in each new book to introduce my past works, which are in conversation with this current one.

This means that, for those readers who have not read any of my previous books, the summary below is helpful.

That clarified—this original contribution of mine to the study of methodology takes the form of an approach known as "the theory of methodological holism" or "methodological holism" in short, as already worked out in *FPHC* (2004).

I have made good use of this methodological approach of mine for all of my previous works.

The Confusion of Holism with Systems Approach

Yet, it is imperative to stress at the outset, as I often emphasized this in all my previous books, that my approach of "methodological holism" does not oppose or exclude "methodological individualism" (as some readers may be tempted to assume, as is conventionally understood) but actually includes it.

The confusion here has to do with equating holism with systems approach in general (be it about systems theory, chaos theory, complexity theory, or else), which I debunked, while learning something from it, as already elaborated in *The Future of Complexity* (2007)—and *The Future of Post-Human Formal Science* (2010) is another addition to this critique of the intellectual fad of systems approach.

In other words, my meta-theory of holism, or sophisticated methodological holism in short, makes a distinction between doing holistic work and doing systems work, for two reasons.

The first reason is that holism, in my theory, does not imply systems approach, since it rejects systems approach because of the myriad problems that the literature on systems approach has been unable to resolve, while learning something from it.

And the second reason is that my theory is related to the other theory of mine, that is, existential dialectics, which does not exclude methodological individualism (but includes it, because it simply goes beyond both reductionism and reverse-reductionism in a dialectic way).

For these reasons (and others too, as summarized in *Table 4.1*), my version of methodological holism is *sophisticated*—not *vulgar* as sometimes used by inapt scholars employing the same term.

The Ontological Constraints

Sophisticated methodological holism is subject to some ontological constraints, and some good examples include "the formalness-informalness principle," "the absoluteness-relativeness principle," "the partiality-totality principle," "the predictability-unpredictability principle," "the explicability-inexplicability principle," and "the fiction-reallity principle" in existential dialectics.

Of course, there can be more than six ontological principles in relation to method, but the examples here are sufficient to illustrate the point in question.

With this caveat in mind—the first two were already analyzed in previous books of mine like *FC* and *FAE*, whereas the third one was addressed in *FPHU*, the fourth and fifth ones were accounted for in *FPHK* and *FPHML*, respectively, and the last one was addressed in *FPHLIT*.

For now, it suffices to show that sophisticted methodological holism, because of these ontological constraints on methodology, targets two major sins of methodology, namely, what I call, in the absence of better words, (a) *reductionism* and (b) *reverse-reductionism*—both of which come in all shapes and sizes, to be summarized below (and also shown in *Table 4.2*).

Against the Varieties of Reductionism

There are four versions of reductionism to be summarized here, which sophisticated methodological holism rejects, namely, (a1) conceptual, (a2) theoretical, (a3) methodological, and (a4) ontological forms of reductionism.

(a1) In conceptual reductionism, a good case in point concerns myriad dualities like mind vs. body, self vs. world, democracy vs non-democracy, and the like (as already addressed in *FHC*, *FPHC*, and *BDPD*, for instance).

(a2) In theoretical reductionism, an illuminating case study is best exemplified by what I originally called "the foundation fallacy" in *FPHST*, in any attempt to naively understand space-time from the physical perspective as the foundation and, consequently, to dangerously dismiss (or belittle) other perspectives.

In *FAE*, I elaborated further these versions of reductionism in the literature on aesthetics (e.g., form vs. content, representation vs. expression, critics vs. artists, and externalism vs. internalism).

In *FIA*, I revealed other forms of reductionism in the literature on information architecture (e.g., the constructivist argument and the representational argument).

In *FPHU*, I showed the persistent legacy of reductionism, this time, in the literature on anomalous experience (e.g., the obsession with physics, chemistry, and biology for explaining anomalous experience).

In *FPHE*, I examined another case of reductionism in action, in the context of engineering (e.g., technical constraints vs. normative constraints).

In *FPHMM*, I elaborated one more version of reductionism, in relation to the three domains of communication (e.g., the competing views on sending, connecting, and receiving).

In *FPHCT*, I explored another version of reductionism, in relation to invention and innovation (e.g., the bio-psychological argument vs. the socio-cultural arguments).

In *FPHG*, I identified another version of reductionism, in relation to infinity, symmetry, and dimensionality (e.g., the Euclidean argument vs. the non-Euclidean arguments).

In *FPHUP*, I analyzed another version of reductionism, in relation to density and void (e.g., the engineering argument and the ecology argument).

In *PFHL*, I examined another version of reductionism, in relation to structure and context (e.g., the structuralist argument and the contextualist argument).

In *PFHO*, I explored another version of reductionism, in relation to communcation, decision-making, and leadership (e.g., the rational-system argument vs. the natural-system argument vs. the open-system argument).

In *PFHMA*, I revealed another version of reductionism, in relation to the martial body and spirit (e.g., the spiritual argument vs. the materialist argument vs. the defensive argument).

In *PFHS*, I analyzed another version of reductionism, in relation to the sexual body and spirit (e.g., the naturalist argument vs. the constructivist argument).

In *PFHLAW*, I examined another version of reductionism, in relation to law (e.g., the necessity argument vs. the contengency argument).

In *FPHWP*, I show another version of reductionism, in relation to war and peace (e.g., the aggressivist argument vs. the pacifist argument).

In *BEPE*, I analyzed another version of reductionism, in relation to morality and immorality (e.g., the objectivist argument vs. the non-objectivist argument vs. the skeptical argument).

In *BCOS*, I examined another version of reductionism, in relation to the contested beginnings and speculative ends of the universe (e.g., the scientific argument vs. the religious argument vs. the esoteric argument vs. the metaphysical argument).

In *FPHP*, I showed another version of reductionism, in relation to normality and abnormality (e.g., the natural argument vs. the social argument vs. the cultural argument vs. the mental argument).

In *FPHGEOL*, I scrutinized another version of reductionism, in relation to statics and dynamics (e.g., the catastrophe argument vs. the uniformity argument vs. the revision argument).

In *FPHCHESS*, I showed another version of reductionism, in relation to tactics and strategy (e.g., the natural argument vs. the social argument vs. the cultural argument vs. the mental argument).

In *FPHR*, I revealed another version of reductionism, in relation to secularness and sacredness (e.g., the critical argument vs. the skeptical argument vs. the theist argument).

In *FPHEDU*, I examined another version of reductionism, in relation to teaching and learning (e.g., the teacher-centered argument vs. the student-centered argument vs. the balanced argument).

In *FPHH*, I explored another version of reductionism, in relation to joking and laughing (e.g., the natural argument vs. the social argument vs. the cultural argument vs. the mental argument).

In *FPHA*, I revealed another version of reductionism, in relation to sound and silence (e.g., the natural argument vs. the social argument vs. the cultural argument vs. the mental argument).

In *FPHGAM*, I revealed another version of reductionism, in relation to risk and caution (e.g., the natural argument vs. the social argument vs. the cultural argument vs. the mental argument).

In *FPHCOM*, I showed another version of reductionism, in relation to hardware and software (e.g., the natural argument vs. the social argument vs. the cultural argument vs. the mental argument).

In *FPHLIT*, I proposed another version of reductionism, in relation to fiction and non-fiction (e.g., the natural argument vs. the social argument vs. the cultural argument vs. the mental argument).

In *FPHCHEM*, I proposed another version of reductionism, in relation to substances and their changes (e.g., the natural argument vs. the social argument vs. the cultural argument vs. the mental argument).

In *FPHDA*, I proposed another version of reductionism, in relation to quantitative and qualitative research (e.g., the natural argument vs. the social argument vs. the cultural argument vs. the mental argument).

In *FPHGEOG*, I proposed another version of reductionism, in relation to environments and their interactions (e.g., the natural argument vs. the social argument vs. the cultural argument vs. the mental argument).

In *FPHPA*, I proposed another version of reductionism, in relation to the body and its presence (e.g., the natural argument vs. the social argument vs. the cultural argument vs. the mental argument).

In *FPHSEM*, I proposed another version of reductionism, in relation to internality and externality (e.g., the natural argument vs. the social argument vs. the cultural argument vs. the mental argument).

In *FPHARCH*, I proposed another version of reductionism, in relation to form and function (e.g., the natural argument vs. the social argument vs. the cultural argument vs. the mental argument).

In *FPHCRIM*, I proposed another version of reductionism, in relation to heroes and villains (e.g., the natural argument vs. the social argument vs. the cultural argument vs. the mental argument).

In *FPHHIST*, I proposed another version of reductionism, in relation to universality and relativity (e.g., the natural argument vs. the social argument vs. the cultural argument vs. the mental argument).

In *FPHMIG*, I proposed another version of reductionism, in relation to Sameness and Otherness (e.g., the natural argument vs. the social argument vs. the cultural argument vs. the mental argument).

In *FPHTRAN*, I proposed another version of reductionism, in relation to networks and operations (e.g., the natural argument vs. the social argument vs. the cultural argument vs. the mental argument).

In *FPHCA*, I proposed another version of reductionism, in relation to ingredients and techniques (e.g., the natural argument vs. the social argument vs. the cultural argument vs. the mental argument).

In *FPHW*, I proposed another version of reductionism, in relation to uselessness and usefulness (e.g., the natural argument vs. the social argument vs. the cultural argument vs. the mental argument).

In *BNR*, I proposed another version of reductionism, in relation to diversity and discontinuity (e.g., the natural argument vs. the social argument vs. the cultural argument vs. the mental argument).

In *FPHSPORT*, I proposed another version of reductionism, in relation to training and winning (e.g., the natural argument vs. the social argument vs. the cultural argument vs. the mental argument).

And in *FPHPROB*, I propose another version of reductionism, in relation to objectivity and subjectivity (e.g., the natural argument vs. the social argument vs. the cultural argument vs. the mental argument).

(a3) In methodological reductionism, a good illustration can be the debate between different versions of qualitative and quantitative methods (as already analyzed in *FC* and also *FHC*).

In *FPHML*, I examined similar reductionism, this time, in the literature on mathematical logic (e.g., the obsession with consistency, soundness, and completeness).

And in *FPHFS*, I explored the problems of reductionism in the context of formal science (e.g., the analytical argument, as opposed to the synthetic argument).

(a4) In ontological reductionism, an excellent instance involves another debate, this time between emergentism and reductionism in complexity theory (as addressed in *FC*) and also in psychology (as elaborated in *FPHC*, in the context of Being and Belonging).

But to target against the varieties of reductionism constitutes only a side of the same coin, and the other side concerns the varieties of reverse-reductionism.

Against the Varieties of Reverse-Reductionism

The other side of the same coin is the reverse version of reductionism, which is what I want to call, in the absence of better words, *reverse-reductionism*.

My sophisticated version of methodological holism targets against the varieties of reverse-reductionism (just as it also rejects the ones of reductionism).

Perhaps nothing expresses better the popularity of reverse-reductionim than the "anything-goes" mentality in postmodernism of our time, as shown in the following four versions.

(b1) In conceptual reverse-reductionism, any concept of "art" (e.g., fine art, applied art, outsider art, junk art) is welcome in postmodernism (as already analyzed in Ch.4 of *FHC*—and also in *FAE*).

(b2) In theoretical reverse-reductionism, a variety of art and literary theories co-exist. Take the case of literary studies, as there are now Literary Structuralism, Marxist Literary Criticism, New Criticism, Phenomenology, Hermeneutics, Language-Game Literary Criticism, Feminist Literary Criticism, Reception Theory, Reader Response Criticism, Poststructuralism, Semiotics, Pyschoanalytic Literary Criticism, just to cite some well-known ones, with no one being said to be better than any others (as detailedly analyzed in Ch.4 of *FHC*). (S. Raman 1997)

In *BNN*, I even introduced "the compromise fallacy" as another good illustration of theoretical reverse-reductionism, in misleadingly treating both genetic and environmental approaches as equally valid.

(b3) In methodological reverse-reductionism, multiple methodologies are deemed as acceptable in postmodernism (e.g., doing art without praxis, doing art with praxis, and doing art by sublation), as analyzed in Ch.4 of *FHC*.

And in *FPHFS*, I exposed the problems of reverse-reductionism in the context of formal science in relation to systems theory (e.g., the synthetic argument).

(b4) In ontological reverse-reductionism, no privileged ontology is allowed, and the door is open practically for anything in postmodernism (e.g., the equal status of the ontology of Being vs. that of Becoming, as already addressed in Ch.4 of *FHC*—and also in *FPHC*).

In *FAE*, I also introduced another version of reverse-reductionism, that is, "the pluralist fallacy," in the context of understanding aesthetic experience, for instance—although this fallacy has been committed not exclusively in relation to the ontological level (but also at the conceptual, theoretical, and methodological ones).

These dual dangers against reverse-reductionism (in this sub-section) and reductionism (in the previous sub-section) are something that sophisticated methodological holism rejects. And in *FPHFS*, I addressed these two problems further.

This point about the dual dangers is important enough, since many scholars often favor one against the other, but it is vital to target both.

The Holistic Organization of an Inquiry

With these dual dangers against reductionism and reverse-reductionism in mind—sophisticated methodological holism suggests that an inquiry of any given phenomenon is more complete, if treated in the context of a comprehensive analysis at all relevant levels, which challengingly encompass all the fields of human knowledge, ranging from the natural sciences and formal sciences through the social sciences to the humanities.

This section is something that I had already stressed in all my previous books and repeat (*often verbatim*) hereafter. But, for those readers who had read my previous books, this serves as a reminder.

With this reminder in mind—there are multiple ways to engage in a holistic inquiry with all relevent levels of analysis. Over the years, I have proposed different ways to fulfill this holistic methodological requirement, as repeated in my previous books.

Hereafter is a summary of four major ways, namely, (a) by discipline, (b) by domain, (c) by subject, and (d) by meta-analysis—all of which fulfill the holistic requirement of sophisticated methodological holism on a given issue.

(a) In a holistic organization by discipline, a good illustration is the multiple levels of analysis, namely, (a1) the micro-physical, (a2) the chemical, (a3) the biological, (a4) the psychological, (a5) the organizational, (a6) the institutional, (a7) the structural, (a8) the systemic, (a9) the cultural, (a10) the macro-physical (cosmological), and (a11) other relevant levels, which are either a combination of all these levels or the practical applications with a combination of them.

This holistic organization by discipline is used in *FCD* and *BCPC*, for instance.

(b) In a holistic organization by domain, some good candidates include the classification of different perspectives of analysis in relation to (b1) nature, (b2) the mind, (b3) society, and (b4) culture, as already worked out in *BCIV*, *FPHST*, *FC*, *FAE*, *FPHU*, *FPHK*, *FPHE*, and *FPHFS*, just to cite some examples.

Culture in (b4), in this re-classification, is the same as culture in (a9) and can be further divided into different sub-categories, with good examples like (a9i) epistemology (methodology), (a9ii) aesthetics, (a9iii) ethics, (a9iv) metaphysics (ontology), and (a9v) religion. For clarification, epistemology (methodology) in (a9i) and metaphysics (ontology) in (a9iv) overlap with the category of "meta-theory."

Society in (b3) includes the organizational in (a5), the institutional in (a6), the structural in (a7), and the systemic in (a8).

The mind (b2) has more to do with the chemical in (a2), the biological in (a3), and the psychological in (a4), although it can overlap with (a1), (b3) and (b4), for instance.

It should be stressed that the "mind" here also includes the categories like the "brain," "body," and the "spirit" (or the "soul")—whenever needed in a given context, so that one does not fall into the intellectual trap of the "mind-body dualism" in the history of modern (and especially, postmodern) philosophy, which I already went to great lengths to explain in *The Future of Human Civilization*. After all, in everyday's usage, the "mind" already implies the "brain" and the "body" (together with their interactions) and has different levels of consciousness which can be both "spiritual" and non-spiritual.

And nature in (b1) refers to the micro-physical in (a1) and the macro-physical (cosmological) in (a10), although it can also overlap with (a2), (a3) and (a4).

(c) In a holistic organization by subject, some excellent examples concern the analysis of consciousness in *FPHC* (i.e., on Having,

Belonging, and Being); of civilization in *FHC* (i.e., the True, the Holy, the Everyday, the Technological, the Beautiful/Sublime, the Good, and the Just); of international politics in *BWT* (i.e., Hyper-Empires, Meso-Empires, and Micro-Empires); of nature and nurture in *BNN* (i.e., genes and memes); of information architecture in *FIA* (i.e., taxonomy and network); of mathematical logic in *FPHML* (i.e., consistency, soundness, and completeness); of communication in *FPHMM* (i.e., sending, connecting, and receiving); of creative thinking in *FPHCT* (i.e., invention and innovation); of geometry in *FPHG* (i.e., infinity, symmetry, and dimensionality); of urban planning in *FPHUP* (i.e., density and void), of language in *FPHL* (i.e., structure and context), of organization in *FPHO* (i.e., communication, decision-making, and leadership), of martial arts in *FPHMA* (i.e., the martial body and the martial spirit), of sexuality in *FPHS* (i.e., the sexual body and the sexual spirit), of law in *FPHLAW* (i.e., necessity and contingency), of war and peace in *FPHWP* (i.e., aggression and pacificity), of personality in *FPHP* (i.e., normality and abnormality), of ethics in *BEPE* (i.e., morality and immorality), of cosmology in *BCOS* (i.e., the beginnings and ends of the universe), of geology in *FPHGEOL* (i.e., statics and dynamics), of chess in *FPHCHESS* (i.e., tactics and strategy), of religion in *FPHR* (i.e., secularness and sacredness), of education in *FPHEDU* (i.e., teaching and learning), of humor in *FPHH* (i.e., laughing and joking), of acoustics in *FPHA* (i.e., sound and silence), of gambling in *FPHGAM* (i.e., risk and caution), of computing in *FPHCOM* (i.e., hardware and software), of literature in *FPHLIT* (i.e., fiction and non-fiction), of chemistry in *FPHCHEM* (i.e., substances and their changes), of data analysis in *FPHDA* (i.e., quantitative and qualitative research), of geography in *FPHGEOG* (i.e., environments and their interactions), of performing arts in *FPHPA* (i.e., the body and its presence), of semantics in *FPHSEM* (i.e., internality and externality), of architecture in *FPHARCH* (i.e., form and function), of criminality in *FPHCRIM* (i.e., heroes and villains), of history in *FPHHIST* (i.e., universality and relativity), of migration in *FPHMIG* (i.e., Sameness and Otherness), of transportation in *FPHTRAN* (i.e., networks and operations), of culinary art in *FPHCA* (i.e., ingredients and techniques), of waste in *FPHW* (i.e., uselessness and usefulness), of natural resources in *BNR* (i.e., diversity and discontinuity), of sports in *FPHSPORT* (i.e., training and winning), and of probability in *FPHPROB* (i.e., objectivity and subjectivity)—by incorporating all the relevent levels of analysis as cited above in each.

(d) And in a holistic organization by meta-analysis, two good cases in point involve the works on post-capitalism in *BCPC* and on authoritarian liberal democracy in *ALD*, with the classification of analysis in theory and meta-theory.

The essential point to remember here, as I thus stress in all my previous books, is that the multiple levels of analysis can be reorganized in many different ways, insofar as none of the levels (if relevant to an inquiry in question) is ignored or dismissed, to avoid the danger of reductionism (and for that matter, the one of reverse-reductionism, in the opposite direction).

Three Clarifications

Three clarifications are needed here, however, to avoid any hasty misunderstanding.

Firstly, the entities in each classification are not mutually exclusive. For instance, in the holisitc organization by domain, nature in (b1) can also be linked to the chemical in (a2), the biological in (a3), and the systemic in (a8). And the mind (b3) can alternatively be related to the micro-physical in (a1) and the chemical in (a2), for example.

Secondly, there is also the important factor of luck (or uncertainty in conventional usage), but it is already implied or allowed in each of the organizations of an inquiry—especially in relation to the predictability-unpredictability principle and the order-chaos principle in existential dialectics. This topic about uncertainty was already addressed in *FHC*—and more extensively, in *FC*, in the context of order and chaos (and later, to an extent, in *FPHK*, *FPHFS*, and *FPHGAM*, for example).

And lastly, the comparison in a classification is not absolute, but relative, as there are often some shades of gray, not exactly black or white (figuratively speaking).

Some further clarifications and qualifications of sophisticated methodological holism are shown in *Table 4.2*.

Some Distinctive Usefulness

In the end, sophisticated methodological holism—when applied, especially though not exclusively, as illustrated in my numerous works—can enrich the understanding of reality in some distinctive ways. Here are four examples (as revealed in each of my books).

Firstly, it provides a comprehensive analysis of a subject matter, from which much can be learned about reality, both in relation to the perspectives of the mind, nature, society, and culture.

Secondly, it offers a new classification of the subject matter.

Thirdly, it suggests some visions of the future in relation to the subject matter in question.

And fourthly, it proposes some insights on meta-theory (e.g., methodology and ontology) in general—with the clear understanding, however, of the dilemma of specific vs. general ontology (as shown in *Table 4.9* on the syntax of existential dialectics in the context of the dilemma of ontology).

For this very reason, all of these ways are important, without reducing one into the analysis of another. (*FPHCT*)

Chapter Outline

With this summary of my two original meta-theories (on methodology and ontology, *almost verbatim* from my previous works) in mind—this book is organized in four main parts, namely, (a) Part I on the introduction to this book, (b) Part II on objectivity, (c) Part III on subjectivity, and (d) Part IV on the conclusion of the book.

The book is thus organized in four chapters.

The introductory chapter, that is, Chapter One titled "Introduction—The Use of Probability," begins with a questioning of probability, provides a definition and history of probability, and then introduces objectivity and subjectivity in probability study, as well as the diverse aspects of probability—together with the theoretical debate on probability, my interpretivist theory of probability, the relationship between theory and meta-theory, existential dialectics, sophisticated methodological holism, and some further clarifications.

Chapter Two titled "Objectivity and its Ambivalence" examines the nature of objectivity in probability study from the four perspectives of the mind, nature, society, and culture, with a constructive assessment of each.

Chapter Three titled "Subjectivity and its Duality" explores the nature of subjectivity in probability study from the four perspectives of the mind, nature, society, and culture, with a deconstructive critique of each.

The last chapter titled "Conclusion—The Future of Probability" summarizes the analysis in the book in the dialectic context of objectivity and subjectivity in probability with thirty-one major theses, namely, (a) the first thesis: the formalness-informalness principle, (b) the second thesis: the absoluteness-relativeness principle, (c) the third thesis: the partiality-totality principle, (d) the fourth thesis: the predictability-unpredictability principle, (e) the fifth thesis: the explicability-inexplicability principle, (f) the sixth thesis: the fiction-reality principle, (g) the seventh thesis: the cognitiveness-noncognitiveness principle, (h) the eighth thesis: the finiteness-transfiniteness principle, (i) the ninth thesis: the preciseness-vagueness principle, (j) the tenth thesis: the simpleness-complicatedness principle, (k) the eleventh thesis: the openness-hiddenness principle, (l) the twelfth thesis: the denseness-emptiness principle, (m) the thirteenth thesis: the change-constancy principle, (n) the fourteenth thesis: the order-chaos principle, (o) the fifteenth thesis: the slowness-quickness principle, (p) the sixteenth thesis: the expansion-contraction principle, (q) the seventeenth thesis: the theory-praxis principle, (r) the eighteenth thesis: the convention-novelty principle, (s) the nineteenth thesis: the evolution-transformation principle, (t) the twentieth thesis: the symmetry-asymmetry principle, (u) the twenty-first thesis: the softness-hardness principle, (v) the twenty-second thesis: the seriousness-playfulness principle, (w) the twenty-third thesis: the activeness-inactiveness principle, (x) the twenty-fourth thesis: the selfness-otherness principle, (y) the twenty-fifth thesis: the regression-progression principle, (z) the twenty-sixth thesis: the sameness-difference principle, (aa) the twenty-seventh thesis: the stability-reaction principle, (bb) the twenty-eighth thesis: the functionality-nonfunctionality principle, (cc) the twenty-ninth thesis: the intentionality-nonintentionality principle, (dd) the thirtieth thesis: the survivability-non-survivability principle, and (ee) the thirty-first thesis: the post-human interpretiveness—to be elaborated in the rest of the book, with a summary in the concluding chapter.

This seminal project here, if successful, will fundamentally change the way that we think about objectivity and subjectivity in probability study (together with other debates as will be discussed in the rest of the book), from the combined perspectives of the mind, nature, society, and culture, with enormous implications for the human future and what I originally called its "post-human" fate.

Some Clarifications

However, some additional (last) clarifications are needed.

These clarifications, lest any potential misunderstanding by the readers occurs, refer to (a) the conversation with my previous books, (b) the illustrative selection of case studies and examples, (c) detailed analysis versus overall analysis, (d) three distinctive features of using quotations, (e) two difficulties with understanding, (f) repetition, (g) chapter organization to counter conventional wisdom, and (h) the use of neologisms.

These clarifications are something that I also regularly repeat in my previous books too, so these points are repeated hereafter, *often verbatim*—albeit in a different context this time.

In Conversation with My Previous Books

Firstly, my theory in this book is constructed on the theoretical foundation of my previous books (i.e., *FHC, FCD, FPHC, BDPD, BCPC, BCIV, FPHST, BNN, BWT, FC, FAE, ALD, FIA, FPHU, FPHK, FPHML, FPHE, FPHMM, FPHCT, FPHG, FPHUP, FPHL, FPHO, FPHMA, FPHS, FPHLAW, FPHFS, FPHWP, FPHP, BCOS, BEPE, FPHGEOL, FPHCHESS, FPHR, FPHEDU, FPHH, FPHA, FPHGAM, FPHCOM, FPHLIT, FPHCHEM, FPHDA, FPHGEOG, FPHPA, FPHSEM, FPHARCH, FPHCRIM, FPHHIST, FPHMIG, FPHTRAN, FPHCA, FPHW, BNR*, and *FPHSPORT*—as explicated in "The List of Abbreviations").

For this reason, many of my books have some familiar words in their titles like "The Future of Post-Human ____" or "Beyond ____to ____," as well as some tables for background information—as they constitute a series of books with the final aim to converge all of my original theories on numerous fields into a unified theory of everything (as already summarized in the section on theory and meta-theory in Chapter One).

Just consider the following two illustrations, namely, (a) in relation to other related theoretical debates and (b) in relation to different visions as worked out in my previous books.

(a) The first illustration is that the theoretical debate here also involves a few other related theoretical issues.

These other related theoretical issues have to do with some deeply contested theoretical debates, which, however, I had already gone to great lengths to analyze in my numerous (previous) books, as summarized below. It should also be stressed, lest misunderstanding occurs, that each of the theoretical debates (as listed below) does not exhaust the issues covered in the books, as each book covers more issues than what is listed below (for illustration only).

- On Probability
  –Ex: *The Future of Post-Human Gambling*
- On Continuousness and Infinity
  –Ex: *The Future of Post-Human Geometry*
- On Quantam Mechanics
  –Ex: *The Future of Post-Human Unconsciousness*

–Ex: *The Future of Post-Human Consciousness*
–Ex: *The Future of Post-Human Geometry*
–Ex: *The Future of Post-Human Space-Time*
–Ex: *The Future of Complexity*

- On Logic and Rationality
  –Ex: *The Future of Post-Human Mathematical Logic*
  –Ex: *The Future of Post-Human Humor*
- On Computation
  –Ex: *The Future of Post-Human Computing*
- On Data Analysis
  –Ex: *The Future of Post-Human Data Analysis*
- On Decision-Making
  –Ex: *The Future of Post-Human Organization*
- On Epistemology and Ontology
  –Ex: *The Future of Post-Human Knowledge*
  –Ex: *The Future of Post-Human Consciousness*
  –Ex: 2 volumes, *The Future of Human Civilization*
- On Complexity
  –Ex: *The Future of Complexity*
- On History
  –Ex: *The Future of Post-Human History*
- On Learning
  –Ex: *The Future of Post-Human Language*
  –Ex: *The Future of Post-Human Education*
- On Cognition
  –Ex: *The Future of Post-Human Unconsciousness*
  –Ex: *The Future of Post-Human Consciousness*
  –Ex: *The Future of Capitalism and Democracy*
  –Ex: *The Future of Post-Human Creative Thinking*
  –Ex: *The Future of Post-Human Personality*
- On Intuition
  –Ex: *The Future of Post-Human Unconsciousness*
  –Ex: *The Future of Post-Human Consciousness*
  –Ex: *Beyond Capitalism to Post-Capitalism*
  –Ex: *The Future of Capitalism and Democracy*
- On Emotion
  –Ex: *The Future of Post-Human Unconsciousness*
  –Ex: *The Future of Post-Human Consciousness*
  –Ex: *The Future of Aesthetic Experience*
  –Ex: *Beyond Capitalism to Post-Capitalism*
  –Ex: *The Future of Capitalism and Democracy*

–Ex: *The Future of Post-Human Personality*
- On Behavior
  –Ex: *The Future of Post-Human Language*
  –Ex: *The Future of Post-Human Unconsciousness*
  –Ex: *The Future of Post-Human Consciousness*
  –Ex: *Beyond Capitalism to Post-Capitalism*
  –Ex: *The Future of Capitalism and Democracy*
  –Ex: *The Future of Post-Human Sexuality*
  –Ex: *The Future of Post-Human Personality*
  –Ex: *The Future of Post-Human Martial Arts*
- On Communication
  –Ex: *The Future of Post-Human Mass Media*
  –Ex: *The Future of Post-Human Organization*
- On Spirituality
  –Ex: 2 volumes, *The Future of Human Civilization*
  –Ex: *The Future of Capitalism and Democracy*
  –Ex: *The Future of Post-Human Religion*
  –Ex: *Beyond Civilization to Post-Civilization*
  –Ex: *The Future of Post-Human Sexuality*
  –Ex: *The Future of Post-Human Martial Arts*
  –Ex: *The Future of Post-Human Unconsciousness*
  –Ex: *The Future of Post-Human Consciousness*
  –Ex: *Beyond Ethics to Post-Ethics*
- On Aesthetics
  –Ex: *The Future of Aesthetic Experience*
  –Ex: *The Future of Post-Human Culinary Art*
  –Ex: *The Future of Post-Human Architecture*
  –Ex: *The Future of Post-Human Martial Arts*
  –Ex: *The Future of Post-Human Performing Art*
  –Ex: *The Future of Post-Human Literature*
- On Civilization
  –Ex: *Beyond Civilization to Post-Civilization*
  –Ex: 2 volumes, *The Future of Human Civilization*
- On Freedom
  –Ex: 2 volumes, *The Future of Human Civilization*
  –Ex: *The Future of Capitalism and Democracy*
  –Ex: *Beyond Capitalism to Post-Capitalism*
  –Ex: 2 volumes, *Beyond Democracy to Post-Democracy*
  –Ex: *Beyond Civilization to Post-Civilization*
- On Being and Belonging
  –Ex: *The Future of Post-Human Consciousness*

–Ex: 2 volumes, *The Future of Human Civilization*
- On Different Dimensions of Human Existence
  –Ex: 2 volumes, *The Future of Human Civilization*
- On Modernity, Post-Modernity, and After-Postmodernity
  –Ex: 2 volumes, *The Future of Human Civilization*
  –Ex: *Beyond Civilization to Post-Civilization*
- On Pre-Modernity
  –Ex: *Beyond Civilization to Post-Civilization*

Therefore, those who are interested in these other theoretical issues (and debates) which are related to the current context can consult these other books of mine for more details.

(b) And the second illustration is that one of the major theses of this book concerns the future, be it human and later post-human, which is closely linked with the different visions that I had worked out in my previous books, in relation to the mind, nature, society, and culture.

Of course, for the convenience of the readers, a summary of some of these visions will be provided whenever necessary, as shown in the two sections on meta-theory above (*often verbatim*) and in different tables at the end of this chapter and the concluding chapter, for instance.

Yet, I still expect the readers to read my previous books directly for more details—as this book is written in conversation with my previous ones.

Case Studies and Examples

Secondly, the case studies and examples in the chapters are not exhaustive but solely illustrative. Nor are they necessarily mutually exclusive, as they can be reclassified in a different way. And exceptions are allowed.

One is tempted to assume, however, that, since the case studies and their examples are not exhaustive, more case studies and examples are needed. But the problem here is that just adding more of them does not necessarily change the conclusions to be drawn and may even lead to redundancy.

After all, the case studies and examples used in this book are chosen with care and deemed sufficient for the purpose at hand, even though they are solely illustrative.

Detailed Analysis versus Overall Synthesis

Thirdly, there is a distinction between detailed analysis and overall synthesis, in that the former is, especially though not exclusively, more for the specialized viewpoint of a specific discipline, whereas the latter is, especially though not exclusively, more for the broad horizon of an interdisciplinary approach—relatively speaking, of course.

Or, by analogy, there is a distinction between depicting a "tree" (with tiny details) in a "forest" and viewing the entire "forest" itself (with the broader horizon).

This book, like all previous ones of mine, tries to strike a balance between the two forms of understanding, although this does not imply that the two forms should be exactly equally offered, in an exact 50-50 balance.

Instead, the book, more often than not, does not allow the tiny details to obscure the larger picture in a given case study, because the entire book is inter-disciplinary in nature and tries to look at the entire "forest," not being bogged down with the detailed analysis of an individual "tree" in the forest, by analogy.

In this sense, although detailed analysis can be provided whenever needed for a given case study, the important point to remember is the overall horizon with its broadness of scope.

Three Distinctive Features of Using Quotations

Fourthly, the way that quotes are used in this book (and others of mine) has three distinctive features, as explained below.

The first distinctive feature is that, at the end of each quotation, sometimes there are more than one reference—in which case the first reference is for the original citation, and the rest of the references are for the convenience of the reader for more information about the issue(s) under quotation.

The second distinctive feature is that quotations are used as often as possible, so as to let others speak for themselves, without the nuisance of putting my own words into their mouths. Even more importantly, this allows me to use the original text as an evidence for critical analysis at the end of each section.

And the third distinctive feature is that all these quotations are used for my own purposes, in that I often creatively (or originally) re-organize them for a new way of analyzing the issue(s) in question.

Two Difficulties with Understanding

Fifthly, to understand this book (or any other book of mine) is difficult for two main reasons.

In regard to the first difficulty, this book, just like many other books of mine, is in conceptual and theoretical conversation with other books of mine (57 books, as of December 13, 2012), so this requires a vast background information on numerous subject matters.

And in regard to the second difficulty, this book also analyzes the subject matter in a way very different from the expectation of a conventional specialist reader, since it examines reality outside the specialist's interest (or outside his or her "comfort zone"), from the four broad perspectives of the mind, nature, society, and culture, such that it covers all domains of knowledge (in the social sciences, the natural sciences, the formal sciences, and the humanities). In other words, those who are truly interested in my project for a unified theory of everything would find many books of mine highly stimulating, but those who are specialists in a subject matter (like physics, geography, chess, and so on, in a strictly specialist sense) will not find them suitable to their specialist taste (or expectation).

Not surprisingly, a conventional specialist reader would find these two difficulties (with understanding) almost practically impossible to overcome, as they almost require the mind of a polymath. Even then, the best advice is that one should still try one's best to understand it.

Repetition

Sixthly, many of my books have something in common, and the reason is to show "the unified theory of everything" in all domains of knowledge.

For this reason, there is certain repetition in all books (especially in regard to the background information, together with the tables, in the introductory chapter and the concluding chapter). Therefore, these books are all titled "The Future of Post-Human ____" (or at times "Beyond ____ to ____") and constitute the "post-human" series.

Chapter Organization to Counter Conventional Wisdom

Seventhly, all of my books aim to deconstruct conventional wisdom and to replace it with new thinking, and the chapters in my books are often organized in two distinctive ways for this purpose

The first distinctive way is to show the polarized thinking in conventional wisdom by devoting two main chapters on opposites and then deconstructing them later. For example, in *Beyond Ethics to Post-Ethics* (2011), Chapter Two first introduces the conventional wisdom for morality but then shows that morality has its negative side often unsaid—and Chapter Three does just the opposite, that is, first introduces the conventional wisdom against immorality but then shows that immorality has its bright side often unspoken of.

And the second distinctive way is to show the opposites (embedded in conventional wisdom) in a dialectic way. For instance, in the above example, the task here is to show how and why the analysis of morality (in Chapter Two) is inadequate without the inclusion of its immorality, just as the examination of immorality (in Chapter Three) is insufficient without the consideration of its morality—to the extent that there is no morality without immorality, and vice versa.

But there are two qualifications.

The first qualification is that the opposites do not have to be always mutually exclusive in the conventional meaning (like morality vs. immorality in *BEPE* and heroes vs. villains in *FPHCRIM*)—but are simply disproportional in focus in other cases (in that some thinkers focus more on one than on the other), and good examples are networks vs. operations (in *FPHTRAN*), ingredients vs. techniques (in *FPHCA*), and form vs. function (in *FPHARCH*). In other words, there are multiple meanings of the opposites.

And the second qualification is that some books (like the 2 volumes of *The Future of Human Civilization*, the 2 volumes of *Beyond Democracy to Post-Democracy*, and *The Future of Capitalism and Democracy*) are very lengthy and analyze many issues in more than two main chapters. Yet, there is a general pattern of introducing and then deconstructing conventional wisdom in my works.

The Use of Neologisms

And finally, I use different neologisms in my books, mostly to introduce my original concepts and theories, and good instances include those here or elsewhere in my previous books (e.g., "the interpretivist theory of probability," "post-democracy," "hyper-spatial consciousness," "post-capitalism," and whatnot). Needless to say, they are used here solely for our current intellectual convenience, as they will be renamed differently in different ways in future history.

As I thus wrote in *FCD* (508-9), "all these terms 'post-capitalism,' 'post-democracy'...and other ones as introduced in...[the] project (e.g.,...'posthuman elitists,' and 'posthuman counter-elitists,' just to cite a few of them) are more for our current intellectual convenience than to the liking of future humans and post-humans, who will surely invent more tasteful neologisms to call their own eras, entities, and everything else, for that matter. But the didactic point here is to use the terms to foretell what the future might be like, not that its eras and entities must be called so exactly and permanently. After all, William Shakespeare (1995: Act II, Scene II, Line 47) well said long ago: 'What is in a name? That which we call a rose by any other name would smell as sweet.'"

As I had also stressed time and again before, each of the neologisms can be re-written as a different "X," only to be re-named differently by the powers that be in different eras of future history.

With these clarifications for thought in mind—I now proceed to Chapter Two on objectivity and its ambivalence.

Table 1.1. Diverse Aspects of Probability

- **Objectivity and Subjectivity**
  - —Objectivity
    - Ex: measurement
    - Ex: mind independence
    - Ex: logical relation
  - —Subjectivity
    - Ex: individual perspective
    - Ex: thought (mind)
    - Ex: personal belief

- **Main Distinctions**
  - —Ex: rational vs. logical
  - —Ex: independent vs. conditional
  - —Ex: discrete vs. continuous
  - —Ex: abstract vs. applied

Notes: The categories are solely illustrative (not exhaustive), and the comparison is relative (not absolute), nor are they necessarily mutually exclusive. And some can be easily re-classified elsewhere. As generalities, they allow exceptions.
Source: From *Sec. 1.3* & *Sec. 1.4* of *FPHPROB.* See book for citations.

Table 1.2. Nine Great Future Transformations of Probability (Part I)

- **In Relation to the Mind**
  —*New Goals*
  - This search search for "new goals" will give future advanced humans and later their successors, namely, the post-humans, different motivations to do probability in a way that current humans cannot—in relation to my previous visions on the future "post-human" supersession of the human mind like "supersession computing," "metamorphic humor," "contrarian personality," "hyper-martial body," "hyper-sexual body," "floating consciousness," "hyper-spatial consciousness," "thinking machines," "unfolding unconsciousness," "thinking robots," "genetically altered superior beings," "cyborgs," etc.

  —*New Means*
  - This quest for "new means" will give future advanced humans and later their successors, namely, the post-humans, the capacity to find novel options to achieve the goals in regard to probability in a way that current humans cannot.

- **In Relation to Nature**
  —*New Comparisons*
  - This search for "new comparisons" will give future advanced humans and later their successors, namely, the post-humans, the ability to learn from animals, plants, and the universe, through comparison and contrast, about probability in a way that current humans cannot.

(continued on next page)

Table 1.2. Nine Great Future Transformations of Probability (Part II)

- **In Relation to Nature (*cont'd*)**
  —*New Integrations*
  - This search for "new integrations" will give future advanced humans and later their successors, namely, the post-humans, the ability to live and interact with nature for probability in a way that current humans cannot. This is especially so, in the context of the great future transformations of geology (in *FPHGEOL*), of geography (in *FPHGEOG*), of migration (in *FPHMIG*), and of transportation (in *FPHTRAN*)—in relation to my previous visions on the future "post-human" adventures into different dimensions of space-time unto multiverses like "the alteration of space-time" (*Table 1.3* & *Table 1.4*), "multiverses" (*Table 1.7* & *Table 1.8*), "hyperspace" (*Table 1.9* & *Table 1.25*), "time travel" (*Table 1.10* & *Table 1.11*), and others.

    This extension is consistent with my other theory, namely, "the theory of the cyclical progression of system integration and fragmentation" in *FCD* (2002)—at the social-systemic level (as already clarified in *Sec. 1.8* and shown below):

    Local → Regional → Global →Solar →
    Galactic → Clustery...→ Multiversal

(continued on next page)

Table 1.2. Nine Great Future Transformations of Probability (Part III)

- **In Relation to Society**
  - —*New Praxes*
    - This search for "new praxes" will give future advanced humans and later their successors, namely, the post-humans, the ability to engage in practice (like activism, politics, personal experience, etc.) for the future of probability in a way that current humans cannot.
  - —*New Technologies*
    - This search for "new technologies" will give future advanced humans and later their successors, namely, the post-humans, the ability to come up with different techniques, materials, and experiments for probability in a way that current humans cannot.
  - —*New Stratifications*
    - This emergence of "new stratifications" will lead future advanced humans and later their successors, namely, the post-humans, to face new inequalities in the context of probability in a way that current humans cannot.

- **In Relation to Culture**
  - —*New Sub-Fields*
    - This search for "new sub-fields" will give future advanced humans and later their successors, namely, the post-humans, the ability to explore new specialties for probability in a way that current humans cannot.
  - —*New Interpretations*
    - This search for "new interpretations" will give future advanced humans and later their successors, namely, the post-humans, the ability to come up with different evaluations and views about probability in a way that current humans cannot.

Notes: The categories and examples are solely illustrative (not exhaustive), and the comparison is relative (not absolute), nor are they necessarily mutually exclusive. And some can be easily re-classified elsewhere. As generalities, they allow exceptions.
Source: From *Ch.4* of *FPHPROB*. See text for citations.

Table 1.3. The Theoretical Debate on Space-Time (Part I)

- **Isaac Newton's Absolutist (Substantivist) Theory of Space-Time**
  —space and time are independent from each other. The structure of space-time is $E3 \times E1$ (with the structure of space, *P*, as the set of spatial locations in a three-dimensional Euclidean space, *E3*, and the structure of time as the set of temporal moments, *T*, in the one-dimensional real time, *E1*).
  —space and time are also independent from the effects of matter and events. The existence of space and time is possible even in a world absent of matter (and, for that matter, even in a world absent of events), as if they were material objects but with their total unchangingness thorough time.

- **Albert Einstein's Relativist Theory of Space-Time**
  —space and time are interchangeable (not absolute), just as matter and energy are equivalent (not independent) with the famous equation, $E = mc^2$ (as in the special theory of relativity in 1905).
  —space-time and matter-energy are also relative in a grand union (as in the general theory of relativity in 1915). Thus, each pair affects the other pair, as "matter 'tells' spacetime how to curve [in a non-Euclidean geometry] and curved spacetime 'tells' matter how to behave….Space contracts near mass and dilates away from it. Time dilates near mass and contracts away from it. Clocks positioned farther away from the mass of the earth run faster than clocks closer to the earth." (L. Shlain 1991: 328-330)

(continued on next page)

Table 1.3. The Theoretical Debate on Space-Time (Part II)

- **Peter Baofu's Perspectival Theory of Space-Time**
  —space and time can be understood from multiple perspectives, be they in relation to culture, society, nature, and the mind, with each perspective revealing something about the nature of space-time and simultaneously delimiting its view. This is subject to "the regression-progression principle" in existential dialectics.
  —each perspective of space and time exists in society and culture with good reasons, with some being more successful and hegemonic (dominant) than others. This is subject to "the symmetry-asymmetry principle" in existential dialectics.
  —space and time will not last, to be eventually superseded (altered) by post-humans in different forms (e.g.,stretching/shrinking space-time, engineering more dimensions of space-time, and manipulating multiverses), be they here in this universe or in multiverses. Thus, even the physical existence of space-time cannot last forever, with ever more transformations in the process. This is subject to "the change-constancy principle" in existential dialectics.
  —the conventional wisdom (especially by physicists) of treating the physical perspective of space and time as the foundation of all other perspectives (of space and time) and of regarding them as much less important is a form of reductionism,committing what I call *the foundation fallacy*, in misleadingly dismissing the multiple perspectives of space and time in relation to culture, society, nature, and the mind.

Notes: The examples in each category are solely illustrative (not exhaustive), and the comparison is relative (not absolute), nor are they necessarily mutually exclusive. Some can be easily re-classified elsewhere. As generalities, they allow exceptions.
Source: A summary of *Sec.1.1*, *Sec.1.2*, and *Sec.1.3* of *FPHST*—and, for that matter, the rest of *FPHST*

Table 1.4. Main Reasons for Altering Space-Time

- **The Need to Make New Energy-Matter**
  —Ex: manipulating molecular bonds for new materials
  —Ex: creating nanotechnologies on the atomic scale
  —Ex: engineering the atomic nucleus
  —Ex: restructuring most elementary particles
  —Ex: inventing new forms of matter and energy

- **The Need to Create New Space-Time**
  —Ex: creating "warp drive" (as in science fiction) for space travel
  —Ex: creating "pocket universes"

- **The Need to Conquer the Cosmos unto Multiverses**
  —Ex: spreading floating consciousness and hyper-spatial consciousness, besides other forms that humans have never known, in the cosmos and beyond unto multiverses for ultimate conquest

Notes: The examples in each category are solely illustrative (not exhaustive) nor necessarily mutually exclusive, and the comparison is relative (not absolute). As generalities, they allow exceptions. Also, it should be stressed that the three reasons are all related, in that they all contribute to the evolution of intelligent life in the cosmos unto multiverses in the most distant future beyond our current knowledge.
Sources: A summary of *Sec.6.2* of *FPHST*. See also *FHC*, *FCD*, *FPHC*, *FC*, *FPHU*, and *FPHG*, for example.

Table 1.5. The Technological Frontiers of the Micro-World

- **Type I-Minus**
  —Ex: building structures and mining
- **Type II-Minus**
  —Ex: playing with the genetic makeups of living things
- **Type III-Minus**
  —Ex: manipulating molecular bonds for new materials
- **Type IV-Minus**
  —Ex: creating nanotechnologies on the atomic scale
- **Type V-Minus**
  —Ex: engineering the atomic nucleus
- **Type VI-Minus**
  —Ex: restructuring most elementary particles
- **Type Ω-Minus**
  —Ex: altering the structure of space-time

Notes: As already indicated in *Sec.4.4.2.2* of *FPHC*, the problem with this micro-classification (from Barrow's work) is that the civilization types (with the exception of Type Ω-Minus, for example) are not quite distinct, since many of them can be achieved more or less in a civilization, to the extent that Type II-minus and Type III-minus, just to cite two plausible types, can be historically contemporaneous, relatively speaking, unlike the vast historical distance between, say, Type O and Type I (or Type I and Type II) civilizations. In other words, the micro-classification here is not very useful to understand civilization types but is revealing to see the technological frontiers of the micro-world.
Sources: A reconstruction from J.Barrow (1998:133), as originally shown in *Table 4.7* of *FPHC*. See *FPHC* for more info.

Table 1.6. Finity, Transfinity, and Infinity (Part I)

- **Infinity**
  —The term "infinity," which is "symbolically represented with ∞," derives from the Latin *infinitas* (or "unboundedness"), to refer to "several distinct concepts—usually linked to the idea of 'without end'—which arise in philosophy, mathematics, and theology." (WK 2008c)
  —In the context of mathematics, "infinity" is often "treated as if it were a number (i.e., it counts or measures things: 'an infinite number of terms') but it is a different type of 'number' from the real numbers. Infinity is related to limits,...large cardinals,...projective geometry, extended real numbers and the absolute Infinite," for instance. (WK 2008c)

- **Transfinity**
  —To avoid confusion, my usage of the word "transfinite" (as used here in the ontological principle) is not the same as the one which was originally coined by Georg Cantor to refer to "cardinal numbers or ordinal numbers that are larger than all finite numbers, yet not necessarily absolutely infinite" (in the mind of God), as opposed to "relative" infinity (in the mind of Man). (WK 2008d) In other words, for Cantor, there are two kinds of infinity, "absolute" and "relative"—with "relative" infinity or "transfinity" (in the mind of Man) being dependent on "absolute" infinity (in the mind of God) for its existence.
  —By contrast, my usage of the word "transfinite" here differs radically from the one used by Cantor (and other mathematicians) and is more limited, in light of the problems confronting any attempt to understand the idea of infinity, be it by intuition, imagination, and conception (as already shown in the prevous three sub-sections).
  —With this clarification in mind, I allow numbers which can be many times larger—or smaller, for that matter—than the finite things that we encounter in daily life, but they do not have to be related to the idea of infinity at all (which may not exist).

(continued on next page)

Table 1.6. Finity, Transfinity, and Infinity (Part II)

- **Transfinity** ***(cont'd)***
  —Of course, there may be some *borderline* cases, in which it is not clear whether the number in question is transfinite (in my usage) or simply a mathematical convenience. A good example of a borderline case is the Planck unit of length for "the smallest space possibly measured in nature," which is "less than billionths of trillionths of trillionths of an inch" (or something like 1.6×10^{-35} meters). (P. Baofu 2006a; N. McAleer 1987: 219; WKV 2008; D. Corbett 2008)
  —The ideas concerning symmetry and dimensions in the examples above were addressed in Chapter Three and Chapter Four of *FPHG*. For now, it suffices to further add that it is mind-boggling to even conceive or imagine extreme numbers like the Planck unit of length (or others). Does it really exist at all? Or is it merely a mathematical convenience to describe a physical situation that the human mind does not understand properly?
  —In any event, "transfinity" (in my usage) can be used as a further extension (of number) to the finite numbers in daily life—albeit with some borderline cases in mind.

- **Finity**
  —It refers to numbers which are bounded (that is, with an end), especially (though not exclusively) in relation to things in everyday life.

Source: From *Sec. 1.4* and *Sec. 2.2.3* of *FPHG*

Table 1.7. Theoretical Speculations of Multiverses

- **"Baby Universes" (Ex: Andre Linde and others)**
  —Ex: In a flat universe theory, "even if our part of it eventually collapses,...some spots in the cosmos would suddenly start inflating on their own, creating brand-new 'baby universes.'" (P. Baofu 2000: 623)

- **"Parallel Universes" (Ex: Stephen Hawking and others)**
  —Ex: In quantum cosmology, there allows the existence of infinite numbers of parallel universes, with tunneling among them. (M. Kaku 1994: 256) Hawking later revised his views on this.

- **"Pocket Universes" (Ex: Alan Guth)**
  —Ex: "As the pocket universes live out their lives and recollapse or dwindle away, new universes are generated to take their place....While life in our pocket universe will presumably die out, life in the universe as a whole will thrive for eternity." (A. Guth 1997: 248; P. Baofu 2002: 482)

- **"Brane Worlds" (Ex: Warren Siegel, Lisa Randall, and others)**
  —Ex: Our universe is stuck on a membrane of space-time embedded in a larger cosmos, with different brane worlds connecting and/or colliding with each other.

Notes: These examples are solely illustrative (not exhaustive), and some of the items can be reclassified somewhere else. Nor are they always mutually exclusive. Since they are generalities, exceptions are expected.
Source: From *Table 4.8* of *FPHC*

Table 1.8. The Confusion between "Many Worlds" and "Multiverse" (Part I)

- **First Problem**

—The first problem in this second confusion concerns a debate between the "many-worlds" interpretation of quantum mechanics by Hugh Everett (1957) and others, and the Copenhagen interpretation by Niels Bohr and Werner Heisenberg on the effects of quantum states after interacting with an external environment (e.g., by a measuring observer).

—On the one hand, the Copenhagen interpretation of quantum mechanics is more indeterministic in its concern with the probabilities of the outcomes of a wave function after interacting with an external environment (e.g., by a measuring observer using light to detect, for example, "an electron in a particular region around the nucleus at a particular time." (WK 2008e) Each wave function for the state of a particle refers to "a mathematical representation used to calculate the probability for it to be found in a location, or state of motion." (WK 2008f) And the "wavefunction collapse" means that "the act of measurement causes the calculated set of probabilities to 'collapse' to the value defined by the measurement." (WK 2008f)

—On the other hand, the Everett's interrpetation is more deterministic, in rejecting "the objective reality of wavefunction collapse" and "instead explaining the subjective appearance of wavefunction collapse with the mechanism of quantum decoherence." (WK 2008g) The words "quantum decoherence" here refers to "the mechanism by which quantum systems interact with their environments to exhibit probabilistically additive behavior," in such a way that "the quantum nature of the system is simply 'leaked' into the environment." (WK 2008g) For example, "[a]s a result of an interaction, the wave functions of the system and the measuring device become entangled with each other. Decoherence happens when different portions of the system's wavefunction become entangled *in different ways* [italic added] with the measuring device...." (WK 2008g) Thus, for Everett, the appearance of "wavefunction collapse" can be explained by "quantum decoherence" for quantum states to branch out in different independent worlds after interacting with an external environment.

(continued on next page)

Table 1.8. The Confusion between "Many Worlds" and "Multiverse" (Part II)

- **First Problem *(cont'd)***
  - —Yet, a major criticism against Everett's interpretation is that these different independent worlds in Everett's sense "will never be accessible to us," because of the interference by the measuring observer in an external environment. (WK 2008g)
  - —For illustration, "once a measurement is done, the measured system becomes entangled with both the physicist who measured it and a huge number of other particles, some of which are photons flying away towards the other end of the universe; in order to prove that the wave function did not collapse one would have to bring all these particles back and measure them again, together with the system that was measured originally. This is completely impractical, but even if one can theoretically do this, it would destroy any evidence that the original measurement took place (including the physicist's memory)." (WK 2008g)
  - —But then, the critics ask, If these independent worlds in Everett's sense are inaccessible to us, how do we know much of anything really about them?

- **Second Problem**
  - —What is relevant at the quantum mechanical level may not necessarily apply to the world at the physicsl level.
  - —In fact, the persistent inability among contemporoary physicists to unifiy the theory of relativity for large objects at the physical level and quantum mechanics for tiny objects at the quantum mechanical level should warn us about the danger of reductionism, by reducing the understanding of different universes ("parallel universes") at the physical level from a generalization of "many worlds" at the quantum mechanical level.

- **Third Problem**
  - —There is no empirical evidence to suggest that there are infinite (or extremely many) possible worlds in the cosmos as the many-worlds model by Everett would have us to believe.

Source: From *Sec.4.4.1* of *FPHG*

Table 1.9. Hyperspace and Its Challenge (Part I)

- **First Problem**
  —Hyperspace for intra- and inter-universal travel is assumed to be possible, on the basis of four main assumptions; although there may be others, these are the often cited ones.
  —The first assumption concerns the existence of "entry" and "exit" points in hyperspace, which is easier said than done to enter, go through, and then exit hyperspace.
  —For instance, using hyperspace "requires complex calculations...of single or multiple hyperspace jumps and the control of the jump....Peturbations such as those experienced by ship in space from the gravitional field around an object such as a planet or even a star are exacerbated in hyperspatial travel, since mass in real space distorts hyperspace in an equal measure. 'Jumping' near to a gravitational mass is likely to make resulting exit from hyperspace to be highly uncertain, with the level of improbability *i* increasing with the square of the distance to the nearest gravitional 'well.'" (WK 2008h)
  —Besides, "[o]nly spaceships equipped with a special force field can enter hyperspace, because exposure to hyperspace even for short period of time is hazardous to unprotected humans." (WK 2008h)

- **Second Problem**
  —The second assumption concerns the faster-than-light speed. It is well known that Einstein argued for the speed of light as the ultimate limit, but nowadays phyiscists do allow the violation of this limit under certain conditions (like the hypothetical existence of tachyons, as will be explained further in a later sub-section on time travel).
  —Even then, there is a challenge to explain in science fiction "why ships can travel faster than light in hyperspace," and two good arguments are that "hyperspace may be smaller than real space and therefore a star ship's propulsion seems to be greatly multiplied, or else the speed of light in hyperspace is not a barrier as it is in real space. Whatever the reasoning, the general effect is that ships traveling in hyperspace seem to have broken the speed of light, appearing at their destinations much more quickly and without the shift in time that the Theory of Relativity would suggest." (WK 2008h)

(continued on next page)

Table 1.9. Hyperspace and Its Challenge (Part II)

- **Third Problem**
  —The third assumption concerns the speed of time (to be measured, say, by a clock), but this is elaborated in a different section on the related issue of "time travel" (as shown in a different table on time travel).

- **Fourth Problem**
  —The fourth assumption concerns spatial shortcut, as it is possible, because of the curvature of space, in that "the manifold of ordinary three-dimensional space is curved in four or more 'higher' spacial dimensions (a 'hyperspace' in the geometric sense…). This curvature causes certain widely separated points in three-dimensional space to nonetheless be 'adjacent' to each other four-dimensionally. Creating an aperture in 4D space (a wormhole) between these locations can allow instantaneous transit between the two locations; a common comparison is that of a folded piece of paper, where a hole punched through two folded sections is more direct than a line drawn between them on the sheet. This idea probably arose out of certain popular descriptions of General Relativity and/or Riemannian manifolds, and may be the original form from which later concepts of hyperspace arose." (WK 2008h)
  —But this does not mean that hyperpsace is always safe. In fact, "in some science fiction, the danger of hyperspace travel is due to the chance that the route through hyperspace may take a ship too close to a celestial body with a large gravitational field, such as a star. In such scenarios, if a starship passes too close to a large gravitational field while in hyperspace, the ship is forcibly pulled out of hyperspace and reverts to normal space. Therefore, certain hyperspace 'routes' may be mapped out that are safe, not passing too close to stars or other dangers." (WJ 2008h)

Source: From *Sec.4.4.1* of *FPHG*

Table 1.10. The Problems of Time Travel into the Future (Part I)

- **First Problem**

—The first problem concerns "time dilation under the Theory of Special Relativity" (which separates space-time from mass-energy as special cases). Suppose you are on a moving train and an observer outside the train is watching you along the tracks. As your train moves along at the speed of light (or something close to it), "time, as measured by your watch, ticked along at a slower pace than time measured by the observer. Not only that, distance changed, too. For the observer, a one-foot ruler whizzing by on the train would have measured less than a foot." (NO 2000)

—In fact, you may not even notice that this slower clicking of the clock; in fact, everything looks normal to you inside the train as it was before: "The weird thing is that, for you on the train, time wouldn't seem to be moving slower and your ruler wouldn't be shorter—all would appear normal." (NO 2008) By contrast, you instead think that "time on the rest of the Earth would appear to be ticking along slower and its rulers would be shorter." (NO 2000) So, both of you insist that the other clock (not his) is clicking more slowly—assuming, of course, an inertial frame of reference.

—Now, change the story a bit, in that, this time, instead of an "inertial reference frame," acceleration is allowed, and you are on your way to board a spaceship. As an illustration, suppose "[y]ou board a spaceship and take off for deep space. The ship approaches the speed of light. Time for you seems to pass as it always has. It takes you about five seconds to tie your shoe. But to an observer on Earth (assuming he or she could watch you), you are moving at a snail's pace. It takes hours to tie your shoe." (NO 2000) In any event, "you continue on your journey. You slow down, stop, and accelerate back to Earth. You arrive home. You have aged two years during your flight. Two hundred years have passed on Earth. You have successfully travelled forward through time." (NO 2000) In this case, both the observer (if he is still alive) and you agree that you have aged less.

(continued on next page)

Table 1.10. The Problems of Time Travel into the Future (Part II)

- **First Problem *(cont'd)***
  —In other words, when you are back to Earth, a lot of things have changed since, and you may find rather strange to be in a new environment which you hardly recognize, when compared with what you used to think as your city, or the Earth (two years ago, well, in accordance to your time of measurement). This means that you are in the future (two hundred years later), even though you are still relatively young (only two years older). But, can you go back? Well, no. The reason is that "[a]ccording to relativity, you can only move through time in one direction." (NO 2000)

- **Second Problem**
  —The second problem concerns "time dilation under the Theory of General Relativity" (which unifies space-time with mass-energy in a general theory), the same result of time dilation occurs.
  —For instance, when "one clock is deeper in a gravity well than the other," a result is that "the clock deeper in the well" will "tick…more slowly; this effect must be taken into account when calibrating the clocks on the satellites of the Global Positioning System, and it could lead to significant differences in rates of aging for observers at different distances from a black hole." (WK 2008i)
  —More precisely, by way of another example, "it has been calculated that, under general relativity, a person could travel forward in time at a rate four times that of distant observers by residing inside a spherical shell with a diameter of 5 meters and the mass of Jupiter. For such a person, every one second of their 'personal' time would correspond to four seconds for distant observers. Of course, squeezing the mass of a large planet into such a structure is not expected to be within our technological capabilities in the near future." (WK 2008i; R. Gott 2002)
  —The point here, however, is that time travel into the future, in this sense, is allowed, theoretically speaking, but in a very limited way, from the theory of relativity. But one can also ask, What if the theory of relativity were wrong? Well, in that case, there would be a need to find a better theory to explain time travel into the future.

Source: From *Sec.4.4.2* of *FPHG*

Table 1.11. The Problems of Time Travel into the Past (Part I)

• **First Problem**

—The first problem concerns the faster-than-light-speed, since, in accordance to the theory of special relativity, when an object goes faster than the speed of light, something weird can happen, in that, instead of having the clock clicking more slowly, it is moving backward, in relation to some inertial frame of reference—which then raisies the issue of whether time travel into the past is possible at all.

—But the problem here is about the violation of the law of casuality. In everyday language, an excellent illustration of causality violation is that, suppose you travel back into the past and then kill your parents, this then means that you could not have been born. But this is not true, since you are alive.

—Yet, "in the case of a hypothetical signal moving faster than light, there would always be some frames in which the signal was received before it was sent, so that the signal could be said to have moved backwards in time." The violation of casuality exists in this case.

—The first part of the problem is that, in accordance to the theory of special relativity, "it would take an infinite amount of energy to accelerate a slower-than-light object to the speed of light"—let alone the energy needed to propel an object to go faster than the speed of light. (WK 2008i) Although some suggest the possibility of "negative energy," it remains to be seen to what extent this is true.

—And the second part of the problem is that, "although relativity does not forbid the theoretical possibility of tachyons which move faster than light at all times, when analyzed using quantum field theory it seems that it would not actually be possible to use them to transmit information faster than light and there is no evidence for their existence." (WK 2008i; S. Chase 1993)

(continued on next page)

Table 1.11. The Problems of Time Travel into the Past (Part II)

- **Second Problem**
  —The second problem concerns cosmic string and black holes, since it is the theory of general relativity which "extends the special theory to cover gravity, illustrating it in terms of curvature in spacetime caused by mass-energy and the flow of momentum." (WK 2008i)
  —On the other hand, "[g]eneral relativity describes the universe under a system of field equations…and there exist solutions to these equations that permit what are called 'closed time-like curves,' and hence time travel into the past….The first of these was proposed by Kurt Gödel, a solution known as the Gödel metric, but his (and many others') example requires the universe to have physical characteristics that it does not appear to have." (WK 2008i; K. Thorne 1994)
  —But then, the problem now becomes: "Whether general relativity forbids closed time-like curves for all realistic conditions is unknown." (WK 2008i)

- **Third Problem**
  —The third problem concerns wormholes and related ones, and "[w]ormholes are a hypothetical warped spacetime which are also permitted by the Einstein field equations of general relativity, although it would be impossible to travel through a wormhole unless it was what is known as a traversable wormhole." (WK 2008i; M. Visser 1995) In other words, "a wormhole is a hypothetical topological feature of spacetime that is fundamentally a 'shortcut' through space and time. Spacetime can be viewed as a 2D surface, and when 'folded' over, a wormhole bridge can be formed. A wormhole has at least two mouths which are connected to a single throat or tube. If the wormhole is traversable, matter can 'travel' from one mouth to the other by passing through the throat." (WK 2008j)
  —But the problem here is three-fold, as shown in what follows.

(continued on next page)

Table 1.11. The Problems of Time Travel into the Past (Part III)

- **Third Problem *(cont'd)***
  —The first part of the problem is that "there is no observational evidence for wormholes." (WK 2008j)
  —The second part of the problem is that it is not clear how exactly wormholes (and black holes, for that matter) could allow time travel into the past at all.
  —And the third part of the problem is that if using wormholes (and black holes, for that matter) to travel into the past requires something to go faster than light too, then this raises another issue concerning the two problems as previously indicated in the sub-section on the faster-than-light-speed.

- **Fourth Problem**
  —The fourth problem is that some physicists have argued that, even should time travel into the past be possible, by way of these three methods, the person still would not return to the same past history that he exactly experienced before, but in a different one. (WK 2008i)
  —In other words, "strictly speaking," time travel into the past is not "really" possible, in light of these criticisms, so there is no need to worry that one could travel into the past and kill one's parent, as a way to violate the law of casuality.

Source: From *Sec.4.4.2* of *FPHG*

Table 1.12. The Conceptual Dimensions of Consciousness (and Other Mental States)

- **On Heredity and Time**
  —Heredity and the Environment
  —The Past and the Present

- **On Layers of Mental States and Abnormality**
  —Consciousness, Unconsciousness, and Preconsciousness
  —Normality and Paranormality

- **On Organicity and Motivation**
  —Mechanicity and Organicity
  —Primary Motivations and Multiple Motivations

- **On Other (Mostly Epistemic) Considerations**
  —Synthesis and Analysis
  —Situation and the Subject
  —Process and Outcome
  —Reasoning and Other Modes of Thinking
  —Meta-Conceptual Nominalism and Realism

Notes:: These examples are solely illustrative (not exhaustive), and some of the items can be reclassified somewhere else. Nor are they always mutually exclusive. Since they are generalities, exceptions are expected.
Sources: From *FPHC*. A re-construction, but with my own contribution, originally from G.Lindzey & C.Hall, *Introduction to Theories of Personality* (NY: John Wiley & Sons, 1985).

Table 1.13. The Theoretical Levels of Consciousness (and Other Mental States) (Part I)

- **At the Micro-Physical Theoretical Level**
  —Quantum-Mechanics
  —Electromagnetism

- **At the Chemical Theoretical Level**
  —Biochemistry

- **At the Biological Theoretical Level**
  —Evolutionary Biology
  —Neuroscience
  —Artificial Intelligence

- **At the Psychological Theoretical Level**
  —Psychodynamic Psychology
    - Psychoanalytical Psychology
    - Analytical Psychology
    - Socially Oriented Psychology

  —Experimental (Behavioral) Psychology
    - Operant Reinforcement Theory
    - Stimulus-Response Theory
    - Social Learning Theory

  —Cognitive (Gestalt) Psychology
    - Humanist Psychology
    - Existential Psychology
    - Field Theory

  —Psychometric Psychology
    - Idiographics
    - Constitutional Psychology

  —Social Psychology
    - Symbolic Interactive Theory
    - Social Exchange Theory

(continued on next page)

Table 1.13. The Theoretical Levels of Consciousness (and Other Mental States) (Part II)

- **At the Organizational Theoretical Level**
  - —Managerial-Bureaucratic Theory
  - —Oligarchic Theory
  - —Network Theory

- **At the Institutional Theoretical Level**
  - —Functionalist Theory
  - —Anomic Theory

- **At the Structural Theoretical Level**
  - —Conflict Theory
    - Marxian Theory
    - Critical Theory
    - Weberian Theory
  - —Games Theory (in Formal Theory)
  - —Feminist Theory
    - Feminist Concerns
    - Feminist Hopes

- **At the Systemic Theoretical Level**
  - —Equilibrium Theory
  - —System Theory
  - —Chaos Theory

- **At the Cosmological Theoretical Level**
  - —Superluminal Model
  - —The Theory of Floating Consciousness

(continued on next page)

Table 1.13. The Theoretical Levels of Consciousness (and Other Mental States) (Part III)

- **At the Cultural Theoretical Level**
  - —Substantive Theories
    - Structuralist Theory
    - Post-Structuralist Theory (in Postmodernism)
  - —Meta-Theories
    - Epistemic Objectivism vs. Epistemic Historicism
    - Epistemic Subjectivism vs. Epistemic Non-Subjectivism
      - –Phenomenology
      - –Ethnomethodology
      - –Hermeneutics
    - Epistemic Relativism vs. Epistemic Absolutism
    - Epistemic Reductionism vs. Epistemic Emergencism

- **At Other Levels**
  - —Structuration Theory
  - —Reflexive Socioanalysis

Notes: These examples are solely illustrative (not exhaustive), and some of the items can be reclassified somewhere else. Nor are they always mutually exclusive. Since they are generalities, exceptions are expected.
Source: From many different sources as indicated in *FPHC*

Table 1.14. The Thematic Issues of Consciousness (and Other Mental States)

- **The Factor of History**
  —Pre-Modernity
  —Modernity
  —Postmodernity
  —After-Postmodernity

- **The Importance of Needs**
  —Having (e.g., the everyday, the technological)
  —Belonging (e.g., the just and the good)
  —Being (e.g., the true, the holy, the sublime/beautiful, and the good)

Notes: These examples are solely illustrative (not exhaustive). Nor are they always mutually exclusive. Since they are generalities, exceptions are expected. Also, some categories overlap with others; for instance, the good is also related with being-ness, just as the everyday can be connected with belong-ness.
Sources: From *FPHC*, based on *FHC* and *FCD*

Table 1.15. Having, Belonging, and Being in Consciousness

- **Having (e.g., About the Everyday and Technological)**
  —Ex: Physiological needs (e.g., thirst, hunger, sex)
  —Ex: Safety (e.g., stability, freedom from fear and anxiety)

- **Belonging (e.g., About the Good and Just)**
  —Ex: Social interaction
  —Ex: Friendship, acquaintance
  —Ex: Love, family
  —Ex: Self-respect, respect from others

- **Being (e.g., About the True, Holy, Beautiful/Sublime, and Good)**
  —Ex: Understanding of reality about self and world
  —Ex: Spiritual quest for holiness, beauty, sublimity, and goodness

Notes: These examples are solely illustrative (not exhaustive), and some of the items can be reclassified somewhere else. Nor are they always mutually exclusive. Since they are generalities, exceptions are expected.
Sources: From *FPHC*, based on *FHC* and *FCD*

Table 1.16. The Having-Ness of Consciousness (and Other Mental States)

- **At the Micro-Physical Level**
  —Ex: The finer physics of nutrition

- **At the Chemical Level**
  —Ex: Food guide pyramid

- **At the Bio-Psychological Level**
  —Ex: The hierarchy of needs

- **At the Institutional Level**
  —Ex: The production of commodity fetishism

- **At the Organizational Level**
  —Ex: Mass standardization (and customization); bureaucratic rationality

- **At the Structural Level**
  —Ex: The divide between the haves and the have-nots

- **At the Cultural Level**
  —Ex: The bondage of greater expectations

- **At the Systemic Level**
  —Ex: The technological transformation of poverty with transvaluations

- **At the Cosmological Level**
  —Ex: Floating consciousness in deep space without the human physiological needs

Notes: These examples are solely illustrative (not exhaustive), and some of the items can be reclassified somewhere else. Nor are they always mutually exclusive. Since they are generalities, exceptions are expected.
Sources: From *FPHC*. See also *FCD* and *FHC*.

Table 1.17. The Belonging-Ness of Consciousness (and Other Mental States)

- **At the Micro-Physical Level**
  —Ex: Statistical physics and social networks

- **At the Chemical Level**
  —Ex: Olfaction and social relations

- **At the Bio-Psychological Level**
  —Ex: Dual human nature on relationships

- **At the Institutional Level**
  —Ex: Private property and social cleavage

- **At the Organizational Level**
  —Ex: Legal formalism and iron cage

- **At the Structural Level**
  —Ex: Class conflicts and social alienation

- **At the Cultural Level**
  —Ex: Ruthless society and the culture of law

- **At the Systemic Level**
  —Ex: Human extinction and post-human relations

- **At the Cosmological Level**
  —Ex: The transcendence of equality in the cosmos and beyond

Notes: These examples are solely illustrative (not exhaustive), and some of the items can be reclassified somewhere else. Nor are they always mutually exclusive. Since they are generalities, exceptions are expected.
Sources: From *FPHC*. See also *FCD* and *FHC*.

Table 1.18. The Being-Ness of Consciousness (and Other Mental States) (Part I)

- **At the Micro-Physical Level**
  —Ex: Space-time and matter-energy in classical mechanics, quantum mechanics, and the theory of relativity

- **At the Chemical Level**
  —Ex: Drugs and well being

- **At the Biological Level**
  —Ex: Neuroscience and mystical experiences

- **At the Psychological Level**
  —Ex: Reductionism and emergencism

- **At the Institutional Level**
  —Ex: The proliferation of cults and sects, and the corporate-art industry

- **At the Organizational Level**
  —Ex: The organization of ideas in the scheme of things

- **At the Structural Level**
  —Ex: Disciplinary power and control in the order of things

- **At the Systemic Level**
  —Ex: Mutualism and ecological psychology

- **At the Cultural Level**
  —Ex: The recurrence of competing spirits, and floating consciousness as a climax of evolution

(continued on next page)

Table 1.18. The Being-Ness of Consciousness (and Other Mental States) (Part II)

- **At the Cosmological Level**
  —Ex: The search for elsewherewhen, and the evolution of other consciousnesses (e.g., the hyper-spatial consciousness)

Notes: These examples are solely illustrative (not exhaustive), and some of the items can be reclassified somewhere else. Nor are they always mutually exclusive. Since they are generalities, exceptions are expected.
Sources: From *FPHC*. See also *FCD* and *FHC*.

Table 1.19. Cognitive Partiality in Different Mental States

- **The Conscious**
  - —*Biased*
    - Ex: Self-overratingness
    - Ex: Self-modesty
    - Ex: Self-fulfilling prophesy
  - —*Shallow*
    - Ex: Attribution error
    - Ex: Heuristics
    - Ex: Illusory thinking
  - —*Short-term*
    - Ex: Deindividuation
  - —*Materialistic*
    - Ex: Hierarchy of needs

- **The Subconscious**
  - Ex: mental reconstruction
  - Ex: stereotyping

- **The Unconscious**
  - Ex: automatic processing
  - Ex: cognitive intuition

Notes: These examples are solely illustrative (not exhaustive), and some of the items can be reclassified somewhere else. Nor are they always mutually exclusive. Since they are generalities, exceptions are expected. And the comparison is relative, not absolute.
Source: From *BCPC*

Table 1.20. Emotional Non-Neutrality and Behavioral Alteration in Different Mental States

- **The Emotional**
  - —*Conscious*
    - Ex: Feelings
  - —*Subconscious*
    - Ex: Moods
  - —*Unconscious*
    - Ex: Empathy
    - Ex: Instincts

- **The Behavioral**
  - —*Conscious*
    - Ex: Foot-in-the-door
    - Ex: Role playing
    - Ex: Saying-is-believing
  - —*Subconscious*
    - Ex: Discrimination
  - —*Unconscious*
    - Ex: Non-Verbal Communication

Notes: These examples are solely illustrative (not exhaustive), and some of the items can be reclassified somewhere else. Nor are they always mutually exclusive. Since they are generalities, exceptions are expected. And the comparison is relative, not absolute.
Source: From *BCPC*

Table 1.21. The Limits of Intuition in Unconsciousness

- **Cognitive Intuition (or Intuitive Thinking)**
  - —*Features*
    - Ex: mental schemas
    - Ex: expertise
    - Ex: blindsight
  - —*Problems*
    - Ex: inaccurate
    - Ex: non-explanatory

- **Emotional Intuition**
  - —*Features*
    - Ex: empathy
    - Ex: instincts
  - —*Problems*
    - Ex: over-generalized
    - Ex: non-explanatory

- **Behavioral Intuition**
  - —*Features*
    - Ex: prosemics
    - Ex: kinesics
  - —*Problems*
    - Ex: unreliable
    - Ex: non-explanatory

Notes: These examples are solely illustrative (not exhaustive), and some of the items can be reclassified somewhere else. Nor are they always mutually exclusive. Since they are generalities, exceptions are expected. And the comparison is relative, not absolute.
Source: From *BCPC*

Table 1.22. The Wealth/Poverty Dialectics in Different Mental States: The Case of Cognition

- **The Conscious**
  - —*Biased*
    - Ex: If richer in self-serving bias, then poorer in understanding others
  - —*Shallow*
    - Ex: If richer in heuristic thinking, then poorer in comprehending reality
  - —*Short-term*
    - Ex: If richer in short-term tendency, then poorer in long-term planning
  - —*Materialistic*
    - Ex: If richer in physiological obsession, then poorer in spiritual enlightenment

- **The Subconscious**
  - Ex: If richer in mental reconstruction, then poorer in accuracy of recall
  - Ex: If richer in stereotyping, then poorer in understanding groups

- **The Unconscious**
  - Ex: If richer in automatic processing, then poorer in depth of analysis
  - Ex: If richer in cognitive intuition, then poorer in reliability of judgments

Notes: These examples are solely illustrative (not exhaustive), and some of the items can be reclassified somewhere else. Nor are they always mutually exclusive. Since they are generalities, exceptions are expected. And the comparison is relative, not absolute.
Source: From *BCPC*

Table 1.23. The Wealth/Poverty Dialectics in Different Mental States: The Case of Emotion and Behavior

- **Emotion**
  - —*Conscious*
    - Ex: If richer in relying on familiarity and similarity for social attraction, then poorer in having diverse social relations
  - —*Subconscious*
    - Ex: If richer in moodiness, then poorer in sober thinking
  - —*Unconscious*
    - Ex: If richer in empathy, then poorer in task-oriented efficiency
    - Ex: If richer in instincts, then poorer in critical thinking

- **Behavior**
  - —*Conscious*
    - Ex: If richer in foot-in-the-door technique, then poorer in instant Gratification
    - Ex: If richer in role playing, then poorer in identity stability
  - —*Subconscious*
    - Ex: If richer in discriminating, then poorer in social harmony
  - —*Unconscious*
    - Ex: If richer in kinesics and proxemics, then poorer in verbal skills, relatively speaking.

Notes: These examples are solely illustrative (not exhaustive), and some of the items can be reclassified somewhere else. Nor are they always mutually exclusive. Since they are generalities, exceptions are expected. And the comparison is relative, not absolute.
Source: From *BCPC*

Table 1.24. The Theoretical Debate on Nature vs. Nurture (Part I)

- **The Environmental Approach**
  —*Thesis*: It focuses, relatively speaking, more on the environment (culture and society) in explaining the achievement gap among individuals and for that matter, countries or regions, when contrasted with the natural factors.
  —*Discourse*: Examples include Jose Ortega y Gasset ("Man has no nature; what he has is history"), Ashley Montagu ("Man is man because he has no instincts, because everything he is and has become he has learned from his culture, from the man-made part of the environment, from other human beings"), Stephen Jay Gould ("[The] brain [is] capable of a full range of behaviors and predisposed to none"), and Jesse Jackson (who blames white racism for the failure of blacks to close the achievement gap between whites and blacks in America). The works on dependency theory in international political economy (with a Marxian influence) and on the Protestant work ethic (by Max Weber) also point to this environmental direction.

- **The Genetic Approach**
  —*Thesis*: It focuses instead, relatively speaking again, on hereditory factors (e.g., genes and evolution) in explaining the achievement gap among individuals and for that matter, countries or regions, when contrasted with the envrionmental factors.
  —*Discourse*: Examples are Hans Eysenck and William Sheldon (in constitutional psychology), Konrad Lorenz (in his work on innate aggressive human nature), Gary Marcus (on the complexities of human thought by a tiny number of genes) and Robert Plomin, together with Michael Owen and Peter McGuffin (on the genetic basis of complex human behaviors).

(continued on next page)

Table 1.24. The Theoretical Debate on Nature vs. Nurture (Part II)

- **The Compromise Approach**
  —*Thesis*: It seeks the middle-of-the-road argument in regard to nature and nurture and regards all differences among individuals and groups as the result of the mixture of both nature and nurture, more or less equally.
  —*Discourse*: Examples include C. Murray and R. Herrnstein ("It seems highly likely to us that both genes and environment have something to do with this issue") and Dan Dennett ("Surely 'everyone knows' that the nature-nurture debate was resolved long ago, and neither side wins since everything – is – a – mixture – of – both – and - it's – all – very -complicated....")

- **The Transcendent Approach**
  —*Thesis*: It goes beyond both nature and nurture (without, however, committing the compromise fallacy) in showing their closely intertwined interactions in producing the behavioral differences as often seen in individual human endeavors on the micro scale, and for that matter, in country (or regional) endeavors on the macro one—in the context of my five theses, namely, (a) the compromise fallacy, (b) no oppression without self-oppression, (c) no success without failure, (d) the factor of randomness, and (e) the post-human vision, to be elaborated in Chapter Six.
  —*Discourse*: Peter Baofu proposed this approach on the basis of his "theory of contrastive advantages" (as an original theoretical contribution to the debate, which was first proposed and analyzed in *The Future of Capitalism and Democracy*). In the end, the human genes will not last, to be eventually superseded by post-human life forms, so the debate between genes and memes has obscured something profoundly important about the future that the world has never known. And the debate is also misleading and faulty in its dichotomy.

Source: From *BNN*. See the book for citations and details.

Table 1.25. Physical Challenges to Hyper-Spatial Consciousness

- **The Understanding of a Higher-Dimensional World of Space-Time**
  —Ex: 4 for traditional aspects of space-time (e.g., length, width, breadth and time) plus 6 more new dimensions in theory of hyper-space, with profound implications for practical applications to new forms of consciousness.

- **The Mastering of Dark Matter and Dark Energy**
  —Ex: "ordinary matter" (e.g., atoms, molecules) as a mere 4.4% of the universe, with 23% made of "cold dark matter" and the rest (about 73%) of mysterious "dark energy," with fundamental significance to questions about the limit of the speed of energy (or info), the availability of energy for use, and the nature of space-time, just to cite some examples.

- **The Exploration of Multiverses**
  —Ex: theoretical speculation of other universes (e.g., "baby universes," "gateways" in black holes, "wave function of the universe," "many worlds," "brane worlds"), with potentially seminary discoveries of different physical laws in relation to matter-energy and space-time, and vital differences to the future of post-human conquest of other universes (for the emergence of new forms of consciousness).

Notes: These examples are solely illustrative (not exhaustive), and some of the items can be reclassified somewhere else. Nor are they always mutually exclusive. Since they are generalities, exceptions are expected. The point here is to give a rough picture of the evolution of consciousness to the hyper-spatial consciousness and others totally unknown to current earthlings. As a note of clarification, it makes no difference to my argument as to whether or not the hyper-spatial consciousness may emerge before, during, and after floating consciousness.
Source: From *Table 4.5* of *FPHC*

Table 1.26. The Theory of Floating Consciousness (Part I)

- **At the Micro-Physical Level**
  —Ex: intelligent life without the human physical-chemical system

- **At the Chemical Level**
  —Ex: space radiation and toxins

- **At the Bio-Psychological Level**
  —Ex: exo-biological evolution in deep space
  —Ex: genetic engineering of new beings

- **At the Institutional Level**
  —Ex: post-capitalism
  —Ex: post-democracy

- **At the Organizational Level**
  —Ex: less legal-formalistic routines

- **At the Structural Level**
  —Ex: alien forms of violence

- **At the Cultural Level**
  —Ex: transcending freedom
  —Ex: transcending equality

(continued on next page)

Table 1.26. The Theory of Floating Consciousness (Part II)

• **At the Cosmological Level**
—Ex: parallel universes
—Ex: pocket universes

• **At the Systemic Level**
—Ex: space habitats (in zero-gravity)

Notes: Each example draws from the works of different scholars in the field. For instance, at the cosmological level, the idea of parallel universes is from the theoretical speculation in quantum cosmology by Stephen Hawking and others, while the one of pocket universes comes from the theoretical work of Allan Guth at MIT. And at the institutional level, I proposed post-capitalism and post-democracy in *FCD* (and later, from *BDPD* and *BCPC*). In addition, the examples are solely illustrative (not exhaustive), and some of the items can be reclassified somewhere else. Nor are they always mutually exclusive. Since they are generalities, exceptions are expected.
Sources: From *FPHC*—and, originally, from *FCD*

Table 1.27. The Potential of Unfolding Unconsciousness

- **Superior Senses**
  - —*Through space*
    - Ex: clair-sentience (in feeling)
    - Ex: clair-voyance (in seeing)
    - Ex: clair-audience (in hearing)
    - Ex: clair-austance (in tasting)
    - Ex: clair-alience (in smelling)
    - Ex: clair-cognizance (in knowing)
  - —*Through time*
    - Ex: pre-cognition (in knowing), pre-sentiment (in feeling)
    - Ex: retro-cognition (in knowing), retro-sentiment (in feeling)

- **Superior Contacts**
  - —*With living minds*
    - Ex: telepathy
  - —*With the dead*
    - Ex: mediumship
  - —*With other-living entities*
    - Ex: telekinesis

Notes: The categories and examples are solely illustrative (not exhaustive), and the comparison is relative (not absolute), nor are they necessarily mutually exclusive. And some can be easily re-classified elsewhere. As generalities, they allow exceptions.
Source: A summary of Ch.5 of *FPHU*

Table 1.28. The Future Exploration of Unfolding Unconsciousness

- **In the Context of Nature**
  - —Ex: the higher-dimensional space-time in hyperspace theory
  - —Ex: the particle-wave interactions in quantum mechanics
  - —Ex: the study of brain waves in electromagnetic theory

- **In the Context of the Mind**
  - —Ex: the unconscious fantasies in psychoanalysis
  - —Ex: the mind-altering drugs in biochemistry
  - —Ex: the manipulation of neural activity in neurobiology

- **In the Context of Culture**
  - —Ex: the correlation between anomalous experience and moral/religious/aesthetic interests in culture studies
  - —Ex: the role of myths and language in epistemology

- **In the Context of Society**
  - —Ex: different organizational agendas in anomalous research
  - —Ex: the scholarly divide on anomalous study
  - —Ex: interests and propaganda in institutioanl research
  - —Ex: the transformative power of technology in research on social systems
  - —Ex: the natural and non-natural selection of other worlds in exo-biology
  - —Ex: the role of the post-human conscious mind in qualitative demography

Notes: The examples in each category are solely illustrative (not exhaustive), and the comparison is relative (not absolute), nor are they necessarily mutually exclusive. And some can be easily re-classified elsewhere. As generalities, they allow exceptions.
Source: A summary of Chs.2-5 of *FPHU*, in conjunction with the rest of the book

Table 1.29. Creative Techniques and Traits

- **Creative Techniques**
  - —*Comprehensiveness*
    - Ex: Broadness of Scope
    - Ex: In-Depth Analysis
    - Ex: Multifaceted Taxonomy
    - Ex: Numerous Theories
  - —*Arrangement*
    - Ex: Separation (or Subtraction)
    - Ex: Combination (or Addition)
    - Ex: Replacement
    - Ex: Accommodation
    - Ex: Reversal
    - Ex: Stretching
  - —*Serendipity*
    - Ex: Play
    - Ex: Chance

- **Creative Traits**
  - —*Openness / Risk*
    - Ex: Ability to welcome new ideas or to do new things, with risk
  - —*Discipline*
    - Ex: Ability to sacrifice whatever it takes to succeed
  - —*Resilience / Confidence*
    - Ex: Ability to take failures to mind, not to heart
  - —*Trust*
    - Ex: Ability to cooperate with others for team work, if necessary

Notes: The examples in the categories are solely illustrative (not exhaustive), and the comparison is relative (not absolute), nor are they necessarily mutually exclusive. And some can be easily re-classified elsewhere. As generalities, they allow exceptions.
Source: From *Sec.4.2* of *FPHCT*. For more details, also read the rest of the book.

Table 1.30. The Desirability of Creativity, and Its Dark Sides (Part I)

- **In Relation to Instrumental Rationality**
  —*Desirability*
  - Ex: For the sake of efficiency and effectiveness in this capitalist era of our time.

  —*Undesirability*
  - Ex: What Max Weber (1930) called the "iron cage" of capitalism, where live the "sensualists without heart, specialists without spirit"—or something which the Frankfurt School has forcefully asked us to be freed from.

- **In Relation to Substantive Rationality**
  —*Desirability*
  - Ex: For the sake of God, the King, Motherland, or other comparable ideals.

  —*Undesirability*
  - Ex: Domination and oppression of various forms—something which the moderns had struggled to free themselves from in the first place, since the dawn of modernity.

- **In Relation to Autonomous Rationality**
  —*Desirability*
  - Ex: For the sake of itself, the autonomy of creative endeavor.

  —*Undesirability*
  - Ex: The decadence of its own degeneration (e.g., the donwsides of postmodernism in our postmodern times, or what Nietzsche once called the unbearable "unreality" and "falseness" of the autonomous artist's "innermost existence"—and, in other cases, the painful suffering from different mental illnesses for those highly creative individuals).

(continued on next page)

Table 1.30. The Desirability of Creativity, and Its Dark Sides (Part II)

Notes: The examples in the categories are solely illustrative (not exhaustive), and the comparison is relative (not absolute), nor are they necessarily mutually exclusive. And some can be easily re-classified elsewhere. As generalities, they allow exceptions.
Also, I already went to great lenghts to explain the desirability and its dark sides of all these three rationalities (although there may be others too) in the two volumes of *FHC*—and, for that matter, *FCD*.
Source: From *Sec.4.9* of *FPHCT*. For more details, also read the rest of the book.

Table 1.31. Posthuman-Ism, Post-Humanism, and Trans-Humanism

- **Post-Humanism**
  —The neologism "post-human" used in my books should *not* be confused with another term which looks similar but has a totally different meaning in the literature of postmodernism, namely, "post-humanism"—which constitutes a critique of "humanism" as tradionally understood (especially, though not exclusively, in relation to the idea of progress in science and reason in the Enlightenment project). (WK 2008)
  —My works reject the project of "postmodernism" and propose the future world of what I orignally called "after-postmodernity" in *FHC* and *FCD*, for instance.

- **Trans-Humanism**
  —Also, the neologism "post-human" used in my books should *not* be confused with a similar term which is used to champion the ideology of technology for the fturue co-existence between humans and cyborgs in "trans-humanism." (WK 2008a)
  —Instead, my term "post-human" in relation to "posthuman-ism" also rejects "trans-humanism" (especially, though not exclusively, in relation to the promises of technology) and refers to something else altogether, that is, the future extinction of humans and its post-human successors in deep space and beyond unto multiverses.
  —Chapter Two of my book on ethics (*BEPE*) provides a more comprehensive critique of trans-humanism.

- **Posthuman-Ism**
  —With these two clarifications in mind, the "post-humans" (as envisioned in my books) can take different forms, and I proposed, in my different books, some of them, such as "unfolding unconsciousness," "floating consciousness," "hyper-spatial consciousness," "thinking machines," "thinking robots," "genetically altered superior beings," "cyborgs," and others (as already analyzed in my previous works).
  —They will evolve in conjunction with other visions of mine in relation to nature, society, and culture.

Sources: From *Sec.1.6* of *FPHE* (and also *BEPE*). For more details, also read the rest of the books (and other books of mine).

Table 1.32. Three Great Future Transformations of the Martial Body (Part I)

- **Virtual Battle**
  - —Ex: virtual fight
  - —Ex: virtual memory

- **The Hyper-Martial Body**
  - —*In relation to the physical dimension*
    - Ex: striking (e.g., punching, kicking, trapping, acupressure-striking, and so on in an amazing way)
    - Ex: grapping (e.g., throwing, pinning, joint-locking, and so on in an amazing way)
    - Ex: running and walking (e.g., on different surfaces)
    - Ex: jumping (e.g., through walls, trees, etc.)
    - Ex: flying (e.g., through space)
    - Ex: enduring pain (e.g., absorbing punches, kicks, and so on in an amazing way)
    - Ex: using extremely advanced hi-tech weapons
    - Ex: healing (e.g., with special energy, etc.)
    - Ex: entering (e.g., into different dimensions of space)
    - Ex: sensing (e.g., in extraordinary ways through space and time, etc.)
    - Ex: wearing hi-tech suits for defensive and offensive purposes (e.g., like a hi-tech armor)
  - —*In relation to the mental dimension*
    - Ex: the ability to relax the martial body in a way that current humans cannot.
    - Ex: the ability to concentrate the martial mind in a way that current humans are not accustomed to.
    - Ex: the ability to breathe and be timely in preparing the martial body for fighting in a way that current humans are not good at.

- **The Post-Martial Floating Mind**
  - —Ex: the mind without the human body
  - —Ex: the quest for spirituality beyond martial arts

(continued on next page)

Table 1.32. Three Great Future Transformations of the Martial Body (Part II)

Notes: The examples in the categories are solely illustrative (not exhaustive), and the comparison is relative (not absolute), nor are they necessarily mutually exclusive. And some can be easily re-classified elsewhere. As generalities, they allow exceptions.
Source: From *Sec. 2.5.1* of *FPHMA*. See text for citations.

Table 1.33. Three Great Future Transformations of the Sexual Body

- **Virtual Pleasure**
  —Ex: virtual Sex
  —Ex: virtual Recall

- **The Hyper-Sexual Body**
  —*In relation to the physical dimension*
  - Ex: the ability to perform differenet sexual positions without the various physical limitations that humans currently have.
  - Ex: the ability to produce unusual amount of sexual energy to endure in sexual acts in a way that humans currently cannot.
  - Ex: the ability to engage in multiple sexual experiences comparable to (but different from) non-human sexuality (e.g., hermaphrodite, monoecious, sex-switching, and many other types) that current humans cannot.

  —*In relation to the mental dimension*
  - Ex: the ability to relax the sexual body in a way that current humans are not accustomed to.
  - Ex: the ability to concentrate the sexual mind in a way that current humans are not good at.
  - Ex: the ability to breathe and be patient in preparing the sexual body for sexuality in a way that current humans are not able to.

- **The Post-Sexual Floating Mind**
  —Ex: the mind without the human body
  —Ex: the quest for spirituality beyond sexuality

Notes: The examples in the categories are solely illustrative (not exhaustive), and the comparison is relative (not absolute), nor are they necessarily mutually exclusive. And some can be easily re-classified elsewhere. As generalities, they allow exceptions.
Source: From *Sec. 3.4.2* of *FPHS*. See text for more info.

Table 1.34. The Trinity of Pre-Modernity

- **Pre-Free-Spirited Pre-Modernity (Pre-Modernism) and Its Internal Split**
  —Competing worldviews and values both within and between linear centric (e.g., Islamic, Christian, Judaic, Imperial Roman) and cyclical-centric (e.g., Confucian, Taoist, Hindu, and Buddhist) orientations
  —Compare modernism with pre-modernism here in relation to the seven dimensions of human existence like the true and the holy (e.g., different versions of epistemic dogmas and religious superstitions), the everyday and the technological (e.g.,different versions of non-technophilism and non-consumerism), the beautiful/sublime (e.g., different versions of aesthetic non-autonomy), and the good and the just (e.g., different versions of moral particularism).

- **Pre-Capitalist Pre-Modernity (Pre-Modernization) and Its Own Discontents**
  —Competing versions of societal arrangements (e.g., feudalism, monarchism, and the holy order)

- **Hegemonic Pre-Modernity and Its Countervailing Forces**
  —Different power centers and their enemies (e.g., the Roman Empire and the "barbarian hordes," the "Holy Crusades" and the Muslims, the Middle Kingdom and the invading tribes, different social castes in India, and warring Greek city-states)

Notes: The examples in each category are solely illustrative (not exhaustive) nor necessarily mutually exclusive, and the comparison is relative (not absolute). As generalities, they allow exceptions. Also, it does not matter what the "base" era is in the analysis of any trinity. And in the present context, the "base" era is modernity (for instance, with its "free-spirited modernity" and the other two parts). So, for pre-modernity, the trinity takes the form of, say, "pre-free-spirited pre-modernity," together with the other two parts.
Sources: From Ch.2 of *BCIV* and also the 2 volumes of *FHC*

Table 1.35. The Trinity of Modernity (Part I)

- **Free-Spirited Modernity (Modernism) and Its Internal Split**
  —*On the True and the Holy*
    - The freedom from the dogmas of the past to the better understanding of, and union with, the world and self (Ch.3 of *FHC*)
    - Alternative discourses: about the true (e.g., anti-science discourses) and the holy (non-mainstream theologies) (Ch.3 of *FHC*)

  —*On the Technological and the Everyday*
    - The freedom from life harshness to the higher technophilic, consumeristic lifeform (Ch.2 of *FHC*)
    - Alternative discourses: about the everyday (e.g., transcendental mindsets) and the technological (e.g., Arcadianism) (Ch.2 of *FHC*)

  —*On the Good and the Just*
    - The freedom from the theo-aristocratic tyranny to the moral universality for a just society (Ch.5 of *FHC*)
    - Alternative discourses: about the just (e.g., Communism, Anarchism) and the good (e.g., Nazism/Fascism, and Zarathustrianism) (Chs.5-6 of *FHC*)

  —*On the Beautiful and the Sublime*
    - The freedom from the external distortion of aesthetic pleasure to the boundless infinity of totality in artistic autonomy (Ch.4 of *FHC*)
    - Alternative discourses: about the beautiful/sublime (e.g., kitsch and historical avant-gardism) (Ch.4 of *FHC*)

- **Capitalist Modernity (Modernization) and Its Own Discontents**
  —*During the Industrial Revolution*
    - Ex: Marx on the institution of inequality (Ch.1 of *FHC*)

  —*During the Modern Rational-Instrumental Epoch*
    - Ex: Weber on the politics of soft liberal institutions (Ch.5 of *FHC*)

  —*During the Great Depression*
    - Ex: Keynes on the myth of the free market (Chs.1-3 of *FHC*)

  —*During the Cold War*
    - Ex: Lasch on the narcissistic culture industry (Chs.2-3 of *FHC*)

(continued on next page)

Table 1.35. The Trinity of Modernity (Part II)

- **Hegemonic Modernity and Its Countervailing Forces**
  —*The Legacies of Colonialism and Imperialism*
    - Ex: European colonization of most of the modern world (Ch.1 of *FHC*)

  —*The Struggle for Decolonialization*
    - Ex: The countervailing forces of resentment, rechantment, and regionalism (Chs.1 & 6 of *FHC*)

Notes: The examples in each category are solely illustrative (not exhaustive) nor necessarily mutually exclusive, and the comparison is relative (not absolute). As generalities, they allow exceptions.
Sources: From the 2 volumes of *FHC*—and also from *FCD*

Table 1.36. The Trinity of Post-Modernity

- **Free-Spirited Postmodernity (Postmodernism) and Its Internal Split**
  —*On the True and the Holy*
  - Postmodern performative turn for knowing and its enemies (Ch.3 of *FHC*)
  - Postmodern comparative theology and its opponents (Ch.3 of *FHC*)

  —*On the Technological and the Everyday*
  - Postmodern corporate technological mindset and its adversaries (Ch.2 of *FHC*)
  - Postmodern postmaterialism and its critics (Ch.2 of *FHC*)

  —*On the Good and the Just*
  - Postmodern politics of difference and its foes (Ch.5 of *FHC*)

  —*On the Beautiful and the Sublime*
  - Postmodern deconstruction and its dissenters (Ch.4 of *FHC*)

- **Capitalist Postmodernity (Postmodernization) and Its Own Discontents**
  —*During the Post-Cold War and Beyond*
  - Ex: post-Fordism and its shortcomings (Ch.6 of *FHC*; Chs.6-7 of *FCD*)

- **Hegemonic Postmodernity and Its Countervailing Forces**
  —*The Debate on the Global Village*
  - Ex: uni-civilizationalism vs. multi-civilizationalism (Ch.6 of *FHC*)

  —*The Resistance Movement*
  - Ex: rechantment and the politics of civilizational claims (e.g., Islamic, Confucian and other ethos in relation to the Same) (Ch.6 of *FHC*; Ch.10 of *FCD*)
  - Ex: resentment and the politics of resurgence (e.g., the rising Chinese superpower, the growing EU, and other players in relation to the U.S. and her allies) (Ch.6 of *FHC*; Ch.8 of *FCD*)
  - Ex: regionalism and the politics of inequality (e.g., trans- or international blocs, the North-South divide, NGO's) (Ch.6 of *FHC*; Ch.5 of *FCD*)

Notes: The examples in each category are solely illustrative (not exhaustive) nor necessarily mutually exclusive, and the comparison is relative (not absolute). As generalities, they allow exceptions.
Sources: From *FCD* and the 2 volumes of *FHC*

Table 1.37. The Trinity of After-Postmodernity

- **Free-Spirited After-Postmodernity (After-Postmodernism) and Its Internal Split**
  —The discourse of naked contingency (Ch.10 of *FCD*; Ch.4 of *FPHC*)

- **Post-Capitalist After-Postmodernity (After-Postmodernization) and Its Own Discontents**
  —Different versions of post-capitalism and post-democracy, and their enemies (Ch.10 of *FCD*; Chs.3-4 of *FPHC*)

- **Hegemonic After-Postmodernity and Its Countervailing Forces**
  —The Cyclical Progression of Hegemony in Multiverses (Chs.9-10 of *FCD*; Ch.4 of *FPHC*)

Notes: The examples in each category are solely illustrative (not exhaustive) nor necessarily mutually exclusive, and the comparison is relative (not absolute). As generalities, they allow exceptions.
Sources: From *FCD* and also *FHC*

• Part Two •

Objectivity

CHAPTER 2
OBJECTIVITY AND ITS AMBIVALENCE

> Objectivity is a false god. We are human beings and we screw up or have flaws that are hidden from us.
> —Ken Auletta (TE 2012)

The Robustness of Objectivity

Probability interpretations involve both objectivity and subjectivity—as already indicated in *Sec. 1.3*.

The relationship between objectivity and subjectivity is treated as dialectic in this book, in that there is no objectivity without subjectivity—and vice versa.

Yet, a common scholarly bias in the literature is to favor one over the other. Consequently, this book is to fix this bias by devoting a whole chapter on each, with the understanding that the two are related.

This dialectic relationship between objectivity and subjectivity can be revealed by showing how and why the analysis of objectivity (in Chapter Two) is inadequate without the inclusion of subjectivity, just as the examination of subjectivity (in Chapter Three) is insufficient without the consideration of objectivity.

The issues concerning objectivity in probability study are to be addressed in this chapter (Chapter Two), whereas the issues concerning subjectivity will be analyzed in the next chapter (Chapter Three) instead.

With this dialectic treatment in mind—a good way to examine objectivity and its ambivalence in probability study is by way of the evaluation of the extent to which it is in fact both possible (or impossible) and desirable (or undesirable).

This can be done by way of a comprehensive analysis of objectivity in the context of probability from the four perspectives of the mind, nature, society, and culture (in accordance to my sophisticated methodological holism, as explained in *Sec. 1.9*).

In other words, this chapter is organized in four main sections, namely, in relation to (2.2) objectivity and the mind, (2.3) objectivity and nature, (2.4) objectivity and society, and (2.5) objectivity and culture—to be addressed hereafter, in that order (and summarized in *Table 2.1*).

Objectivity and the Mind

Objectivity, when examined from the perspective of the mind, can illuminate its possibility (or impossibility) and desirability (or undesirability), and this can be shown by way of two case studies, namely, (2.2.1) objectivity, imagination, and mental models, and (2.2.2) objectivity, emotion, and rationality—to be addressed in what follows, respectively.

Objectivity, Imagination, and Mental Models

A good way to evaluate the possibility (or impossibility) and desirability (or undesirability) of objectivity from the perspective of the mind is to explore objectivity, imagination, and mental models.

An argument in the literature is that, as Kenneth Craik suggested in 1943,…the mind constructs 'small-scale models' of reality that it uses to anticipate events." (WK 2012p)

In psychology, "the term 'mental models' is sometimes used to refer to mental representations or mental simulation generally….People infer that a conclusion is valid if it holds in all the possibilities….They reject a conclusion if they find a counterexample, i.e., a possibility in which the premises hold, but the conclusion does not." (WK 2012p) (W. Schroyens 2003; N. Verschueren 2005)

History of Mental Models

The history of mental models is short, as "the term 'mental model' is believed to have originated with Kenneth Craik in his 1943 book *The Nature of Explanation.* Georges-Henri Luquet in *Le dessin enfantin* (*Children's Drawings*), published in 1927 by Alcan, Paris, argued that children construct internal models, a view that influenced, among others, child psychologist Jean Piaget." (WK 2012p)

Then, "Philip Johnson-Laird published *Mental Models: Towards a Cognitive Science of Language, Inference and Consciousness* in 1983. In the same year, Dedre Gentner and Albert Stevens edited a collection of chapters in a book also titled *Mental Models*. The first line of their book explains the idea further: 'One function of this chapter is to belabor the obvious; people's views of the world, of themselves, of their own capabilities, and of the tasks that they are asked to perform, or topics they are asked to learn, depend heavily on the conceptualizations that they bring to the task.'" (WK 2012p)

Later, "there has been much discussion and use of the idea in human-computer interaction and usability by researchers including Donald Norman and Steve Krug (in his book *Don't Make Me Think*). Walter Kintsch and Teun A. van Dijk, using the term *situation model* (in their book *Strategies of Discourse Comprehension*, 1983), showed the relevance of mental models for the production and comprehension of discourse." (WK 2012p)

Characteristics of Mental Models

A mental model has distinctive characteristics, and hereafter are some main ones for illustration (as summarized in *Table 2.2*), in that it is commonly: (WK 2012p; D. Ford 2009)

- "founded on hardly qualifiable, impugnable, obscure, or incomplete facts."
- "flexible" and "considerably variable in positive as well as in negative sense."
- used for its "effects as information filter," as it "causes selective perception, perception of only selected parts of information."
- allowed to use any "sources of information, which one cannot find anywhere else," whenever they are are available.
- reliant on "deeply held images of thinking and acting," which are "so basic to understanding of the world that people are hardly conscious of them."

Of course, there can be other characteristics, so the examples above are illustrative, not exhaustive.

Forms of Mental Models

Mental models have some basic forms to make learning easier, and hereafter are three basic forms (as summarized in *Table 2.2*): (WK 2012p)

- "Polygons"—are those "whose vertices sharing the edge represent related items."

- "Causal-Loop Diagrams"—are to "display…tendency and a direction of information connections and the resulting causality."
- "Flow Diagrams"—are "a more appropriate way to express a dynamic system."

Of course, there are also written forms of mental models, which "can help remember more details, such as more items in a long list." (WK 2012p)

Principles of Mental Models

Mental models follow certain principles in regard to their construction, and hereafter are some main ones for illustration (as summarized in *Table 2.2*): (WK 2012p)

- "Each mental model represents a possibility. A mental model represents one possibility, capturing what is common to all the different ways in which the possibility may occur." (WK 2012p; P. Johnson-Laird 2002)
- "Mental models are iconic, i.e., each part of a model corresponds to each part of what it represents." (WK 2012p; P. Johnson-Laird 2006)
- "Mental models are based on a principle of truth: they typically represent only those situations that are possible, and each model of a possibility represents only what is true in that possibility according to the proposition."
- "Mental models can represent what is false, temporarily assumed to be true, for example, in the case of counterfactual conditionals and counterfactual thinking." (WK 2012p; R. Byrne 2005)

Again, there can be other principles, so the examples above are illustrative, not exhaustive.

Reasoning with Mental Models

The reasoning with mental models focuses more on counter-examples, and the reason for this is that "people infer that a conclusion is valid if it holds in all the possibilities," and therefore "procedures for reasoning with mental models rely on counter-examples to refute invalid inferences; they establish validity by ensuring that a conclusion holds over all the models of the premises." (WK 2012p)

In this way, "reasoners" with mental models then "focus on a subset of the possible models of multiple-model problems, often just a single model....They reject a conclusion if they find a counterexample, i.e., a possibility in which the premises hold, but the conclusion does not." (WK 2012p; W. Schroyens 2003; N. Verschueren 2005)

One good example of reasoning with a mental model is "the following description from physicist Richard Feynman," as he wrote: "I had a scheme, which I still use today when somebody is explaining something that I'm trying to understand: I keep making up examples. For instance, the mathematicians would come in with a terrific theorem, and they're all excited. As they're telling me the conditions of the theorem, I construct something which fits all the conditions. You know, you have a set (one ball)—disjoint (two balls). Then the balls turn colors, grow hairs, or whatever, in my head as they put more conditions on. Finally they state the theorem, which is some dumb thing about the ball which isn't true for my hairy green ball thing, so I say, 'False!'" (WK 2012p)

Problems with Mental Models

However, for the critics, this argument on imagination and mental models has some major problems. Consider, for illustration, a few examples below (as summarized in *Table 2.2*).

Firstly, one main criticism of the argument on imagination and mental models is that there are alternative explanations of human reasoning.

For instance, for the critics, there are alternative views on human reasoning, which are not based on mental models, and good examples are those based on "formal rules of inference (e.g., O'Brien, 2009), domain-specific rules of inference (e.g., Cheng & Holyoak, 2008; Cosmides, 2005), or probabilities (e.g., Oaksford and Chater, 2007)." (WK 2012p)

Secondly, another main criticism of the argument on imagination and mental models is that there is no general consensus of whether mental models are better than other models in explaining human reasoning.

For example, for the critics, "scientific debate continues about whether human reasoning is based on mental models....Many empirical comparisons of the different theories have been carried out." (WK 2012p; K. Oberauer 2006)

Thirdly, still another main criticism of the argument on imagination and mental models is that mental models are simplistic.

For instance, for the critics, "compared with the complexities surrounding the world, it [a mental model] is very limited, and even when the scientific model is extensive and in accordance with a certain reality in the derivation of logical consequences of it," and in this way, "in the simplification of reality, creating a model can find a sense of reality, seeking to overcome systemic thinking and system dynamics," albeit with its bias. (WK 2012p)

Fourthly, an additional main criticism of the argument on imagination and mental models is that they have their own limits.

For instance, for the critics, a mental model "must take into account such restrictions" on different factors like "age" and "working memory; i.e., well-known rules on the maximum number of elements that people are suddenly able to remember, gestaltismus or failure of the principles of logic, etc." (WK 2012p; P. Barrouillet 2000)

Fifthly, another additional main criticism of the argument on imagination and mental models is that a counter-example does not necessarily refute an inductive conclusion totally.

For instance, for the critics, a counter-example can weaken an inductive conclusion but does not necessarily refute it totally.

And sixthly, still another additional main criticism of the argument on imagination and mental models is that there are complex situations which require optimal decision making (not on the basis of the simplification of reality as in mental models).

For instance, for the critics, "optimal decision" can be "used when the decision is important enough to motivate the time it takes to analyze it, or when it is too complex to solve with more simple intuitive approaches." (WK 2012t)

Beyond Mental Models

These criticisms of the argument on imagination and mental models should not be misunderstood as a total rejection of its validity but serve to show us the two opposing sides of the debate, such that the possibility (or impossibility) and desirability (or undesirability) of objectivity (from the perspective of the mind with objectivity, imagination, and mental models as a case study here) are not to the extent that the respective defenders would like us to believe.

In addition, the analysis of objectivity, imagination, and mental models sheds some light on the ontological principles in existential dialectics, and good examples include the formalness-informalness principle, the absoluteness-relativeness principle, the partiality-totality principle, the predictability-unpredictability principle, the explicability-inexplicability principle, the fiction-reality principle, the cognitiveness-noncognitiveness principle, the finiteness-transfiniteness principle, the preciseness-vagueness principle, the simpleness-complicatedness principle, the openness-hiddenness principle, the denseness-emptiness principle, the change-constancy principle, the order-chaos principle, the slowness-quickness principle, the expansion-contraction principle, the theory-praxis principle, the convention-novelty principle, the evolution-transformation principle, the symmetry-asymmetry principle, the softness-hardness principle, the seriousness-playfulness principle, the activeness-inactivess principle, the selfness-otherness principle, the regression-progression principle, the same-difference principle, the stability-reaction principle, the functionality-nonfunctionality principle, and the intentionality-nonintentionality principle.

For instance, in relation to the formalness-informalness principle, if there is formalness (e.g., the formal logical requirement of soundness, completeness, and consistency in a system of ideas, as in the argument on imagination and mental models), there is informalness (e.g., the non-formal existence of unsoundness, incompleteness, and inconsistency in the argument on imagination and mental models, as shown in the criticisms that there are alternative explanations of human reasoning, that there is no general consensus of whether mental models are better than other models in explaining human reasoning, that mental models are simplistic, that they have their own limits, that a counter-example does not necessarily refute an inductive conclusion totally, and that there are complex situations which require optimal decision making [not on the basis of the simplification of reality as in mental models]). And the reverse direction also holds true.

In relation to the absoluteness-relativeness principle, if there is absoluteness (e.g., the absolute view by those who favor mental models), there is relativeness (e.g., what is right for those who favor mental models is not necessarily so for those who argue that there are alternative views on human reasoning, which are not based on mental models, and good examples are those based on "formal rules of inference" and "domain-specific rules of inference," for example). And the reverse direction also holds true.

In relation to the partiality-totality principle, if there is partiality (e.g., the partial view on event anticipation by those who favor mental models), there is totality (e.g., the more holistic view on event anticipation, such that the whole is not the sum of its parts, that is, it cannot be reduced to the sum of opposing views by the advocates of "mental models," "formal rules of inference," "domain-specific rules of inference," and "probabilities," etc., since there will emerge new views in the future not yet known today, just as there are already alternative views nowadays which do not side exclusively with any of them, as is the analysis here). And the reverse direction also holds true.

In relation to the predictability-unpredictability principle, if there is predictability (e.g., the predictable tendency of "reasoners" with mental models to reject a conclusion if they find a counterexample), there is unpredictability (e.g., the more difficult task to predict exactly the extent to which "reasoners" with mental models who focus on a subset of a single model will end up rejecting a particular conclusion on a particular occasion). And the reverse direction also holds true.

In relation to the explicability-inexplicability principle, if there is explicability (e.g., the explanation by "reasoners" with mental models that these models are desirable on the basis of their provision of a sense of reality about a given possibility, that is, about "what is common to all the different ways in which the possibility may occur"), there is inexplicability (e.g., the lack of sufficient explanation by "reasoners" with mental models of why these models are necessarily desirable, since, for the critics, there is no general consensus of whether mental models are better than other models in explaining human reasoning). And the reverse direction also holds true.

In relation to the fiction-reality principle, if there is fiction (e.g., the fictional aspect of mental models, insofar as "mental models can represent what is false, temporarily assumed to be true, for example, in the case of counterfactual conditionals and counterfactual thinking"), there is reality (e.g., the realistic aspect of mental mdoels, insofar as "mental models are based on a principle of truth: they typically represent only those situations that are possible, and each model of a possibility represents only what is true in that possibility according to the proposition"). And the reverse direction also holds true.

In relation to the cognitiveness-noncognitiveness principle, if there is cognitiveness (e.g., a view of things on the basis of reason and evidences, as shown in the reliance of mental models on "deeply held images of thinking and acting" which are "so basic to understanding of the world," as well as "on a principle of truth"), there is noncognitiveness (e.g., a view of things on the basis of non-cognitive factors like envy, jealousy, power, nationality, race, gender, age, class, greed, lust, status, faith, anger, sadness, joy, fear, wish, etc., as shown in the "hope" that mental models are better than other models in explaining human reasoning, even though "scientific debate continues" about this in the field, with no consensus). And the reverse direction also holds true.

In relation to the finiteness-transfiniteness principle, if there is finiteness (e.g., the finite bits of information used in mental models for the simplication of reality), there is transfiniteness (e.g., the transfinite bits of information used in optimal decision-making models when the situation in question is extremely complex). And the reverse direction also holds true.

In relation to the preciseness-vagueness principle, if there is preciseness (e.g., the precise identification of 4 principles of mental models in the article aforecited), there is vagueness (e.g., the vagueness in the identification of the 4 principles of mental models in the article aforecited, since it is not clear why there must be only 4, not 6, 7, 8, and so on). And the reverse direction also holds true.

In relation to the simpleness-complicatedness principle, if there is simpleness (e.g., the simplification of reality in mental models), there is complicatedness (e.g., the complexity of reality in optimal decision-making models). And the reverse direction also holds true.

In relation to the openness-hiddenness principle, if there is openness (e.g., the open exploration of reality by being "flexible" in mental models), there is hiddenness (e.g., the hidden bias in mental models, because of its "effects as information filter" so as to "cause…selective perception, perception of only selected parts of information"). And the reverse direction also holds true.

In relation to the denseness-emptiness principle, if there is denseness (e.g., the relatively denser concentration of info used in optimal decisin-making models), there is emptiness (e.g., the relatively less dense, or more empty, concentration of info used in mental models). And the reverse direction also holds true.

In relation to the change-constancy principle, if there is change (e.g., the ever new ways to deal with info, as shown in mental models), there is constancy (e.g., the ever constant existence of different problems with the ever new ways to deal with info, as shown in the problems with mental models pointed out by the critics). And the reverse direction also holds true.

In relation to the order-chaos principle, if there is order (e.g., the more or less orderly view of reality, if looked from the vantage point of the "effects as information filter" in mental models, so as to "cause…selective perception, perception of only selected parts of information" for a simplified sense of reality), there is chaos (e.g., the more or less chaotic view of reality, if lookd from "hardly qualifiable, impugnable, obscure, or incomplete facts" in mental models, well before they are even "selected" and "filtered"). And the reverse direction also holds true.

In relation to the slowness-quickness principle, if there is slowness (e.g., the relatively slower speed of understadning reality if "hardly qualifiable, impugnable, obscure, or incomplete facts" in mental models are not "selected" and "filtered"), there is quickness (e.g., the relatively quicker speed of understadning reality if "hardly qualifiable, impugnable, obscure, or incomplete facts" in mental models are carefully "selected" and "filtered" to make sense of reality). And the reverse direction also holds true.

In relation to the expansion-contraction principle, if there is expansion (e.g., the relatively more developed ability of those in mental models to deal with the simplification of reality), there is contraction (e.g., the relatively less developed ability of those in mental models to deal with the complexity of reality as in optimal decision-making models). And the reverse direction also holds true.

In relation to the theory-praxis principle, if there is theory (e.g., the theoretical construction of the "principles" of mental models), there is praxis (e.g., the practical application of the "principles" of mental models to the field of "system dynamics"). (WK 2012p) And the reverse direction also holds true.

In relation to the convention-novelty principle, if there is convention (e.g., the conventional wisdom about the formal rules of inference in Aristotle's logic), there is novelty (e.g., the alternative novel challenge to the conventional wisdom about the formal rules of inference, by the new idea of mental models by Kenneth Craik in 1943). And the reverse direction also holds true.

In relation to the evolution-transformation principle, if there is evolution (e.g., the natural evolution of humans to process information in the state of nature), there is transformation (e.g., the technical transformation of human ability to process information by the invention of mental models). And the reverse direction also holds true.

In relation to the symmetry-asymmetry principle, if there is symmetry (e.g., the co-existence of animals and humans), there is asymmetry (e.g., humans can create systematic rules of inference, princples of mental models, etc.—but animals cannot). And the reverse direction also holds true.

In relation to the softness-hardness principle, if there is softness (e.g., the inclusion of certain bits of informaton in mental models, when they are selected), there is hardness (e.g., the exclusion of certain bits of information in mental models, when they are ignored in the process of "filtering" info). And the reverse direction also holds true.

In relation to the seriousness-playfulness principle, if there is seriousness (e.g., the serious business of those who try to figure out how humans anticipate events), there is playfulness (e.g., the playful part of those who try to figure out how humans anticipate events, when they play around with different options over time, like "formal rules of inference," "domain-specific rules of inference," "mental models," "probabilities," etc.). And the reverse direction also holds true.

In relation to the activeness-inactivess principle, if there is activeness (e.g., the relative activeness of those in mental models when they filter info to make sense of reality), there is inactiveness (e.g., the relative inactiveness of those who suffer from such "restrictions" like old age and failing working memory, so that they cannot effectively use mental models). And the reverse direction also holds true.

In relation to the selfness-otherness principle, if there is selfness (e.g., the tendency to be mostly centered on one's self or group, as shown by those in mental models to simplify reality so as to make sense of it for themselves), there is otherness (e.g., the tendency to be mostly centered on others, as shown by those scholars of optimal decision-making models to take the big trouble of going through complex information for a better understanding of the world, as an advance in human knowledge). And the reverse direction also holds true.

In relation to the regression-progression principle, if there is regression (e.g., the regression made by mental models, as shown in the problems pointed out by the critics), there is progression (e.g., the progress made by mental models, as shown in its contribution to the effort of making sense of reality). And the reverse direction also holds true.

In relation to the same-difference principle, if there is similarity in outcome (e.g., the contribution to the molding and control of beliefs, values, and behaviors, regardless of whether this be done by way of mental models, or by way of optimal decision-making models), there is difference in outcome (e.g., the contribution to the molding and control of beliefs, values, and behaviors by way of mental models for a relatively more satisfying view of the world—but the contribution to the molding and control of beliefs, values, and behaviors by way of optimal decision-making models for a relatively more maximizing view of the world). And the reverse direction also holds true.

In relation to the stability-reaction principle, if there is stability (e.g., the relatively more stable condition of formal rules of inference in the older days, like Aristotle's logic), there is reaction (e.g., formal rules of inference in the older days then led to further changes later on, when alternative ways of reasoning like mental models are in fashion nowadays). And the reverse direction also holds true.

In relation to the functionality-nonfunctionality principle, if there is functionality (e.g., the presence of action for which something is fitted, as shown in the creation of formal rules of inference in Aristotle's logic in the older days, when they served the function of providing a relatively more deterministic, logical worldview), there is nonfunctionality (e.g., the relative lack of action for which something is fitted, as shown in the use of mental models nowadays, such that the relatively more deterministic, logical worldview in Aristotle's logic is no longer as much functional in our times as it was before, when the fast-paced, flexible nature of mental models is attractive to some people who wants to simplify the world as a way to make sense of it). And the reverse direction also holds true.

In relation to the intentionality-nonintentionality principle, if there is intentionality (e.g., the planning or design of something for a certain outcome, as shown in the creation of formal rules of inference in Aristotle's logic in the older days, for an understanding of the world in a relatively more deterministic, logical worldview), there is nonintentionality (e.g., the relative absence of planning or design of something for a certain outcome, such that the outcome can even be counter-intentional in being the opposite of what was originally planned, as shown in the rise of mental models nowadays, for the simplification of reality as a way to make sense of it, contrary to the relatively more deterministic, logical worldview in Aristotle's logic). And the reverse direction also holds true.

And in relation to the survivability-nonsurvivability principle, if there is survivability (e.g., the survival of those conclusions which pass the tests of counterexamples in mental models), there is non-survivability (e.g., the non-survival of those conclusions which do not pass the tests of counterexamples in mental models). And the reverse direction also holds true.

Objectivity, Emotion, and Rationality

Another way to evaluate the possibility (or impossibility) and desirability (or undesirability) of objectivity from the perspective of the mind is to explore objectivity, emotion, and rationality.

An argument in the literature is that, as some philosophers like A.C. Grayling argued, "a good rationale must be independent of emotions, personal feelings or any kind of instincts." (WK 2012v)

Theories of Rationality

The word "rationality" refers to "the characteristic of any action, belief, or desire, that makes their choice a necessity. It is a normative concept of reasoning in the sense that rational people should derive conclusions in a consistent way given the information at disposal. It refers to the conformity of one's beliefs with one's reasons to believe, or...with one's reasons for action….A rational decision is one that is not just reasoned, but is also optimal for achieving a goal or solving a problem. Determining optimality for rational behavior requires a quantifiable formulation of the problem, and the making of several key assumptions. When the goal or problem involves making a decision, rationality factors in how much information is available (e.g. complete or incomplete knowledge)." (WK 2012v)

Scholars over time have proposed different theories of rationality. For example, "the German sociologist Max Weber proposed an interpretation of social action that distinguished between four different idealized types of rationality," as shown below (and summarized in *Table 2.3*): (WK 2012v)

- "*Zweckrational* or Purposive/Instrumental" Rationality—"is related to the expectations about the behavior of other human beings or objects in the environment. These expectations serve as means for a particular actor to attain ends, ends which Weber noted were 'rationally pursued and calculated.'"
- "*Wertrational* or Value/Belief-Oriented" Rationality—allows an "action" to be "undertaken for what one might call reasons intrinsic to the actor: some ethical, aesthetic, religious or other motive, independent of whether it will lead to success."
- "Affectual" Rationality—is "determined by an actor's specific affect, feeling, or emotion,…to which Weber himself said that this was a kind of rationality that was on the borderline of what he considered 'meaningfully oriented.'"
- "Traditional" Rationality—is "determined by ingrained habituation."

But "Weber emphasized that it was very unusual to find only one of these orientations: combinations were the norm. His usage also makes clear that he considered the first two as more significant than the others, and it is arguable that the third and fourth are subtypes of the first two." (WK 2012v)

Now, there is an "advantage" in Weber's interpretation, in that "it avoids a value-laden assessment, say, that certain kinds of beliefs are irrational. Instead, Weber suggests that a ground or motive can be given—for religious or affect reasons, for example—that may meet the criterion of explanation or justification even if it is not an explanation that fits the *Zweckrational* orientation of means and ends. The opposite is therefore also true: some means-ends explanations will not satisfy those whose grounds for action are '*Wertrational*.'" (WK 2012v)

Levels of Rationality

Scholars of rationality have also classified rationality in terms of two main levels, that is, at the theoretical level and at the practical level (as summarized in *Table 2.3*).

For instance, "Kant had distinguished theoretical from practical reason. Rationality theorist Jesús Mosterín [2008] makes a parallel distinction between theoretical and practical rationality, although, according to him, reason and rationality are not the same: reason would be a psychological faculty, whereas rationality is an optimizing strategy." (WK 2012v)

At the theoretical level, "rationality has a formal component that reduces to logical consistency and a material component that reduces to empirical support, relying on our inborn mechanisms of signal detection and interpretation. Mosterín [2002] distinguishes between involuntary and implicit belief, on the one hand, and voluntary and explicit acceptance, on the other. Theoretical rationality can more properly be said to regulate our acceptances than our beliefs." (WK 2012v)

And at the practical level, "rationality is the strategy for living one's best possible life, achieving your most important goals and your own preferences in as far as possible. Practical rationality has also a formal component, that reduces to Bayesian decision theory, and a material component, rooted in human nature (lastly, in our genome)." (WK 2012v)

Emotion, and the Quality of Rationality

These theoretical and practical levels aside—rationality is "believed by some philosophers (notably A.C. Grayling) that a good rationale must be independent of emotions, personal feelings or any kind of instincts. Any process of evaluation or analysis, that may be called rational, is expected to be highly objective, logical and 'mechanical.' If these minimum requirements are not satisfied i.e. if a person has been, even slightly, influenced by personal emotions, feelings, instincts or culturally specific, moral codes and norms, then the analysis may be termed irrational, due to the injection of subjective bias." (WK 2012v)

However, not all scholars are as extreme as Grayling, but some suggested that this quality of rationality "may productively appeal to the idealization as a point of reference." (WK 2012v)

Examples of Rationality

The study of rationality has been used to understand different fields. For instance, in organization studies, "individuals or organizations are called rational if they make optimal decisions in pursuit of their goals. It is in these terms that one speaks, for example, of a rational allocation of resources, or of a rational corporate strategy. For such 'rationality,' the decision maker's goals are taken as part of the model, and not made subject to criticism, ethical or otherwise." (WK 2012v)

And in the field of artificial intelligence, "a *rational agent* is one that maximizes its expected utility, given its current knowledge. Utility is the usefulness of the consequences of its actions. The utility function is arbitrarily defined by the designer, but should be a function of *performance*, which is the directly measurable consequences, such as winning or losing money. In order to make a safe agent that plays defensively, a nonlinear function of performance is often desired, so that the reward for winning is lower than the punishment for losing." (WK 2012v)

Problems with the Argument on Emotion and Rationality

However, for the critics, this argument on emotion and rationality has some major problems. Consider, for illustration, a few examples below (as summarized in *Table 2.3*).

Firstly, one main criticism of the argument on emotion and rationality is that there is no general consensus on the usage of "rationality" in different fields.

For example, for the critics, "the term 'rationality' tends to be used differently in different disciplines, including specialized discussions of economics, sociology, psychology, and political science." (WK 2012v)

Secondly, another main criticism of the argument on emotion and rationality is that the embedded assumptions in each rationality make it relative in relation to others.

For instance, for the critics, "collectively, the formulation and background assumptions are the model within which rationality applies," and this is "the relativity of rationality: if one accepts a model in which benefiting oneself is optimal, then rationality is equated with behavior that is self-interested to the point of being selfish; whereas if one accepts a model in which benefiting the group is optimal, then purely selfish behavior is deemed irrational." (WK 2012v)

Thirdly, still another main criticism of the argument on emotion and rationality is that the conventional views about rationality do not give sufficient weight to the issue of power.

For instance, for the critics, "Weber's constructions of rationality have been critiqued both from a Habermasian (1984) perspective (as devoid of social context and under-theorised in terms of social power) and also from a feminist perspective…whereby Weber's rationality constructs are viewed as imbued with masculine values and oriented toward the maintenance of male power." (WK 2012v; M. Eagleton 2003)

Fourthly, an additional main criticism of the argument on emotion and rationality is that the theories of rationality are mixed up in classification.

For instance, for the critics, one good illustration concerns the four types of Weber's rational constructs, since "it is arguable that the third and fourth are subtypes of the first two." (WK 2012v)

Fifthly, another additional main criticism of the argument on emotion and rationality is that it is limited in practice.

For instance, for the critics, some scholars like Philip Johnson-Laird and Ruth M.J. Byrne (2009) argued that "humans are rational in principle but they err in practice, that is, humans have the competence to be rational but their performance is limited by various factors." (WK 2012v)

Sixthly, still another additional main criticism of the argument on emotion and rationality is that it assumes a false conception of human nature.

For instance, for the critics, "modern cognitive science and neuroscience show that studying the role of emotion in mental function (including topics ranging from flashes of scientific insight to making future plans), that no human has ever satisfied this criterion, except perhaps a person with no affective feelings, for example an individual with a massively damaged amygdala or severe psychopathy. Thus, such an idealized form of rationality is best exemplified by computers, and not people." (WK 2012v)

And seventhly, one last main criticism of the argument on emotion and rationality is that there are alternative explanations of rationality.

For instance, for the critics, "debates arise…about whether or not people or organizations are 'really' rational, as well as whether it make sense to model them as such in formal models. Some have argued that a kind of bounded rationality makes more sense for such models." (WK 2012v)

Beyond Emotion and Rationality

These criticisms of the argument on emotion and rationality, however, do not mean a total rejection of its usefulness but serve to show us the two opposing sides of the debate, such that the possibility (or impossibility) and desirability (or undesirability) of objectivity (from the perspective of the mind with objectivity, emotion, and rationality as a case study here) are not to the extent that the respective defenders would like us to believe.

In fact, the analysis of objectivity, emotion, and rationality can teach us something important about the ontological principles in existential dialectics, and good examples include the formalness-informalness principle, the absoluteness-relativeness principle, the partiality-totality principle, the predictability-unpredictability principle, the explicability-inexplicability principle, the fiction-reality principle, the cognitiveness-noncognitiveness principle, the finiteness-transfiniteness principle, the preciseness-vagueness principle, the simpleness-complicatedness principle, the openness-hiddenness principle, the denseness-emptiness principle, the change-constancy principle, the order-chaos principle, the slowness-quickness principle, the expansion-contraction principle, the theory-praxis principle, the convention-novelty principle, the evolution-transformation principle, the symmetry-asymmetry principle, the softness-hardness principle, the seriousness-playfulness principle, the activeness-inactivess principle, the selfness-otherness principle, the regression-progression principle, the same-difference principle, the stability-reaction principle, the functionality-nonfunctionality principle, and the intentionality-nonintentionality principle.

For instance, in relation to the formalness-informalness principle, if there is formalness (e.g., the formal logical requirement of soundness, completeness, and consistency in a system of ideas, as in the argument on emotion and rationality), there is informalness (e.g., the non-formal existence of unsoundness, incompleteness, and inconsistency in the argument on emotion and rationality, as shown in the criticisms that there is no general consensus on the usage of "rationality" in different fields, that the embedded assumptions in each rationality make it relative in relation to others, that the conventional views about rationality do not give sufficient weight to the issue of power, that the theories of rationality are mixed up in classification, that it is limited in practice, that it assumes a false conception of human nature, and that there are alternative explanations of rationality). And the reverse direction also holds true.

In relation to the absoluteness-relativeness principle, if there is absoluteness (e.g., the absolute view within a model of rationality, like "a model in which benefiting oneself is optimal"), there is relativeness (e.g., what is right for a model of rationality, like "a model in which benefiting oneself is optimal," is not necessarily so for a different rationality "in which benefiting the group is optimal," since "purely selfish behavior is deemed irrational" in the latter). And the reverse direction also holds true.

In relation to the partiality-totality principle, if there is partiality (e.g., the partial view about rationality, if looked from the vantage point of a model of rationality, like "a model in which benefiting oneself is optimal"), there is totality (e.g., the more holistic view of about rationality, such that the whole is not the sum of its parts, that is, it cannot be reduced to the sum of opposing views like the four rational constructs by Weber, since there will emerge new views in the future not yet known today, just as there are already alternative views nowadays which do not side exclusively with any of them, as is the analysis here). And the reverse direction also holds true.

In relation to the predictability-unpredictability principle, if there is predictability (e.g., the predictable tendency of "humans" to be "rational in principle, according to some scholars like Philip Johnson-Laird and Ruth M.J. Byrne), there is unpredictability (e.g., the more difficult task to predict exactly the extent to which humans are rational on a particular occasion in practice, since "humans…err in practice, that is,…their performance is limited by various factors," according to some scholars like Philip Johnson-Laird and Ruth M.J. Byrne). And the reverse direction also holds true.

In relation to the explicability-inexplicability principle, if there is explicability (e.g., the explanation by those of benefiting oneself as rational on the basis of "a model in which benefiting oneself is optimal"), there is inexplicability (e.g., the lack of sufficient explanation by those accepting "a model in which benefiting oneself is optimal" of why benefiting oneself is necessrily desirable, since, for the critics, there are alternative models "in which benefiting the group is optimal" and "purely selfish behavior is deemed irrational"). And the reverse direction also holds true.

In relation to the fiction-reality principle, if there is fiction (e.g., the fictional aspect of a rationality in individuals or organizatins which "are called rational if they make optimal decisions in pursuit of their goals," insofar as, "for such 'rationality,' the decision maker's goals are taken as part of the model, and not made subject to criticism, ethical or otherwise"), there is reality (e.g., the realistic aspect of a rationality in individuals or organizatins, insofar as they are allowed to "make optimal decisions in pursuit of their goals" or to be selfish as rational in this case). And the reverse direction also holds true.

In relation to the cognitiveness-noncognitiveness principle, if there is cognitiveness (e.g., a view of things on the basis of reason and evidences, as shown in the use of information to make a decision in a model of rationality, depending on "how much information is available" for the purpose at hand), there is noncognitiveness (e.g., a view of things on the basis of non-cognitive factors like envy, jealousy, power, nationality, race, gender, age, class, greed, lust, status, faith, anger, sadness, joy, fear, wish, etc., as shown in "the background model assumptions describing how the problem is framed and formulated" in a model of rationality, so that they are to be "taken as part of the model, and not made subject to criticism, ethical or otherwise"). And the reverse direction also holds true.

In relation to the finiteness-transfiniteness principle, if there is finiteness (e.g., the finite amount of time and information needed to find a satisfying rational decision for simple problems, as in bounded rationality), there is transfiniteness (e.g., the transfinite amount of time and information needed to "find...the rational decision for arbitrarily complex problems"). (WK 2012v) And the reverse direction also holds true.

In relation to the preciseness-vagueness principle, if there is preciseness (e.g., the precise identification of 4 types of rationality by Weber in the article aforecited), there is vagueness (e.g., the vagueness in the identification of the 4 types of rationality by Weber in the article aforeicted, since it is not clear why there must be only 4, not 2, 3, 5, 6, 7, and so on). And the reverse direction also holds true.

In relation to the simpleness-complicatedness principle, if there is simpleness (e.g., the relatively simple analysis of the argument on emotion and rationality by the advocates), there is complicatedness (e.g., the relatively more complicated analysis of the argument on emotion and rationality, by challenging the claims and assumptions, as shown in the problems pointed out by the critics as cited). And the reverse direction also holds true.

In relation to the openness-hiddenness principle, if there is openness (e.g., the open exploration of 4 types of rationality by Weber), there is hiddenness (e.g., the hidden bias in the 4 types of rationality by Weber, since, for the critics like the Habermasians and feminists, they do not give sufficient weight to the issue of power). And the reverse direction also holds true.

In relation to the denseness-emptiness principle, if there is denseness (e.g., the relatively denser concentration of concerns with selfishness by those accepting "a model in which benefiting oneself is optimal"), there is emptiness (e.g., the relatively less dense, or more empty, concentration of concerns with selfishness by those accepting a model "in which benefiting the group is optimal" and "purely selfish behavior is deemed irrational"). And the reverse direction also holds true.

In relation to the change-constancy principle, if there is change (e.g., the ever new ways to deal with reasoning, as shown in the argument on emotion and rationality), there is constancy (e.g., the ever constant existence of different problems with the ever new ways to deal with reasoning, as shown in the problems with the argument on emotion and rationality pointed out by the critics). And the reverse direction also holds true.

In relation to the order-chaos principle, if there is order (e.g., the more or less orderly view about rationality, if looked only from the vantage point of a particular type of rationality for a particular goal, like "a model in which benefiting oneself is optimal"), there is chaos (e.g., the more or less chaotic view abour rationality, if looked from the conflicting views that there is no general consensus on the usage of "rationality" in different fields, that the embedded assumptions in each rationality make it relative in relation to others, that the theories of rationality are mixed up in classification, etc.). And the reverse direction also holds true.

In relation to the slowness-quickness principle, if there is slowness (e.g., the relatively slower speed of decision making, if the goal is to "find…the rational decision for arbitrarily complex problems" in an optimal way). there is quickness (e.g., the relatively quicker speed of decision making, if the goal is merely to find a satisfying rational decision for simple problems, as in bounded rationality). And the reverse direction also holds true.

In relation to the expansion-contraction principle, if there is expansion (e.g., the relative expansion of the use of the argument for non-emotional rationality in the field of artificial intelligence), there is contraction (e.g., the relative decline of the use of the argument for non-emotional rationality in the field of feminism). And the reverse direction also holds true.

In relation to the theory-praxis principle, if there is theory (e.g., the theoretical construction of an "objective" account of rationality without the distortion of emotion in philosophy), there is praxis (e.g., the practical application of an "objective" account of rationality without the distortion of emotion in philosophy to the field of artificial intelligence). And the reverse direction also holds true.

In relation to the convention-novelty principle, if there is convention (e.g., the conventional wisdom about an "objective" account of rationality without the distortion of emotion in philosophy), there is novelty (e.g., the alternative novel challenge to the conventional wisdom about an "objective" account of rationality without the distortion of emotion in philosophy, by the new idea of the Habermasians and feminists that the conventional views about rationality do not give sufficient weight to the issue of power). And the reverse direction also holds true.

In relation to the evolution-transformation principle, if there is evolution (e.g., the natural evolution of humans to have reasoning and emotion in the state of nature), there is transformation (e.g., the technical transformation of human ability to reason without emotion by the invention of an "objective" account of rationality without the distortion of emotion in philosophy). And the reverse direction also holds true.

In relation to the symmetry-asymmetry principle, if there is symmetry (e.g., the co-existence of animals and humans), there is asymmetry (e.g., humans can propose different systematic models of rationality—but animals cannot). And the reverse direction also holds true.

In relation to the softness-hardness principle, if there is softness (e.g., the inclusion of emotion in the understanding of rationality by the feminists), there is hardness (e.g., the exclusion of emotion in the understanding of rationality by philosophers like A. C. Grayling). And the reverse direction also holds true.

In relation to the seriousness-playfulness principle, if there is seriousness (e.g., the serious business of those who try to understand the nature of rationality), there is playfulness (e.g., the playful part of those who try to understand the nature of rationality, when they play around with different views about rationality, in relation to the theories, levels, interpretations, etc.). And the reverse direction also holds true.

In relation to the activeness-inactivess principle, if there is activeness (e.g., the relative activeness of those experts who excel in making good use of rationality in practice), there is inactiveness (e.g., the relative inactiveness of those humans who cannot be rational "in practice, in that, although they "have the competence to be rational but their performance is limited by various factors"). (WK 2012v) And the reverse direction also holds true.

In relation to the selfness-otherness principle, if there is selfness (e.g., the tendency to be mostly centered on one's self or group, as shown by those who follow a model of rationality in which benefiting oneself is optimal"), there is otherness (e.g., the tendency to be mostly centered on others, as shown by those who follow a different model rationality "in which benefiting the group is optimal"). And the reverse direction also holds true.

In relation to the regression-progression principle, if there is regression (e.g., the regression made by the argument on emotion and rationality, as shown in the problems pointed out by the critics aforecited), there is progression (e.g., the progress made by the argument on emotion and rationality, as shown in its contribution to the understanding of human reasoning aforecited). And the reverse direction also holds true.

In relation to the same-difference principle, if there is similarity in outcome (e.g., the contribution to the molding and control of beliefs, values, and behaviors, regardless of whether this be done by way of

Weberian rationality, or by way of Habermasian rationality), there is difference in outcome (e.g., the contribution to the molding and control of beliefs, values, and behaviors by way of Weberian rationality for a relatively more value-neutral worldview, as in the relativity of rationality—but the contribution to the molding and control of beliefs, values, and behaviors by way of Habermasian rationality for a relatively more context-specific worldview, as in the critique of social power). And the reverse direction also holds true.

In relation to the stability-reaction principle, if there is stability (e.g., the relatively more stable condition of the fashionable view about rationality, as shown in the influential Weberian account of 4 types of rationality), there is reaction (e.g., the influential Weberian account of 4 types of rationality then led to further changes later on, when nowadays, the feminist perspective rejects it as "imbued with masculine values and oriented toward the maintenance of male power"). And the reverse direction also holds true.

In relation to the functionality-nonfunctionality principle, if there is functionality (e.g., the presence of action for which something is fitted, as shown in the influential Weberian account of 4 types of rationality in the older days, when they served the function of propounding "masculine values" and being "oriented toward the maintenance of male power"), there is nonfunctionality (e.g., the relative lack of action for which something is fitted, as shown in the increasing critique of the Weberian account of 4 types of rationality nowadays, such that they are no longer as much functional as they used to be, due to different feminist expectations and campaigns for a different voice on rationality with a feminine focus on equality, affect, and so on in our times). And the reverse direction also holds true.

In relation to the intentionality-nonintentionality principle, if there is intentionality (e.g., the planning or design of something for a certain outcome, as shown in the Weberian account of 4 types of rationality for a better understanding of social action), there is nonintentionality (e.g., the relative absence of planning or design of something for a certain outcome, such that the outcome can even be counter-intentional in being the opposite of what was originally planned, as shown in the critique of the Weberian account of 4 types of rationality by [A.] Etzioni [1988], who reframes thought on decision-making to argue for a reversal of the position put forward by Weber," in that "purposive/instrumental reasoning is subordinated by normative considerations [ideas on how people 'ought' to behave] and affective considerations [as a support system for the development of human relationships]," contrary to the original Weberian construction). (WK 2012v) And the reverse direction also holds true.

And in relation to the survivability-nonsurvivability principle, if there is survivability (e.g., the survival of those notions of rationality when they pass the test of what Richard Brandt proposed as "a form of cognitive-psychotherapy"), there is non-survivability (e.g., the non-survival of those notions of rationality when they fail the test of what Richard Brandt proposed as "a form of cognitive-psychotherapy"). (WK 2012v; R. Brandt 1959) And the reverse direction also holds true.

Objectivity and Nature

Objectivity, when examined from the perspective of nature, can show us in a controversial way its possibility (or impossibility) and desirability (or undesirability), and this can be shown by way of two case studies, namely, (2.3.1) objectivity, the physical world, and discrete probability, and (2.3.2) objectivity, complexity, and uncertainty—to be addressed in what follows, respectively.

Objectivity, the Physical World, and Discrete Probability

A good way to evaluate the possibility (or impossibility) and desirability (or undesirability) of objectivity from the perspective of nature is to explore objectivity, the physical world, and discrete probability.

An argument in the literature is that discreteness in the physical world can be predicted in a precise way, and a good example is the game of rolling a dice (or more than one dice). In other words, "for a single roll of a fair *s*-sided die, the probability of rolling each value is exactly 1/*s*; this is an example of a discrete uniform distribution." (WK 2012q)

Examples of Discreteness

In science, the word "discrete" refers to "the opposite of continuous" and means "something that is separate; distinct; individual"—and good examples of discreteness in the physical world are shown below (and summarized in *Table 2.4*): (WK 2012z)

- "Discrete Particle or Quantum in Physics"—"for example in quantum theory"
- "Discrete Mathematics"—"the study of structures without continuity"

- "Discrete Optimization"—"a branch of optimization in applied mathematics and computer science"
- "Discrete Signal"—"a time series consisting of a sequence of quantities"
- "Discrete Space"—"a simple example of a topological space"
- "Discrete Time"—"a non-continuous time, which results in discrete-time samples"
- "Discrete Pitch"—"a pitch with a steady frequency, rather than an indiscrete gliding, glissando or portamento, pitch"

Of course, there can be other examples, so the ones above are illustrative, not exhaustive.

Properties of Discreteness

Discreteness has some distinctive properties to be understood. For illustration, consider the following examples (as summarized in *Table 2.4*): (WK 2012aa)

- "Every discrete group is totally disconnected."
- "Every subgroup of a discrete group is discrete."
- "Every quotient of a discrete group is discrete."
- "The product of a finite number of discrete groups is discrete."
- "Discrete group is compact if and only if it is finite."
- "Every discrete group is locally compact."

Again, there can be other examples, so the ones above are illustrative, not exhaustive.

Discreteness in Probability Theory

The idea of discreteness is important in probability theory, and a good illustration is about "discrete probability theory," which "deals with events that occur in countable sample spaces," and good practical examples of discrete events include "throwing dice, experiments with decks of cards, and random walk." (WK 2012c)

In probability theory, some well-known discrete probability distributions are "the Poisson distribution, the Bernoulli distribution, the binomial distribution, the geometric distribution, and the negative binomial distribution." (WK 2012h)

However, there are two different definitions (or interpretations) of discrete probability. (WK 2012c & 2012gg)

Firstly, in the older days, there was the "classical definition" of discrete probability, in that "the probability of an event to occur was defined as number of cases favorable for the event, over the number of total outcomes possible in an equiprobable sample space." (WK 2012c; E. Jaynes 2003) Here is an example: "if the event is 'occurrence of an even number when a die is rolled,' the probability is given by *3/6 = 1/2*, since 3 faces out of the 6 have even numbers and each face has the same probability of appearing." (WK 2012c)

And secondly, nowadays, there is the "modern definition" of discrete probability, in that "the modern definition starts with a finite or countable set called the sample space, which relates to the set of all possible outcomes in classical sense, denoted by Ω. It is then assumed that for each element $x \in \Omega$, an intrinsic 'probability' value $f(x)$ is attached, which satisfies the following properties": (WK 2012c)

$$f(x) \in [0, 1] \text{ for all } x \in \Omega\,;$$

$$\sum_{x \in \Omega} f(x) = 1\,.$$

This means that "the probability function $f(x)$ lies between zero and one for every value of x in the sample space Ω, and the sum of $f(x)$ over all values x in the sample space Ω is equal to 1." (WK 2012c)

Thus, "an event is defined as any subset E, of the sample space Ω. The probability of the event E is defined as": (WK 2012c)

$$P(E) = \sum_{x \in E} f(x)\,.$$

In this way, "the probability of the entire sample space is 1, and the probability of the null event is 0." (WK 2012c)

And "the function $f(x)$ mapping a point in the sample space to the 'probability' value is called a probability mass function abbreviated as pmf." (WK 2012c)

A Case Study of Discreteness

For illustration, consider a case study of discrete probability distribution in action, that is, the example of throwing a dice.

For instance, "for a single roll of a fair *s*-sided die, the probability of rolling each value is exactly $1/s$; this is an example of a discrete uniform distribution. For n multiple rolls, with a *s*-sided die the possibility space is equal to s^n. So, for *n* rolls of an *s*-sided die the probability of any result is $1/s^n$." (WK 2012bb)

But "as the number of dice increases, the distribution of the sum of all numbers tends to normal distribution by the central limit theorem," and the normal distribution looks like "a bell-shaped probability density function." (WK 2012bb & 2012cc)

Applications of Discreteness

Discrete probability distributions have been applied for different purposes.

For instance, in dice games, "the fantasy role-playing game *Dungeons & Dragons* (D&D) is largely credited with popularizing dice in such games. Some games use only one type, like *Exalted* which uses only ten-sided dice. Others use numerous types for different game purposes, such as D&D, which makes use of all common polyhedral dice. Dice are used to determine the outcome of events; such usage is called a *check*. Games typically determine results either as a total on one or more dice above or below a fixed number, or a certain number of rolls above a certain number on one or more dice." (WK 2012bb)

And in the game of cleromancy, "dice can be used for divination....A pair of common dice is usual, though other forms of polyhedra can be used. Tibetan Buddhists sometimes use this method of divination. It is highly likely that the Pythagoreans used the Platonic solids as dice; they referred to such dice as 'the dice of the gods,' and they sought to understand the universe through an understanding of geometry in polyhedra. Astrological dice are a specialized set of three 12-sided dice for divination; the first die represents planets, the Sun, the Moon, and the nodes of the Moon; the second die represents the 12 zodiac signs; and the third represents the 12 houses. An icosahedron provides the answers of the Magic 8-Ball, conventionally used to provide answers to yes-or-no questions." (WK 2012bb; K. Guthrie 1988)

Problems with Discrete Probability

However, for the critics, this argument on the physical world and discrete probability has some major problems. Consider, for illustration, a few examples below (as summarized in *Table 2.4*).

Firstly, one main criticism of the argument on the physical world and discrete probability is that mathematics assumes certain forms of perfection which are not true in the real world.

For instance, for the critics, a dice in the real world is not perfectly manufactured, as "dice for role-playing games are usually plastic; early polyhedral dice from the 1970s and 1980s were made of a soft plastic that would easily wear as the die was used, and wear would gradually render the die unusable." (WK 2012bb) Besides this imperfection, there are also other factors such as "circumstances or character skill" (like "movements in the thrower's hand," the texture and shape of the table, and so on). (WK 2012p) Therefore, the probability of rolling each value is not $1/s$ in practice.

Secondly, another main criticism of the argument on the physical world and discrete probability is that there are alternative probability distributions which are not discrete.

For instance, for the critics, a non-discrete probability distribution is known as "a probability distribution that has a probability density function. Mathematicians also call such a distribution absolutely continuous....There are many examples of continuous probability distributions: normal, uniform, chi-squared, and others." (WK 2012h)

Thirdly, still another main criticism of the argument on the physical world and discrete probability is that mathematics use certain constructs of infinity which are not possible to apply in the real world.

For instance, "in probability theory, the central limit theorem (CLT) states that, given certain conditions, the mean of a sufficiently large number of independent random variables, each with finite mean and variance, will be approximately normally distributed...as n approaches infinity" (WK 2012dd) But the problem here is that, in real life, one does not roll a dice or flip a coin in a game so many times to the point of being close to infinity; in other words, "[i]n actual practice..., we cannot flip a coin an infinite number of times; so in general, this formula most accurately applies to situations in which we have already assigned an a priori probability to a particular outcome (in this case, our assumption that the coin was a 'fair' coin)," among other assumptions. (WK 2012ff) In *FPHG* (2009), I already challenged the conventional understanding of infinity and proposed "transfinity" instead.

Fourthly, an additional main criticism of the argument on the physical world and discrete probability is that discreteness is very difficult to be computed in certain situations.

For instance, for the critics, "there exist groups for which computing discrete logarithms is apparently difficult. In some cases (e.g. large prime order subgroups of groups $(Z_p)^\times$) there is not only no efficient algorithm known for the worst case, but also the average-case complexity can be shown to be about as hard as the worst case using random self-reducibility." (WK 2012ee)

And fifthly, another additional main criticism of the argument on the physical world and discrete probability is that probability theory does not explain how probability mass functions come into existence.

For instance, for the critics, "the modern definition [of discrete probability distribution] does not try to answer how probability mass functions are obtained; instead it builds a theory that assumes their existence." (WK 2012c)

Beyond Discrete Probability

These criticisms of the argument on the physical world and discrete probability do not entail a total rejection of its usefulness but serve to show us the two opposing sides of the debate, such that the possibility (or impossibility) and desirability (or undesirability) of objectivity (from the perspective of nature with objectivity, the physical world, and discrete probability as a case study here) are not to the extent that the respective defenders would like us to believe.

Indeed, the analysis of objectivity, the physical world, and discrete probability can illuminate important insights about the ontological principles in existential dialectics, and good examples include the formalness-informalness principle, the absoluteness-relativeness principle, the partiality-totality principle, the predictability-unpredictability principle, the explicability-inexplicability principle, the fiction-reality principle, the cognitiveness-noncognitiveness principle, the finiteness-transfiniteness principle, the preciseness-vagueness principle, the simpleness-complicatedness principle, the openness-hiddenness principle, the denseness-emptiness principle, the change-constancy principle, the order-chaos principle, the slowness-quickness principle, the expansion-contraction principle, the theory-praxis principle, the convention-novelty principle, the evolution-transformation principle, the symmetry-asymmetry principle, the softness-hardness principle, the seriousness-playfulness principle, the activeness-inactivess principle, the selfness-otherness principle, the regression-progression principle, the same-difference principle, the stability-reaction principle, the functionality-nonfunctionality principle, and the intentionality-nonintentionality principle.

For instance, in relation to the formalness-informalness principle, if there is formalness (e.g., the formal logical requirement of soundness and completeness in a system of ideas, as in the argument on the physical world and discrete probability), there is informalness (e.g., the non-formal existence of unsoundness and incompleteness in the argument on the physical world and discrete probability, as shown in the criticisms that mathematics assumes certain forms of perfection which are not true in the real world, that there are alternative probability distributions which are not discrete, that mathematics use certain constructs of infinity which are not possible to apply in the real world, that discreteness is very difficult to be computed in certain situations, and that probability theory does not explain how probability mass functions come into existence). And the reverse direction also holds true.

In relation to the absoluteness-relativeness principle, if there is absoluteness (e.g., the absolute focus on discrete events by the advocates), there is relativeness (e.g., what is appropriate for the advocates of discrete events is not necessarily so for the advocates of continuous events). And the reverse direction also holds true.

In relation to the partiality-totality principle, if there is partiality (e.g., the partial view by those who focus on discreteness), there is totality (e.g., the more holistic view of probability distributions, such that the whole is not the sum of its parts, that is, it cannot be reduced to the sum of opposing views about discrete probability distributions, continuous probability distributions, etc., since there will emerge new views in the future not yet known today, just as there are already alternative views nowadays which do not side exclusively with any of them, as is the analysis here). And the reverse direction also holds true.

In relation to the predictability-unpredictability principle, if there is predictability (e.g., the predictable tendency of a dice to have a probability of rolling each number in terms of $1/s$ in principle), there is unpredictability (e.g., the more difficult task to predict exactly the extent to which rolling a dice on a particular occasion will have a particular number as its outcome in practice). And the reverse direction also holds true.

In relation to the explicability-inexplicability principle, if there is explicability (e.g., the explanation by those in probability theory that "the function $f(x)$ mapping a point in the sample space to the 'probability' value" is "a probability mass function"), there is inexplicability (e.g., the lack of sufficient explanation by those in probability theory of "how probability mass functions are obtained," since the theory "instead…assumes their existence"). (WK 2012c) And the reverse direction also holds true.

In relation to the fiction-reality principle, if there is fiction (e.g., the fictional aspect of the central limit theorem about a normal distribution "as *n* approaches infinity," insofar as "[i]n actual practice..., we cannot flip a coin an infinite number of times"), there is reality (e.g., the realistic aspect of the central limit theorem, insofar as a normal distribution can be approximated, although it cannot be exact, as *n* approaches "a sufficiently large number"). And the reverse direction also holds true.

In relation to the cognitiveness-noncognitiveness principle, if there is cognitiveness (e.g., a view of things on the basis of reason and evidences, as shown in the reasoning of the central limit theorem about a normal distribution "as *n* approaches infinity"), there is noncognitiveness (e.g., a view of things on the basis of non-cognitive factors like envy, jealousy, power, nationality, race, gender, age, class, greed, lust, status, faith, anger, sadness, joy, fear, wish, etc., as shown in the hopeful assumptions in regard to a "fair" coin and the existence of infinity, since "we cannot flip a coin an infinite number of times; so in general, this formula most accurately applies to situations in which we have already assigned an a priori probability to a particular outcome [in this case, our assumption that the coin was a 'fair' coin]," among other assumptions, which are not proved and are therefore neither true nor false within the formula). And the reverse direction also holds true.

In relation to the finiteness-transfiniteness principle, if there is finiteness (e.g., the finite number of times in flipping a coin or rolling a dice as *n* approaches "a sufficiently large number," so that a normal distribution can be approximated, though not exact), there is transfiniteness (e.g., the transfinite idea of "*n*" as its approaches "infinity" in the central limit theorem). And the reverse direction also holds true.

In relation to the preciseness-vagueness principle, if there is preciseness (e.g., the precise identification of 6 properties of discreteness in the article aforecited), there is vagueness (e.g., the vagueness in the identification of the 6 properties of discreteness in the article aforecited, since it is not clear why there must be only 6, not 7, 8, 9, 10, 11, 13, 14, 17, and so on). And the reverse direction also holds true.

In relation to the simpleness-complicatedness principle, if there is simpleness (e.g., the relatively simple analysis of discrete probability and the physical world by the advocates), there is complicatedness (e.g., the relatively more complicated analysis of discrete probability and the physical world, by challenging the claims and assumptions, as shown in the problems pointed out by the critics as cited). And the reverse direction also holds true.

In relation to the openness-hiddenness principle, if there is openness (e.g., the open exploration of a normal distribution "as *n* approaches infinity" in the central limit theorem), there is hiddenness (e.g., the hidden bias in the central limit theorem, since, in practice, "we cannot flip a coin an infinite number of times," so "this formula most accurately applies to situations in which we have already assigned an a priori probability to a particular outcome [in this case, our assumption that the coin was a 'fair' coin]," among other assumptions). And the reverse direction also holds true.

In relation to the denseness-emptiness principle, if there is denseness (e.g., the relatively denser concentration of concerns with discreteness in the Poisson distribution), there is emptiness (e.g., the relatively less dense, or more empty, concentration of concerns with discreteness in the chi-squared distribution). And the reverse direction also holds true.

In relation to the change-constancy principle, if there is change (e.g., the ever new ways to understand mathematical structures, as shown in the construction of discreteness), there is constancy (e.g., the ever constant existence of different problems with the ever new ways to understand mathematical structures, as shown in the problems with discreteness pointed out by the critics aforecited). And the reverse direction also holds true.

In relation to the order-chaos principle, if there is order (e.g., the more or less orderly view about discreteness, if looked from the vantage point of a dice with a probability of obtaining each number in terms of $1/s$ in principle), there is chaos (e.g., the more or less chaotic view about discreteness, if looked from the vantage point of the difficult case when "there exist groups for which computing discrete logarithms is apparently difficult," like cases concerning "large prime order subgroups of groups $(Z_p)^\times$ when "there is not only no efficient algorithm known for the worst case, but also the average-case complexity can be shown to be about as hard as the worst case using random self-reducibility"). (WK 2012ee) And the reverse direction also holds true.

In relation to the slowness-quickness principle, if there is slowness (e.g., the relatively slower speed of finishing the task of rolling a dice many times before the probability of each number will approach $1/s$ in practice, not just in principle), there is quickness (e.g., the relatively quicker speed of calculating the probability of obtaining each number of a dice in terms of $1/s$ in principle). And the reverse direction also holds true.

In relation to the expansion-contraction principle, if there is expansion (e.g., the relative spread of the "modern definition" of probability nowadays), there is contraction (e.g., the relative decline of the "classical definition" of probability nowadays). (WK 2012gg) And the reverse direction also holds true.

In relation to the theory-praxis principle, if there is theory (e.g., the theoretical construction of discrete probability in mathematics), there is praxis (e.g., the practical application of discrete probability in mathematics to the field of games, like dice games and the game of cleromancy). And the reverse direction also holds true.

In relation to the convention-novelty principle, if there is convention (e.g., the "classical definition" of probability by James Bernoulli and Pierre-Simon Laplace in the older days), there is novelty (e.g., the alternative novel challenge to the "classical definition" of probability, by the "modern definition" of probability by John Venn and George Boole). And the reverse direction also holds true.

In relation to the evolution-transformation principle, if there is evolution (e.g., the natural evolution of humans to distinguish "separate" or "distinct" events in the state of nature), there is transformation (e.g., the technical transformation of human ability to distinguish "separate" or "distinct" events by the invention of discrete probability theory). And the reverse direction also holds true.

In relation to the symmetry-asymmetry principle, if there is symmetry (e.g., both animals and humans can perceive separate or distinct things in nature), there is asymmetry (e.g., human can invent discrete probability theory to understand discrete probability distributions—but animals cannot). And the reverse direction also holds true.

In relation to the softness-hardness principle, if there is softness (e.g., the defense of the "classical definition" of probability by James Bernoulli and Pierre-Simon Laplace), there is hardness (e.g., the critique of the "classical definition" of probability by John Venn and George Boole). And the reverse direction also holds true.

In relation to the seriousness-playfulness principle, if there is seriousness (e.g., the serious business of those who try to understand probability), there is playfulness (e.g., the playful part of those who try to understand probability, when they play around with different "interpretations" over time, like the "classical" and "modern" definitions aforecited). And the reverse direction also holds true.

In relation to the activeness-inactivess principle, if there is activeness (e.g., the relative activeness of a dice when it is still new and can be used by players for role-playing games, for example), there is inactiveness (e.g., the relative inactiveness of a dice for role-playing games when it wears so much that it renders "the die unusable").

In relation to the selfness-otherness principle, if there is selfness (e.g., the tendency to be mostly centered on one's self or group, as shown by those who play dice games like *Dungeons & Dragons* for their personal enjoyment), there is otherness (e.g., the tendency to be mostly centered on others, as shown by the Pythagoreans who "used the Platonic solids as dice" in their quest for understanding "the universe through an understanding of geometry," as an advancement in knowledge for humanity). And the reverse direction also holds true.

In relation to the regression-progression principle, if there is regression (e.g., the regression made by discrete probability theory, as shown in the problems pointed out by the critics), there is progression (e.g., the progress made by discrete probability theory, as shown in its contribution to the understanding of discreteness aforecited). And the reverse direction also holds true.

In relation to the same-difference principle, if there is similarity in outcome (e.g., the contribution to the molding and control of beliefs, values, and behaviors, regardless of whether this be done by way of playing dice games like *Dungeons & Dragons*, or by way of playing dice games like cleromancy), there is difference in outcome (e.g., the contribution to the molding and control of beliefs, values, and behaviors by way of playing dice games like *Dungeons & Dragons* for a relatively more entertaining lifeworld—but the contribution to the molding and control of beliefs, values, and behaviors by way of playing dice games like cleromancy for a relatively more spiritual lifeworld). And the reverse direction also holds true.

In relation to the stability-reaction principle, if there is stability (e.g., the relatively more stable condition of the "classical" interpretation of probability in the older days), there is reaction (e.g., the "classical" interpretation of probability in the older days then led to further changes later on, when nowadays it is replaced by the "modern" interpretation). And the reverse direction also holds true.

In relation to the functionality-nonfunctionality principle, if there is functionality (e.g., the presence of action for which something is fitted, as shown in the use of dice by the Tibetans, which serve the important function of being a form of divination to make decisions about "health, work" and the like), there is nonfunctionality (e.g., the relative lack of action for which something is fitted, as shown in the use of modern health care and education in the U.S. nowadays, such that the form of divination is not as functional there as it could in Tibet, because of different expectations and standards based on a relatively advanced, hi-tech society in America). (WK 2012hh) And the reverse direction also holds true.

In relation to the intentionality-nonintentionality principle, if there is intentionality (e.g., the planning or design of something for a certain outcome, as shown in the work on the "classical" interpretation of probability by James Bernoulli and Pierre-Simon Laplace in the older days), there is nonintentionality (e.g., the relative absence of planning or design of something for a certain outcome, such that the outcome can even be counter-intentional in being the opposite of what was originally planned, as shown in the attempt to revive the "classical" interpretation by some people nowadays, by relating it to "the general interest in Bayesian probability," contrary to the original intention of James Bernoulli and Pierre-Simon Laplace in the older days). And the reverse direction also holds true.

And in relation to the survivability-nonsurvivability principle, if there is survivability (e.g., the relative survival of the "modern" interpretation of probability nowadays), there is non-survivability (e.g., the relative non-survival of the "classical" interpretation of probaiblity nowadays). And the reverse direction also holds true.

Objectivity, Complexity, and Uncertainty

Another good way to evaluate the possibility (or impossibility) and desirability (or undesirability) of objectivity from the perspective of nature is to explore objectivity, complexity, and uncertainty.

An argument in the literature is that there is emergent uncertainty in the prediction of complex phenomena.

Meanings of Complexity

In common usage, the term "complexity" refers to "something with many parts in intricate arrangement"—but one should not confuse the term "complex" with the related term "complicated," in that "complex is the opposite of independent, while complicated is the opposite of simple." (WK 2012ii)

In science, "there are at this time a number of approaches to characterizing complexity....Neil Johnson [2007] describes complexity science as the study of the phenomena which emerge from a collection of interacting objects." (WK 2012ii; J. Zayed 2010)

But "complexity" has different precise meanings, depending on contexts, as shown below: (WK 2012ii)

- "In Computational Complexity Theory"—the term refers to "the amounts of resources required for the execution of algorithms....The most popular types of computational complexity are the time complexity of a problem equal to the number of steps that it takes to solve an instance of the problem as a function of the size of the input (usually measured in bits), using the most efficient algorithm, and the space complexity of a problem equal to the volume of the memory used by the algorithm (e.g., cells of the tape) that it takes to solve an instance of the problem as a function of the size of the input (usually measured in bits), using the most efficient algorithm. This allows to classify computational problems by complexity class (such as P, NP...)."
- "In Information Processing"—the term refers to "a measure of the total number of properties transmitted by an object and detected by an observer. Such a collection of properties is often referred to as a state."
- "In Software Engineering"—the terms refers to "a measure of the interactions of the various elements of the software. This differs from the computational complexity described above in that it is a measure of the design of the software."

Of course, there can be other contexts, so the examples above are illustrative, not exhaustive.

Attributes of Complexity

However, in general, a complex system has some or all of the following attributes (as summarized in *Table 2.5*): (WK 2012ii; N. Johnson 2007)

- "The number of parts (and types of parts) in the system and the number of relations between the parts is non-trivial—however, there is no general rule to separate 'trivial' from 'non-trivial.'"
- "The system has memory or includes feedback."
- "The system can adapt itself according to its history or feedback."
- "The relations between the system and its environment are...non-linear."
- "The system can be influenced by, or can adapt itself to, its environment."
- "The system is highly sensitive to initial conditions."

Other scholars proposed different views. For example, in *The Future of Complexity* (2007), I already introduced the work by David Green (2001) and David Newth, who proposed the following criteria of what constitutes complexity: (WK 2006c)

- "Self-organization"
- "Connectivity"
- "Criticality"
- "Novelty"
- "Diversity"
- "Emergence"

Of course, there can be other criteria, so the examples above are illustrative, not exhaustive.

Types of Complexity

These meanings and attributes of complexity aside—"Warren Weaver posited in 1948 that the complexity of a particular system is the degree of difficulty in predicting the properties of the system...given the properties of the system's parts....In Weaver's view, complexity comes in two forms: disorganized complexity, and organized complexity," as explained below (and summarized in *Table 2.5*): (WK 2012ii; J. Jacobs 1961)

- "Disorganized Complexity"—"results from the particular system having a very large number of parts, say millions of parts, or many more. Though the interactions of the parts in a 'disorganized complexity' situation can be seen as largely random, the properties of the system as a whole can be understood by using probability and statistical methods. A prime example of disorganized complexity is a gas in a container, with the gas molecules as the parts. Some would suggest that a system of disorganized complexity may be compared, for example, with the (relative) simplicity of the planetary orbits—the latter can be known by applying Newton's laws of motion, though this example involved highly correlated events."
- "Organized Complexity"—"resides in nothing else than the non-random, or correlated, interaction between the parts. These correlated relationships create a differentiated structure that can, as a system, interact with other systems. The coordinated system manifests properties not carried or dictated by individual parts. The organized aspect of this form of complexity vis a vis to other systems than the

subject system can be said to 'emerge,' without any 'guiding hand.' The number of parts does not have to be very large for a particular system to have emergent properties. A system of organized complexity may be understood in its properties (behavior among the properties) through modeling and simulation, particularly modeling and simulation with computers. An example of organized complexity is a city neighborhood as a living mechanism, with the neighborhood people among the system's parts."

Of course, there can be other types, so the examples above are illustrative.

Sources of Complexity

In general, "there are generally rules which can be invoked to explain the origin of complexity in a given system." (WK 2012ii)

For instance, in the case of "disorganized complexity"—"the source...is the large number of parts in the system of interest, and the lack of correlation between elements in the system." (WK 2012ii)

And in the case of "organized complexity"—the source is "from beneficially mutated organisms being selected to survive by their environment for their differential reproductive ability or at least success over inanimate matter or less organized complex organisms," especially in "self-organizing living systems." (WK 2012ii; R. Ulanowicz 1997)

Problems with the Study of Complexity

However, for the critics, this study of complexity has some major problems. Consider, for illustration, a few examples below (as summarized in *Table 2.5*).

Firstly, one main criticism of the study of complexity is that there is a tradeoff between solvability and size (e.g., the larger the size of a problem is, the less solvable the problem becomes).

For example, for the critics, "some problems are difficult to solve, while others are easy. For example, some difficult problems need algorithms that take an exponential amount of time in terms of the size of the problem to solve. Take the travelling salesman problem, for example. It can be solved in time $O\ (n^2 2^n)$ (where n is the size of the network to visit—let's say the number of cities the travelling salesman must visit exactly once). As the size of the network of cities grows, the time needed to find the route grows (more than) exponentially." (WK 2012ii)

Secondly, another main criticism of the study of complexity is that solving complex problems is not practical in certain cases.

For instance, for the critics, "even though a problem may be computationally solvable in principle, in actual practice it may not be that simple. These problems might require large amounts of time or an inordinate amount of space....Time and space are two of the most important and popular considerations when problems of complexity are analyzed. For instance, there exist a certain class of problems that although they are solvable in principle they require so much time or space that it is not practical to attempt to solve them. These problems are called intractable." (WK 2012ii)

Thirdly, still another main criticism of the study of complexity is that there is the problem of Bremermann's fundamental limit on computation.

For instance, H. Bremermann (1977) showed some fundamental constraints on any further improvement of computation (as already discussed in my book titled *The Future of Complexity*), in that "there exists a fundamental limit for the speed and efficiency of computing machines which cannot be overcome. Such a fundamental limit stems from the idea that the maximum speed of signal traveling between the inner components of the computing machine is constrained by the speed of light, i.e., 3.108 m/second. The time-lag of signal traveling is determined by the distance between the computing machine inner devices," and this time-lag is "in turn constrained by the so-called commutation time. Commutation time is the time-interval involved in processing in-formation (signals) through discrete devices....Even if we supposed the (technological) possibility of building a quite small computer and minimizing/optimizing the trajectory of signal-traveling, such a fundamental limit cannot be overcome [since]...there would remain problems whose complexity can be said to be transcomputable." (J. Teixeira 1998)

And fourthly, an additional main criticism of the study of complexity is that some complex problems cannot be predicted.

For instance, for the critics, "a recent study based on molecular simulations and compliance constants describes molecular recognition as a phenomenon of organisation. Even for small molecules like carbohydrates, the recognition process can not be predicted or designed even assuming that each individual hydrogen bond's strength is exactly known." (WK 2012ii)

Beyond the Study of Complexity

These criticisms of the study of complexity, however, should not be misconstrued as a total rejection of its usefulness but serve to show us the two opposing sides of the debate, such that the possibility (or impossibility) and desirability (or undesirability) of objectivity (from the perspective of nature with objectivity, complexity, and uncertainty as a case study here) are not to the extent that the respective defenders would like us to believe.

Furthermore, the analysis of objectivity, complexity, and uncertainty is to enrich our understanding of the ontological principles in existential dialectics, and good examples include the formalness-informalness principle, the absoluteness-relativeness principle, the partiality-totality principle, the predictability-unpredictability principle, the explicability-inexplicability principle, the fiction-reality principle, the cognitiveness-noncognitiveness principle, the finiteness-transfiniteness principle, the preciseness-vagueness principle, the simpleness-complicatedness principle, the openness-hiddenness principle, the denseness-emptiness principle, the change-constancy principle, the order-chaos principle, the slowness-quickness principle, the expansion-contraction principle, the theory-praxis principle, the convention-novelty principle, the evolution-transformation principle, the symmetry-asymmetry principle, the softness-hardness principle, the seriousness-playfulness principle, the activeness-inactivess principle, the selfness-otherness principle, the regression-progression principle, the same-difference principle, the stability-reaction principle, the functionality-nonfunctionality principle, and the intentionality-nonintentionality principle.

For instance, in relation to the formalness-informalness principle, if there is formalness (e.g., the formal logical requirement of soundness and completeness in a system of ideas, as in the argument on complexity and emergence), there is informalness (e.g., the non-formal existence of unsoundness in the argument on complexity and emergence, as shown in the criticisms that there is a tradeoff between solvability and size [e.g., the larger the size of a problem is, the less solvable the problem becomes], that solving complex problems is not practical in certain cases, that there is the problem of Bremermann's fundamental limit on computation, and that some complex problems cannot be predicted, etc.). And the reverse direction also holds true.

In relation to the absoluteness-relativeness principle, if there is absoluteness (e.g., a system operates within its own "initial conditions"), there is relativeness (e.g., different systems have different "initial conditions," so each of these systems is relative to others, since each "is highly sensitive" to its "initial conditions"). (WK 2012ii) And the reverse direction also holds true.

In relation to the partiality-totality principle, if there is partiality (e.g., the partial view on complexity from the vantage point of a particular meaning of complexity in a particular field), there is totality (e.g., the more holistic view of complexity, such that the whole is not the sum of its parts, that is, it cannot be reduced to the sum of opposing views about the meanings of complexity in different fields, since there will emerge new views in the future not yet known today, just as there are already alternative views nowadays which do not side exclusively with any of them, as is the analysis here). And the reverse direction also holds true.

In relation to the predictability-unpredictability principle, if there is predictability (e.g., the predictable tendency of "the properties of the system as a whole" in organized complexity, since they "can be understood by using probability and statistical methods"), there is unpredictability (e.g., the more difficult task to predict exactly the extent to which "the interactions of the parts in a 'disorganized complexity' situation" will behave on a particular future occasion, since they "can be seen as largely random" and there is "the lack of correlation between elements in the system"). And the reverse direction also holds true.

In relation to the explicability-inexplicability principle, if there is explicability (e.g., the explanation, by the advocates of cybernetics in the early history of complexity study, of the research on "the interaction of parts in systems and the structure of systems" on the basis of its contribution to the understanding of the constraints and possibilities of systems), there is inexplicability (e.g., the lack of sufficient explanation by the advocates of cybernetics in the early history of complexity study, of why this way of studying systems is necessarily desirable, since, for the critics in the later history of complexity study, there are such unresolvable problems as the problem of Bremermann's fundamental limit on computation and the problem of intractability, among others). (WK 2012jj) And the reverse direction also holds true.

In relation to the fiction-reality principle, if there is fiction (e.g., the fictional aspect of complexity study, insofar as, for the critics in the later history of complexity study, there are such unresolvable problems such as the problem of Bremermann's fundamental limit on computation and the problem of intractability), there is reality (e.g., the realistic aspect of complexity study, insofar as it contributes to the resolution of some problems which can be solved). And the reverse direction also holds true.

In relation to the cognitiveness-noncognitiveness principle, if there is cognitiveness (e.g., a view of things on the basis of reason and evidences, as shown inWeaver's research on the distinction between "disorganized complexity" and "organized complexity"), there is noncognitiveness (e.g., a view of things on the basis of non-cognitive factors like envy, jealousy, power, nationality, race, gender, age, class, greed, lust, status, faith, anger, sadness, joy, fear, wish, etc., as shown in the hope of those in complexity study to bridge the gap between "disorganized complexity" and "organized complexity" one day, as Weaver had addressed the problem only in "a preliminary way"). (WK 2012ii) And the reverse direction also holds true.

In relation to the finiteness-transfiniteness principle, if there is finiteness (e.g., the finite number of parts in a small system of "organized complexity"), there is transfiniteness (e.g., the transfinite number of parts in an extremely large system of "disorganized complexity"). And the reverse direction also holds true.

In relation to the preciseness-vagueness principle, if there is preciseness (e.g., the precise identification of 6 attributes of complexity in the article aforecited), there is vagueness (e.g., the vagueness in the identification of the 6 attributes of complexity in the article aforecited, since it is not clear why there must be only 6, not 7, 8, 9, 10, and so on). And the reverse direction also holds true.

In relation to the simpleness-complicatedness principle, if there is simpleness (e.g., the relatively simple analysis of complexity study by the advocates), there is complicatedness (e.g., the relatively more complicated analysis of complexity study, by challenging the claims and assumptions, as shown in the problems pointed out by the critics as cited). And the reverse direction also holds true.

In relation to the openness-hiddenness principle, if there is openness (e.g., the open exploration of the relationships among different parts in a system by the advocates of complexity study), there is hiddenness (e.g., the hidden bias in complexity study, since, for the critics, there are different unresolvable problems such as the problem of Bremermann's fundamental limit on computation, the problem of intractability, and so on). And the reverse direction also holds true.

In relation to the denseness-emptiness principle, if there is denseness (e.g., the relatively denser concentration of correlations among different parts in an organized complex system), there is emptiness (e.g., the relatively less dense, or more empty, concentration of correlations among different parts in a disorganized complex system). And the reverse direction also holds true.

In relation to the change-constancy principle, if there is change (e.g., the ever new ways to understand the linkages among different parts, as shown in complexity study), there is constancy (e.g., the ever constant existence of different problems with the ever new ways to understand the linkages among different parts, as shown in the problems with complexity study pointed out by the critics). And the reverse direction also holds true.

In relation to the order-chaos principle, if there is order (e.g., the more or less orderly view of reality, if looked from the vantage point of highly correlated parts in a small organized complex system), there is chaos (e.g., the more or less chaotic view of reality, if looked from the vantage point of many uncorrelated parts in an extremely large disorganized complex system). And the reverse direction also holds true.

In relation to the slowness-quickness principle, if there is slowness (e.g., the relatively slower speed of a problem to be solved, if the size is extremely large), there is quickness (e.g., the relatively quicker speed of of a problem to be solved, if the size is extremely small). And the reverse direction also holds true.

In relation to the expansion-contraction principle, if there is expansion (e.g., the relative expansion of "web science" in the research on complexity study nowadays), there is contraction (e.g., the relative decline of "cybernetics" in the research on complexity study nowadays). And the reverse direction also holds true.

In relation to the theory-praxis principle, if there is theory (e.g., the theoretical construction of "emergence" in complexity study), there is praxis (e.g., the practical application of "emergence" in complexity study to the field of software engineering, like the issue of "programming complexity"). (WK 2012ii) And the reverse direction also holds true.

In relation to the convention-novelty principle, if there is convention (e.g., the conventional wisdom about complexity study to understand the linkages among different parts in a system), there is novelty (e.g., the alternative novel challenge to the conventional wisdom about complexity study to understand the linkages among different parts in a system, by the new idea that there are different unresolvable problems such as the problem of Bremermann's fundamental limit on computation, the problem of intractability, and so on). And the reverse direction also holds true.

In relation to the evolution-transformation principle, if there is evolution (e.g., the natural evolution of humans to observe different parts of things in the state of nature), there is transformation (e.g., the technical transformation of human ability to understand different parts of things by the invention of complexity theory). And the reverse direction also holds true.

In relation to the symmetry-asymmetry principle, if there is symmetry (e.g., the co-existence of different problems studied by those in complexity study), there is asymmetry (e.g., some problems in complexity study are "tractable"—but some others are "intractable"). And the reverse direction also holds true.

In relation to the softness-hardness principle, if there is softness (e.g., the defense of complexity study by the advocates), there is hardness (e.g., the critique of complexity study by the critics). And the reverse direction also holds true.

In relation to the seriousness-playfulness principle, if there is seriousness (e.g., the serious business of those who try to understand complexity), there is playfulness (e.g., the playful part of those who try to understand complexity, when they play around with different topics and/or different fields over time, like "complex behaviour," "complex mechanisms," "complex simulations," "complex systems," "complexity in data," "complexity in molecular recognition," "complexity management," etc.). (WK 2012ii) And the reverse direction also holds true.

In relation to the activeness-inactivess principle, if there is activeness (e.g., the relative activeness of complexity study when it can be used to solve some problems which are tractable), there is inactiveness (e.g., the relative inactiveness of complexity study when it cannot be used to solve some problems which are intractable).

In relation to the selfness-otherness principle, if there is selfness (e.g., the tendency to be mostly centered on one's self or group, as shown by those in business who uses "complexity management" so as to "efficiently control value-adding complexity" for profits), there is otherness (e.g., the tendency to be mostly centered on others, as shown in the work by Weaver on the distinction between "disorganized complexity" and "organized complexity," as an advancement in knowledge for humanity). And the reverse direction also holds true.

In relation to the regression-progression principle, if there is regression (e.g., the regression made by complexity study, as shown in the problems pointed out by the critics aforecited), there is progression (e.g., the progress made by complexity study, as shown in its contribution to the understanding of the linkages among different parts in a system aforecited). And the reverse direction also holds true.

In relation to the same-difference principle, if there is similarity in outcome (e.g., the contribution to the molding and control of beliefs, values, and behaviors, regardless of whether this be done by way of the use of complexity study in mathematics, or by way of the use of complexity study in business), there is difference in outcome (e.g., the contribution to the molding and control of beliefs, values, and behaviors by way of the use of complexity study in mathematics for a relatively more logical-formalistic worldview—but the contribution to the molding and control of beliefs, values, and behaviors by way of the use of complexity study in business for a relatively more materialstic worldview). And the reverse direction also holds true.

In relation to the stability-reaction principle, if there is stability (e.g., the relatively more stable condition of cybernetics in the early history of complexity study, when it was widely studied), there is reaction (e.g., cybernetics in the early history of complexity study led to further changes later on, when nowadays it is superseded by different new sub-fields, like "web science" and "e-science" in the current history of complexity study). (WK 2012ii) And the reverse direction also holds true.

In relation to the functionality-nonfunctionality principle, if there is functionality (e.g., the presence of action for which something is fitted, as shown in the use of creation theory in the older days, when it served the function of explaining the complexity of things in the world in terms of the "guiding hand" of God), there is nonfunctionality (e.g., the relative lack of action for which something is fitted, as shown in the rise of complexity science nowadays, such that it is used to understand the complexity of things in the world without appealing to a divine design, that is, in terms of "emergence" without any "guiding hand," such that creation theory is no longer as much functional in our time as it was dominant in Medieval Europe). (WK 2012ii) And the reverse direction also holds true.

In relation to the intentionality-nonintentionality principle, if there is intentionality (e.g., the planning or design of something for a certain outcome, as shown in the use of complexity study for the understanding of linkages among different parts in a system), there is nonintentionality (e.g., the relative absence of planning or design of something for a certain outcome, such that the outcome can even be counter-intentional in being the opposite of what was originally planned, as shown by R. Tulip [2012], who used complexity study for "a rational foundation for systematic Biblical theology"—contrary to the original mission of complexity study in science). And the reverse direction also holds true.

And in relation to the survivability-nonsurvivability principle, if there is survivability (e.g., the relative survival of complexity study in modern science since its beginnng in the 1940s for the understanding of linkages among different parts in a system, without the help of any "guiding hand" or God), there is non-survivability (e.g., the relative non-survival of creation theory in modern science, when it tried to understand the linkages among different parts in a system with the help of the "guiding hand" of a supernatural being, although it remains relevant in Western theology). And the reverse direction also holds true.

Objectivity and Society

Objectivity, when examined from the perspective of society, can reveal to us some powerful features in relation to its possibility (or impossibility) and desirability (or undesirability), and this can be shown by way of two case studies, namely, (2.4.1) objectivity, consumption, and reliability engineering, and (2.4.2) objectivity, organizational behavior, and optimal decision—to be addressed hereafter.

Objectivity, Consumption, and Reliability Engineering

A good way to evaluate the possibility (or impossibility) and desirability (or undesirability) of objectivity from the perspective of society is to explore objectivity, consumption, and reliability engineering.

An argument in the literature is that the reliability (or the probability of failure) in regard to products and services used by consumers can be predicted for an interval of time.

Reliability Theory (in Reliability Engineering)

The term "reliability" refers to the "probability of failure" in reliability theory, which "describes the probability of a system completing its expected function during an interval of time. It is the basis of reliability engineering, which is an area of study focused on optimizing the reliability, or probability of successful functioning, of systems, such as airplanes, linear accelerators, and any other product. It developed apart from the mainstream of probability and statistics. It was originally a tool to help nineteenth century maritime insurance and life insurance companies compute profitable rates to charge their customers." (WK 2012kk)

In technical terms, "reliability" is defined as "the probability that a device will perform its intended function during a specified period of time under stated conditions," as expressed below: (WK 2012ll)

$$R(t) = Pr\{T > t\} = \int_t^{\infty} f(x)\,dx$$

Here, "*f(x)* is the failure probability density function and *t* is the length of the period of time (which is assumed to start from time zero)." (WK 2012ll)

There are four main issues about reliability in reliability theory, which serves as "the foundation of reliability engineering" (as summarized in *Table 2.6*): (WK 2012ll)

- "Reliability is a probability"—which means "that failure is regarded as a random phenomenon: it is a recurring event, and we do not express any information on individual failures, the causes of failures, or relationships between failures, except that the likelihood for failures to occur varies over time according to the given probability function. Reliability engineering is concerned with meeting the specified probability of success, at a specified statistical confidence level."
- "Reliability is predicated on 'intended function'"—which means "operation without failure. However, even if no individual part of the system fails, but the system as a whole does not do what was intended, then it is still charged against the system reliability. The system requirements specification is the criterion against which reliability is measured."
- "Reliability applies to a specified period of time"—which means that "a system has a specified chance that it will operate without failure before time *t*. Reliability engineering ensures that components and materials will meet the requirements during the specified time. Units other than time may sometimes be used. The automotive industry might specify reliability in terms of miles, the military might specify reliability of a gun for a certain number of rounds fired. A piece of mechanical equipment may have a reliability rating value in terms of cycles of use."
- "Reliability is restricted to operation under stated (or explicitly defined) conditions"—which means that "it is impossible to design a system for unlimited conditions. A Mars Rover will have different specified conditions than a family car. The operating environment must be addressed during design and testing. Also, that same rover may be required to operate in varying conditions requiring additional scrutiny."

Of course, there can be other issues, so the ones above are illustrative, not exhaustive.

Reliability Requirements

Theory is one thing, but implementation is another. Thus, in reliability engineering, it is important "to adequately specify the reliability and maintainability requirements as defined by the stakeholders in terms of their overall availability needs." (WK 2012ll)

Consider some main reliability requirements below (as summarized in *Table 2.6*): (WK 2012ll)

- "Systems Engineering Logic"—which is to "drive a (system or part) design to incorporate features and functions that prevent failures from occurring or limit consequences from failure in the first place....The maintainability requirements address the costs of repairs as well as repair time. Testability requirements provide the link between reliability and maintainability and should address detectability of failure modes (on a particular system level), isolation levels and the creation of diagnostics (procedures)."
- "Reliability Tasks and Documentation"—which is to "address requirements for various reliability tasks and documentation during system development, test, production, and operation. These requirements are generally specified in the contract statement of work and depend on how much leeway the customer wishes to provide to the contractor. Reliability tasks include various analyses, planning, and failure reporting. Task selection depends on the criticality of the system as well as cost. A critical system may require a formal failure reporting and review process throughout development, whereas a non-critical system may rely on final test reports."

Of course, there can be other requirements, so the ones above are illustrative, not exhaustive.

Reliability Prediction

In general, works in reliability engineering are supposed to deal with some or all of the following predictions (as summarized in *Table 2.6*): (WK 2012ll)

- Ex: "help assess the effect of product reliability on maintenance activity and the quantity of spare units required for acceptable field performance of any particular system. For example, predictions of the frequency of unit level maintenance actions can be obtained. Reliability prediction can be used to size spare populations."

- Ex: "provide necessary input to system-level reliability models. System-level reliability models can subsequently be used to predict, for example, frequency of system outages in steady-state, frequency of system outages during early life, expected downtime per year, and system availability."
- Ex: "provide necessary input to unit and system-level Life Cycle Cost Analyses. Life cycle cost studies determine the cost of a product over its entire life. Therefore, how often a unit will have to be replaced needs to be known. Inputs to this process include unit and system failure rates. This includes how often units and systems fail during the first year of operation as well as in later years."
- Ex: "assist in deciding which product to purchase from a list of competing products. As a result, it is essential that reliability predictions be based upon a common procedure."
- Ex: "can be used to set factory test standards for products requiring a reliability test. Reliability predictions help determine how often the system should fail."
- Ex: "are needed as input to the analysis of complex systems such as switching systems and digital cross-connect systems. It is necessary to know how often different parts of the system are going to fail even for redundant components."
- Ex: "can be used in design trade-off studies. For example, a supplier could look at a design with many simple devices and compare it to a design with fewer devices that are newer but more complex. The unit with fewer devices is usually more reliable."
- Ex: "can be used to set achievable in-service performance standards against which to judge actual performance and stimulate action."

Again, there can be other predictions, so the ones above are illustrative, not exhaustive.

Reliability Modeling

Reliability predictions require some forms of modeling for the process of "predicting or understanding the reliability of a component or system prior to its implementation." (WK 2012ll)

For illustration, hereafter are two commonly used models for "part level predictions" (as summarized in *Table 2.6*): (WK 2012ll)

- "The Physics of Failure Approach"—"uses an understanding of physical failure mechanisms involved, such as mechanical crack propagation or chemical corrosion degradation or failure."
- "The Parts Stress Modeling Approach"—"is an empirical method for prediction based on counting the number and type of components of the system, and the stress they undergo during operation."

Surely, there can be other approaches, so the ones above are illustrative, not exhaustive.

Reliability Testing

Reliability modeling allows testing later, and "the purpose of reliability testing is to discover potential problems with the design as early as possible and, ultimately, provide confidence that the system meets its reliability requirements." (WK 2012ll)

There are some important issues of reliability testing to be considered, as shown below: (WK 2012ll)

- "Test Levels"—"Reliability testing may be performed at several levels and there are different types of testing. Complex systems may be tested at component, circuit board, unit, assembly, subsystem and system levels....For example, performing environmental stress screening tests at lower levels, such as piece parts or small assemblies, catches problems before they cause failures at higher levels. Testing proceeds during each level of integration through full-up system testing, developmental testing, and operational testing, thereby reducing program risk."
- "Risk Levels"—"Statistical confidence is increased by increasing either the test time or the number of items tested. Reliability test plans are designed to achieve the specified reliability at the specified confidence level with the minimum number of test units and test time. Different test plans result in different levels of risk to the producer and consumer. The desired reliability, statistical confidence, and risk levels for each side influence the ultimate test plan. The customer and developer should agree in advance on how reliability requirements will be tested."
- "Test Strategy"—"As part of the requirements phase, the reliability engineer develops a test strategy with the customer. The test strategy makes trade-offs between the needs of the reliability organization, which wants as much data as possible, and constraints such as cost,

> schedule and available resources. Test plans and procedures are developed for each reliability test, and results are documented."

Again, there can be other issues, so the ones above are illustrative, not exhaustive.

Problems with Reliability Engineering

However, for the critics, reliability engineering (with reliability theory as its foundation) has some major problems. Consider, for illustration, a few examples below (as summarized in *Table 2.6*).

Firstly, one main criticism of reliability engineering is that there is a trade-off between "reliability" and "safety."

For example, for the critics, "safety engineering and reliability engineering may even have contradicting requirements," because "reliability engineering is in the end only concerned with cost," but "safety engineering…is more specific and regulated" and "relates to only very specific and system Safety Hazards that could potentially lead to severe accidents." (WK 2012ll)

Secondly, another main criticism of reliability engineering is that reliability testing can be infeasible in practice.

For instance, for the critics, "it is not always feasible to test all system requirements. Some systems are prohibitively expensive to test; some failure modes may take years to observe; some complex interactions result in a huge number of possible test cases; and some tests require the use of limited test ranges or other resources." (WK 2012ll)

Thirdly, still another main criticism of reliability engineering is that there is the problem of definition in testing.

For instance, for the critics, "a key aspect of reliability testing is to define 'failure.' Although this may seem obvious, there are many situations where it is not clear whether a failure is really the fault of the system. Variations in test conditions, operator differences, weather and unexpected situations create differences between the customer and the system developer." (WK 2012ll)

Fourthly, an additional main criticism of reliability engineering is that testing does not always need to be formal.

For instance, for the critics, "for small, non-critical systems, reliability engineering may be informal. As complexity grows, the need arises for a formal reliability function. Because reliability is important to the customer, the customer may even specify certain aspects of the reliability organization." (WK 2012ll)

Fifthly, another additional main criticism of reliability engineering is that the bias of perception can screw up the understanding of reliability.

For instance, for the critics, "when reliability is considered from the perspective of the consumer of a technology or service, actual reliability measures may differ dramatically from perceived reliability. One bad experience can be magnified in the mind of the customer, inflating the perceived unreliability of the product. One plane crash where hundreds of passengers die will immediately instill fear in a large percentage of the flying consumer population, regardless of actual reliability data about the safety of air travel." (WK 2012kk)

Sixthly, still another additional main criticism of reliability engineering is that the validity of testing can be problematic due to questionable assumptions.

For instance, for the critics, "to perform a proper quantitative reliability prediction for systems may be difficult....On part level, results can be obtained often with higher confidence as many samples might be used for the available testing financial budget, however unfortunately these tests might lack validity on system level due to the assumptions that had to be made for part level testing." (WK 2012ll)

And seventhly, one last main criticism of reliability engineering is that test errors are hard to avoid.

For instance, for the critics, "with each test both a statistical type 1 and type 2 error could be made and depends on sample size, test time, assumptions and the needed discrimination ratio. There is risk of incorrectly accepting a bad design (type 2 error) and the risk of incorrectly rejecting a good design (type 1 error)." (WK 2012ll)

Beyond Reliability Engineering

These criticisms of reliability engineering, however, do not mean a total rejection of its usefulness but serve to show us the two opposing sides of the debate, such that the possibility (or impossibility) and desirability (or undesirability) of objectivity (from the perspective of society with objectivity, consumption, and reliability engineering as a case study here) are not to the extent that the respective defenders would like us to believe.

Furthermore, the analysis of objectivity, consumption, and reliability engineering can teach us something important about the ontological principles in existential dialectics, and good examples include the formalness-informalness principle, the absoluteness-relativeness principle, the partiality-totality principle, the predictability-unpredictability principle, the explicability-inexplicability principle, the fiction-reality principle, the cognitiveness-noncognitiveness principle, the finiteness-transfiniteness principle, the preciseness-vagueness principle, the simpleness-complicatedness principle, the openness-hiddenness principle, the denseness-emptiness principle, the change-constancy principle, the order-chaos principle, the slowness-quickness principle, the expansion-contraction principle, the theory-praxis principle, the convention-novelty principle, the evolution-transformation principle, the symmetry-asymmetry principle, the softness-hardness principle, the seriousness-playfulness principle, the activeness-inactivess principle, the selfness-otherness principle, the regression-progression principle, the same-difference principle, the stability-reaction principle, the functionality-nonfunctionality principle, and the intentionality-nonintentionality principle.

For instance, in relation to the formalness-informalness principle, if there is formalness (e.g., the formal logical requirement of soundness, completeness, and consistency in a system of ideas, as in the study of reliability engineering), there is informalness (e.g., the non-formal existence of unsoundness, incompleteness, and inconsistency in the study of reliability engineering, as shown in the criticisms that there is a trade-off between "reliability" and "safety," that reliability testing can be infeasible in practice, that there is the problem of definition in testing, that testing does not always need to be formal, that the bias of perception can screw up the understanding of reliability, that the validity of testing can be problematic due to questionable assumptions, and that test errors are hard to avoid). And the reverse direction also holds true.

In relation to the absoluteness-relativeness principle, if there is absoluteness (e.g., the absolute view of those who focus on "reliability" in reliability engineering), there is relativeness (e.g., what is right for those who focus on "reliability" in reliability engineering is not necessarily so for those who focus on "safety" instead). And the reverse direction also holds true.

In relation to the partiality-totality principle, if there is partiality (e.g., the partial view of "failure" if looked from the vantage point of "reliability"), there is totality (e.g., the more holistic view of "failure," such that the whole is not the sum of its parts, that is, it cannot be reduced to the sum of opposing views like "reliability," "safety," "tradeoff," "definition," etc., since there will emerge new views in the future not yet known today, just as there are already alternative views nowadays which do not side exclusively with any of them, as is the analysis here). And the reverse direction also holds true.

In relation to the predictability-unpredictability principle, if there is predictability (e.g., the predictable tendency of those in reliability engineering to make certain predictions aforecited), there is unpredictability (e.g., the more difficult task to predict exactly the extent to which a particular reliability model will succeed in making correct predictions on a particular occasion, because of the problems pointed out by the critics aforecited). And the reverse direction also holds true.

In relation to the explicability-inexplicability principle, if there is explicability (e.g., the explanation by those in reliability engineering of why it is desirable on the basis of being concerned with "cost," etc.), there is inexplicability (e.g., the lack of sufficient explanation by those in "reliability engineering of why it is necessarily desirable to focus on "cost," since, for the critics, there are other important concerns like "safety," and so on). And the reverse direction also holds true.

In relation to the fiction-reality principle, if there is fiction (e.g., the fictional aspect of reliability about air travel, insofar as "one bad experience can be magnified in the mind of the customer, inflating the perceived unreliability of the product"), there is reality (e.g., the realistic aspect of reliability about air travel, insofar there are actual cases of "plane crash where hundreds of passengers died"). And the reverse direction also holds true.

In relation to the cognitiveness-noncognitiveness principle, if there is cognitiveness (e.g., a view of things on the basis of reason and evidences, as shown in the facts about some accidents of "plane crash" which did occur in the history of air travel), there is noncognitiveness (e.g., a view of things on the basis of non-cognitive factors like envy, jealousy, power, nationality, race, gender, age, class, greed, lust, status, faith, anger, sadness, joy, fear, wish, etc., as shown in the "fear" in "a large percentage of the flying consumer population," when they heard about a "plane crash where hundreds of passengers died,... regardless of actual reliability data about the safety of air travel"). And the reverse direction also holds true.

In relation to the finiteness-transfiniteness principle, if there is finiteness (e.g., the finite number of "test levels" like "component, circuit board, unit, assembly, subsystem and system levels"), there is transfiniteness (e.g., the transfinite number of all the individual items in history which can be included for testing at any of the levels in reliability engineering, in principle). And the reverse direction also holds true.

In relation to the preciseness-vagueness principle, if there is preciseness (e.g., the precise identification of 2 main approaches to reliability modeling in the article aforecited, namely, "the physics of failure approach" and "the parts stress modeling approach"), there is vagueness (e.g., the vagueness in the identification of the 2 main approaches to reliability modeling in the article aforecited, since it is not clear why there must be only 2, not 4, 7, 8, 12, 13, and so on). And the reverse direction also holds true.

In relation to the simpleness-complicatedness principle, if there is simpleness (e.g., the relatively simple analysis of reliability engineering by the practitioners), there is complicatedness (e.g., the relatively more complicated analysis of reliability engineering, by challenging the claims and assumptions, as shown in the problems pointed out by the critics as cited). And the reverse direction also holds true.

In relation to the openness-hiddenness principle, if there is openness (e.g., the open exploration of "reliability" by those in reliability engineering), there is hiddenness (e.g., the hidden bias in the focus on "reliability" by those in reliability engineering, since there are other important issues not emphasized, like "safety" as stressed by those in safety engineering). And the reverse direction also holds true.

In relation to the denseness-emptiness principle, if there is denseness (e.g., the relatively denser concentration of concerns with "reliability" by those in reliability engineering), there is emptiness (e.g., the relatively less dense, or more empty, concentration of concerns with "reliability" by those in safety engineering). And the reverse direction also holds true.

In relation to the change-constancy principle, if there is change (e.g., the ever new ways to cope with failure, as shown in reliability engineering), there is constancy (e.g., the ever constant existence of different problems with the ever new ways to cope with failure, as shown in the problems with reliability engineering pointed out by the critics). And the reverse direction also holds true.

In relation to the order-chaos principle, if there is order (e.g., the more or less orderly view about products and services, if looked from the vantage point of their "successes" like good performances), there is chaos (e.g., the more or less chaotic view about products and services, if looked from the vantage point of their "failures" or unpredicted accidents). And the reverse direction also holds true.

In relation to the slowness-quickness principle, if there is slowness (e.g., the relatively slower speed of a product to fail in the laboratory if a normal life testing, not an "accelerated life testing," is used for testing instead), there is quickness (e.g., the relatively quicker speed of a product to fail in the laboratory if "accelerated life testing" is used for testing). (WK 2012ll) And the reverse direction also holds true.

In relation to the expansion-contraction principle, if there is expansion (e.g., the relatively more developed ability to deal with "stress" in the physics of failure approach to reliability engineering), there is contraction (e.g., the relatively less developed ability to deal with "redundancy" in the physics of failure approach to reliability engineering). And the reverse direction also holds true.

In relation to the theory-praxis principle, if there is theory (e.g., the theoretical construction of "probability" in mathematics like mainstream probability study and statistics), there is praxis (e.g., the practical application of probability" in mathematics like mainstream probability study and statistics to the field of engineering, like "reliability engineering"). And the reverse direction also holds true.

In relation to the convention-novelty principle, if there is convention (e.g., the mainstream of probability study and statistics), there is novelty (e.g., the alternative novel challenge to the mainstream of probability study and statistics, by the new idea of "reliability theory," which "developed apart from the mainstream of probability and statistics"). (WK 2012kk) And the reverse direction also holds true.

In relation to the evolution-transformation principle, if there is evolution (e.g., the natural evolution of humans to experience failures in the state of nature), there is transformation (e.g., the technical transformation of human ability to cope with failures by the invention of reliability engineering). And the reverse direction also holds true.

In relation to the symmetry-asymmetry principle, if there is symmetry (e.g., the co-existence of different risks in reliability testing), there is asymmetry (e.g., "the risk of incorrectly accepting a bad design" in "type 2 error"—but "the risk of incorrectly rejecting a good design" in type 1 error"). And the reverse direction also holds true.

In relation to the softness-hardness principle, if there is softness (e.g., the defense of focusing on "reliability" by the advocates), there is hardness (e.g., the critique of focusing on reliability by the critics, like those who focus on "safety" instead). And the reverse direction also holds true.

In relation to the seriousness-playfulness principle, if there is seriousness (e.g., the serious business of those who work on reliability models), there is playfulness (e.g., the playful part of those who work on reliability models, when they play around with different approaches to modeling over time, like "the physics of failure approach," "the parts stress modeling approach," etc.). And the reverse direction also holds true.

In relation to the activeness-inactivess principle, if there is activeness (e.g., the relative activeness of those products which perform well in reliability testing), there is inactiveness (e.g., the relative inactiveness of those products which fail in reliability testing). And the reverse direction also holds true.

In relation to the selfness-otherness principle, if there is selfness (e.g., the tendency to be mostly centered on one's self or group, as shown in the focus on reliability by those in a company which is "only concerned with cost" to minmize the "loss of revenue for the company"), there is otherness (e.g., the tendency to be mostly centered on others, as shown in the focus on "safety" by those in government for the welfare of the larger "society"). And the reverse direction also holds true.

In relation to the regression-progression principle, if there is regression (e.g., the regression made by reliability engineering, as shown in the problems pointed out by the critics), there is progression (e.g., the progress made by reliability engineering, as shown in its contribution to the enhancement of reliability aforecited). And the reverse direction also holds true.

In relation to the same-difference principle, if there is similarity in outcome (e.g., the contribution to the molding and control of beliefs, values, and behaviors, regardless of whether this be done by way of focusing on reliability, or by way of focusing on safety), there is difference in outcome (e.g., the contribution to the molding and control of beliefs, values, and behaviors by way of focusing on reliability for a relatively more efficient lifeworld—but the contribution to the molding and control of beliefs, values, and behaviors by way of focusing on safety for a relatively more effective lifeworld). And the reverse direction also holds true.

In relation to the stability-reaction principle, if there is stability (e.g., the relatively more stable condition of companies which focused more on cost in the early days of industrialization), there is reaction (e.g., the stable condition of companies which focused more on cost in the early days of industrialization then led to further changes later on, when in our times, there are laws governing "consumer protection" for safety, like the establishment of the Federal Trade Commission in 1914). (WK 2012mm) And the reverse direction also holds true.

In relation to the functionality-nonfunctionality principle, if there is functionality (e.g., the presence of action for which something is fitted, as shown in the focus on cost in the older days, which served the function of laissez-faire capitalism in the early days of industrialization), there is nonfunctionality (e.g., the relative lack of action for which something is fitted, as shown in the rise of laws and organizations for consumer protection nowadays, such that the selfish focus on cost in the older days is no longer much functional in our times, because of different standards and expectations for safety, etc.). And the reverse direction also holds true.

In relation to the intentionality-nonintentionality principle, if there is intentionality (e.g., the planning or design of something for a certain outcome, as shown by those in reliability engineering who work on the exact measurement of the performance of "a device and its intended function during a specified period of time under stated conditions"), there is nonintentionality (e.g., the relative absence of planning or design of something for a certain outcome, such that the outcome can even be counter-intentional in being the opposite of what was originally planned, as shown in the distortion of perception by consumers, when "one bad experience can be magnified in the mind of the customer, inflating the perceived unreliability of the product,...regardless of actual reliability data"—contrary to the original mission to measure actual reliability). And the reverse direction also holds true.

And in relation to the survivability-nonsurvivability principle, if there is survivability (e.g., the survival of those devices which pass reliability testing), there is non-survivability (e.g., the non-survival of those devices which do not pass reliability testing). And the reverse direction also holds true.

Objectivity, Organizational Behavior, and Optimal Decision

Another good way to evaluate the possibility (or impossibility) and desirability (or undesirability) of objectivity from the perspective of society is to explore objectivity, organizational behavior, and optimal decision.

An argument in the literature is that optimality can be achieved in decision making within an organization.

Decision Making in Organizatioal Behavior

The term "organizational behavior" refers to "a field of study that investigates the impact that individuals, groups and structures have on behavior within an organization. It is an interdisciplinary field that includes sociology, psychology, communication, and management; and it complements the academic studies of organizational theory (which is focused on organizational and intra-organizational topics) and human resource studies (which is more applied and business-oriented). It may also be referred to as organizational studies or organizational science. The field has its roots in industrial and organizational psychology." (WK 2012nn)

One major distinction in American academia is to differentiate three types of organizational behavior, as shown below (and summarized in *Table 2.7*): (WK 2012nn)

- "Micro" Organizational Behaviour—"refers to individual and group dynamics in an organizational setting."
- "Meso" Organizational Behavior—studies "power, culture, and the networks of individuals," "units in organizations," and "'field' level analysis which study how whole populations of organizations interact."
- "Macro" Organizational Behavior—"studies whole organizations and industries, how they adapt, and the strategies, structures and contingencies that guide them."

Of course, there can be other distinctions; for instance, a different distinction concerns major issues in organizational behavior, as shown below: (WK 2012nn)

- Ex: "systems framework"
- Ex: "decision making"
- Ex: "organization structures and dynamics"
- Ex: "personality traits"
- Ex: "control and stress"
- Ex: "motivation in organizations"

An interesting issue (on the list) to be considered here concerns "decision making" in organizational behavior.

The term "decision making" here refers to "the mental processes (cognitive process) resulting in the selection of a course of action among several alternative scenarios. Every decision making process produces a final choice. The output can be an action or an opinion of choice." (WK 2012oo; J. Reason 1990)

Decision Making Stages

Decision making in an organization can be understood in terms of different stages. For instance, "developed by B. Aubrey Fisher, there are four stages that should be involved in all group decision making. These stages, or sometimes called phases, are important for the decision making process to begin," as shown below (and summarized in *Table 2.7*): (WK 2012oo)

- "Orientation Stage"—"is where members meet for the first time and start to get to know each other."

- "Conflict Stage"—is where "once group members become familiar with each other, disputes, little fights and arguments occur. Group members eventually work it out."
- "Emergence Stage"—is where "the group begins to clear up vague opinions by talking about them."
- "Reinforcement Stage"—is where "members finally make a decision, while justifying themselves that it was the right decision."

Of course, this does not always mean that "all group members fully agree—they may not want argue further just to be liked by other group members or to 'fit in.'" (WK 2012oo; T. Postmes 2001)

Decision Making Steps

Decision-making can follow a series of steps as shown below: (WK 2012pp)

- "Verify, define, and detail the problem."
- "Generate all possible solutions."
- "Generate objective assessment criteria."
- "Choose the best solution generated."
- "Implement the preferred alternative."
- "Monitor and evaluate outcomes and results."
- "Feedback."

Surely, these decision making steps are relative to a specific decision making model, like "the rational planning model" (or "the rational decision making model"). (WK 2012pp)

Decision Making Optimality

Then a challenging question to ask is how a decision making within an organization can reach optimality.

An "optimal decision" in organizatoinal behavior refers to "a decision such that no other available decision options will lead to a better outcome"; "in order to compare the different decision outcomes, one commonly assigns a relative utility to each of them. If there is uncertainty in what the outcome will be, the optimal decision maximizes the expected (average) utility. Sometimes, the equivalent problem of minimizing loss is considered, particularly in financial situations, where the utility is defined as economic gain....For example, it may well be the optimal decision for someone to buy a sports car rather than a station wagon, if the outcome in terms of another criterion (e.g., effect on personal image) is more desirable, even given the higher cost and lack of versatility of the sports car." (WK 2012t)

In terms of mathematical description, an optimal decision can be defined. First, "each decision *d* in a set *D* of available decision options will lead to an outcome $o = f(d)$. All possible outcomes form the set *O*. Assigning a utility $U_O(o)$ to every outcome, we can define the utility of a particular decision *d* as": (WK 2012t)

$$U_D(d) = U_O(f(d))$$

Then, an optimal decision d_{opt} is one that maximizes $U_D(d)$, as shown below: (WK 2012t)

$$d_{\mathbf{opt}} = \arg\max_{d \in D} U_D(d)$$

Then, an optimal decision can be found in the following three steps: (WK 2012t)

- "predicting the outcome *o* for every decision *d*"
- "assigning a utility $U_O(o)$ to every outcome *o*
- "finding the decision *d* that maximizes $U_D(d)$

In case of uncertainty in outcome, that is, when "it is not possible to predict with certainty what will be the outcome of a particular decision, a probabilistic approach is necessary. In its most general form, it can be expressed as follows: given a decision *d*, we know the probability distribution for the possible outcomes described by the conditional probability density $p(o|d)$. We can then calculate the expected utility of decision *d*," as shown below: (WK 2012t; M. DeGroot 1970)

$$U_D(d) = \int p(o|d)U(o)do$$

where "the integral is taken over the whole set *O*," but the rest of the mathematical expression, that is, the optimal decision d_{opt}, is the same as the one (afore-expressed) above. (WK 2012t)

Optimizing vs. Satisficing Decision

But an "optimal" decision (which "attempts to find the best option available") should not be confused with a "satisficing" decision—which "attempts to meet an acceptability threshold," even though "a satisficing strategy may often be (near) optimal if the costs of the decision-making process itself, such as the cost of obtaining complete information, are considered in the outcome calculus." (WK 2012u)

For instance, one way to express "satisficing" is "just constraint satisfaction, the process of finding a solution satisfying a set of constraints, without concern for finding an optimum. Any such satisficing problem can be formulated as an (equivalent) optimization problem using the Indicator function of the satisficing requirements as an objective function." (WK 2012u)

Mathematically speaking, "if X denotes the set of all options and $S \in X$ denotes the set of 'satisficing' options, then selecting a satisficing solution (an element of S) is equivalent to the following optimization problem": (WK 2012u)

$$\max_{s \in X} I_S(s)$$

where "I_S denotes the indicator function of S," as shown below: (WK 2012u)

$$I_S(s) := \begin{cases} 1 \ , & s \in S \\ 0 \ , & s \notin S \end{cases} \quad , \ s \in X$$

In this way, "a solution $s \in X$ to this optimization problem is optimal if, and only if, it is a satisficing option (an element of S)." (WK 2012u)

Problems with Optimal Decision

However, for the critics, this argument on optimal decision has some major problems. Consider, for illustration, a few examples below (as summarized in *Table 2.7*).

Firstly, one main criticism of the argument on optimal decision is that optimization is seldom achieved in real life, as decision making can be more intuitive and less formal.

For instance, for the critics, "in practice, few people verify that their decisions are optimal, but instead use...approaches to make decisions that are 'good enough.'" (WK 2012t) Hereafter are some good examples of everyday decision techniques which often do not yield optimal outcomes: (WK 2012oo & 2012qq; W. Bell 1956; R. Batley 2006; A. Tversky 1972 & 1979)

- "Pros and Cons"—"listing the advantages and disadvantages of each option, popularized by Plato and Benjamin Franklin. Contrast the costs and benefits of all alternatives."
- "Elimination by Aspects": "choosing between alternatives using Mathematical Psychology Technique was introduced by Amos Tversky in 1972. It is a covert elimination process that involves comparing all available alternatives by aspects. The decision-maker chooses an aspect; any alternatives without that aspect are eliminated. The decision-maker repeats this process with as many aspects as needed until there remains only one alternative."
- "Preference Trees"—"In 1979 Amos Tversky and Shmuel Sattach updated the elimination by aspects technique by presenting a more ordered and structured way of comparing the available alternatives. This technique compared the alternatives by presenting the aspects in a decided and sequential order. It became a more hierarchical system in which the aspects are ordered from general to specific."
- "Authority"—"Acquiesce to a person in authority or an 'expert,' just following orders"
- "Flipism"—"flipping a coin, cutting a deck of playing cards, and other random or coincidence methods"
- "Divination"—"Prayer, tarot cards, astrology, augurs, revelation, or other forms of divination"
- "Mistrusted Authorities"—"Taking the most opposite action compared to the advice of mistrusted authorities (parents, police officers, partners)"
- "Opportunity Cost"—"calculating the opportunity cost of each options and decide the decision"
- "Bureaucratic"—"set up criteria for automated decisions."
- "Political"—"negotiate choices among interest groups."
- "Heuristic"—"where an exhaustive search is impractical, heuristic methods are used to speed up the process of finding a satisfactory solution. Examples of this method include using a rule of thumb, an educated guess, an intuitive judgment, or common sense."

Of course, there can be other everyday techniques, so the ones above are illustrative, not exhaustive.

Secondly, another main criticism of the argument on optimal decision is that optimal and non-optimal decision-making strategies are relative to different personality traits.

For instance, for the critics, "some research has suggested that satisficing/maximizing and other decision-making strategies, like personality traits, have a strong genetic component and endure over time. This genetic influence on decision-making behaviors has been found through classical twin studies, in which decision-making tendencies are self-reported by pairs of twins and then compared between monozygotic and dizygotic twins. This implies that people can be categorized into 'maximizers' and 'satisficers,' with some people landing in between." (WK 2012u; I. Simonson 2011)

Thirdly, still another main criticism of the argument on optimal decision is that its scientific objectivity is questionable.

For instance, "Robert A. Dahl wrote in 1947 that a science of organizations was impossible due to cultural differences among organizations. Similarly, a 1957 review of the second edition questioned whether scientific studies of important aspects of organizational decision-making 'under controlled conditions' could ever be undertaken." (WK 2012rr; R. Golembiewski 1988; E. Banfield 1957)

Fourthly, an additional main criticism of the argument on optimal decision is that it is often taken out of context.

For instance, for the critics, "while the rational...model was innovative at its conception, the concepts are controversial and questionable processes today.....Rather than conceptualising human agents as rational planners, Lucy Suchman [2007] argues, agents can better be understood as engaging in situated action. Going further, Guy Benveniste [1994] argued that the rational model could not be implemented without taking the political context into account." (WK 2012pp)

Fifthly, another additional main criticism of the argument on optimal decision is that it has many questionable assumptions.

For instance, for the critics, hereafter are some examples of questionable assumptions embedded in an optimal decision model: (WK 2012pp)

- Ex: "requires a great deal of time."
- Ex: "requires great deal of information."

- Ex: "assumes rational, measurable criteria are available and agreed upon."
- Ex: "assumes accurate, stable and complete knowledge of all the alternatives, preferences, goals and consequences."
- Ex: "assumes a rational, reasonable, non-political world."

Of course, these are just some examples, as there can be more.

Sixthly, still another additional main criticism of the argument on optimal decision is that it does not give sufficient attention to non-rational factors.

For instance, for the critics, it is also important to include such non-rational factors like "the role of emotions and habit in organizations." (WK 2012rr; M. Cohen 2007)

And seventhly, one last main criticism of the argument on optimal decision is that it ignores power and corruption.

For instance, for the critics, the argument for optimal decision often ignores "how immoral behavior, such as police corruption, can occur within organizations." (WK 2012rr; J. Gow 2003)

Beyond Optimal Decision

These criticisms of optical decision, however, do not imply a total rejection of its usefulness but serve to show us the two opposing sides of the debate, such that the possibility (or impossibility) and desirability (or undesirability) of objectivity (from the perspective of nature with objectivity, organizational behavior, and optimal decision as a case study here) are not to the extent that the respective defenders would like us to believe.

Furthermore, the analysis of objectivity, organizational behavior, and optimal decision sheds some light on the ontological principles in existential dialectics, and good examples include the formalness-informalness principle, the absoluteness-relativeness principle, the partiality-totality principle, the predictability-unpredictability principle, the explicability-inexplicability principle, the fiction-reality principle, the cognitiveness-noncognitiveness principle, the finiteness-transfiniteness principle, the preciseness-vagueness principle, the simpleness-complicatedness principle, the openness-hiddenness principle, the denseness-emptiness principle, the change-constancy principle, the order-chaos principle, the slowness-quickness principle, the expansion-contraction principle, the theory-praxis principle, the convention-novelty principle, the evolution-transformation principle, the symmetry-asymmetry principle, the softness-hardness principle, the seriousness-playfulness principle, the activeness-inactivess principle, the selfness-otherness principle, the regression-progression principle, the same-difference principle, the stability-reaction principle, the functionality-nonfunctionality principle, and the intentionality-nonintentionality principle.

For instance, in relation to the formalness-informalness principle, if there is formalness (e.g., the formal logical requirement of soundness and completeness in a system of ideas, as in the argument for optimal decision), there is informalness (e.g., the non-formal existence of unsoundness and incompleteness in the argument for optimal decision, as shown in the criticisms that optimization is seldom achieved in real life because decision making can be more intuitive and less formal, that optimal and non-optimal decision-making strategies are relative to different personality traits, that its scientific objectivity is questionable, that it is often taken out of context, that it has many questionable assumptions, that it does not give sufficient attention to non-rational factors, and that it ignores power and corruption). And the reverse direction also holds true.

In relation to the absoluteness-relativeness principle, if there is absoluteness (e.g., the absolute view on optimal decision making by its advocates), there is relativeness (e.g., what is right for the advocates of optimal decision making is not necessarily so for those critics who prefer satisficing decision making). And the reverse direction also holds true.

In relation to the partiality-totality principle, if there is partiality (e.g., the partial view about decision making, if looked from the vantage point of optimal decision), there is totality (e.g., the more holistic view about decision making, such that the whole is not the sum of its parts, that is, it cannot be reduced to the sum of opposing views by those for optimal decision, satisficing decision, and so on, since there will emerge new views in the future not yet known today, just as there are already alternative views nowadays which do not side exclusively with any of them, as is the analysis here). And the reverse direction also holds true.

In relation to the predictability-unpredictability principle, if there is predictability (e.g., the predictable tendency of an optimal decision in situation of uncertainty to make use of a probabilistic approach, with "the probability distribution for the possible outcomes described by the conditional probability density *p [o|d]*), there is unpredictability (e.g., the more difficult task to predict exactly the extent to which "the outcome of a particular decision" will be on a particular occasion, in a situation of uncertainty). And the reverse direction also holds true.

In relation to the explicability-inexplicability principle, if there is explicability (e.g., the explanation by the advocates that an optimal decision is desirable because it gives "the best option available"), there is inexplicability (e.g., the lack of sufficient explanation by the advocates of why an optimal decision is necessarily desirable, because, for the critics, optimization is seldom achieved in real life because decision making can be more intuitive and less formal, and non-optimal decision-making strategies are relative to different personality traits, its scientific objectivity is questionable, and it has many questionable assumptions). And the reverse direction also holds true.

In relation to the fiction-reality principle, if there is fiction (e.g., the fictional aspect of optimal decision, insofar as its scientific objectivity is questionable and it has many questionable assumptions), there is reality (e.g., the realistic aspect of optimal decision, insofar as it "attempts to find the best option available," at least as an ideal point of reference). And the reverse direction also holds true.

In relation to the cognitiveness-noncognitiveness principle, if there is cognitiveness (e.g., a view of things on the basis of reason and evidences, as shown in the different steps of a rational decision making model to follow, aforecited), there is noncognitiveness (e.g., a view of things on the basis of non-cognitive factors like envy, jealousy, power, nationality, race, gender, age, class, greed, lust, status, faith, anger, sadness, joy, fear, wish, etc., as shown in the dependence on "authority" by those who use it as an everyday decision technique and "acquiesce to a person in authority or an 'expert,' just following orders," not on the basis of reason and evidences). And the reverse direction also holds true.

In relation to the finiteness-transfiniteness principle, if there is finiteness (e.g., the finite number of bits of information needed for a satisficing decision), there is transfiniteness (e.g., the transfinite number of bits of information needed for an optimal decision, by those who follow a strict interpretation of the rational decision making model). And the reverse direction also holds true.

In relation to the preciseness-vagueness principle, if there is preciseness (e.g., the precise identification of 7 decision making steps in the article aforecited), there is vagueness (e.g., the vagueness in the identification of the 7 decision making steps in the article aforecited, since it is not clear why there must be only 7, not 8, 9, 10, and so on). And the reverse direction also holds true.

In relation to the simpleness-complicatedness principle, if there is simpleness (e.g., the relatively simple analysis of optimal decision by the advocates), there is complicatedness (e.g., the relatively more complicated analysis of optimal decision, by challenging the claims and assumptions, as shown in the problems pointed out by the critics as cited). And the reverse direction also holds true.

In relation to the openness-hiddenness principle, if there is openness (e.g., the open exploration of achieving optimality in decision making within organizational behavior), there is hiddenness (e.g., the hidden bias in the quest for achieving optimality in decision making within organizational behavior, because its scientific objectivity is questionable, it is often taken out of context, it has many questionable assumptions, it does not give sufficient attention to non-rational factors, and it ignores power and corruption). And the reverse direction also holds true.

In relation to the denseness-emptiness principle, if there is denseness (e.g., the relatively denser concentration of concerns with optimality for those in the rational decision making model), there is emptiness (e.g., the relatively less dense, or more empty, concentration of concerns with optimality for those in the satisficing decision making model). And the reverse direction also holds true.

In relation to the change-constancy principle, if there is change (e.g., the ever new ways to deal with decision making, as shown by optimal decision making in organizational behavior), there is constancy (e.g., the ever constant existence of different problems with the ever new ways to deal with decision making, as shown in the problems with optimal decision making in organizational behavior pointed out by the critics aforecited). And the reverse direction also holds true.

In relation to the order-chaos principle, if there is order (e.g., the more or less orderly view about decision making, if looked from the vantage point of following the mathematical equations for optimal decision aforecited), there is chaos (e.g., the more or less chaotic view about decision making, if looked from the vantage point of "flipism," "divination," or other chanceful techniques). And the reverse direction also holds true.

In relation to the slowness-quickness principle, if there is slowness (e.g., the relatively slower speed of making a decision if an optimal decision making model is used), there is quickness (e.g., the relatively quicker speed of making a decision if a satisficing decision making model is used). And the reverse direction also holds true.

In relation to the expansion-contraction principle, if there is expansion (e.g., the relatively more developed ability to deal with optimality for the practitioners of the rational decision making model), there is contraction (e.g., the relatively less developed ability to deal with authority as an everyday decision making for the practitioners of the rational decision making model). And the reverse direction also holds true.

In relation to the theory-praxis principle, if there is theory (e.g., the theoretical construction of "probability distribution" in mathematics), there is praxis (e.g., the practical application of "probability distribution" in mathematics to the field of organizational behavior for optimal decision making). And the reverse direction also holds true.

In relation to the convention-novelty principle, if there is convention (e.g., the conventional wisdom about optimal decision making in organizational behavior in the middle of the 20th century), there is novelty (e.g., the alternative novel challenge to the conventional wisdom about optimal decision making in organizational behavior in the middle of the 20th century, by the new idea of "bounded rationality" in the 1960s and 1970s). And the reverse direction also holds true.

In relation to the evolution-transformation principle, if there is evolution (e.g., the natural evolution of humans to make decision in the state of nature), there is transformation (e.g., the technical transformation of human ability to make decision by the invention of the mathematical formulation of optimal decision). And the reverse direction also holds true.

In relation to the symmetry-asymmetry principle, if there is symmetry (e.g., the co-existence of different everyday decision techniques, like "authority" and "mistrusted authorites" in the article aforecited), there is asymmetry (e.g., some individuals follow "authority"—but others "mistrust authorities"). And the reverse direction also holds true.

In relation to the softness-hardness principle, if there is softness (e.g., some individuals "acquiesce to a person in authority or an 'expert,' just following orders"), there is hardness (e.g., some individuals take "the most opposite action compared to the advice of mistrusted authorities [parents, police officers, partners]"). And the reverse direction also holds true.

In relation to the seriousness-playfulness principle, if there is seriousness (e.g., the serious business of those who try to make good decisions), there is playfulness (e.g., the playful part of those who try to make good decisions, when they play around with different "everyday decision techniques" over time as cited earlier). And the reverse direction also holds true.

In relation to the activeness-inactivess principle, if there is activeness (e.g., the relative activeness of those who are able to make a lot of decisions based on satisficing techniques), there is inactiveness (e.g., the relative inactiveness of those who are not able to make a lot of decisions, because they follow the optimal decision making approach which requires "a great deal of time" and a "great deal of information," etc.). And the reverse direction also holds true.

In relation to the selfness-otherness principle, if there is selfness (e.g., the tendency to be mostly centered on one's self or group, as shown by those who follow the "political" everyday technique to negotiate choices for their own interest groups), there is otherness (e.g., the tendency to be mostly centered on others, as shown by those scholars who contribute to the rational decision making model, as an advancement in the knowledge of organizational behavior for humanity). And the reverse direction also holds true.

In relation to the regression-progression principle, if there is regression (e.g., the regression made by optimal decision, as shown in the problems pointed out by the critics), there is progression (e.g., the progress made by optimal decision, as shown in its contribution to the understanding of finding the best option available in decision making aforecited). And the reverse direction also holds true.

In relation to the same-difference principle, if there is similarity in outcome (e.g., the contribution to the molding and control of beliefs, values, and behaviors, regardless of whether this be done by way of rational decision-making model, or by way of bounded rationality), there is difference in outcome (e.g., the contribution to the molding and control of beliefs, values, and behaviors by way of rational decision-making model for a relatively more maximizing personalitiy type—but the contribution to the molding and control of beliefs, values, and behaviors by way of bounded rationality for a relatively more satisficing personality type). And the reverse direction also holds true.

In relation to the stability-reaction principle, if there is stability (e.g., the relatively more stable condition of the "scientific management" approach to organizational behavior in the early part of the 20th century), there is reaction (e.g., the "scientific management" approach to organizational behavior in the early part of the 20th century then led to further changes later on, when nowadays, the satisficing approach and the cultural approach gain more attention). (WK 2012nn) And the reverse direction also holds true.

In relation to the functionality-nonfunctionality principle, if there is functionality (e.g., the presence of action for which something is fitted, as shown in the rise of the "scientific management" approach to organizational behavior in the early part of the 20th century, which served the function of scientific optimism at the time, like the belief that "rationalizing the organization with precise sets of instructions and time-motion studies would lead to increased productivity"), there is nonfunctionality (e.g., the relative lack of action for which something is fitted, as shown in the rise of bounded rationality in the 1960s and 1970s, such that scientific optimism is no longer functional to the extent that it was in earlier times, due to different expectations and beliefs in our time, with a different interest in the cultural and social contexts of organizational behavior). And the reverse direction also holds true.

In relation to the intentionality-nonintentionality principle, if there is intentionality (e.g., the planning or design of something for a certain outcome, as shown in the creation of the "rationality as optimization" in the early part of the 20th century in organizational behavior for "increased productivity"), there is nonintentionality (e.g., the relative absence of planning or design of something for a certain outcome, such that the outcome can even be counter-intentional in being the opposite of what was originally planned, as shown in the rise of "bounded rationality" in our time, which proposes the opposite view that "rationality of individuals is limited by the information they have, the cognitive limitations of their minds, and the finite amount of time they have to make a decision," contrary to the original mission of "rationality as optimization"). (WK 2012ss) And the reverse direction also holds true.

And in relation to the survivability-nonsurvivability principle, if there is survivability (e.g., the survival of "bounded rationality" since 1960s unto now), there is non-survivability (e.g., the non-survival of "the rational planning model," which "has fallen out of mass use as of the last decade"). (WK 2012pp) And the reverse direction also holds true.

Objectivity and Culture

Objectivity, when examined from the perspective of culture, can cast an illuminating light on its possibility (or impossibility) and desirability (or undesirability), and this can be shown by way of two case studies, namely, (2.5.1) objectivity, propositional calculus, and inference rules, and (2.5.2) objectivity, history, and common causes—to be addressed hereafter, in that order.

Objectivity, Propositional Calculus, and Inference Rules

An excellent case study here to understand objectivity from the perspective of culture has to do with objectivity, propositional calculus, and inference rules.

An argument in the literature is that a system of inference rules and axioms (in propositional calculus) can yield true propositions.

In mathematical logic, the term "propositional calculus" is also known as "propositional logic," "sentential calculus," or "sentential logic"—and refers to "a formal system in which formulas of a formal language may be interpreted as representing propositions. A system of inference rules and axioms allows certain formulas to be derived, called theorems; which may be interpreted as true propositions....Truth-functional propositional logic is a propositional logic whose interpretation limits the truth values of its propositions to two, usually true and false. Truth-functional propositional logic and systems isomorphic to it are considered to be zeroth-order logic." (WK 2012tt)

History of Propositional Calculus

Historically, in Western philosophy, propositional logic was developed from the earlier "syllogistic logic which was focused on terms" and "was developed into a formal logic by Chrysippus and expanded by the Stoics. The logic was focused on propositions....However, later in antiquity, the propositional logic developed by the Stoics was no longer understood. As a result, the system was essentially reinvented by Peter Abelard." (WK 2012tt; SEP 2012; J. Marenbon 2007)

Later, "propositional logic was eventually refined using symbolic logic. Gottfried Leibniz has been credited with being the founder of symbolic logic for his work with the calculus ratiocinator. Although his work was the first of its kind, it was unknown to the larger logical community. As a result, many of the advances achieved by Leibniz were reachieved by logicians like George Boole and Augustus De Morgan completely independent of Leibniz." (WK 2012tt)

Then, "just as propositional logic can be seen as an advancement from the earlier syllogistic logic, Gottlob Frege's predicate logic was an advancement from the earlier propositional logic. Predicate logic has been described as combining 'the distinctive features of syllogistic logic and propositional logic.' As a result, it ushered a new era in the history of logic. However, advances in propositional logic were still made after Frege. These include Natural Deduction, Truth-Trees and Truth-Tables. Natural deduction was invented by Jan Łukasiewicz. Truth-Trees were invented by Evert Willem Beth. The invention of truth-tables, however, is of controversial attribution." (WK 2012tt; P. Hurley 2007)

Basic Concepts of Propositional Calculus

In general, "a calculus is a formal system that consists of a set of syntactic expressions (*well-formed formulæ* or *wffs*), a distinguished subset of these expressions (axioms), plus a set of formal rules that define a specific binary relation, intended to be interpreted as logical equivalence, on the space of expressions. When the formal system is intended to be a logical system, the expressions are meant to be interpreted as statements, and the rules, known as inference rules, are typically intended to be truth-preserving. In this setting, the rules (which may include axioms) can then be used to derive ('infer') formulæ representing true statements from given formulæ representing true statements....In addition a semantics may be given which defines truth and valuations (or interpretations)." (WK 2012tt)

Then, "the language of a propositional calculus consists of" the following: (WK 2012tt)

- "a set of primitive symbols, variously referred to as atomic formulae, placeholders, proposition letters, or variables"
- "a set of operator symbols, variously interpreted as logical operators or logical connectives"

Although there are "many different formulations" of "a standard propositional calculus," they "differ in the details of" the following (as summarized in *Table 2.8*): (WK 2012tt)

- "the...language"—"that is, the particular collection of primitive symbols and operator symbols"
- "the set of axioms"—that is, "distinguished formulæ"
- "the set of inference rules"—that is, the entailment of a conclusion based on a set of premises

Then, there are basic concepts, as shown below (and summarized in *Table 2.8*):

- "Truth-Values"—"all propositions have exactly one of two truth-values: true or false [in propositional calculus]. To take an example, let P be the proposition that it is raining outside. This will be true if it is raining outside and false otherwise." (WK 2012tt)
- "Truth-Functional Operators"—For instance, "$\neg P$" is "the negation of P, which can be thought of as the denial of P. In the example above, expresses that it is not raining outside, or by a more standard reading: 'It is not the case that it is raining outside.' When P is true,

$\neg P$ is false; and when P is false, $\neg P$ is true. $\neg\neg P$ always has the same truth-value as P." (WK 2012tt)

- "Conjunction"—"is a truth-functional connective which forms a proposition out of two simpler propositions, for example, P and Q. The conjunction of P and Q is written $P \wedge Q$, and expresses that each are true. We read $P \wedge Q$ as 'P and Q.' For any two propositions, there are four possible assignments of truth values": (WK 2012tt)

"P is true and Q is true."
"P is true and $Q$is false."
"P is false and Q is true."
"P is false and Q is false."

Thus, "the conjunction of P and Q is true in case 1 and is false otherwise." (WK 2012tt)

- "Disjunction"—"resembles conjunction in that it forms a proposition out of two simpler propositions. We write it $P \vee Q$, and it is read 'P or Q.' It expresses that either P or Q is true. Thus, in the cases listed above, the disjunction of P or Q is true in all cases except 4." (WK 2012tt)
- "Material Conditional"—"joins two simpler propositions, and we write $P \rightarrow Q$, which is read 'if P then Q.' The proposition to the left of the arrow is called the antecedent and the proposition to the right is called the consequent….It expresses that Q is true whenever P is true. Thus it is true in every case above except case 2, because this is the only case when P is true but Q is not….The material conditional is often confused with physical causation. The material conditional, however, only relates two propositions by their truth-values—which is not the relation of cause and effect. It is contentious in the literature whether the material implication represents logical causation." (WK 2012tt)
- "Biconditional" —"joins two simpler propositions, and we write $P \leftrightarrow Q$, which is read 'P if and only if Q.' It expresses that P

and Q have the same truth-value, thus P if and only if Q is true in cases 1 and 4, and false otherwise." (WK 2012tt)

Surely, this summary is merely introductory for illustration, as things can be more complicated.

Generic Description of Propositional Calculus

A propositional calculus can be described generally as "a formal system $\mathcal{L} = \mathcal{L}(\mathrm{A}, \Omega, \mathrm{Z}, \mathrm{I})$," which means the following: (WK 2012tt)

- "*The Alpha Set* A"—"is a finite set of elements called *proposition symbols* or *propositional variables*. Syntactically speaking, these are the most basic elements of the formal language $\mathcal{L}$, otherwise referred to as *atomic formulæ* or *terminal elements*. In the examples to follow, the elements of A are typically the letters p, q, r, and so on." (WK 2012tt)
- "*The Omega Set* Ω"—"is a finite set of elements called operator symbols or logical connectives….In the more familiar propositional calculi, Ω is typically partitioned as follows": (WK 2012tt)

$$\Omega_1 = \{\neg\},$$
$$\Omega_2 \subseteq \{\wedge, \vee, \rightarrow, \leftrightarrow\}.$$

And "a frequently adopted convention treats the constant logical values as operators of arity zero," as shown below: (WK 2012tt)

$$\Omega_0 = \{0, 1\}.$$

But "some writers use the tilde (~), or N, instead of $\neg$; and some use the ampersand (&), the prefixed K,…instead of $\wedge$. Notation varies even more for the set of logical values, with symbols like {false, true}, {F, T}, or $\{\bot, \top\}$ all being seen in various contexts instead of {0, 1}." (WK 2012tt)

- "*The Zeta Set* Z"—"is a finite set of transformation rules that are called *inference rules* when they acquire logical applications."

- "*The Iota Set* $\mathcal{I}$"—"is a finite set of *initial points* that are called *axioms* when they receive logical interpretations." (WK 2012tt)

In addition, "the language of $\mathcal{L}$, also known as its set of *formulæ*, *well-formed formulas* or *wffs*, is inductively defined by the following rules": (WK 2012tt)

- Rule 1—"Base: Any element of the alpha set A is a formula of $\mathcal{L}$."
- Rule 2—"If $p_1, p_2, \ldots, p_j$ are formulæ and f is in Ω_j, then $(f(p_1, p_2, \ldots, p_j))$ is a formula."
- Rule 3—"Closed: Nothing else is a formula of $\mathcal{L}$."

Then, "repeated applications of these rules permits the construction of complex formulæ," and hereafter are some examples: (WK 2012tt)

- "By rule 1, p is a formula."
- "By rule 2, $\neg p$ is a formula."
- "By rule 1, q is a formula."
- "By rule 2, $(\neg p \vee q)$ is a formula."

Again, these rules are illustrative, not exhaustive.

Soundness and Completeness of Inference Rules

Rules like this have "crucial properties," like "soundness" and "completeness," in that "they are sound and complete. Informally this means that the rules are correct and that no other rules are required." (WK 2012tt)

A formalization is possible as to "what it means for a formula ϕ to be implied by a certain set S of formulae. Informally this is true if in all worlds that are possible given the set of formulae S the formula ϕ also holds. This leads to the following formal definition: We say that a set S of wffs semantically entails (or implies) a certain wff ϕ if all truth assignments that satisfy all the formulae in S also satisfy ϕ." (WK 2012tt)

Then, "we define syntactical entailment such that ϕ is syntactically entailed by S if and only if we can derive it with the inference rules that were presented above in a finite number of steps. This allows us to formulate exactly what it means for the set of inference rules to be sound and complete" in what follows: (WK 2012tt)

- "Soundness"—"If the set of wffs $\mathcal{S}$ syntactically entails wff ϕ, then $\mathcal{S}$, semantically entails ϕ."
- "Completeness"—"If the set of wffs $\mathcal{S}$ semantically entails wff ϕ, then $\mathcal{S}$ syntactically entails ϕ."

Again, this summary is introductory for illustration.

Interpretation of Truth-Functional Propositional Calculus

In the end, "a truth-functional propositional calculus $\mathcal{P}$" can be interpreted by "an assignment to each propositional symbol of $\mathcal{P}$ of one or the other (but not both) of the truth values truth (T) and falsity (F), and an assignment to the connective symbols of $\mathcal{P}$ of their usual truth-functional meanings. An interpretation of a truth-functional propositional calculus may also be expressed in terms of truth tables." (WK 2012tt; G. Hunter 1971)

For instance, "for *n* distinct propositional symbols there are 2^n distinct possible interpretations. For any particular symbol *a*, for example, there are $2^1=2$ possible interpretations," as shown below: (WK 2012tt)

- "*a* is assigned T, or"
- "*a* is assigned F."

And "for the pair *a, b* there are $2^2=4$ possible interpretations," as shown below: (WK 2012tt)

- "both are assigned T,"
- "both are assigned F,"
- "*a* is assigned T and *b* is assigned F, or"
- "*a* is assigned F and *b* is assigned T."

Because "$\mathcal{P}$ has $\aleph_0$, that is, denumerably many propositional symbols, there are $2^{\aleph_0} = \mathfrak{c}$, and therefore uncountably many distinct possible interpretations of $\mathcal{P}$." (WK 2012tt)

Problems with Propositonal Calculus

However, for the critics, propositional calculus has some major problems. Consider, for illustration, a few examples below (as summarized in *Table 2.8*).

Firstly, one main criticism of propositional calculus is that the quest for completeness in mathematical logic is problematic.

For instance, "Gödel's incompleteness theorems are two theorems of mathematical logic that establish inherent limitations of all but the most trivial axiomatic systems capable of doing arithmetic. The theorems, proven by Kurt Gödel in 1931, are important both in mathematical logic and in the philosophy of mathematics. The two results are widely, but not universally, interpreted as showing that Hilbert's program to find a complete and consistent set of axioms for all mathematics is impossible, giving a negative answer to Hilbert's second problem." (WK 2012uu)

Secondly, another main criticism of propositional calculus is that it does not include modal logic.

For instance, for the critics, "modal logic also offers a variety of inferences that cannot be captured in propositional calculus." (WK 2012tt) The reason is that "modal logic is a type of formal logic primarily developed in the 1960s that extends classical propositional and predicate logic to include operators expressing modality. Modals—words that express modalities—qualify a statement. For example, the statement 'John is happy' might be qualified by saying that John is usually happy, in which case the term 'usually' is functioning as a modal. The traditional alethic modalities, or modalities of truth, include possibility ('Possibly, p,' 'It is possible that p'), necessity ('Necessarily, p,' 'It is necessary that p'), and impossibility ('It is impossible that p'). Other modalities that have been formalized in modal logic include temporal modalities, or modalities of time (notably, 'It was the case that p,' 'It has always been that p,' 'It will be that p,' 'It will always be that p'), deontic modalities (notably, 'It is obligatory that p,' and 'It is permissible that p'), epistemic modalities, or modalities of knowledge ('It is known that p') and doxastic modalities, or modalities of belief ('It is believed that p')." (WK 2012uu; A. Prior 1955 & 1967; N. Rescher 1968 & 1971; J. Hinntikka 1962)

Thirdly, still another main criticism of propositional calculus is that it does not include alternative many-valued logics.

For instance, for the critics, "many-valued logics are those allowing sentences to have values other than *true* and *false*....For example, *neither* and *both* are standard 'extra values'; 'continuum logic' allows each sentence to have any of an infinite number of 'degrees of truth' between *true* and *false*....These logics often require calculational devices quite distinct from propositional calculus." (WK 2012tt) In *FPHML* (and other books of mine), I already introduced my version of many-value logic known as "existential dialectics."

Fourthly, an additional main criticism of propositional calculus is that it does not pay attention to context-specific rules of inference.

For instance, for the critics, there is an alternative theory of "the cognitive processes that human reasoning is based on," which is that "people rely on domain-specific or content-sensitive rules of inference." (WK 2012vv; L. Cosmides 2005)

Fifthly, another additional main criticism of propositional calculus is that it does not give weight to the factors of power and social action.

For instance, the Habermasian perspective is to introduce "social context" in conjunction with "social power" in the discourse on rationality—just as and the feminist perspective is to criticize "masculine values," which often distort the social construction of inference rules for "the maintenance of male power." (WK 2012v)

Sixthly, still another additional main criticism of propositional calculus is that it is too formal and thus ignores less formal and more intuitive approaches to human thinking.

For instance, as already discussed in *Sec. 4.2* on optimal decision making, for the critics, "in practice, few people verify that their decisions are optimal, but instead use...approaches to make decisions that are 'good enough,'" especially with "everyday decision techniques." (WK 2012t, 2012oo & 2012qq; W. Bell 1956; R. Batley 2006; A. Tversky 1972 & 1979)

And seventhly, one last main criticism of propositional calculus is that it is too simplistic, because there are are higher-order logics.

For instance, for the critics, "first-order logic (aka first-order predicate logic) results when the 'atomic sentences' of propositional logic are broken up into terms, variables, predicates, and quantifiers, all keeping the rules of propositional logic with some new ones introduced....For example, from 'All dogs are mammals' we may infer 'If Rover is a dog then Rover is a mammal'....With the tools of first-order logic it is possible to formulate a number of theories, either with explicit axioms or by rules of inference, that can themselves be treated as logical calculi. Arithmetic is the best known of these; others include set theory and mereology. Second-order logic and other higher-order logics are formal extensions of first-order logic. Thus, it makes sense to refer to propositional logic as 'zeroth-order logic,' when comparing it with these logics." (WK 2012tt)

Beyond Propositional Calculus

These criticisms of propositional calculus, however, do not reject its usefulness but serve to show us the two opposing sides of the debate, such that the possibility (or impossibility) and desirability (or undesirability) of objectivity (from the perspective of culture with objectivity, propositional calculus, and inference rules as a case study here) are not to the extent that the respective defenders would like us to believe.

Furthermore, the analysis of objectivity, propositional calculus, and inference rules can show us valuable information about the ontological principles in existential dialectics, and good examples include the formalness-informalness principle, the absoluteness-relativeness principle, the partiality-totality principle, the predictability-unpredictability principle, the explicability-inexplicability principle, the fiction-reality principle, the cognitiveness-noncognitiveness principle, the finiteness-transfiniteness principle, the preciseness-vagueness principle, the simpleness-complicatedness principle, the openness-hiddenness principle, the denseness-emptiness principle, the change-constancy principle, the order-chaos principle, the slowness-quickness principle, the expansion-contraction principle, the theory-praxis principle, the convention-novelty principle, the evolution-transformation principle, the symmetry-asymmetry principle, the softness-hardness principle, the seriousness-playfulness principle, the activeness-inactivess principle, the selfness-otherness principle, the regression-progression principle, the same-difference principle, the stability-reaction principle, the functionality-nonfunctionality principle, and the intentionality-nonintentionality principle.

For instance, in relation to the formalness-informalness principle, if there is formalness (e.g., the formal logical requirement of soundness and completeness in a system of ideas, as in propositional calculus), there is informalness (e.g., the non-formal existence of unsoundness and incompleteness in propositional calculus, as shown in the criticisms that the quest for completeness in mathematical logic is problematic, that it does not include modal logic, that it does not include alternative many-valued logics, that it does not pay attention to context-specific rules of inference, that it does not give weight to the factors of power and social action, that it is too formal and thus ignores less formal and more intuitive approaches to human thinking, and that it is too simplistic because there are are higher-order logics). And the reverse direction also holds true.

In relation to the absoluteness-relativeness principle, if there is absoluteness (e.g., the absolute "assignment to each propositional symbol of $\mathcal{P}$ of one or the other [but not both] of the truth values truth [T] and falsity [F]" in propositional calculus), there is relativeness (e.g., what is right for propositional calculus to assign "one or the other [but not both] of the truth values truth [T] and falsity [F]" is not necessarily so for "many-valued logics" which "allow...values other than *true* and *false*"). And the reverse direction also holds true.

In relation to the partiality-totality principle, if there is partiality (e.g., the partial view about inference rules in propositional calculus), there is totality (e.g., the more holistic view about inference rules, such that the whole is not the sum of its parts, that is, it cannot be reduced to the sum of opposing views in propositional calculus, in higher-order logics, in modal logic, in many-valued logics, in incompleteness theorem, in everyday decision techniques, etc., since there will emerge new views in the future not yet known today, just as there are already alternative views nowadays which do not side exclusively with any of them, as is the analysis here). And the reverse direction also holds true.

In relation to the predictability-unpredictability principle, if there is predictability (e.g., the predictable tendency of propositional calculus to be refined by Predicate Logic, Natural Deduction, Truth-Trees, Truth-Tables, and so on), there is unpredictability (e.g., the more difficult task to predict exactly the extent to which propositional calculus will be refined in a particular way on a particular future occasion, since there will like be new logics in the future not yet knonw today). And the reverse direction also holds true.

In relation to the explicability-inexplicability principle, if there is explicability (e.g., the explanation by those in propositional calculus of the importance of "assignment to each propositional symbol of $\mathcal{P}$ of one or the other [but not both] of the truth values truth [T] and falsity [F]," on the basis of classical laws of thought for two-valued logic), there is inexplicability (e.g., the lack of sufficient explanation by those in propositional calculus of why the two-value logic is necessarily desirable, since, for the critics, there are alternative "many-valued logics" which "allow...values other than *true* and *false*"). And the reverse direction also holds true.

In relation to the fiction-reality principle, if there is fiction (e.g., the fictional aspect of completeness in propositional calculus, insofar as "Gödel's incompleteness theorems are two theorems of mathematical logic that establish inherent limitations of all but the most trivial axiomatic systems capable of doing arithmetic"), there is reality (e.g., the realistic aspect of completeness in propositional calculus, insofar as the 3 rules aforecited are complete). And the reverse direction also holds true.

In relation to the cognitiveness-noncognitiveness principle, if there is cognitiveness (e.g., a view of things on the basis of reason and evidences, as shown in the use of axioms and inference rules in propositional calculus), there is noncognitiveness (e.g., a view of things on the basis of non-cognitive factors like envy, jealousy, power, nationality, race, gender, age, class, greed, lust, status, faith, anger, sadness, joy, fear, wish, etc., as shown in the assumption of two-valued logic in propositional calculus, even though there are alternative many-valued logics). And the reverse direction also holds true.

In relation to the finiteness-transfiniteness principle, if there is finiteness (e.g., the finite number of possible interpretations, when $2^1=2$), there is transfiniteness (e.g., the transfinite number of possible interpretations, when $2^{\aleph_0} = \mathfrak{c}$). And the reverse direction also holds true.

In relation to the preciseness-vagueness principle, if there is preciseness (e.g., the precise identification of 3 inference rules in the article aforecited), there is vagueness (e.g., the vagueness in the identification of the 3 inference rules in the article aforecited, since it is not clear why there must be only 3, not 4, 6, 7, 9, 13, and so on). And the reverse direction also holds true.

In relation to the simpleness-complicatedness principle, if there is simpleness (e.g., the relatively simple analysis of logic from the vantage point of propositional logic as "zeroth-order logic"), there is complicatedness (e.g., the relatively more complicated analysis of logic from the vantage point of higher-order logics). And the reverse direction also holds true.

In relation to the openness-hiddenness principle, if there is openness (e.g., the open exploration of two truth values in propositional calculus), there is hiddenness (e.g., the hidden bias in the two truth values in propositional calculus, since there are "values other than *true* and *false*" in many-valued logics). And the reverse direction also holds true.

In relation to the denseness-emptiness principle, if there is denseness (e.g., the relatively denser concentration of concerns with "zeroth-order logic" in propositional logic), there is emptiness (e.g., the relatively less dense, or more empty, concentration of concerns with "zeroth-order logic" in higher-order logics). And the reverse direction also holds true.

In relation to the change-constancy principle, if there is change (e.g., the ever new ways to understand logic, as shown in propositional logic), there is constancy (e.g., the ever constant existence of different problems with the ever new ways to understand logic, as shown in the problems with propositional logic pointed out by the critics). And the reverse direction also holds true.

In relation to the order-chaos principle, if there is order (e.g., the more or less orderly view about logic, if looked only from the vantage point of propositional logic), there is chaos (e.g., the more or less chaotic view about logic, if looked from the multiple, often conflicting, vantage points like propositional calculus, higher-order logics, modal logic, many-valued logics, incompleteness theorem, everyday decision techniques, mental models, etc.). And the reverse direction also holds true.

In relation to the slowness-quickness principle, if there is slowness (e.g., the relatively slower speed to calculate possible interpretations if when $2^{\aleph_0} = \mathfrak{c}$), there is quickness (e.g., the relatively quicker speed to calculate possible interpretations if $2^1=2$). And the reverse direction also holds true.

In relation to the expansion-contraction principle, if there is expansion (e.g., the relative spread of propositional calculus in the 19th and 20th centuries), there is contraction (e.g., the relative decline of traditional syllogistic logic in the 19th and 20th centuries). And the reverse direction also holds true.

In relation to the theory-praxis principle, if there is theory (e.g., the theoretical construction of "graphs" as a subfield of mathematics), there is praxis (e.g., the practical application of "graphs" as a subfield of mathematics to another subfield like propositional logic for "graphical calculi"). And the reverse direction also holds true.

In relation to the convention-novelty principle, if there is convention (e.g., the tradition of syllogistic logic in ancient Greece, like the Aristotelian syllogisms), there is novelty (e.g., the alternative novel challenge to the tradition of syllogistic logic in ancient Greece, by the new idea of propositional logic in modern times). And the reverse direction also holds true.

In relation to the evolution-transformation principle, if there is evolution (e.g., the natural evolution of humans to think in the state of nature in a primitive way), there is transformation (e.g., the technical transformation of human ability to think by the invention of propositional calculus). And the reverse direction also holds true.

In relation to the symmetry-asymmetry principle, if there is symmetry (e.g., the co-existence of humans and animals), there is asymmetry (e.g., humans have the intellectual power to create propositional logic—but animals do not). And the reverse direction also holds true.

In relation to the softness-hardness principle, if there is softness (e.g., the defense of syllogistic logic by Aristotle in ancient Greece), there is hardness (e.g., the critique of syllogistic logic by Gottlob Frege for propositional calculus in modern times). And the reverse direction also holds true.

In relation to the seriousness-playfulness principle, if there is seriousness (e.g., the serious business of those who try to understand inference rules), there is playfulness (e.g., the playful part of those who try to understand inference rules, when they play around with different options over time, like propositional calculus, higher-order logics, modal logic, many-valued logics, incompleteness theorem, everyday decision techniques, mental models, etc.). And the reverse direction also holds true.

In relation to the activeness-inactivess principle, if there is activeness (e.g., the relative activeness of those followers of propositional calculus in the community of its advocates), there is inactiveness (e.g., the relative inactiveness of those followers of propositional calculus, when they are not welcome in the community of those who follow the alternative theory of "domain-specific or content-sensitive rules of inference"). And the reverse direction also holds true.

In relation to the selfness-otherness principle, if there is selfness (e.g., the tendency to be mostly centered on one's self or group, as shown in the "controversy" surrounding the credit for "the invention of truth-tables," as there has been disputes as to who should receive the honor to invent it first), there is otherness (e.g., the tendency to be mostly centered on others, as shown in the efforts of Gottfried Leibniz, who "has been credited with being the founder of symbolic logic for his work with the calculus ratiocinator" but "it was unknown to the larger logical community" at the time and was not much appreciated then). (WK 2012tt) And the reverse direction also holds true.

In relation to the regression-progression principle, if there is regression (e.g., the regression made by propositional calculus, as shown in the problems pointed out by the critics), there is progression (e.g., the progress made by propositional calculus, as shown in its contribution to the understanding of inference rules aforecited). And the reverse direction also holds true.

In relation to the same-difference principle, if there is similarity in outcome (e.g., the contribution to the molding and control of beliefs, values, and behaviors, regardless of whether this be done by way of two-valued logic, or by way of many-valued logics), there is difference in outcome (e.g., the contribution to the molding and control of beliefs, values, and behaviors by way of two-valued logic for a relatively more simple worldview—but the contribution to the molding and control of beliefs, values, and behaviors by way of many-valued logics for a relatively more complex worldview). And the reverse direction also holds true.

In relation to the stability-reaction principle, if there is stability (e.g., the relatively more stable condition of syllogistic logic in the older days, when it was influential in ancient Greece), there is reaction (e.g., syllogistic logic in the older days then led to further changes later on, when nowadays it is superseded by propositional calculus and other logics). And the reverse direction also holds true.

In relation to the functionality-nonfunctionality principle, if there is functionality (e.g., the presence of action for which something is fitted, as shown in the efforts by mathematicians in "the logicist period from the Begriffsschrift of Frege to the *Principia Mathematica* of Russell and Whitehead," which served the function to "incorporate the logic of all mathematical and scientific discourse in a single unified system"), there is nonfunctionality (e.g., the relative lack of action for which something is fitted, as shown in "Gödel's incompleteness theorems," which later challenged Hilbert's impossible task to find "a complete and consistent set of axioms for all mathematics," such that this futile quest for "the logic of all mathematical and scientific discourse in a single unified system" is no longer much functional in our time, because of a more critical rejection of any attempt to find "the foundations of mathematics" nowadays). (WK 2012ww) And the reverse direction also holds true.

In relation to the intentionality-nonintentionality principle, if there is intentionality (e.g., the planning or design of something for a certain outcome, as shown in "Hilbert's program to find a complete and consistent set of axioms for all mathematics"), there is nonintentionality (e.g., the relative absence of planning or design of something for a certain outcome, such that the outcome can even be counter-intentional in being the opposite of what was originally planned, as shown in "Gödel's incompleteness theorems…that establish inherent limitations of all but the most trivial axiomatic systems capable of doing arithmetic" and show that "Hilbert's program to find a complete and consistent set of axioms for all mathematics is impossible, giving a negative answer to Hilbert's second problem," contrary to the original mission of Hilbert). And the reverse direction also holds true.

And in relation to the survivability-nonsurvivability principle, if there is survivability (e.g., the survival of propositional logic in modern times), there is non-survivability (e.g., the non-survival of traditional syllogistic logic in modern times, as "propositional logic" is now "seen as an advancement from the earlier syllogistic logic"). (WK 2012tt) And the reverse direction also holds true.

Objectivity, History, and Common Causes

Another way to evaluate the possibility (or impossibility) and desirability (or undesirability) of objectivity from the perspective of culture concerns objectivity, history, and common causes.

An argument in the literature is that there are certain historical causes which can be quantitied and predicted, and these causes are known as "common causes," which are "the usual, historical, quantifiable variation in a system…, being recognised and discussed as early as 1703 by Gottfried Leibniz." (WK 2012x) This common-cause variation is "predictable, at least approximately in frequency. This variation is evident from the experience base." (WK 2012x)

Concepts of Common Cause

Back in 1703, "Jacob Bernoulli wrote to Gottfried Leibniz to discuss their shared interest in applying mathematics and probability to games of chance. Bernoulli speculated whether it would be possible to gather mortality data from gravestones and thereby calculate, by their existing practice, the probability of a man currently aged 20 years outliving a man aged 60 years." (WK 2012x)

This exchange "captures the central idea that some variation is predictable, at least approximately in frequency. This common-cause variation is evident from the experience base." (WK 2012x)

Definitions of Common Cause

Common-cause can be defined in different ways. For instance, one good definition is to characterise it with the following features (as summarized in *Table 2.9*): (WK 2012x)

- "Phenomena constantly active within the system"
- "Variation predictable probabilistically"
- "Irregular variation within an historical experience base"
- "Lack of significance in individual high or low values"

For instance, "the outcomes of a perfectly balanced roulette wheel are a good example of common-cause variation. Common-cause variation is the noise within the system." (WK 2012x)

And "Walter A. Shewhart originally used the term *chance cause*. The term *common cause* was coined by Harry Alpert in 1947. The Western Electric Company used the term *natural pattern*. Shewhart called a process that features only common-cause variation as being *in statistical control.* This term is deprecated by some modern statisticians who prefer the phrase stable and predictable." (WK 2012x; W. Shewhart 1931; WEC 1956)

Examples of Common Cause

There are plenty of examples of common cause in history, and hereafter are some of them (as summarized in *Table 2.9*): (WK 2012x)

- Ex: "inappropriate procedures"
- Ex: "poor design"
- Ex: "poor maintenance of machines"
- Ex: "lack of clearly defined Standard operating procedures"
- Ex: "poor working conditions, e.g. lighting, noise, dirt, temperature, ventilation"
- Ex: "substandard raw materials"
- Ex: "assurement error"
- Ex: "quality control error"
- Ex: "vibration in industrial processes"
- Ex: "ambient temperature and humidity"
- Ex: "normal wear and tear"
- Ex: "variability in settings"
- Ex: "computer response time"

Surely, there can be other examples, so the ones above are illustrative only.

Importance of Common Cause

The study of common cause is important for the understanding of the history of different fields.

For instance, in the history of engineering, "common mode, or common cause, failure has a more specific meaning....It refers to events which are not statistically independent. That is, failures in multiple parts of a system caused by a single fault, particularly random failures due to environmental conditions or aging. An example is when all of the pumps for a fire sprinkler system are located in one room. If the room becomes too hot for the pumps to operate, they will all fail at essentially the same time, from one cause (the heat in the room)." (WK 2012x)

In an electronic system, "a fault in a power supply injects noise onto a supply line, causing failures in multiple subsystems. This is particularly important in safety-critical systems using multiple redundant channels. If the probability of failure in one subsystem is p, then it would be expected that an N channel system would have a probability of failure of p^N. However, in practice, the probability of failure is much higher because they are not statistically independent; for example ionizing radiation or electromagnetic interference (EMI) may affect both channels." (WK 2012x; W. Shewhart 1931)

For this reason, "the principle of redundancy states that, when events of failure of a component are statistically independent, the probabilities of their joint occurrence multiply. Thus, for instance, if the probability of failure of a component of a system is one in one thousand per year, the probability of the joint failure of two of them is one in one million per year, provided that the two events are statistically independent. This principle favors the strategy of the redundancy of components. One place this strategy is implemented is in RAID 1, where two hard disks store a computer's data redundantly." (WK 2012x)

But there is a drawback; for instance, "there can be many common modes: consider a RAID1 where two disks are purchased online and are installed in a computer, there can be many common modes": (WK 2012x)

- Ex: "the disks are likely to be from the same manufacturer and of the same model, therefore they share the same design flaws."
- Ex: "the disks are likely to have similar serial numbers, thus they may share any manufacturing flaws affecting production of the same batch."
- Ex: "the disks are likely to have been shipped at the same time, thus they are likely to have suffered from the same transportation damage."
- Ex: "as installed both disks are attached to the same power supply, making them vulnerable to the same power supply issues."
- Ex: "as installed both disks are in the same case, making them vulnerable to the same overheating events."
- Ex: "they will be both attached to the same card or motherboard, and driven by the same software, which may have the same bugs."
- Ex: "because of the very nature of RAID1, both disks will be subjected to the same workload and very closely similar access patterns, stressing them in the same way."

In addition, "if the events of failure of two components are maximally statistically dependent, the probability of the joint failure of both is identical to the probability of failure of them individually. In such a case, the advantages of redundancy are negated. Strategies for the avoidance of common mode failures include keeping redundant components physically isolated." (WK 2012x)

An excellent example of "redundancy with isolation is a nuclear power plant. The new ABWR has three divisions of Emergency Core Cooling Systems, each with its own generators and pumps and each isolated from the others. The new European Pressurized Reactor has two containment buildings, one inside the other. However, even here it is not impossible for a common mode failure to occur (for example, caused by a highly-unlikely Richter 9 earthquake)." (WK 2012x)

Problems with History and Common Causes

However, for the critics, the argument on history and common causes has some major problems. Consider, for illustration, a few examples below (as summarized in *Table 2.9*).

Firstly, one main criticism of the argument on history and common causes is that there are "special causes" outside the historical experience base which are not predictable.

For instance, for the critics, "new, unanticipated, emergent or previously neglected phenomena (e.g. 'new diseases') result in variation outside the historical experience base. [Walter] Shewhart and [Edwards] Deming argued that such special-cause variation is fundamentally unpredictable in frequency of occurrence or in severity." (WK 2012x)

Special-causes have the following characteristics: (WK 2012x)

- Ex: "new, unanticipated, emergent or previously neglected phenomena within the system"
- Ex: "variation inherently unpredictable, even probabilistically"
- Ex: "variation outside the historical experience base"
- Ex: "evidence of some inherent change in the system or our knowledge of it"
- Ex: "special-cause variation always arrives as a surprise. It is the signal within a system."

Hereafter are good examples of "special cause" in history: (WK 2012x; W. Shewhart 2012)

- Ex: "poor adjustment of equipment"
- Ex: "operator falls asleep"
- Ex: "faulty controllers"
- Ex: "machine malfunction"
- Ex: "computer crashes"

- Ex: "poor batch of raw material"
- Ex: "power surges"
- Ex: "high healthcare demand from elderly people"
- Ex: "broken part"
- Ex: "abnormal traffic (click-fraud) on web ads"
- Ex: "extremely long lab testing turnover time due to switching to a new computer system"
- Ex: "operator absent"

And in the history of economics, the idea of special cause is referred to "under the rubric of 'Knightian uncertainty.' John Maynard Keynes and Frank Knight both discussed the inherent unpredictability of economic systems in their work and used it to criticise the mathematical approach to economics, in terms of expected utility, developed by Ludwig von Mises and others. Keynes in particular argued that economic systems did not automatically tend to the equilibrium of full employment owing to their agents' inability to predict the future. As he remarked in *The General Theory of Employment, Interest and Money*": "as living and moving beings, we are forced to act…[even when] our existing knowledge does not provide a sufficient basis for a calculated mathematical expectation." (WK 2012x) However, "Keynes' thinking was at odds with the classical liberalism of the Austrian school of economists, but G. L. S. Shackle recognised the importance of Keynes's insight and sought to formalise it within a free-market philosophy." (WK 201x)

Secondly, another main criticism of the argument on history and common causes is that there are causes which can be within the historical experience base but are subjective in orientation.

For instance, for the critics, one excellent example is "the subjective Bayesian probability," which lies outside both common cause and special cause. (WK 2012x)

And thirdly, still another main criticism of the argument on history and common causes is that common causes are not necessarily desirable and any attempt to correct them could lead to more undesirable outcomes.

For instance, for the critics, "a common-cause failure is equivalent to noise in the system and specific actions cannot be made to prevent the failure," as Harry Alpert thus explained: "A riot occurs in a certain prison. Officials and sociologists turn out a detailed report about the prison, with a full explanation of why and how it happened here, ignoring the fact that the causes were common to a majority of prisons, and that the riot could have happened anywhere." (WK 2012x) This then means that "there is a temptation to react to an extreme outcome and to see it as significant, even where its causes are common to many situations and the distinctive circumstances surrounding its occurrence," which are "the results of mere chance. Such behaviour has many implications within management, often leading to ad-hoc interventions that merely increase the level of variation and frequency of undesirable outcomes." (WK 2012x)

Beyond Common and Special Causes

These criticisms of the study of history and common causes, however, should not be misunderstood as a total rejection of its usefulness but serve to show us the two opposing sides of the debate, such that the possibility (or impossibility) and desirability (or undesirability) of objectivity (from the perspective of culture with objectivity, history, and common causes as a case study here) are not to the extent that the respective defenders would like us to believe.

After all, the analysis of objectivity, history, and common causes can teach us something insightful about the ontological principles in existential dialectics, and good examples include the formalness-informalness principle, the absoluteness-relativeness principle, the partiality-totality principle, the predictability-unpredictability principle, the explicability-inexplicability principle, the fiction-reality principle, the cognitiveness-noncognitiveness principle, the finiteness-transfiniteness principle, the preciseness-vagueness principle, the simpleness-complicatedness principle, the openness-hiddenness principle, the denseness-emptiness principle, the change-constancy principle, the order-chaos principle, the slowness-quickness principle, the expansion-contraction principle, the theory-praxis principle, the convention-novelty principle, the evolution-transformation principle, the symmetry-asymmetry principle, the softness-hardness principle, the seriousness-playfulness principle, the activeness-inactivess principle, the selfness-otherness principle, the regression-progression principle, the same-difference principle, the stability-reaction principle, the functionality-nonfunctionality principle, and the intentionality-nonintentionality principle.

For instance, in relation to the formalness-informalness principle, if there is formalness (e.g., the formal logical requirement of soundness and completeness in a system of ideas, as in the argument on history and common causes), there is informalness (e.g., the non-formal existence of unsoundness and incompleteness in the argument on history and common causes, as shown in the criticisms that there are "special causes" outside the historical experience base which are not predictable, that there are causes which can be within the historical experience base but are subjective in orientation, and that common causes are not necessarily desirable and any attempt to correct them could lead to more undesirable outcomes). And the reverse direction also holds true.

In relation to the absoluteness-relativeness principle, if there is absoluteness (e.g., the absolute view by those who are tempted "to react to an extreme outcome and to see it as significant"), there is relativeness (e.g., what is right for those who are tempted "to react to an extreme outcome and to see it as significant" is not necessarily so for the critics, who pointed out that "its causes are common to many situations and the distinctive circumstances surrounding its occurrence," which are "the results of mere chance"). And the reverse direction also holds true.

In relation to the partiality-totality principle, if there is partiality (e.g., the partial view about a prison riot in history by those who are tempted "to react to an extreme outcome and to see it as significant"), there is totality (e.g., the more holistic view about a prison riot in history, such that the whole is not the sum of its parts, that is, it cannot be reduced to the sum of opposing views by the officials, the critics, etc., since there will emerge new views in the future not yet known today, just as there are already alternative views nowadays which do not side exclusively with any of them, as is the analysis here). And the reverse direction also holds true.

In relation to the predictability-unpredictability principle, if there is predictability (e.g., the predictable tendency of the existence of special causes in history), there is unpredictability (e.g., the more difficult task to predict exactly the extent to which a particular special cause will occur on a particular future occasion, since special causes are "new, unanticipated, emergent or previously neglected phenomena within the system" and their "variation" is "inherently unpredictable, even probabilistically"). And the reverse direction also holds true.

In relation to the explicability-inexplicability principle, if there is explicability (e.g., the explanation by those officials of the importance to intervene in a "prison riot" on the basis of its being "an extreme outcome" which requires urgent action), there is inexplicability (e.g., the lack of sufficient explanation by those officials of why it is necessarily desirable to intervene a "prison riot" on the basis of its being "an extreme outcome," since, for the critics, there is a temptation "to react to an extreme outcome and to see it as significant, even where its causes are common to many situations and the distinctive circumstances surrounding its occurrence," which are "the results of mere chance," so this "temptation to react to an extreme outcome...often lead...to ad-hoc interventions that merely increase the level of variation and frequency of undesirable outcomes"). And the reverse direction also holds true.

In relation to the fiction-reality principle, if there is fiction (e.g., the fictional aspect of a response to a prison riot, insofar as there is a temptation by officials "to react to an extreme outcome and to see it as significant, even where its causes are common to many situations and the distinctive circumstances surrounding its occurrence," which are "the results of mere chance"), there is reality (e.g., the realistic aspect of a response to a prison riot, insofar as there is a need for officials to deal with a prison riot, whenever it happens). And the reverse direction also holds true.

In relation to the cognitiveness-noncognitiveness principle, if there is cognitiveness (e.g., a view of things on the basis of reason and evidences, as shown in the empirical research by Keynes that "economic systems did not automatically tend to the equilibrium of full employment owing to their agents' inability to predict the future"), there is noncognitiveness (e.g., a view of things on the basis of non-cognitive factors like envy, jealousy, power, nationality, race, gender, age, class, greed, lust, status, faith, anger, sadness, joy, fear, wish, etc., as shown in the ideological politics against Keynes' finding by those in "the classical liberalism of the Austrian school of economics" at the time, who did not appreciate it). And the reverse direction also holds true.

In relation to the finiteness-transfiniteness principle, if there is finiteness (e.g., the finite number of the categories of causes in history as identified by Keynes, namely, "frequency probability," "subjective or Bayesian probability," and "special causes"), there is transfiniteness (e.g., the transfinite number of all the individual phenomena in history that can be classified under "frequency probability," "subjective or Bayesian probability," or "special causes"). And the reverse direction also holds true.

In relation to the preciseness-vagueness principle, if there is preciseness (e.g., the precise identification of 4 features of common causes in the article aforecited), there is vagueness (e.g., the vagueness in the identification of the 4 features of common causes in the article aforecited, since it is not clear why there must be only 4, not 5, 6, 7, 8, and so on). And the reverse direction also holds true.

In relation to the simpleness-complicatedness principle, if there is simpleness (e.g., the relatively simple analysis of history and common causes by the advocates), there is complicatedness (e.g., the relatively more complicated analysis of history and common causes, by challenging the claims and assumptions, as shown in the problems pointed out by the critics as cited). And the reverse direction also holds true.

In relation to the openness-hiddenness principle, if there is openness (e.g., the open exploration of common causes in history), there is hiddenness (e.g., the hidden bias in the analysis of common causes in history, because there are alternative causes like "special causes"). And the reverse direction also holds true.

In relation to the denseness-emptiness principle, if there is denseness (e.g., the relatively denser concentration of concerns with "the equilibrium of full employment" in the free market by those in the Austrial School of Economics), there is emptiness (e.g., the relatively less dense, or more empty, concentration of concerns with "the equilibrium of full employment" in the free market by those critics like Keynes, who challenged the idea). And the reverse direction also holds true.

In relation to the change-constancy principle, if there is change (e.g., the ever new ways to study history, as shown in the argument on history and common causes), there is constancy (e.g., the ever constant existence of different problems with the ever new ways to study history, as shown in the problems with the argument on history and common causes pointed out by the critics). And the reverse direction also holds true.

In relation to the order-chaos principle, if there is order (e.g., the more or less orderly view about the free market, if looked from the vantage point of those in the Austrian School of Economics, who have strong faith on the ability of the free market to reach "the equilibrium of full employment"), there is chaos (e.g., the more or less chaotic view about the free market, if looked from the vantage point of Keynes, who argued that "economic systems did not automatically tend to the equilibrium of full employment owing to their agents' inability to predict the future" and "the inherent unpredictability of economic systems"). And the reverse direction also holds true.

In relation to the slowness-quickness principle, if there is slowness (e.g., the relatively slower speed of the free market to ever reach "the equilibrium of full employment," for Keynes), there is quickness (e.g., the relatively quicker speed of the free market to ever reach "the equilibrium of full employment," if government intervention is undertaken, for Keynes). And the reverse direction also holds true.

In relation to the expansion-contraction principle, if there is expansion (e.g., "the probability of the joint failure of two" components in a system increase to "one in one million per year, provided that the two events are statistically independent," in the example about the principle of redundancy in the article aforecited), there is contraction (e.g., "the probability of failure of a component of a system" decreases to "one in one thousand per year," in the example about the principle of redundancy in the article aforecited). And the reverse direction also holds true.

In relation to the theory-praxis principle, if there is theory (e.g., the theoretical construction of "common cause" and "special cause" in mathematics), there is praxis (e.g., the practical application of "common cause" and "special cause" in mathematics to the field of engineering in the article aforecited). And the reverse direction also holds true.

In relation to the convention-novelty principle, if there is convention (e.g., the conventional wisdom about "common cause" by Jacob Bernoulli in 1703), there is novelty (e.g., the alternative novel challenge to the conventional wisdom about "common cause" by Jacob Bernoulli, by the new idea of "special cause" by Keynes and others in the 20th century). And the reverse direction also holds true.

In relation to the evolution-transformation principle, if there is evolution (e.g., the natural evolution of humans to experience variation of events in the state of nature in a primitive way), there is transformation (e.g., the technical transformation of human ability to understand variation of events by the invention of "common cause" and "special cause" in mathematics and statistics). And the reverse direction also holds true.

In relation to the symmetry-asymmetry principle, if there is symmetry (e.g., the co-existence of different views on uncertainty and prediction in economics), there is asymmetry (e.g., the view of the free market to reach "the equilibrium of full employment" is more influential among those in the Austrian School —but the view about the "inherent unpredictability of economic systems" is more influential among those like Keynes). And the reverse direction also holds true.

In relation to the softness-hardness principle, if there is softness (e.g., the defense of the free market to reach "the equilibrium of full employment" by those in the Austrian School), there is hardness (e.g., the critique of the free market to reach "the equilibrium of full employment" by those in the Keynesian School). And the reverse direction also holds true.

In relation to the seriousness-playfulness principle, if there is seriousness (e.g., the serious business of those who try to understand the causes of history), there is playfulness (e.g., the playful part of those who try to understand the causes of history, when they play around with different interpretations over time, like "frequency probability," "subjective or Bayesian probability," and "special causes"). And the reverse direction also holds true.

In relation to the activeness-inactivess principle, if there is activeness (e.g., the relative activeness of those who are lucky to be able to work, even when the free market does not reach "the equilibrium of full employment" due to "special causes" contributing to the "inherent unpredictability of economic systems"), there is inactiveness (e.g., the relative inactiveness of those who are not lucky and cannot find work when the free market does not reach "the equilibrium of full employment" due to "special causes" contributing to the "inherent unpredictability of economic systems"). And the reverse direction also holds true.

In relation to the selfness-otherness principle, if there is selfness (e.g., the tendency to be mostly centered on one's self or group, as shown in the emphasis by those in the Austrian School on the individual choices of economic agents to pursue their own self-interests in the free market), there is otherness (e.g., the tendency to be mostly centered on others, as shown in the call for government intervention for the benefits of larger society by those in the Keynesian School to fix the problems of the free market, when it cannot reach "the equilibrium of full employment"). And the reverse direction also holds true.

In relation to the regression-progression principle, if there is regression (e.g., the regression made by the study on history and common causes, as shown in the problems pointed out by the critics), there is progression (e.g., the progress made by the study on history and common causes, as shown in its contribution to the understanding of "the usual, historical, quantifiable variation in a system" aforecited). And the reverse direction also holds true.

In relation to the same-difference principle, if there is similarity in outcome (e.g., the contribution to the molding and control of beliefs, values, and behaviors, regardless of whether this be done by way of the Keynesian School, or by way of the Austrial School), there is difference in outcome (e.g., the contribution to the molding and control of beliefs, values, and behaviors by way of the Keynesian School for a relatively more government-interventionist lifeworld—but the contribution to the molding and control of beliefs, values, and behaviors by way of the Austrial School for a relatively more free-market lifeworld). And the reverse direction also holds true.

In relation to the stability-reaction principle, if there is stability (e.g., the relatively more stable condition of the free market ideology in the older days, when people believed in the free market to eventually reach "the equilibrium of full employment"), there is reaction (e.g., the free market ideology in the older days led to further changes later on, when nowadays, there are those in the Keynesian School who advocate "the inherent unpredictability of economic systems," which requires some government intervention). And the reverse direction also holds true.

In relation to the functionality-nonfunctionality principle, if there is functionality (e.g., the presence of action for which something is fitted, as shown in the belief in the free market ideology in the older days, which served the function of supporting laissez-faire economics at the time), there is nonfunctionality (e.g., the relative lack of action for which something is fitted, as shown in the rise of Keynsian economics in modern times, such that the ideology of free market is no longer dominant or functional nowadays, due to different expectations and beliefs in "the inherent unpredictability of economic systems," especially since the Great Depression and government intervention in American economic life ever since, which is the remaining baston of world capitalism). And the reverse direction also holds true.

In relation to the intentionality-nonintentionality principle, if there is intentionality (e.g., the planning or design of something for a certain outcome, as shown in the pursuit of each individual's interests in the free market), there is nonintentionality (e.g., the relative absence of planning or design of something for a certain outcome, such that the outcome can even be counter-intentional in being the opposite of what was originally planned, as shown in the failure of the free market to reach "the equilibrium of full employment" during the Great Depression, when each individual pursued his own self-interests in the free market, contrary to the original design of "the Invisible Hand" in the free-market ideology). And the reverse direction also holds true.

And in relation to the survivability-nonsurvivability principle, if there is survivability (e.g., the survival of those components in an electronic system which perform as designed), there is non-survivability (e.g., the non-survival of those components in an electronic system which fail to the point that it has to be discarded). And the reverse direction also holds true.

The Dogmaticalness of Objectivity

This comprehensive analysis of the possibility (or impossibility) and desirability (or undesirability) of objectivity in the context of probablity study, from the perspectives of the mind, nature, society, and culture, is important to show us the different ways in which objectivity is both possible and desirable, but not to the extent that the spokespersons from each side would like us to believe.

Objectivity thus has its dogmaticalness, just as it has its robustness too, as the other side of the same coin.

But this is only one part of a larger story, since there is the other side of the story, which concerns subjectivity, the opposite of objectivity. Since this chapter already deals with objectivity, the next chapter will therefore address the issue of subjectivity.

These dual analyses should not be underestimated, because those on the side of objectivity often dismiss subjectivity merely as the other side in the study of probability without really appreciating it from the vantage point of subjectivity too—and vice versa.

With this in mind, let's now turn to Chapter Three for the study of subjectivity and its duality—for which we now turn to chapter three.

Table 2.1. Objectivity and Its Ambivalence

- **Objectivity and the Mind**
  —Ex: objectivity, imagination, and mental models
  —Ex: objectivity, emotion, and rationality

- **Objectivity and Nature**
  —Ex: objectivity, the physical world, and discrete probability
  —Ex: objectivity, complexity, and uncertainty

- **Objectivity and Society**
  —Ex: objectivity, consumption, and reliability engineering
  —Ex: objectivity, organizational behavior, and optimal decision

- **Objectivity and Culture**
  —Ex: objectivity, propositional calculus, and inference rules
  —Ex: objectivity, history, and common causes

Notes: The examples in the categories are solely illustrative (not exhaustive), and the comparison is relative (not absolute), nor are they necessarily mutually exclusive. And some can be easily re-classified elsewhere. As generalities, they allow exceptions.
Source: A summary of Ch.2 of *FPHPROB*

Table 2.2. Objectivity, Imagination, and Mental Models

- **Characteristics of Mental Models**
  - —Ex: used for its "effects as information filter," as it "causes selective perception, perception of only selected parts of information."
  - —Ex: allowed to use any "sources of information, which one cannot find anywhere else," whenever they are are available.
  - —Ex: reliant on "deeply held images of thinking and acting."

- **Forms of Mental Models**
  - —Ex: "polygons"
  - —Ex: "causal-loop diagrams"
  - —Ex: "flow diagrams"

- **Principles of Mental Models**
  - —Ex: "each mental model represents a possibility."
  - —Ex: "mental models are iconic."
  - —Ex: "mental models are based on a principle of truth."
  - —Ex: "mental models can represent what is false, temporarily assumed to be true."

- **Problems with Mental Models**
  - —Firstly, there are alternative explanations of human reasoning.
  - —Secondly, there is no general consensus of whether mental models are better than other models in explaining human reasoning.
  - —Thirdly, mental models are simplistic.
  - —Fourthly, they have their own limits.
  - —Fifthly, a counter-example does not necessarily refute an inductive conclusion totally.
  - —And sixthly, there are complex situations which require optimal decision making (not on the basis of the simplification of reality as in mental models).

Notes: The examples in the categories are solely illustrative (not exhaustive), and the comparison is relative (not absolute), nor are they necessarily mutually exclusive. And some can be easily re-classified elsewhere. As generalities, they allow exceptions.
Source: From *Sec. 2.2.1* of *FPHPROB*. See book for citations.

Table 2.3. Objectivity, Emotion, and Rationality

- **Theories of Rationality**
  - —Ex: "*Zweckrational* or purposive/instrumental" rationality
  - —Ex: "*Wertrational* or value/belief-oriented" rationality

- **Levels of Rationality**
  - —Ex: "at the theoretical level, "rationality has a formal component that reduces to logical consistency and a material component that reduces to empirical support, relying on our inborn mechanisms of signal detection and interpretation."
  - —Ex: "at the practical level, "rationality is the strategy for living one's best possible life, achieving your most important goals and your own preferences in as far as possible."

- **Problems with the Argument on Emotion and Rationality**
  - —Firstly, there is no general consensus on the usage of "rationality" in different fields.
  - —Secondly, the embedded assumptions in each rationality make it relative in relation to others.
  - —Thirdly, the conventional views about rationality do not give sufficient weight to the issue of power.
  - —Fourthly, the theories of rationality are mixed up in classification.
  - —Fifthly, it is limited in practice.
  - —Sixthly, it assumes a false conception of human nature.
  - —And seventhly, there are alternative explanations of rationality.

Notes: The examples in the categories are solely illustrative (not exhaustive), and the comparison is relative (not absolute), nor are they necessarily mutually exclusive. And some can be easily re-classified elsewhere. As generalities, they allow exceptions.
Source: From *Sec. 2.2.2* of *FPHPROB.* See book for citations.

Table 2.4. Objectivity, the Physical World, and Discrete Probability

- **Examples of Discreteness**
  —Ex: "discrete particle or quantum in physics"
  —Ex: "discrete mathematics"
  —Ex: "discrete signal"
  —Ex: "discrete space"
  —Ex: "discrete time"

- **Properties of Discreteness**
  —Ex: "every discrete group is totally disconnected."
  —Ex: "every subgroup of a discrete group is discrete."
  —Ex: "every quotient of a discrete group is discrete."
  —Ex: "the product of a finite number of discrete groups is discrete."

- **Discreteness in Probability Theory**
  —Ex: "a good illustration is about "discrete probability theory," which "deals with events that occur in countable sample spaces," and good practical examples of discrete events include "throwing dice, experiments with decks of cards, and random walk." (WK 2012c)

- **Problems with Discrete Probability**
  —Firstly, mathematics assumes certain forms of perfection which are not true in the real world.
  —Secondly, there are alternative probability distributions which are not discrete.
  —Thirdly, mathematics use certain constructs of infinity which are not possible to apply in the real world.
  —Fourthly, discreteness is very difficult to be computed in certain situations.
  —And fifthly, probability theory does not explain how probability mass functions come into existence.

Notes: The examples in the categories are solely illustrative (not exhaustive), and the comparison is relative (not absolute), nor are they necessarily mutually exclusive. And some can be easily re-classified elsewhere. As generalities, they allow exceptions.
Source: From *Sec. 2.3.1* of *FPHPROB*. See book for citations.

Table 2.5. Objectivity, Complexity, and Uncertainty

- **Attributes of Complexity**
  - —Ex: "the number of parts (and types of parts) in the system and the number of relations between the parts is non-trivial—however, there is no general rule to separate 'trivial' from 'non-trivial.'"
  - —Ex: "the system has memory or includes feedback."
  - —Ex: "the system can adapt itself according to its history or feedback."
  - —Ex: "the relations between the system and its environment are…non-linear."
  - —Ex: "the system can be influenced by, or can adapt itself to, its environment."
  - —Ex: "the system is highly sensitive to initial conditions."

- **Types of Complexity**
  - —Ex: disorganized complexity
  - —Ex: organized complexity

- **Problems with the Study of Complexity**
  - —Firstly, there is a tradeoff between solvability and size (e.g., the larger the size of a problem is, the less solvable the problem becomes).
  - —Secondly, solving complex problems is not practical in certain cases.
  - —Thirdly, there is the problem of Bremermann's fundamental limit on computation.
  - —And fourthly, some complex problems cannot be predicted.

Notes: The examples in the categories are solely illustrative (not exhaustive), and the comparison is relative (not absolute), nor are they necessarily mutually exclusive. And some can be easily re-classified elsewhere. As generalities, they allow exceptions.
Source: From *Sec. 2.3.2* of *FPHPROB*. See book for citations.

Table 2.6. Objectivity, Consumption, and Reliability Engineering

- **Foundation of Reliability Engineering**
  —Ex: "reliability is a probability"
  —Ex: "reliability is predicated on 'intended function'"
  —Ex: "reliability applies to a specified period of time"
  —Ex: "reliability is restricted to operation under stated (or explicitly defined) conditions"

- **Reliability Requirements**
  —Ex: "systems engineering logic"
  —Ex: "reliability tasks and documentation"

- **Reliability Prediction**
  —Ex: to "provide necessary input to system-level reliability models."
  —Ex: "to set factory test standards for products requiring a reliability test."
  — Ex: to be used "as input to the analysis of complex systems such as switching systems and digital cross-connect systems."

- **Reliability Modeling**
  —Ex: "the physics of failure approach"
  —Ex: "the parts stress modeling approach"

- **Problems with Reliability Engineering**
  —Firstly, there is a trade-off between "reliability" and "safety."
  —Secondly, reliability testing can be infeasible in practice.
  —Thirdly, there is the problem of definition in testing.
  —Fourthly, testing does not always need to be formal.
  —Fifthly, the bias of perception can screw up the understanding of reliability.
  —Sixthly, the validity of testing can be problematic due to questionable assumptions.
  —And seventhly, test errors are hard to avoid.

Notes: The examples in the categories are solely illustrative (not exhaustive), and the comparison is relative (not absolute), nor are they necessarily mutually exclusive. And some can be easily re-classified elsewhere. As generalities, they allow exceptions.
Source: From *Sec. 2.4.1* of *FPHPROB.* See book for citations.

Table 2.7. Objectivity, Organizational Behavior, and Optimal Decision

- **Decision Making in Organizatioal Behavior**
  - —Ex: "micro" organizational behaviour
  - —Ex: "meso" organizational behavior
  - —Ex: "macro" organizational behavior

- **Decision Making Stages**
  - —Ex: "orientation stage"
  - —Ex: "conflict stage"
  - —Ex: "emergence stage"
  - —Ex: "reinforcement stage"

- **Decision Making Goals**
  - —Ex: "optimizing"
  - —Ex: "satisficing"

- **Problems with Optimal Decision**
  - —Firstly, optimization is seldom achieved in real life, as decision making can be more intuitive and less formal.
  - —Secondly, optimal and non-optimal decision-making strategies are relative to different personality traits.
  - —Thirdly, its scientific objectivity is questionable.
  - —Fourthly, it is often taken out of context.
  - —Fifthly, it has many questionable assumptions.
  - —Sixthly, it does not give sufficient attention to non-rational factors.
  - —And seventhly, it ignores power and corruption.

Notes: The examples in the categories are solely illustrative (not exhaustive), and the comparison is relative (not absolute), nor are they necessarily mutually exclusive. And some can be easily re-classified elsewhere. As generalities, they allow exceptions.
Source: From *Sec. 2.4.2* of *FPHPROB.* See book for citations.

Table 2.8. Objectivity, Propositional Calculus, and Inference Rules

- **Components of Propositional Calculus**
  —Ex: "the…language"—"that is, the particular collection of primitive symbols and operator symbols"
  —Ex: "the set of axioms"—that is, "distinguished formulæ"
  —Ex: "the set of inference rules"—that is, the entailment of a conclusion based on a set of premises

- **Concepts of Propositional Calculus**
  —Ex: "truth-values"
  —Ex: "truth-functional operators"
  —Ex: "conjunction"
  —Ex: "disjunction"
  —Ex: "material conditional"
  —Ex: "biconditional"

- **Problems with Propositonal Calculus**
  —Firstly, the quest for completeness in mathematical logic is problematic.
  —Secondly, it does not include modal logic.
  —Thirdly, it does not include alternative many-valued logics.
  —Fourthly, it does not pay attention to context-specific rules of inference.
  —Fifthly, it does not give weight to the factors of power and social action.
  —Sixthly, it is too formal and thus ignores less formal and more intuitive approaches to human thinking.
  —And seventhly, it is too simplistic, because there are are higher-order logics.

Notes: The examples in the categories are solely illustrative (not exhaustive), and the comparison is relative (not absolute), nor are they necessarily mutually exclusive. And some can be easily re-classified elsewhere. As generalities, they allow exceptions.
Source: From *Sec. 2.5.1* of *FPHPROB*. See book for citations.

Table 2.9. Objectivity, History, and Common Causes

- **Charcteristics of Common Cause**
  —Ex: "phenomena constantly active within the system"
  —Ex: "variation predictable probabilistically"
  —Ex: "irregular variation within an historical experience base"
  —Ex: "lack of significance in individual high or low values"

- **Examples of Common Cause**
  —Ex: "inappropriate procedures"
  —Ex: "poor maintenance of machines"
  —Ex: "substandard raw materials"
  —Ex: "assurement error"
  —Ex: "quality control error"
  —Ex: "vibration in industrial processes"
  —Ex: "variability in settings"

- **Problems with History and Common Causes**
  —Firstly, one main criticism of the argument on history and common causes is that there are "special causes" outside the historical experience base which are not predictable.
  —Secondly, another main criticism of the argument on history and common causes is that there are causes which can be within the historical experience base but are subjective in orientation.
  —And thirdly, still another main criticism of the argument on history and common causes is that common causes are not necessarily desirable and any attempt to correct them could lead to more undesirable outcomes.

Notes: The examples in the categories are solely illustrative (not exhaustive), and the comparison is relative (not absolute), nor are they necessarily mutually exclusive. And some can be easily re-classified elsewhere. As generalities, they allow exceptions.
Source: From *Sec. 2.5.2* of *FPHPROB.* See book for citations.

• PART THREE •

Subjectivity

CHAPTER 3
SUBJECTIVITY AND ITS DUALITY

> Does not the very word 'creative' mean to build, to initiate, to give out, to act—rather than to be acted upon, to be subjective?
> —Berenice Abbott (TE 2012a)

The Flexibility of Subjectivity

Probability interpretations involve both objectivity and subjectivity—as already indicated in *Sec. 1.3* and *Sec. 2.1*.

The relationship between objectivity and subjectivity is treated as dialectic in this book, in that there is no objectivity without subjectivity—and vice versa.

There is a good reason for this dialectic treatment, in that a common scholarly bias in the literature is to favor one over the other. Therefore, this book is to fix this bias by devoting a whole chapter on each, with the understanding that the two are related.

This dialectic relationship between objectivity and subjectivity can be revealed by showing how and why the analysis of objectivity (in Chapter Two) is inadequate without the inclusion of subjectivity, just as the examination of subjectivity (in Chapter Three) is insufficient without the consideration of objectivity.

The issues concerning subjectivity in probability study are to be addressed in this chapter (Chapter Three), whereas the issues concerning objectivity were already analyzed in the previous chapter (Chapter Two).

With this dialectic treatment in mind—a good way to examine subjectivity and its duality in probability study is by way of the evaluation of the extent to which it is in fact both possible (or impossible) and desirable (or undesirable).

This can be done by way of a comprehensive analysis of subjectivity in the context of probability from the four perspectives of the mind, nature, society, and culture (in accordance to my sophisticated methodological holism, as explained in *Sec. 1.9*).

In other words, this chapter is organized in four main sections, namely, in relation to (3.2) subjectivity and the mind, (3.3) subjectivity and nature, (3.4) subjectivity and society, and (3.5) subjectivity and culture—to be addressed hereafter, in that order (and summarized in *Table 3.1*).

Subjectivity and the Mind

Subjectivity, when examined from the perspective of the mind, can illuminate its possibility (or impossibility) and desirability (or undesirability), and this can be shown by way of two case studies, namely, (3.2.1) subjectivity, Bayesian probability, and personal belief, and (3.2.2) subjectivity, Q analysis, and the association in thinking—to be addressed in what follows, respectively.

Subjectivity, Bayesian Probability, and Personal Belief

A good way to evaluate the possibility (or impossibility) and desirability (or undesirability) of subjectivity from the perspective of the mind is to explore subjectivity, Bayesian probability, and personal belief.

An argument in the literature is that, as Thomas Bayes once argued, "the Bayesian interpretation of probability can be seen as an extension of logic that enables reasoning with propositions whose truth or falsity is uncertain. To evaluate the probability of a hypothesis, the Bayesian probabilist specifies some prior probability, which is then updated in the light of new, relevant data." (WK 2012n; J. Paulos 2011)

According to the subjectivist interpretation, in this way, "probability quantifies a 'personal belief.'" (WK 2012n; B. de Finetti 1974)

History of Bayesian Probability

Historically, "the term *Bayesian* refers to Thomas Bayes (1702–1761), who proved a special case of what is now called Bayes' theorem in a paper titled 'An Essay towards solving a Problem in the Doctrine of Chances.' In that special case, the prior and posterior distributions were Beta distributions and the data came from Bernoulli trials. It was Pierre-Simon Laplace (1749–1827) who introduced a general version of the theorem and used it to approach problems in celestial mechanics, medical statistics, reliability, and jurisprudence." (WK 2012n; S. McGrayne 2011; S. Stigler 1986)

In the 20th century, "there was a dramatic growth in research and applications of Bayesian methods, mostly attributed to the discovery of Markov chain Monte Carlo methods, which removed many of the computational problems, and an increasing interest in nonstandard, complex applications....Bayesian methods are...accepted and used, such as in the fields of machine learning and talent analytics." (WK 2012n; C. Bishop 2007; R. Wolpert 2004)

Characteristics of Bayesian Probability

In general, "the Bayesian interpretation provides a standard set of procedures and formulae to perform this calculation. Bayesian probability interprets the concept of probability as 'an abstract concept, a quantity that we assign theoretically, for the purpose of representing a state of knowledge, or that we calculate from previously assigned probabilities.'" (WK 2012n; E. Jaynes 1986)

For example, "Bayesian methods are characterized by the following concepts and procedures" (as summarized in *Table 3.2*): (WK 2012n)

- "The sequential use of the Bayes' formula: when more data become available after calculating a posterior distribution, the posterior becomes the next prior."
- "For the frequentist, a hypothesis is a proposition (which must be either true or false), so that the frequentist probability of a hypothesis is either one or zero. In Bayesian statistics, a probability can be assigned to a hypothesis that can differ from 0 or 1 if the truth value is uncertain."
- "The use of hierarchical models and marginalization over the values of nuisance parameters [which are not of immediate interest]. In most cases, the computation is intractable, but good approximations can be obtained using Markov chain Monte Carlo methods."

Of course, there can be other characteristics, so the examples above are illustrative, not exhaustive.

Justifications of Bayesian Probability

Bayesian probability has been defended on the basis of some main justifications, and three main arguments are introduced below (and summarized in *Table 3.2*). (WK 2012n)

(a) Axiomatic Approach

The first argument to justify Bayesian probability is known as the "axiomatic approach."

This approach was based on the work of Richard T. Cox (2001), who "showed that Bayesian updating follows from several axioms, including two functional equations and a controversial hypothesis of differentiability." (WK 2012n)

However, "it is known that Cox's 1961 development (mainly copied by Jaynes) is non-rigorous, and in fact a counterexample has been found by Halpern [1999]. The assumption of differentiability or even continuity is questionable since the Boolean algebra of statements may only be finite. Other axiomatizations have been suggested by various authors to make the theory more rigorous." (WK 2012n; M. Dupré 2009)

(b) Dutch Book Approach

The second argument to justify Bayesian probability is known as the "Dutch book approach."

This approach was based on the work of de Finetti on betting, in that "a Dutch book is made when a clever gambler places a set of bets that guarantee a profit, no matter what the outcome is of the bets. If a bookmaker follows the rules of the Bayesian calculus in the construction of his odds, a Dutch book cannot be made." (WK 2012n)

But the critic like Ian Hacking begged to differ, since he argued that "traditional Dutch book arguments did not specify Bayesian updating: they left open the possibility that non-Bayesian updating rules could avoid Dutch books," as Hacking (1967) thus wrote: "And neither the Dutch book argument, nor any other in the personalist arsenal of proofs of the probability axioms, entails the dynamic assumption. Not one entails Bayesianism. So the personalist requires the dynamic assumption to be Bayesian. It is true that in consistency a personalist could abandon the Bayesian model of learning from experience. Salt could lose its savour." (WK 2012n)

In addition, "there are non-Bayesian updating rules that also avoid Dutch books (as discussed in the literature on 'probability kinematics' following the publication of Richard C. Jeffrey's rule, which is itself regarded as Bayesian). The additional hypotheses sufficient to (uniquely) specify Bayesian updating are substantial, complicated, and unsatisfactory." (WK 2012n; B. van Frassen 1989; J. Joyce 2003)

(c) Decision Theory Approach

And the third argument to justify Bayesian probability is known as the "decision theory approach."

This approach was based on the work of Abraham Wald on decision theory, who "proved that every admissible statistical procedure is either a Bayesian procedure or a limit of Bayesian procedures. Conversely, every Bayesian procedure is admissible." (WK 2012n; A. Wald 1950; J. Bernardo 1994)

Problems with Bayesian Probability

However, for the critics, the argument on Bayesian probability and personal belief has some major problems. Consider, for illustration, a few examples below (as summarized in *Table 3.2*).

Firstly, one main criticism of the argument on Bayesian probability and personal belief is that there is an alternative interpretation of Bayesian probability.

For instance, for the critics, "according to the *objectivist* view [of Bayesian probability], the rules of Bayesian statistics can be justified by requirements of rationality and consistency and interpreted as an extension of logic"—unlike "the subjectivist view," in which "probability quantifies a 'personal belief.'" (WK 2012n; E. Jaynes 1986; R. Cox 2001; B. de Finetti 1974) In other words, "in the objectivist stream, the statistical analysis depends on only the model assumed and the data analysed. No subjective decisions need to be involved. In contrast, 'subjectivist' statisticians deny the possibility of fully objective analysis for the general case." (WK 2012n)

Secondly, another main criticism of the argument on Bayesian probability and personal belief is that the probability of a random event can be quantified without the need of Bayesian statistics.

For instance, the proponents of "frequentist probability...claim...that the probability of a random event denotes the *relative frequency of occurrence* of an experiment's outcome, when repeating the experiment. This interpretation considers probability to be the relative frequency 'in the long run' of outcomes." (WK 2012n; I. Hacking 1965)

Thirdly, still another main criticism of the argument on Bayesian probability and personal belief is that there can be an eclectic approach which is neither Bayesian (subjectivist) nor objectivist.

For instance, "an alternative solution is the eclectic view, which accepts both interpretations: depending on the situation, one selects one of the two interpretations for pragmatic, or principled, reasons." (WK 2012; D. Cox 2006)

Fourthly, an additional main criticism of the argument on Bayesian probability and personal belief is that different individual Bayesians disagree with each other about a particular probability.

For instance, for the critics, "Bayesian methods would suggest that one hypothesis was more probable than the other, but individual Bayesians might differ about which was the more probable and by how much, by virtue of having used different priors; but that's the same thing as disagreeing on significance levels, except significance levels are just an *ad hoc* device which are not really a probability, while priors are not only justified by the rules of probability but there is definitely a normative methodology to define beliefs…." (WK 2012) It is thus no wonder that some critics had argued that "the individual agent's probability distribution could be objectively studied in experiments," and "the objectivity of science lies not in the psychology of individual scientists, but in the process of science and especially in statistical methods." (WK 2012n; S. Stigler 1978)

Fifthly, another additional main criticism of the argument on Bayesian probability and personal belief is that its influence is still limited.

For instance, for the critics, "despite the growth of Bayesian research, most undergraduate teaching is still based on frequentist statistics." (WK 2012n)

And sixthly, still another additional main criticism of the argument on Bayesian probability and personal belief is that its justifications are problematic.

For instance, the justifications of Bayseian probability as introduced above are problematic, as explained in the previous sub-section.

Beyond Bayesian Probability

These criticisms against Bayesian probability do not mean a total rejection of its validity but serve to show us the two opposing sides of the debate, such that the possibility (or impossibility) and desirability (or undesirability) of subjectivity (from the perspective of the mind with subjectivity, Bayesian probability, and personal belief as a case study here) are not to the extent that the respective defenders would like us to believe.

Furthermore, the analysis of subjectivity, Bayesian probability, and personal belief reveals something important about the ontological principles in existential dialectics, and good examples include the formalness-informalness principle, the absoluteness-relativeness principle, the partiality-totality principle, the predictability-unpredictability principle, the explicability-inexplicability principle, the fiction-reality principle, the cognitiveness-noncognitiveness principle, the finiteness-transfiniteness principle, the preciseness-vagueness principle, the simpleness-complicatedness principle, the openness-hiddenness principle, the denseness-emptiness principle, the change-constancy principle, the order-chaos principle, the slowness-quickness principle, the expansion-contraction principle, the theory-praxis principle, the convention-novelty principle, the evolution-transformation principle, the symmetry-asymmetry principle, the softness-hardness principle, the seriousness-playfulness principle, the activeness-inactivess principle, the selfness-otherness principle, the regression-progression principle, the same-difference principle, the stability-reaction principle, the functionality-nonfunctionality principle, and the intentionality-nonintentionality principle.

For instance, in relation to the formalness-informalness principle, if there is formalness (e.g., the formal logical requirement of soundness, completeness, and consistency in a system of ideas, as in the argument for Bayesian probability), there is informalness (e.g., the non-formal existence of unsoundness, incompleteness, and inconsistency in the argument for Bayesian probability, as shown in the criticisms that there is an alternative interpretation of Bayesian probability, that the probability of a random event can be quantified without the need of Bayesian statistics, that there can be an eclectic approach which is neither Bayesian [subjectivist] nor objectivist, that different individual Bayesians disagree with each other about a particular probability, that its influence is still limited, and that its justifications are problematic). And the reverse direction also holds true.

In relation to the absoluteness-relativeness principle, if there is absoluteness (e.g., the absolute view by those for the objectivist interpretation of probability, like the "frequentist probability…claim…that the probability of a random event denotes the *relative frequency of occurrence* of an experiment's outcome, when repeating the experiment," such that the probability of a random event can be quantified without the need of Bayesian statistics), there is relativeness (e.g., what is right for those in favor of the objectivist interpretation of probability is not necessarily so for those in favor of the subjectivist version of Bayesian probability). And the reverse direction also holds true.

In relation to the partiality-totality principle, if there is partiality (e.g., the partial view on probability by those in favor of the subjectivist version of Bayesian probability), there is totality (e.g., the more holistic view on probability, such that the whole is not the sum of its parts, that is, it cannot be reduced to the sum of opposing views like the objectivist interpretation of probability, the two opposing versions of Bayesian probability, etc., since there will emerge new views in the future not yet known today, just as there are already alternative views nowadays which do not side exclusively with any of them, as shown in the argument by some that there can be an eclectic approach which is neither Bayesian [subjectivist] nor objectivist). And the reverse direction also holds true.

In relation to the predictability-unpredictability principle, if there is predictability (e.g., the predictable tendency by those for Bayesian probability to argue that one hypothesis can be "more probable than the other"), there is unpredictability (e.g., the more difficult task to predict exactly which "individual Bayesians might differ about which was the more probable and by how much, by virtue of having used different priors"). And the reverse direction also holds true.

In relation to the explicability-inexplicability principle, if there is explicability (e.g., the explanation by those like Cox in favor of Bayesian probability on the basis of the "axiomatic approach," in that Bayesian updating follows from several axioms"), there is inexplicability (e.g., the lack of sufficient explanation, by those in favor of Bayesian probability on the basis of the "axiomatic approach," of why the "axiomatic approach" is necessarily desirable, since, for the critics, "Cox's 1961 development [mainly copied by Jaynes] is non-rigorous, and in fact a counterexample has been found by Halpern [1999]"). And the reverse direction also holds true.

In relation to the fiction-reality principle, if there is fiction (e.g., the fictional aspect of the Dutch book approach, insofar as the critic like Ian Hacking argued that "traditional Dutch book arguments did not specify Bayesian updating: they left open the possibility that non-Bayesian updating rules could avoid Dutch books"), there is reality (e.g., the realistic aspect of the Dutch book approach, insofar as, under some specified conditions, "a Dutch book is made when a clever gambler places a set of bets that guarantee a profit, no matter what the outcome is of the bets," but "if a bookmaker follows the rules of the Bayesian calculus in the construction of his odds, a Dutch book cannot be made"). And the reverse direction also holds true.

In relation to the cognitiveness-noncognitiveness principle, if there is cognitiveness (e.g., a view of things on the basis of reason and evidences, as shown in the "Bayesian methods," in that "one hypothesis" is "more probable than the other... by virtue of having used different priors"), there is noncognitiveness (e.g., a view of things on the basis of non-cognitive factors like envy, jealousy, power, nationality, race, gender, age, class, greed, lust, status, faith, anger, sadness, joy, fear, wish, etc., as shown in the "Bayesian methods," in which "individual Bayesians might differ about which was the more probable and by how much," due to differences in "personal belief"). (WK 2012n; B. de Finetti 1974) And the reverse direction also holds true.

In relation to the finiteness-transfiniteness principle, if there is finiteness (e.g., the finite number of procedures which characterize "Bayesian methods," as described earlier in regard to the 3 examples), there is transfiniteness (e.g., the transfinite amount of time in the computation of "intractable" problems in some Bayesian procedures, like "the use of hierarchical models and marginalization over the values of nuisance parameters [which are not of immediate interest]"). And the reverse direction also holds true.

In relation to the preciseness-vagueness principle, if there is preciseness (e.g., the precise identification of 3 justifications of Bayesian probability, namely, "axiomatic approach," "Dutch book approach," and "decision theory approach"), there is vagueness (e.g., the vagueness in the identification of the `3 justifications of Bayesian probability, since it is not clear why there must be only 3, not 4, 5, 6, 7, and so on). And the reverse direction also holds true.

In relation to the simpleness-complicatedness principle, if there is simpleness (e.g., the relatively simple analysis of Bayesian probability by its advocates), there is complicatedness (e.g., the relatively more complicated analysis of Bayesian probability, by challenging the claims and assumptions, as shown in the problems pointed out by the critics as cited). And the reverse direction also holds true.

In relation to the openness-hiddenness principle, if there is openness (e.g., the open exploration of calculating probability on the basis of using "different priors" in the Bayesian methods), there is hiddenness (e.g., the hidden bias in the Bayesian methods, since "individual Bayesians might differ about which was the more probable and by how much, by virtue of having used different priors"). And the reverse direction also holds true.

In relation to the denseness-emptiness principle, if there is denseness (e.g., the relatively denser concentration of concerns with "personal belief" in the Bayesian probability interpretation), there is emptiness (e.g., the relatively less dense, or more empty, concentration of concerns with "personal belief" in the objectivist probability interpretation). And the reverse direction also holds true.

In relation to the change-constancy principle, if there is change (e.g., the ever new ways to study probability, as shown in the construction of the Bayesian methods), there is constancy (e.g., the ever constant existence of different problems with the ever new ways to study probability, as shown in the problems with the Bayesian methods pointed out by the critics). And the reverse direction also holds true.

In relation to the order-chaos principle, if there is order (e.g., the more or less orderly view on probability, if looked from the vantage point of "the frequentist probability of a hypothesis," which must be "either one or zero," in an orderly way), there is chaos (e.g., the more or less chaotic view on probability, if looked from the vantage point of Bayesian statistics, in which "a probability can be assigned to a hypothesis that can differ from 0 or 1," in that "the truth value is uncertain" or unknown, such that it can go in any direction). And the reverse direction also holds true.

In relation to the slowness-quickness principle, if there is slowness (e.g., the relatively slower speed of computation to be completed, if it is "intractable," in cases involving "the use of hierarchical models and marginalization over the values of nuisance parameters" in Bayesian methods), there is quickness (e.g., the relatively quicker speed of computation to be completed, if "good approximations can be obtained using Markov chain Monte Carlo methods," in cases involving "the use of hierarchical models and marginalization over the values of nuisance parameters" in Bayesian methods). And the reverse direction also holds true.

In relation to the expansion-contraction principle, if there is expansion (e.g., the relative "dramatic growth in research and applications of Bayesian methods, mostly attributed to the discovery of Markov chain Monte Carlo methods," in the 1980s), there is contraction (e.g., the relative decline of Bayesian statistics in "most undergraduate teaching," which "is still based on frequentist statistics"). And the reverse direction also holds true.

In relation to the theory-praxis principle, if there is theory (e.g., the theoretical construction of "Bayesian methods" in the study of probability), there is praxis (e.g., the practical application of "Bayesian methods" in the study of probability to the fields of "machine learning and talent analytics"). And the reverse direction also holds true.

In relation to the convention-novelty principle, if there is convention (e.g., the conventional understanding of Bayesian probability as worked out by Bayes and Laplace in the older days of the 18th and 19th centuries), there is novelty (e.g., the alternative novel challenge to the conventional understanding of Bayesian probability as worked out by Bayes and Laplace in the older days of the 18th and 19th centuries, by the new ideas in the 20th century, when "the ideas of Laplace were further developed in two different directions, giving rise to objective and subjective currents in Bayesian practice"). (WK 2012n) And the reverse direction also holds true.

In relation to the evolution-transformation principle, if there is evolution (e.g., the natural evolution of humans to hold personal beliefs in the state of nature), there is transformation (e.g., the technical transformation of human ability to understand personal beliefs by the invention of Bayesian methods in the study of probability). And the reverse direction also holds true.

In relation to the symmetry-asymmetry principle, if there is symmetry (e.g., the existence of different ways to study probability, like Bayesian statistics and frequentist statistics), there is asymmetry (e.g., the popularity of Bayesian statistics in the fields of "machine learning and talent analytics"—but the popularity of frequentist statistics in "most undergraduate teaching"). And the reverse direction also holds true.

In relation to the softness-hardness principle, if there is softness (e.g., the defense of the Dutch book approach by de Finetti), there is hardness (e.g., the critique of the Dutch book approach by Ian Hacking). And the reverse direction also holds true.

In relation to the seriousness-playfulness principle, if there is seriousness (e.g., the serious business of those who try to understand the nature of probability), there is playfulness (e.g., the playful part of those who try to understand the nature of probability, when they play around with different interpretations and versions over time, like the "objectivist" interpretation, the "subjectivist" interpretation, the two different versions of Bayesian methods, the "eclectic" alternative, etc.). And the reverse direction also holds true.

In relation to the activeness-inactivess principle, if there is activeness (e.g., the relative activeness of those for Bayesian methods, when they actively practice them in the fields of "machine learning and talent analytics"), there is inactiveness (e.g., the relative inactiveness of those for Bayesian methods, when they are not welcome in "most undergraduate teaching," which "is still based on frequentist statistics"). And the reverse direction also holds true.

In relation to the selfness-otherness principle, if there is selfness (e.g., the tendency to be mostly centered on one's self or group, as shown in "a clever gambler," who "places a set of bets that guarantee a profit, no matter what the outcome is of the bets," such that "a Dutch book is made"), there is otherness (e.g., the tendency to be mostly centered on others, as shown in the Dutch book argument proposed by de Finetti, as a contribution to the advance in the scientific knowledge of probability for humanity). And the reverse direction also holds true.

In relation to the regression-progression principle, if there is regression (e.g., the regression made by Bayesian methods, as shown in the problems pointed out by the critics), there is progression (e.g., the progress made by Bayesian methods, as shown in their contribution to the fields of "machine learning and talent analytics" aforecited). And the reverse direction also holds true.

In relation to the same-difference principle, if there is similarity in outcome (e.g., the contribution to the molding and control of beliefs, values, and behaviors, regardless of whether this be done by way of the objectivist interpretation of probability, or by way of the Bayesian interpretation of probability), there is difference in outcome (e.g., the contribution to the molding and control of beliefs, values, and behaviors by way of the objectivist interpretation of probability for a relatively more scientific worldview—but the contribution to the molding and control of beliefs, values, and behaviors by way of of the Bayesian interpretation of probability for a relatively more psychological personal worldview). And the reverse direction also holds true.

In relation to the stability-reaction principle, if there is stability (e.g., the relatively more stable condition of the ideas of probability as worked out by Bayes and Laplace in the older days of the 18th and 19th centuries), there is reaction (e.g., the ideas of probability as worked out by Bayes and Laplace in the older days of the 18th and 19th centuries led to further changes later on, when "the ideas of Laplace were further developed in two different directions, giving rise to objective and subjective currents in Bayesian practice" in the 20th century). And the reverse direction also holds true.

In relation to the functionality-nonfunctionality principle, if there is functionality (e.g., the presence of action for which something or someone is fitted, as shown in the use of the Bible in medieval Europe, which served the function of understanding reality along the hermeneutic line), there is nonfunctionality (e.g., the relative lack of action for which something or someone is fitted, as shown in the use of the objectivist version of Bayesian methods nowadays, such that the biblical hermenuetics is no longer functional "in the process of science and especially in statistical methods" in our time, since "to meet the needs of science…, Bayesian statisticians have developed 'objective' methods for specifying prior probabilities"). (WK 2012n) And the reverse direction also holds true.

In relation to the intentionality-nonintentionality principle, if there is intentionality (e.g., the planning or design of something for a certain outcome, as shown in the work of de Finetti on the "Dutch book approach" as a way to justify Bayesian probability), there is nonintentionality (e.g., the relative absence of planning or design of something for a certain outcome, such that the outcome can even be counter-intentional in being the opposite of what was originally planned, as shown in the work of Ian Hacking, who showed that "traditional Dutch book arguments did not specify Bayesian updating: they left open the possibility that non-Bayesian updating rules could avoid Dutch books," and "neither the Dutch book argument, nor any other in the personalist arsenal of proofs of the probability axioms, entails the dynamic assumption. Not one entails Bayesianism"—contrary to the original intention of de Finetti to justify Bayesian probability with the "Dutch book approach"). And the reverse direction also holds true.

And in relation to the survivability-nonsurvivability principle, if there is survivability (e.g., the survival of the "axiomatic approach," insofar as "other axiomatizations have been suggested by various authors to make the theory more rigorous" nowadays), there is non-survivability (e.g., the non-survival of the "axiomatic approach," insofar as "Cox's 1961 development [mainly copied by Jaynes] is non-rigorous, and in fact a counterexample has been found by Halpern"). (WK 2012n; M. Dupré 2009) And the reverse direction also holds true.

Subjectivity, Q Analysis, and the Association in Thinking

Another way to evaluate the possibility (or impossibility) and desirability (or undesirability) of subjectivity from the perspective of the mind is to explore subjectivity, Q analysis, and the association in thinking.

An argument in the literature is that "people think about ideas in relation to other ideas, rather than in isolation." (WK 2012xx)

Influence of R Analysis

Historically, "Q Methodology is a research method used in psychology and other social sciences to study people's 'subjectivity'—that is, their viewpoint. Q was developed by psychologist William Stephenson." (WK 2012xx)

But, because Stephenson studied "at University College London under Charles Spearman, a pioneer of factor analysis," it is therefore important to understand the relationship between factor analysis (like R analysis) and Q analysis. (WK 2012yy)

After all, "the name 'Q' comes from the form of factor analysis that is used to analyze the data. Normal factor analysis, called 'R method,' involves finding correlations between variables (say, height and age) across a sample of subjects." (WK 2012xx)

For illustration, here is an example of how R analysis works: "Suppose a psychologist proposes a theory that there are two kinds of intelligence, 'verbal intelligence' and 'mathematical intelligence,' neither of which is directly observed. Evidence for the theory is sought in the examination scores from each of 10 different academic fields of 1000 students. If each student is chosen randomly from a large population, then each student's 10 scores are random variables. The psychologist's theory may say that for each of the 10 academic fields, the score averaged over the group of all students who share some common pair of values for verbal and mathematical 'intelligences' is some constant times their level of verbal intelligence plus another constant times their level of mathematical intelligence, i.e., it is a linear combination of those two 'factors.' The numbers for a particular subject, by which the two kinds of intelligence are multiplied to obtain the expected score, are posited by the theory to be the same for all intelligence level pairs, and are called 'factor loadings' for this subject." (WK 2012zz)

For instace," the theory may hold that the average student's aptitude in the field of taxonomy is {10 × the student's verbal intelligence} + {6 × the student's mathematical intelligence}. The numbers 10 and 6 are the factor loadings associated with taxonomy. Other academic subjects may have different factor loadings. Two students having identical degrees of verbal intelligence and identical degrees of mathematical intelligence may have different aptitudes in taxonomy because individual aptitudes differ from average aptitudes. That difference is called the 'error' — a statistical term that means the amount by which an individual differs from what is average for his or her levels of intelligence." (WK 2012zz)

So, "the observable data that go into factor analysis would be 10 scores of each of the 1000 students, a total of 10,000 numbers. The factor loadings and levels of the two kinds of intelligence of each student must be inferred from the data." (WK 2012zz)

Distinctiveness of Q Analysis

With this influence of R analysis in mind—Q analysis differs from R analysis, in that, while R analysis "involves finding correlations between variables (say, height and age) across a sample of subjects," Q analysis "looks for correlations between subjects across a sample of variables." (WK 2012xx)

In other words, "Q factor analysis reduces the many individual viewpoints of the subjects down to a few 'factors,' which represent shared ways of thinking. It is sometimes said that Q factor analysis is R factor analysis with the data table turned sideways." (WK 2012xx)

In general, "the data for Q factor analysis come from a series of 'Q sorts' performed by one or more subjects. A Q sort is a ranking of variables—typically presented as statements printed on small cards—

according to some 'condition of instruction.' For example, in a Q study of people's views of a celebrity, a subject might be given statements like 'He is a deeply religious man' and 'He is a liar,' and asked to sort them from 'most like how I think about this celebrity' to 'least like how I think about this celebrity.' The use of ranking, rather than asking subjects to rate their agreement with statements individually, is meant to capture the idea that people think about ideas in relation to other ideas, rather than in isolation." (WK 2012xx)

Q analysis depends on the technique of "concourse," which means that "the sample of statements for a Q sort is drawn from…the sum of all things people say or think about the issue being investigated. Since concourses do not have clear membership lists (as would be the case in the population of subjects), statements cannot be drawn randomly. Commonly Q methodologists use a structured sampling approach in order to ensure that they include the full breadth of the concourse." (WK 2012xx)

A distinctive advantage of Q analysis is that, when contrasted with "other social science research methodologies, such as surveys,…it typically uses many fewer subjects. This can be a strength, as Q is sometimes used with a single subject. In such cases, a person will rank the same set of statements under different conditions of instruction. For example, someone might be given a set of statements about personality traits and then asked to rank them according to how well they describe herself, her ideal self, her father, her mother, etc." (WK 2012xx)

Application of Q Analysis

Q analysis has been applied in different fields, as it has been used "as a research tool in a wide variety of disciplines including nursing, veterinary medicine, public health, transportation, education, rural sociology, hydrology and mobile communication. The methodology is particularly useful when researchers wish to understand and describe the variety of subjective viewpoints on an issue." (WK 2012xx; S. Bracken 2006; J. Previte 2007; G. Raadgever 2008)

For illustration, "in studies of intelligence, Q factor analysis can generate Consensus based assessment (CBA) scores as direct measures. Alternatively, the unit of measurement of a person in this context is his factor loading for a Q-sort he or she performs. Factors represent norms with respect to schemata. The individual who gains the highest factor loading on an Operant factor is the person most able to conceive the norm for the factor." (WK 2012xx)

In the older days, "the 'Q sort' data collection procedure is…done using a paper template and the sample of statements or other stimuli printed on individual cards. However, there are also computer software applications for conducting online Q sorts. For example, nQue is a web-based commercial software application that uses a drag-and-drop, graphical user interface to conduct online Q sorts that mimic the analog, paper-based sorting procedure. UC Riverside's International Situations Project is a similar, university-developed web-based application." (WK 2012xx)

Problems with Q Methodology

However, for the critics, Q methodology has some major problems. Consider, for illustration, a few examples below.

Firstly, one main criticism of Q methodology is that its analysis is theory-dependent.

For example, for the critics, "naming factors may require knowledge of theory because seemingly dissimilar attributes can correlate strongly for unknown reasons." (WK 2012zz)

Secondly, another main criticism of Q methodology is that it does not explain why the association of ideas in someone's mind exists in the way that it does.

For instance, for the critics, in Q analysis, "the individual who gains the highest factor loading on an Operant factor is the person most able to conceive the norm for the factor," but "what the norm means is a matter, always, for conjecture and refutation (Popper). It may be indicative of the wisest solution, or the most responsible, the most important, or an optimized-balanced solution"; alternatively, it may be interpreted as meaning something else altogether, so "these are all matters for future determination." (WK 2012xx)

Thirdly, still another main criticism of Q methodology is that there are alternative approaches.

For instance, for the critics, "an alternative method that determines the similarity among subjects somewhat like Q methodology, as well as the cultural 'truth' of the statements used in the test, is Cultural Consensus Theory." (WK 2012xx)

Fourthly, an additional main criticism of Q methodology is that the validity of its analysis is relative to the data collected about an individual.

For instance, for the critics, Q analysis suffers from the same problem of R analysis, in that "factor analysis can be only as good as the data allows. In psychology, where researchers often have to rely on less valid and reliable measures such as self-reports, this can be problematic." (WK 2012zz)

And fifthly, another additional main criticism of Q methodology is that its overall evaluation as a method has yet to be fully accessed.

For instance, for the critics, "little information on validation of the methods is available currently and the retrospective process using the methodology for any researches needs to be taken." (WK 2012xx; M. Buckley 2002)

Beyond Q Methodology

These criticisms of Q methodology should not be misunderstood as a total rejection of its usefulness but serve to show us the two opposing sides of the debate, such that the possibility (or impossibility) and desirability (or undesirability) of subjectivity (from the perspective of the mind with subjectivity, Q analysis, and the association in thinking as a case study here) are not to the extent that the respective defenders would like us to believe.

Besides, the analysis of subjectivity, Q analysis, and the association in thinking can cast a bright light on the ontological principles in existential dialectics, and good examples include the formalness-informalness principle, the absoluteness-relativeness principle, the partiality-totality principle, the predictability-unpredictability principle, the explicability-inexplicability principle, the fiction-reality principle, the cognitiveness-noncognitiveness principle, the finiteness-transfiniteness principle, the preciseness-vagueness principle, the simpleness-complicatedness principle, the openness-hiddenness principle, the denseness-emptiness principle, the change-constancy principle, the order-chaos principle, the slowness-quickness principle, the expansion-contraction principle, the theory-praxis principle, the convention-novelty principle, the evolution-transformation principle, the symmetry-asymmetry principle, the softness-hardness principle, the seriousness-playfulness principle, the activeness-inactivess principle, the selfness-otherness principle, the regression-progression principle, the same-difference principle, the stability-reaction principle, the functionality-nonfunctionality principle, and the intentionality-nonintentionality principle.

For instance, in relation to the formalness-informalness principle, if there is formalness (e.g., the formal logical requirement of soundness, completeness, and consistency in a system of ideas, as in Q methodology), there is informalness (e.g., the non-formal existence of unsoundness, incompleteness, and inconsistency in Q methodology, as shown in the criticisms that its analysis is theory-dependent, that it does not explain why the association of ideas in someone's mind exists in the way that it does, that there are alternative approaches, that the validity of its analysis is relative to the data collected about an individual, and that its overall evaluation as a method has yet to be fully accessed). And the reverse direction also holds true.

In relation to the absoluteness-relativeness principle, if there is absoluteness (e.g., the absolute view on analysis by those for R analysis, with its focus on "correlations between variables [say, height and age] across a sample of subjects"), there is relativeness (e.g., what is right for those for R analysis, with its focus on "correlations between variables…across a sample of subjects," is not necessarily so for those in favor of Q analysis, which looks "for correlations between subjects across a sample of variables," since "for mathematical reasons no one data matrix would be suitable for analysis with both Q and R"). (WK 2012xx) And the reverse direction also holds true.

In relation to the partiality-totality principle, if there is partiality (e.g., the partial view on analysis by those for R analysis), there is totality (e.g., the more holistic view on analysis, such that the whole is not the sum of its parts, that is, it cannot be reduced to the sum of opposing views like R analysis, Q analysis, Cultural Consensus Theory, etc., since there will emerge new views in the future not yet known today, just as there are already alternative views nowadays which do not side exclusively with any of them, as is the analysis here). And the reverse direction also holds true.

In relation to the predictability-unpredictability principle, if there is predictability (e.g., the predictable tendency of those in Q analysis to argue that "the individual who gains the highest factor loading on an Operant factor is the person most able to conceive the norm for the factor"), there is unpredictability (e.g., the more difficult task to predict exactly the extent to which, in Q analysis, "the norm for the factor" is to mean, since "what the norm means is a matter, always, for conjecture and refutation [Popper],…so "these are all matters for future determination"). And the reverse direction also holds true.

In relation to the explicability-inexplicability principle, if there is explicability (e.g., the explanation, by those for Q analysis, of its desirability on the basis of its focus on "correlations between subjects across a sample of variables," so as to show that "people think about ideas in relation to other ideas, rather than in isolation"), there is inexplicability (e.g., the lack of sufficient explanation, by those for Q analysis, of why it is necessarily desirable, since, for the critics, there are alternatives like R methodology, Cultural Consensus Theory, etc.). And the reverse direction also holds true.

In relation to the fiction-reality principle, if there is fiction (e.g., the fictional aspect of Q analysis, insofar as "the data collected" are not valid or reliable since its analysis "can be only as good as the data allows"), there is reality (e.g., the realistic aspect of Q analysis, insofar as "the data collected" are indeed valid or reliable). And the reverse direction also holds true.

In relation to the cognitiveness-noncognitiveness principle, if there is cognitiveness (e.g., a view of things on the basis of reason and evidences, as shown in Q analysis, since its analysis "can be only as good as the data allows"), there is noncognitiveness (e.g., a view of things on the basis of non-cognitive factors like envy, jealousy, power, nationality, race, gender, age, class, greed, lust, status, faith, anger, sadness, joy, fear, wish, etc., as shown in the good faith in Q analysis, insofar as "little information on validation of the methods is available currently"). And the reverse direction also holds true.

In relation to the finiteness-transfiniteness principle, if there is finiteness (e.g., the finite number of subjects used in Q analysis, since "one salient difference between Q and other social science research methodologies, such as surveys, is that it typically uses many fewer subjects"), there is transfiniteness (e.g., the transfinite number of all the indiviudal subjects in history that have ever existed and can be used in Q analysis, at least in principle). And the reverse direction also holds true.

In relation to the preciseness-vagueness principle, if there is preciseness (e.g., the precise identification of 8 applications of Q analysis in the article aforecited, namely, "nursing, veterinary medicine, public health, transportation, education, rural sociology, hydrology and mobile communication"), there is vagueness (e.g., the vagueness in the identification of the 8 applications of Q analysis in the article aforecited, since it is not clear why there must be only 8, not, 9, 13, 15, and so on). And the reverse direction also holds true.

In relation to the simpleness-complicatedness principle, if there is simpleness (e.g., the relatively simple analysis of Q methodology by its advocates), there is complicatedness (e.g., the relatively more complicated analysis of Q methodology, by challenging the claims and assumptions, as shown in the problems pointed out by the critics as cited). And the reverse direction also holds true.

In relation to the openness-hiddenness principle, if there is openness (e.g., the open exploration of how "people think about ideas in relation to other ideas, rather than in isolation" by Q analysis), there is hiddenness (e.g., the hidden bias in Q analysis, as shown in the problems that its analysis is theory-dependent, that the validity of its analysis is relative to the data collected about an individual, and that its overall evaluation as a method has yet to be fully accessed). And the reverse direction also holds true.

In relation to the denseness-emptiness principle, if there is denseness (e.g., the relatively denser concentration of concerns with "correlations between variables [say, height and age] across a sample of subjects" by those in R analysis), there is emptiness (e.g., the relatively less dense, or more empty, concentration of concerns with "correlations between variables [say, height and age] across a sample of subjects" by those in Q analysis, since they focus on "correlations between subjects across a sample of variables" instead). And the reverse direction also holds true.

In relation to the change-constancy principle, if there is change (e.g., the ever new ways to study analysis, as shown in Q analysis), there is constancy (e.g., the ever constant existence of different problems with the ever new ways to study analysis, as shown in the problems with Q analysis pointed out by the critics). And the reverse direction also holds true.

In relation to the order-chaos principle, if there is order (e.g., the more or less orderly view on the ideas of an individual, if looked from the vantage point of Q analysis, which tries to show that "people think about ideas in relation to other ideas, rather than in isolation"), there is chaos (e.g., the more or less chaotic view on the ideas of an individual, if looked from the vantage point of understanding ideas in isolation in the mindset of an individual, such that they are not closely linked with each other). And the reverse direction also holds true.

In relation to the slowness-quickness principle, if there is slowness (e.g., the relatively slower speed of doing Q sorts in Q analysis, if the traditional way of "a paper template and the sample of statements or other stimuli printed on individual cards" is used instead), there is quickness (e.g., the relatively quicker speed of doing Q sorts in Q analysis, if sophisticated computer software applications are used). And the reverse direction also holds true.

In relation to the expansion-contraction principle, if there is expansion (e.g., the relative expansion of using computer software applications for doing Q sorts nowadays), there is contraction (e.g., the relative decline of using "a paper template and the sample of statements or other stimuli printed on individual cards" for doing Q sorts nowadays). And the reverse direction also holds true.

In relation to the theory-praxis principle, if there is theory (e.g., the theoretical construction of Q methodology), there is praxis (e.g., the practical application of Q methodology to the applied fields like "nursing, veterinary medicine, public health, transportation, education, rural sociology, hydrology and mobile communication"). And the reverse direction also holds true.

In relation to the convention-novelty principle, if there is convention (e.g., the conventional wisdom about using R analysis in the older days), there is novelty (e.g., the alternative novel challenge to the conventional wisdom about using R analysis in the older days, by the new idea of using Q analysis nowadays). And the reverse direction also holds true.

In relation to the evolution-transformation principle, if there is evolution (e.g., the natural evolution of humans to observe and analyze things in the state of nature, albeit in a primitive way), there is transformation (e.g., the technical transformation of human ability to observe and analyze things by the invention of Q methodology). And the reverse direction also holds true.

In relation to the symmetry-asymmetry principle, if there is symmetry (e.g., the co-existence of different analyses, like Q analysis, Cultural Consensus Theory, etc.), there is asymmetry (e.g., the popularity of Q analysis among those who research on how "people think about ideas in relation to other ideas, rather than in isolation"—but the popularity of Cultural Consensus Theory among those who research on "the cultural 'truth' of the statements used in the test"). And the reverse direction also holds true.

In relation to the softness-hardness principle, if there is softness (e.g., the defense of Q methodology by William Stephenson), there is hardness (e.g., the critique of Q methodology by the critics, as shown in the problems aforecited). And the reverse direction also holds true.

In relation to the seriousness-playfulness principle, if there is seriousness (e.g., the serious business of those who research on the nature of analysis), there is playfulness (e.g., the playful part of those who research on the nature of analysis, when they play around with different approaches over time, like R analysis, Q analysis, Cultural Consensus Theory, etc.). And the reverse direction also holds true.

In relation to the activeness-inactivess principle, if there is activeness (e.g., the relative activeness of those in favor of Q methodology when it is widely used in fields like "nursing, veterinary medicine, public health, transportation, education, rural sociology, hydrology and mobile communication"), there is inactiveness (e.g., the relative inactiveness of those in favor of Q methodology, when it is not used in other fields like "marketing, product management, operations research, and other applied sciences that deal with large quantities of data," where R analysis is more widely used instead). (WK 2012zz) And the reverse direction also holds true.

In relation to the selfness-otherness principle, if there is selfness (e.g., the tendency to be mostly centered on one's self or group, as shown in Q analysis, when the "norm" for a factor by an individual "may be indicative of the wisest solution, or the most responsible, the most important, or an optimized-balanced solution," which he most desires for his own well-being), there is otherness (e.g., the tendency to be mostly centered on others, as shown in Q analysis, when it contributes to the understanding of how "people think about ideas in relation to other ideas, rather than in isolation," as an advance in the knowledge of psychology for humanity). And the reverse direction also holds true.

In relation to the regression-progression principle, if there is regression (e.g., the regression made by Q methodology, as shown in the problems pointed out by the critics), there is progression (e.g., the progress made by Q methodology, as shown in its contribution to the understanding of how "people think about ideas in relation to other ideas, rather than in isolation"). And the reverse direction also holds true.

In relation to the same-difference principle, if there is similarity in outcome (e.g., the contribution to the molding and control of beliefs, values, and behaviors, regardless of whether this be done by way of Q methodology, or by way of Cultural Consensus Theory), there is difference in outcome (e.g., the contribution to the molding and control of beliefs, values, and behaviors by way of Q methodology for a relatively more psychological worldview—but the contribution to the molding and control of beliefs, values, and behaviors by way of Cultural Consensus Theory for a relatively more cultural worldview). And the reverse direction also holds true.

In relation to the stability-reaction principle, if there is stability (e.g., the relatively more stable condition of factor analysis like R analysis, when it was pioneered by Charles Spearman and others in the older days), there is reaction (e.g., the development of factor analysis led to further changes later on, when it further evolved into Q analysis by William Stephenson and others). And the reverse direction also holds true.

In relation to the functionality-nonfunctionality principle, if there is functionality (e.g., the presence of action for which something or someone is fitted, as shown in the use of "a paper template and the sample of statements or other stimuli printed on individual cards" for doing Q sorts in the older days, which served the function of a relatively less technologically advanced society at the time), there is nonfunctionality (e.g., the relative lack of action for which something or someone is fitted, as shown in the use of sophisticated computer software applications for doing Q sorts nowadays, such that the use of "a paper template and the sample of statements or other stimuli printed on individual cards" in the older days is no longer much functional in our computer age). And the reverse direction also holds true.

In relation to the intentionality-nonintentionality principle, if there is intentionality (e.g., the planning or design of something for a certain outcome, as shown in the construction of factor analysis like R analysis by Charles Spearman and others in the older days), there is nonintentionality (e.g., the relative absence of planning or design of something for a certain outcome, such that the outcome can even be counter-intentional in being the opposite of what was originally planned, as shown in the construction of Q analysis by William Stephenson and others later on, such that its focus is on "correlations between subjects across a sample of variables"—contrary to the original intention of Spearman for factor analysis, which is more on "correlations between variables [say, height and age] across a sample of subjects" instead). And the reverse direction also holds true.

And in relation to the survivability-nonsurvivability principle, if there is survivability (e.g., the relative survival of using sophisticated computer software applications for doing Q sorts nowadays), there is non-survivability (e.g., the relative non-survival of the use of "a paper template and the sample of statements or other stimuli printed on individual cards" for doing Q sorts nowadays). And the reverse direction also holds true.

Subjectivity and Nature

Subjectivity, when examined from the perspective of nature, can show us in a controversial way its possibility (or impossibility) and desirability (or undesirability), and this can be shown by way of two case studies, namely, (3.3.1) subjectivity, measurement, and natural condition, and (3.3.2) subjectivity, quantum mechanics, and uncertainty—to be addressed in what follows, respectively.

Subjectivity, Measurement, and Natural Condition

A good way to evaluate the possibility (or impossibility) and desirability (or undesirability) of subjectivity from the perspective of nature is to explore subjectivity, measurement, and natural condition.

An argument in the literature is that "all measurements are subject to uncertainty," such that there is an element of subjectivity in measurement (due to the effects of "the measuring system, the measurement procedure, the skill of the operator, the environment, and other effects"), so this measurement uncertainty becomes a natural condition, regardless of any particular culture and society. (WK 2012aaa)

Background of Measurement Uncertainty

The issue about measurement uncertainty is a major topic in "metrology" (or "the science of measurement"), and measurement uncertainty refers to "a non-negative parameter characterizing the dispersion of the values attributed to a measured quantity. The uncertainty has a probabilistic basis and reflects incomplete knowledge of the quantity. All measurements are subject to uncertainty and a measured value is only complete if it is accompanied by a statement of the associated uncertainty." (WK 2012aaa)

The reason for uncertainty to exist in measurement has to do with the fact that "when a quantity is measured, the outcome depends on the measuring system, the measurement procedure, the skill of the operator, the environment, and other effects. Even if the quantity were to be measured several times, in the same way and in the same circumstances, a different measured value would in general be obtained each time, assuming that the measuring system has sufficient resolution to distinguish between the values." (WK 2012aaa)

So, "the dispersion of the measured values would relate to how well the measurement is made. Their average would provide an estimate of the true value of the quantity that generally would be more reliable than an individual measured value. The dispersion and the number of measured values would provide information relating to the average value as an estimate of the true value." (WK 2012aaa)

Even then, however, "this information would not generally be adequate. The measuring system may provide measured values that are not dispersed about the true value, but about some value offset [or correction] from it. Take a domestic bathroom scale. Suppose it is not set to show zero when there is nobody on the scale, but to show some value offset from zero. Then, no matter how many times the person's mass were re-measured, the effect of this offset would be inherently present in the average of the values." (WK 2012aaa)

Types of Measurement Uncertainty

In general, "there are two types of measurement error," namely, "systematic error and random error" (as summarized in *Table 3.3*). (WK 2012aaa)

(a) Systematic Error

The first type of measurement uncertainty has to do with "systematic error," which is a measurement bias "associated with the fact that a measured value contains an offset [or a correction]. In general, a systematic error, regarded as a quantity, is a component of error that remains constant or depends in a specific manner on some other quantity." (WK 2012aaa)

(b) Random Error

And the second type of measurement uncertainty has to do with "random error," which "is associated with the fact that when a measurement is repeated, it will generally provide a measured value that is different from the previous value. It is random in that the next measured value cannot be predicted exactly from previous such values." (WK 2012aaa)

Models of Measurement Uncertainty

In real life, "the direct measurement of a quantity…occurs rarely," so models are needed.

But "there are many types of measurement in practice and therefore many models. A simple measurement model (for example for a scale, where the mass is proportional to the extension of the spring) might be sufficient for everyday domestic use. Alternatively, a more sophisticated model of a weighing, involving additional effects such as air buoyancy, is capable of delivering better results for industrial or scientific purposes. In general there are often several different quantities, for example temperature, humidity and displacement, that contribute to the definition of the measurand, and that need to be measured." (WK 2012aaa)

In addition, "correction terms should be included in the measurement model when the conditions of measurement are not exactly as stipulated. These terms correspond to systematic errors. Given an estimate of a correction term, the relevant quantity should be corrected by this estimate. There will be an uncertainty associated with the estimate, even if the estimate is zero, as is often the case. Instances of systematic errors arise in height measurement, when the alignment of the measuring instrument is not perfectly vertical, and the ambient temperature is different from that prescribed. Neither the alignment of the instrument nor the ambient temperature is specified exactly, but information concerning these effects is available, for example the lack of alignment is at most 0.001° and the ambient temperature at the time of measurement differs from that stipulated by at most 2 °C." (WK 2012aaa)

In general, "the items required by a measurement model to define a measurand are known as input quantities in a measurement model. The model is often referred to as a functional relationship. The output quantity in a measurement model is the measurand. Formally, the output quantity, denoted by Y, about which information is required, is often related to input quantities, denoted by $X_1, \ldots, X_N$, about which information is available, by a measurement model in the form" as shown below: (WK 2012aaa)

$$Y = f(X_1, \ldots, X_N),$$

with f above to refer to "the measurement function. A general expression for a measurement model" is shown below: (WK 2012aaa)

$$h(Y, X_1, \ldots, X_N) = 0.$$

This means that "a procedure exists for calculating Y given $X_1, \ldots, X_N$ and that Y is uniquely defined by this equation." (WK 2012aaa)

Probability Distributions of Measurement Uncertainty

However, "the true values of the input quantities $X_1, \ldots, X_N$ are unknown," so "$X_1, \ldots, X_N$ are characterized by probability distributions and treated mathematically as random variables. These distributions describe the respective probabilities of their true values lying in different intervals, and are assigned based on available knowledge concerning $X_1, \ldots, X_N$. Sometimes, some or all of $X_1, \ldots, X_N$ are interrelated and the relevant distributions, which are known as joint, apply to these quantities taken together." (WK 2012aaa)

Then, "consider estimates $x_1, \ldots, x_N$, respectively, of the input quantities $X_1, \ldots, X_N$, obtained from certificates and reports, manufacturers' specifications, the analysis of measurement data, and so on. The probability distributions characterizing $X_1, \ldots, X_N$ are chosen such that the estimates $x_1, \ldots, x_N$, respectively, are the expectations of $X_1, \ldots, X_N$. Moreover, for the ith input quantity, consider a so-called standard uncertainty, given the symbol $u(x_i)$, defined as the standard deviation of the input quantity X_i. This standard uncertainty is said to be associated with the (corresponding) estimate X_i." (WK 2012aaa)

Of course, "the use of available knowledge to establish a probability distribution to characterize each quantity of interest applies to the X_i and also to Y. In the latter case, the characterizing probability distribution for Y is determined by the measurement model together with the probability distributions for the X_i. The determination of the probability distribution for Y from this information is known as the *propagation of distributions*." (WK 2012aaa)

In addition, "prior knowledge about the true value of the output quantity Y can also be considered. For the domestic bathroom scale, the fact that the person's mass is positive, and that it is the mass of a person, rather than that of a motor car, that is being measured, both constitute prior knowledge about the possible values of the measurand in this example. Such additional information can be used to provide a probability distribution for Y that can give a smaller standard deviation for Y and hence a smaller standard uncertainty associated with the estimate of Y." (WK 2012aaa; J. Bernardo 1994; C. Elster 2007)

Evaluation of Measurement Uncertainty

There are two main types of evaluating measurement uncertainty, namely, Type A and Type B (as summarized in *Table 3.3*).

(a) Type A Evaluation

In Type A evaluation, "knowledge about an input quantity X_i is inferred from repeated measured values," and "the assumption is often made that the distribution best describing an input quantity X given repeated measured values of it (obtained independently) is a Gaussian distribution. X then has expectation equal to the average measured value and standard deviation equal to the standard deviation of the average. When the uncertainty is evaluated from a small number of measured values (regarded as instances of a quantity characterized by a Gaussian distribution), the corresponding distribution can be taken as a t-distribution." (WK 2012aaa)

(b) Type B Evaluation

in Type B evaluation, knowledge about an input quantity X_i is inferred from "scientific judgement or other information concerning the possible values of the quantity," and "often the only available information is that X lies in a specified interval [a, b]. In such a case, knowledge of the quantity can be characterized by a rectangular probability distribution[8] with limits a and b." (WK 2012aaa; K. Weise 1992)

Stages of the Evaluation of Measurement Uncertainty

Ther are two main stages of evaluating measurement uncertainty, namely, (a) formulation stage and (b) calculation stage (as summarized in *Table 3.3*). (WK 2012aaa)

(a) Formulation Stage

In the formulation stage, the following tasks are carried out: (WK 2012aaa)

- "defining the output quantity Y (the measurand)"
- "identifying the input quantities on which Y depends"
- "developing a measurement model relating Y to the input quantities"
- "on the basis of available knowledge, assigning probability distributions—Gaussian, rectangular, etc.—to the input quantities (or a joint probability distribution to those input quantities that are not independent)"

Of course, there can be other tasks, so these are just examples.

(b) Calculation Stage

In the calculation stage, the focus is on "propagation and summarizing," that is, "propagating the probability distributions for the input quantities through the measurement model to obtain the probability distribution for the output quantity Y, and summarizing by using this distribution to obtain" the following info: (WK 2012aaa)

- "the expectation of Y, taken as an estimate y of Y"
- "the standard deviation of Y, taken as the standard uncertainty $u(y)$ associated with y"
- "a coverage interval containing Y with a specified coverage probability"

In fact, "the propagation stage of uncertainty evaluation is known as the propagation of distributions, various approaches for which are available." (WK 2012aaa)

Problems with the Study of Measurement Uncertainty

However, for the critics, this study of measurement uncertainty has some major problems. Consider, for illustration, a few examples below (as summarized in *Table 3.3*).

Firstly, one main criticism of the study of measurement uncertainty is that there are alternative perspectives to understand uncertainty.

For example, for the critics, this study of measurement uncertainty claims that "random variables are proper mathematical models for uncertain quantities and simple probability distributions are sufficient for representing all forms of measurement uncertainties. In some situations, however, a mathematical interval rather than a probability distribution might be a better model of uncertainty. This may include situations involving periodic measurements, binned data values, censoring, detection limits, or plus-minus ranges of measurements where no particular probability distribution seems justified or where one cannot assume that the errors among individual measuresments are completely independent." (WK 2012aaa)

Secondly, another main criticism of the study of measurement uncertainty is that its assumptions can be questioned.

For instance, for the critics, in Type A evaluation, an assumption here is that "the distribution best describing an input quantity X given repeated measured values of it (obtained independently) is a Gaussian distribution," but the problem here is that "other considerations" are needed in other cases, especially "when the measured values are not obtained independently." (WK 2012aaa) And in Type B evaluation, an assumption here is that knowledge about an input quantity X_i is inferred from "scientific judgement or other information concerning the possible values of the quantity," and "often the only available information is that X lies in a specified interval [a, b]"; but the problem here is that "different information" can be needed in other cases, so "a probability distribution consistent with that information would be used" instead. (WK 2012aaa; K. Weise 1992)

Thirdly, still another main criticism of the study of measurement uncertainty is that there are more robust ways to do measurement uncertainty.

For instance, for the critics, this study of measurement uncertainty is not robust enough, and there are alternative ways. An excellent example of "a more robust representation of measurement uncertainty" is the use of "interval [*a*,*b*]," which is "different from a rectangular or uniform probability distribution over the same range in that the latter suggests that the true value lies inside the right half of the range $[(a+b)/2, b]$ with probability one half, and within any subinterval of $[a,b]$ with probability equal to the width of the subinterval divided by $b–a$. The interval makes no such claims, except simply that the measurement lies somewhere within the interval. Distributions of such measurement intervals can be summarized as probability boxes and Dempster-Shafer structures over the real numbers, which incorporate both aleatoric and epistemic uncertainties." (WK 2012aaa)

Fourthly, an additional main criticism of the study of measurement uncertainty is that it has its own inherent limits.

For instance, for the critics, in 1997, a Joint Committee for Guides in Metrology admitted that, in evaluating measurement uncertainty, "it is not possible to state how well the true value of the measurand is known, but only how well it is believed to be known. Measurement uncertainty can therefore be described as a measure of how well one believes one knows the true value of the measurand. This uncertainty reflects the incomplete knowledge of the measurand." (WK 2012aaa)

And fifthly, another additional main criticism of the study of measurement uncertainty is that there are conflicting interpretations of probability in measurement uncertainty.

For instance, for the critics, the Joint Committee for Guides in Metrology is more inclined towards a "subjectivist" interpretation when it treated "measurement uncertainty…as a measure of how well one believes one knows the true value of the measurand….The notion of 'belief' is an important one, since it moves metrology into a realm where results of measurement need to be considered and quantified in terms of probabilities that express degrees of belief." (WK 2012aaa) But the problem here is that there is an opposing interpretation, known as the "objectivist" interpretation of probability (as already addressed in Chapter Two).

Beyond Measurement Uncertainty

These criticisms of the study of measurement uncertainty do not claim a total rejection of its usefulness but serve to show us the two opposing sides of the debate, such that the possibility (or impossibility) and desirability (or undesirability) of subjectivity (from the perspective of nature with subjectivity, measurement, and natural condition as a case study here) are not to the extent that the respective defenders would like us to believe.

Furthermore, the analysis of subjectivity, measurement, and natural condition illuminates the insights of the ontological principles in existential dialectics, and good examples include the formalness-informalness principle, the absoluteness-relativeness principle, the partiality-totality principle, the predictability-unpredictability principle, the explicability-inexplicability principle, the fiction-reality principle, the cognitiveness-noncognitiveness principle, the finiteness-transfiniteness principle, the preciseness-vagueness principle, the simpleness-complicatedness principle, the openness-hiddenness principle, the denseness-emptiness principle, the change-constancy principle, the order-chaos principle, the slowness-quickness principle, the expansion-contraction principle, the theory-praxis principle, the convention-novelty principle, the evolution-transformation principle, the symmetry-asymmetry principle, the softness-hardness principle, the seriousness-playfulness principle, the activeness-inactivess principle, the selfness-otherness principle, the regression-progression principle, the same-difference principle, the stability-reaction principle, the functionality-nonfunctionality principle, and the intentionality-nonintentionality principle.

For instance, in relation to the formalness-informalness principle, if there is formalness (e.g., the formal logical requirement of soundness, completeness, and consistency in a system of ideas, as in the study of measurement uncertainty), there is informalness (e.g., the non-formal existence of unsoundness, incompleteness, and inconsistency in the study of measurement uncertainty, as shown in the criticisms that there are alternative perspectives to understand uncertainty, that its assumptions can be questioned, that there are more robust ways to do measurement uncertainty, that it has its own inherent limits, and that there are conflicting interpretations of probability in measurement uncertainty). And the reverse direction also holds true.

In relation to the absoluteness-relativeness principle, if there is absoluteness (e.g., the absolute view by those in mainstream metrology who claimed that "random variables are proper mathematical models for uncertain quantities and simple probability distributions are sufficient for representing all forms of measurement uncertainties"), there is relativeness (e.g., what is right for those in mainstream metrology who claimed that "random variables are proper mathematical models for uncertain quantities and simple probability distributions are sufficient for representing all forms of measurement uncertainties" is not necessarily so for the critics who offered alternative perspectives to understand uncertainty, like the idea that "a mathematical interval rather than a probability distribution might be a better model of uncertainty"). And the reverse direction also holds true.

In relation to the partiality-totality principle, if there is partiality (e.g., the partial view on measurement uncertainty by those in the Joint Committee for Guides in Metrology, when they proposed a "subjectivist" interpretation in treating "measurement uncertainty…as a measure of how well one believes one knows the true value of the measurand," and "the notion of 'belief' is an important one, since it moves metrology into a realm where results of measurement need to be considered and quantified in terms of probabilities that express degrees of belief"), there is totality (e.g., the more holistic view on measurement uncertainty, such that the whole is not the sum of its parts, that is, it cannot be reduced to the sum of opposing views like the "subjectivist" interpretation, the "objectivist" interpretation, etc., since there will emerge new views in the future not yet known today, just as there are already alternative views nowadays which do not side exclusively with any of them, as is the analysis here). And the reverse direction also holds true.

In relation to the predictability-unpredictability principle, if there is predictability (e.g., the predictable tendency of a "random error" in measurement uncertainty to be "associated with the fact that when a measurement is repeated, it will generally provide a measured value that is different from the previous value"), there is unpredictability (e.g., the more difficult task to predict exactly what the next measured value is on a particular occasion, since a "random error" in measurement uncertainty is "random in that the next measured value cannot be predicted exactly from previous such values"). And the reverse direction also holds true.

In relation to the explicability-inexplicability principle, if there is explicability (e.g., the explanation, by those in mainstream metrology, of the desirability of using "random variables" and "simple probability distributions" for understanding measurement uncertainties"), there is inexplicability (e.g., the lack of sufficient explanation, by those in mainstream metrology, of why using "random variables" and "simple probability distributions" are necessarily sufficient for understanding measurement uncertainties, since, for the critics, "in some situations, however, a mathematical interval rather than a probability distribution might be a better model of uncertainty"). And the reverse direction also holds true.

In relation to the fiction-reality principle, if there is fiction (e.g., the fictional aspect of Type A evaluation, insofar as the assumption that "the distribution best describing an input quantity X given repeated measured values of it [obtained independently] is a Gaussian distribution" is problematic in other cases, especially "when the measured values are not obtained independently"), there is reality (e.g., the realistic aspect of Type

A evaluation, insofar as the assumption that "the distribution best describing an input quantity X given repeated measured values of it (obtained independently) is a Gaussian distribution" is valid in some situations, especially when the measured values are obtained independently). And the reverse direction also holds true.

In relation to the cognitiveness-noncognitiveness principle, if there is cognitiveness (e.g., a view of things on the basis of reason and evidences, as shown in the 2 stages of evaluating measurement uncertainty, in regard to output quantity Y and input quantities $X_1, \ldots, X_N$), there is noncognitiveness (e.g., a view of things on the basis of non-cognitive factors like envy, jealousy, power, nationality, race, gender, age, class, greed, lust, status, faith, anger, sadness, joy, fear, wish, etc., as shown in the confession by the Joint Committee for Guides in Metrology in 1997 when it treated "measurement uncertainty…as a measure of how well one believes one knows the true value of the measurand," on the basis of "degrees of belief" or faith). And the reverse direction also holds true.

In relation to the finiteness-transfiniteness principle, if there is finiteness (e.g., the finite number of the categories of situation, in which "a mathematical interval rather than a probability distribution might be a better model of uncertainty," like the situations involving "periodic measurements, binned data values, censoring, detection limits, or plus-minus ranges of measurements where no particular probability distribution seems justified or where one cannot assume that the errors among individual measuresments are completely independent"), there is transfiniteness (e.g., the transfinite number of all the differentiable units in a normal or Gaussian distribution in a Type A evaluation of measurement uncertainty, since "the function is supersmooth…, implying that it is infinitely differentiable"). (WK 2012aaa & 2012cc) And the reverse direction also holds true.

In relation to the preciseness-vagueness principle, if there is preciseness (e.g., the precise classification of 2 stages of evaluating measurement uncertainty in the article aforecited), there is vagueness (e.g., the vagueness in the classification of 2 stages of evaluating measurement uncertainty in the article aforecited, since it is not clear why there must be only 2, not 3, 4, 5, and so on). And the reverse direction also holds true.

In relation to the simpleness-complicatedness principle, if there is simpleness (e.g., the relatively simple analysis of measurement uncertainty by its advocates), there is complicatedness (e.g., the relatively more complicated analysis of measurement uncertainty, by challenging the claims and assumptions, as shown in the problems pointed out by the critics as cited). And the reverse direction also holds true.

In relation to the openness-hiddenness principle, if there is openness (e.g., the open exploration of evaluating measurement uncertainty by Type A and Type B), there is hiddenness (e.g., the hidden bias in Type A and Type B evaluations of measurement uncertainty, because of its questionable assumptions afore-discussed). And the reverse direction also holds true.

In relation to the denseness-emptiness principle, if there is denseness (e.g., the relatively denser concentration of concerns with "random variables" and "simple probability distributions" for evaluating measurement uncertainties by those in mainstream metrology), there is emptiness (e.g., the relatively less dense, or more empty, concentration of concerns with "random variables" and "simple probability distributions" for evaluating measurement uncertainties by the critics who offered an alternative perspective, that is, the idea that "a mathematical interval rather than a probability distribution might be a better model of uncertainty"). And the reverse direction also holds true.

In relation to the change-constancy principle, if there is change (e.g., in a random error, "when a measurement is repeated, it will generally provide a measured value that is different from the previous value"), there is constancy (e.g., the existence of "quantities representing physical constants," like "modulus of elasticity and specific heat" in measurement uncertainty). (WK 2012aaa) And the reverse direction also holds true.

In relation to the order-chaos principle, if there is order (e.g., the more or less orderly view on random error, if looked from the vantage point that, as is well expected, "when a measurement is repeated it will generally provide a measured value that is different from the previous value"), there is chaos (e.g., the more or less chaotic view on random error, if looked from the vantage point that "the next measured value cannot be predicted exactly from previous such values," such that it can go in any direction). And the reverse direction also holds true.

In relation to the slowness-quickness principle, if there is slowness (e.g., the relatively slower speed of evaluating measurement uncertainty, if the task is complicated by including difficult situations like a huge number of output quantities and input quantities, "periodic measurements, binned data values, censoring, detection limits,...plus-minus ranges of measurements," etc.), there is quickness (e.g., the relatively quicker speed of evaluating measurement uncertainty, if the task is simplified by focusing only on situations involving "random variables" and "simple probability distributions," together with the inherent restrictive assumptions). And the reverse direction also holds true.

In relation to the expansion-contraction principle, if there is expansion (e.g., the expansion of the size of a mathematical model to evaluate measurement uncertainty, if it includes a huge number of output quantities and input quantities, "periodic measurements, binned data values, censoring, detection limits,...plus-minus ranges of measurements," etc.), there is contraction (e.g., the relative shrinkage of the size of a mathematical model to evaluate measurement uncertainty, if it makes things easier by focusing only on simple situations involving only a few "random variables" and "simple probability distributions," together with the inherent restrictive assumptions). And the reverse direction also holds true.

In relation to the theory-praxis principle, if there is theory (e.g., the theoretical construction of the models of evaluating measurement uncertainty in metrology), there is praxis (e.g., the practical application of the models of evaluating measurement uncertainty by such applied organizations like the International Electrotechnical Commission, the International Federation of Clinical Chemistry and Laboratory Medicine, the International Laboratory Accreditation Cooperation, the International Organization for Standardization, the International Union of Pure and Applied Chemistry, etc.). (WK 2012aaa) And the reverse direction also holds true.

In relation to the convention-novelty principle, if there is convention (e.g., the conventional wisdom about using only "random variables" and "simple probability distributions" for evaluating measurement uncertainties by those in mainstream metrology), there is novelty (e.g., the alternative novel challenge to the conventional wisdom about using only "random variables" and "simple probability distributions" for evaluating measurement uncertainties by those in mainstream metrology, by the new idea of the critics, in that "a mathematical interval rather than a probability distribution might be a better model of uncertainty"). And the reverse direction also holds true.

In relation to the evolution-transformation principle, if there is evolution (e.g., the natural evolution of humans to measure things in the state of nature in a primitive way), there is transformation (e.g., the technical transformation of human ability to understand measurement by the invention of "metrology" for studying measurement uncertainty). And the reverse direction also holds true.

In relation to the symmetry-asymmetry principle, if there is symmetry (e.g., the coexistence of different ways for evaluating measurement uncertainties), there is asymmetry (e.g., the popularity of the simple way of using only "random variables" and "simple probability distributions" for evaluating measurement uncertainties among those in mainstream metrology—but the popularity of the more complicated way of using "a mathematical interval rather than a probability distribution," together with multivariate mathematical modeling, for evaluating measurement uncertainties among the critics). And the reverse direction also holds true.

In relation to the softness-hardness principle, if there is softness (e.g., the defense of using only "random variables" and "simple probability distributions" for evaluating measurement uncertainties by those in mainstream metrology aforecited), there is hardness (e.g., the critique of using only "random variables" and "simple probability distributions" for evaluating measurement uncertainties by the critics aforecited). And the reverse direction also holds true.

In relation to the seriousness-playfulness principle, if there is seriousness (e.g., the serious business of those who try to understand the business of evaluating uncertainty), there is playfulness (e.g., the playful part of those who try to understand the business of evaluating uncertainty, when they play around with different proposals and interpretations over time, like the conventional way of using only "random variables" and "simple probability distributions," the more robust way of using "interval [*a*,*b*]," the "subjectivist" interpretation, the "objectivist" interpretation, etc.). And the reverse direction also holds true.

In relation to the activeness-inactivess principle, if there is activeness (e.g., the relative activeness of those in favor of using "random variables" and "simple probability distributions" for evaluating measurement uncertainties, when they try to apply them in simple situations), there is inactiveness (e.g., the relative inactiveness of those in favor of using "random variables" and "simple probability distributions" for evaluating measurement uncertainties, when they confront difficult situations which require different practices, like a huge number of output quantities and input quantities, "periodic measurements, binned data values, censoring, detection limits,...plus-minus ranges of measurements," etc., which then require alternative ways of evaluating measurement uncertainty). And the reverse direction also holds true.

In relation to the selfness-otherness principle, if there is selfness (e.g., the tendency to be mostly centered on one's self or group, as shown in those manufacturing firms, which use metrology for their products, like "the bathroom scale" which "convert[s] a measured extension of a spring into an estimate of the measurand, the mass of the person on the scale"), there is otherness (e.g., the tendency to be mostly centered on others, as shown in the Joint Committee for Guides in Metrology to provide guidelines for the advance in the knowledge of measurement uncertainty for humanity). And the reverse direction also holds true.

In relation to the regression-progression principle, if there is regression (e.g., the regression made by the study of measurement uncertainty, as shown in the problems pointed out by the critics), there is progression (e.g., the progress made by the study of measurement uncertainty, as shown in the contribution to the understanding of systematic and random errors aforecited). And the reverse direction also holds true.

In relation to the same-difference principle, if there is similarity in outcome (e.g., the contribution to the molding and control of beliefs, values, and behaviors, regardless of whether this be done by way of "modern standards" for measurement, or by way of "arbitrary standards set up by regional or local authorities" in antiquity), there is difference in outcome (e.g., the contribution to the molding and control of beliefs, values, and behaviors by way of "modern standards" for measurement for a relatively more scientific worldview—but the contribution to the molding and control of beliefs, values, and behaviors by way of "arbitrary standards set up by regional or local authorities" in antiquity for a relatively more pre-scientific worldview). (WK 2012bbb) And the reverse direction also holds true.

In relation to the stability-reaction principle, if there is stability (e.g., the relatively more stable condition of "arbitrary standards set up by regional or local authorities" for measurement in antiquity), there is reaction (e.g., "arbitrary standards set up by regional or local authorities" for measurement in antiquity led to further changes later on, when "modern standards" for measurement were set up since the scientific revolution). (WK 2012aaa) And the reverse direction also holds true.

In relation to the functionality-nonfunctionality principle, if there is functionality (e.g., the presence of action for which something or someone is fitted, as shown in the use of "arbitrary standards set up by regional or local authorities" for measurement in antiquity, which served the function of the pre-scientific lifeworld at the time), there is nonfunctionality (e.g., the relative lack of action for which something or someone is fitted, as shown in the use of "modern standards" for measurement set up since the scientific revolution, such that the older "arbitrary standards set up by regional or local authorities" for measurement in antiquity are no longer much functional in our relatively more scientific age). And the reverse direction also holds true.

In relation to the intentionality-nonintentionality principle, if there is intentionality (e.g., the planning or design of something for a certain outcome, as shown in the "arbitrary standards set up by regional or local authorities" in antiquity, for measurement as a way to facilitate local commerce), there is nonintentionality (e.g., the relative absence of planning or design of something for a certain outcome, such that the outcome can even be counter-intentional in being the opposite of what was originally planned, as shown in the "modern standards" for measurement set up since the scientific revolution, along the more scientific line, contrary to the original pre-scientific "arbitrary standards" in antiquity). (WK 2012bbb) And the reverse direction also holds true.

And in relation to the survivability-nonsurvivability principle, if there is survivability (e.g., the survival, in our time, of many "modern standards" for measurement set up since the scientific revolution, like mathematical modeling aforecited), there is non-survivability (e.g., the non-survival, in our time, of many "arbitrary standards set up by regional or local authorities" in antiquity). And the reverse direction also holds true.

Subjectivity, Quantum Mechanics, and Uncertainty

Another good way to evaluate the possibility (or impossibility) and desirability (or undesirability) of subjectivity from the perspective of nature is to explore subjectivity, quantum mechanics, and uncertainty.

An argument in the literature is that probability theory contributes to the understanding of uncertainty in the quantum mechanical world—with an element of subjectivity, in that "there is no way to say what the state of a system fundamentally is, only what the result of observations might be….Niels Bohr believed that the probability distributions…depend on which measurements we choose to perform." (WK 2012r)

Uncertainty and Related Concepts

As a start, the term "uncertainty" is sometimes confused with the related term "risk," so "many specialists in decision theory, statistics and other quantitative fields have defined uncertainty, risk, and their measurement" in different ways, as shown below (and summarized in *Table 3.4*): (WK 2012ccc)

- "Uncertainty"—"the lack of certainty, A state of having limited knowledge where it is impossible to exactly describe the existing state, a future outcome, or more than one possible outcome."
- "Risk"—"a state of uncertainty where some possible outcomes have an undesired effect or significant loss"

- "Measurement of Uncertainty"—"a set of possible states or outcomes where probabilities are assigned to each possible state or outcome—this also includes the application of a probability density function to continuous variables"
- "Measurement of Risk"—"a set of measured uncertainties where some possible outcomes are losses, and the magnitudes of those losses—this also includes loss functions over continuous variables"

Of course, there can be other definitions, so the ones above are illustrative, not exhaustive.

History of the Uncertainty Principle

This clarification of the term "uncertainty" aside—the issue of uncertainty is important in the subfield of physics known as "quantum mechanics," where there exists "the uncertainty principle," which is "any of a variety of mathematical inequalities asserting a fundamental limit to the precision with which certain pairs of physical properties of a particle, such as position x and momentum p, can be known simultaneously. The more precisely the position of some particle is determined, the less precisely its momentum can be known, and vice versa"—and "according to the Copenhagen interpretation of quantum mechanics, there is no fundamental reality that the quantum state describes, just a prescription for calculating experimental results. There is no way to say what the state of a system fundamentally is, only what the result of observations might be....Niels Bohr believed that the probability distributions...depend on which measurements we choose to perform." (WK 2012r)

In 1925, Werner Heisenberg "formulated the Uncertainty Principle at Niels Bohr's institute in Copenhagen, while working on the mathematical foundations of quantum mechanics," and, "following pioneering work with Hendrik Kramers, Heisenberg developed matrix mechanics, which replaced the ad-hoc old quantum theory with modern quantum mechanics. The central assumption was that the classical concept of motion does not fit at the quantum level, and that electrons in an atom do not travel on sharply defined orbits. Rather, the motion is smeared out in a strange way: the Fourier transform of time only involve those frequencies that could be seen in quantum jumps." (WK 2012r; D. Cassidy 2012)

At the time, "Heisenberg's paper did not admit any unobservable quantities like the exact position of the electron in an orbit at any time; he only allowed the theorist to talk about the Fourier components of the motion. Since the Fourier components were not defined at the classical frequencies, they could not be used to construct an exact trajectory, so that the formalism could not answer certain overly precise questions about where the electron was or how fast it was going. In March 1926, working in Bohr's institute, Heisenberg realized that the non-commutativity implies the uncertainty principle. This implication provided a clear physical interpretation for the non-commutativity, and it laid the foundation for what became known as the Copenhagen interpretation of quantum mechanics. Heisenberg showed that the commutation relation implies an uncertainty, or in Bohr's language a complementarity. Any two variables that do not commute cannot be measured simultaneously—the more precisely one is known, the less precisely the other can be known." (WK 2012r; N. Bohr 1958)

As Heisenberg himself, put it, "it can be expressed in its simplest form as follows: One can never know with perfect accuracy both of those two important factors which determine the movement of one of the smallest particles—its position and its velocity. It is impossible to determine accurately both the position and the direction and speed of a particle at the same instant." (WK 2012r)

For historical clarification, it is important to note that, "throughout the main body of his original 1927 paper, written in German, Heisenberg used the word, 'Unbestimmtheit' ('indeterminacy'), to describe the basic theoretical principle. Only in the endnote did he switch to the word, 'Unsicherheit' ('uncertainty'). When the English-language version of Heisenberg's textbook, *The Physical Principles of the Quantum Theory*, was published in 1930, however, the translation 'uncertainty' was used, and it became the more commonly used term in the English language thereafter." (WK 2012r; D. Cassidy 2009)

Formulations of the Uncertainty Principle

Over the decades, there are multiple formulations of the uncertainty principle. Consider, for illustration, two main ones hereafter, namely, (a) "wave mechanics" formulation and (b) "matrix mechanics" formulation (as summarized in *Table 3.4*).

(a) Wave Mechanics Formulation

In "wave mechanics" formulation, in accordance to the de Broglie hypothesis, "every object in the universe is a wave, a situation which gives rise to this phenomenon. The position of the particle is described by a wave function $\Psi(x,t)$....The Born rule states that this should be interpreted as a probability density function in the sense that the probability of finding the particle between *a* and *b* is" shown below: (WK 2012r)

$$\mathrm{P}[a \le X \le b] = \int_a^b |\psi(x)|^2 \, \mathrm{d}x$$

Now, "in the case of the single-moded plane wave, $|\psi(x)|^2$ is a uniform distribution. In other words, the particle position is extremely uncertain in the sense that it could be essentially anywhere along the wave packet." (WK 2012r)

Since the wave function consists of many possible waves, "adding together all of these plane waves comes at a cost, namely the momentum has become less precise, having become a mixture of waves of many different momenta. One way to quantify the precision of the position and momentum is the standard deviation σ. Since $|\psi(x)|^2$ is a probability density function for position, we calculate its standard deviation." (WK 2012r)

Here is the trade-off: "We improved the precision of the position, i.e. reduced σ_x, by using many plane waves, thereby weakening the precision of the momentum, i.e. increased σ_p. Another way of stating this is that σ_x and σ_p have an inverse relationship or are at least bounded from below. This is the uncertainty principle…." (WK 2012r)

(b) Matrix Mechanics Formulation

In "the matrix mechanics" formulation, it all starts with the understanding that "observables such as position and momentum are represented by self-adjoint operators. When considering pairs of observables, one of the most important quantities is the commutator. For a pair of operators $\hat{A}$ and $\hat{B}$, we may define their commutator" as shown below: (WK 2012r)

$$[\hat{A}, \hat{B}] = \hat{A}\hat{B} - \hat{B}\hat{A}.$$

Then, "in the case of position and momentum, the commutator is the canonical commutation relation" as shown below: (WK 2012r)

$$[\hat{x}, \hat{p}] = i\hbar.$$

Now, "the physical meaning of the non-commutativity can be understood by considering the effect of the commutator on position and momentum eigenstates. Let $|\psi\rangle$ be a right eigenstate of position with a constant eigenvalue x_0....Suppose for the sake of proof by contradiction that $|\psi\rangle$ is also a right eigenstate of momentum, with constant eigenvalue p_0." (WK 2012r)

But this contradicts the canonical commutation relation requirement as shown below: (WK 2012r)

$$[\hat{x}, \hat{p}]|\psi\rangle = i\hbar|\psi\rangle \neq 0.$$

Therefore, "this implies that no quantum state can be simultaneously both a position and a momentum eigenstate. When a state is measured, it is projected onto an eigenstate in the basis of the observable. For example, if a particle's position is measured, then the state exists at least momentarily in a position eigenstate. This means that the state is not in a momentum eigenstate, however, but rather exists as a sum of multiple momentum basis eigenstates. In other words the momentum must be less precise. The precision may be quantified by the standard deviations, defined by" the following equations: (WK 2012r)

$$\sigma_x = \sqrt{\langle \hat{x}^2 \rangle - \langle \hat{x} \rangle^2}$$
$$\sigma_p = \sqrt{\langle \hat{p}^2 \rangle - \langle \hat{p} \rangle^2}.$$

Like "the wave mechanics interpretation above, we see a tradeoff between the precisions of the two, given by the uncertainty principle." (WK 2012r)

Heisenberg's Illustration of the Uncertainty Principle

These mathematical equations may look intimidating, so Heisenberg provided a simple illustration of the uncertainty principle by asking the reader to imagine "an experimenter trying to measure the position and momentum of an electron by shooting a photon at it," but there are two problems here, as Heisenberg himself explained them below (as summarized in *Table 3.4*): (WK 2012r)

- "Problem 1"—"If the photon has a short wavelength, and therefore, a large momentum, the position can be measured accurately. But the photon scatters in a random direction, transferring a large and uncertain amount of momentum to the electron. If the photon has a long wavelength and low momentum, the collision does not disturb the electron's momentum very much, but the scattering will reveal its position only vaguely."

- "Problem 2"—"If a large aperture is used for the microscope, the electron's location can be well resolved...; but by the principle of conservation of momentum, the transverse momentum of the incoming photon and hence, the new momentum of the electron resolves poorly. If a small aperture is used, the accuracy of both resolutions is the other way around."

So, Heisenberg's point is that "the combination of these trade-offs imply that no matter what photon wavelength and aperture size are used, the product of the uncertainty in measured position and measured momentum is greater than or equal to a lower limit, which is (up to a small numerical factor) equal to Planck's constant. Heisenberg did not care to formulate the uncertainty principle as an exact limit..., and preferred to use it instead, as a heuristic quantitative statement, correct up to small numerical factors, which makes the radically new noncommutativity of quantum mechanics inevitable." (WK 2012r)

Problems with the Uncertainty Principle

However, for the critics, this argument on quantum mechanics and uncertainty has some major problems. Consider, for illustration, a few examples below (as summarized in *Table 3.4*).

Firstly, one main criticism of the argument on quantum mechanics and uncertainty is that the loss of certainty in quantum mechanical measurements can be due to our ignorance of some hidden variables in the real world.

For instance, "Albert Einstein believed that randomness is a reflection of our ignorance of some fundamental property of reality" and "disliked this loss of determinism" in quantum mechanics, so he held that there must be some "hidden variables" to be understood and, "consequently, that the present theory [of quantum mechanics] was incomplete." (WK 2012r & 2012s)

Secondly, another main criticism of the argument on quantum mechanics and uncertainty is that quantum mechanics itself has gone through different revisions and interpretations over time.

For instance, "the earliest versions of quantum mechanics were formulated in the first decade of the 20th century....[Then] quantum theory was significantly reformulated in the mid-1920s by Werner Heisenberg, Max Born and Pascual Jordan, who created matrix mechanics;

Louis de Broglie and Erwin Schrodinger (Wave Mechanics); and Wolfgang Pauli and Satyendra Nath Bose (statistics of subatomic particles). And the Copenhagen interpretation of Niels Bohr became widely accepted. By 1930, quantum mechanics had been further…formalized by the work of David Hilbert, Paul Dirac and John von Neumann, with a greater emphasis placed on measurement in quantum mechanics, the statistical nature of our knowledge of reality, and philosophical speculation about the role of the observer. Quantum mechanics has since branched out into almost every aspect of 20th century physics and other disciplines, such as quantum chemistry, quantum electronics, quantum optics, and quantum information science." (WK 2012s) And this changing history led "the renowned physicist Richard Feynman" to once say, "I think I can safely say that nobody understands quantum mechanics." (WK 2012s; W. Greiner 1994)

Thirdly, still another main criticism of the argument on quantum mechanics and uncertainty is that it can be falsified.

For instance, the critic like Karl Popper (1959) "approached the problem of indeterminacy as a logician and metaphysical realist. He disagreed with the application of the uncertainty relations to individual particles rather than to ensembles of identically prepared particles, referring to them as 'statistical scatter relations.' In this statistical interpretation, a particular measurement may be made to arbitrary precision without invalidating the quantum theory. This directly contrasts the Copenhagen interpretation of quantum mechanics, which is non-deterministic but lacks local hidden variables." (WK 2012r; I. Jarvie 2006)

And fourthly, an additional main criticism of the argument on quantum mechanics and uncertainty is that there can be an alternative in-between perspective.

For instance, for the critics, "the many-worlds interpretation originally outlined by Hugh Everett III in 1957 is partly meant to reconcile the differences between the Einstein and Bohr's views by replacing Bohr's wave function collapse with an ensemble of deterministic and independent universes whose distribution is governed by wave functions and the Schrödinger equation. Thus, uncertainty in the many-worlds interpretation follows from each observer within any universe having no knowledge of what goes on in the other universes." (WK 2012r) Therefore, "many-worlds implies that all possible alternative histories and futures are real, each representing an actual 'world' (or 'universe')." (WK 2012ddd)

Beyond the Uncertainty Principle

These criticisms of quantum mechanics and uncertainty do not entail, however, a total rejection of its usefulness but serve to show us the two opposing sides of the debate, such that the possibility (or impossibility) and desirability (or undesirability) of subjectivity (from the perspective of nature with subjectivity, quantum mechanics, and uncertainty as a case study here) are not to the extent that the respective defenders would like us to believe.

Besides, the analysis of subjectivity, quantum mechanics, and uncertainty reveals something important about the ontological principles in existential dialectics, and good examples include the formalness-informalness principle, the absoluteness-relativeness principle, the partiality-totality principle, the predictability-unpredictability principle, the explicability-inexplicability principle, the fiction-reality principle, the cognitiveness-noncognitiveness principle, the finiteness-transfiniteness principle, the preciseness-vagueness principle, the simpleness-complicatedness principle, the openness-hiddenness principle, the denseness-emptiness principle, the change-constancy principle, the order-chaos principle, the slowness-quickness principle, the expansion-contraction principle, the theory-praxis principle, the convention-novelty principle, the evolution-transformation principle, the symmetry-asymmetry principle, the softness-hardness principle, the seriousness-playfulness principle, the activeness-inactivess principle, the selfness-otherness principle, the regression-progression principle, the same-difference principle, the stability-reaction principle, the functionality-nonfunctionality principle, and the intentionality-nonintentionality principle.

For instance, in relation to the formalness-informalness principle, if there is formalness (e.g., the formal logical requirement of soundness and consistency in a system of ideas, as in the uncertainty principle), there is informalness (e.g., the non-formal existence of unsoundness and inconsistency in the uncertainty principle, as shown in the criticisms that the loss of certainty in quantum mechanical measurements can be due to our ignorance of some hidden variables in the real world, that quantum mechanics itself has gone through different formulations over time, that it can be falsified, and that there can be an alternative in-between perspective). And the reverse direction also holds true.

In relation to the absoluteness-relativeness principle, if there is absoluteness (e.g., the absolute view by Einstein on the loss of certainty in quantum mechanical measurements due to our ignorance of some hidden variables in the real world), there is relativeness (e.g., what is right for

Einstein in regard to the loss of certainty in quantum mechanical measurements due to our ignorance of some hidden variables in the real world is not necessarily so for Heisenberg because of the uncertainty principle). And the reverse direction also holds true.

In relation to the partiality-totality principle, if there is partiality (e.g., the partial view on quantum mechanics by Heisenberg), there is totality (e.g., the more holistic view on quantum mechanics, such that the whole is not the sum of its parts, that is, it cannot be reduced to the sum of opposing views by Heisenberg, Einstein, Everett, etc., since there will emerge new views in the future not yet known today, just as there are already alternative views nowadays which do not side exclusively with any of them, as is the analysis here). And the reverse direction also holds true.

In relation to the predictability-unpredictability principle, if there is predictability (e.g., the predictable tendency by the advocates of the uncertainty principle that "the more precisely the position of some particle is determined, the less precisely its momentum can be known"), there is unpredictability (e.g., the more difficult task to predict exactly "which certain pairs of physical properties of a particle, such as position x and momentum p, can be known simultaneously," which is not possible in accordance to the uncertainty principle). And the reverse direction also holds true.

In relation to the explicability-inexplicability principle, if there is explicability (e.g., the explanation, by those in favor of the Copenhagen interpretation of quantum mechanics, that "there is no way to say what the state of a system fundamentally is, only what the result of observations might be"), there is inexplicability (e.g., the lack of sufficient explanation, by those in favor of the Copenhagen interpretation of quantum mechanics, of why the uncertainty principle is necessarily true, since the critic like Einstein argued that "randomness is a reflection of our ignorance of some fundamental property of reality," so "the present theory [of quantum mechanics] was incomplete"). And the reverse direction also holds true.

In relation to the fiction-reality principle, if there is fiction (e.g., the fictional aspect of the uncertainty principle, insofar as Popper pointed out that, "in…his statistical interpretation, a particular measurement may be made to arbitrary precision without invalidating the quantum theory" and "this directly contrasts the Copenhagen interpretation of quantum mechanics, which is non-deterministic but lacks local hidden variables"), there is reality (e.g., the realistic aspect of the uncertainty principle, insofar as Heisenberg pointed out that there is "a tradeoff between the precisions" of the position and momentum of a sub-atomic particle under certain conditions). And the reverse direction also holds true.

In relation to the cognitiveness-noncognitiveness principle, if there is cognitiveness (e.g., a view of things on the basis of reason and evidences, as shown in the 2 mathematical formulations of the uncertainty principle aforecited), there is noncognitiveness (e.g., a view of things on the basis of non-cognitive factors like envy, jealousy, power, nationality, race, gender, age, class, greed, lust, status, faith, anger, sadness, joy, fear, wish, etc., as shown in the wishful belief by some people that the uncertainty principle can be applied for the understanding of free will, even though the critics had showed that "nontrivial biological mechanisms requiring quantum mechanics are unlikely, due to the rapid decoherence time of quantum systems at room temperature"). (WK 2012r; P. Davies 2004) And the reverse direction also holds true.

In relation to the finiteness-transfiniteness principle, if there is finiteness (e.g., the finite number of the categories of worlds, like "histories," "futures," etc. in the "many-worlds interpretation" of quantum mechanics), there is transfiniteness (e.g., the transfinite number of all the possible individual "alternative histories and futures," according to the "many-worlds interpretation" of quantum mechanics). And the reverse direction also holds true.

In relation to the preciseness-vagueness principle, if there is preciseness (e.g., the precise identification of 2 formulations of the uncertainty principle in the article aforecited, namely, "wave mechanics" and "matrix mechanics"), there is vagueness (e.g., the vagueness in the identification of the 2 formulations of the uncertainty principle in the article aforecited, namely, "wave mechanics" and "matrix mechanics," since it is not clear why there must be only 2, not 3, 4, 5, and so on). And the reverse direction also holds true.

In relation to the simpleness-complicatedness principle, if there is simpleness (e.g., the relatively simple analysis of the uncertainty principle, by its advocates), there is complicatedness (e.g., the relatively more complicated analysis of the uncertainty principle, by challenging the claims and assumptions, as shown in the problems pointed out by the critics as cited). And the reverse direction also holds true.

In relation to the openness-hiddenness principle, if there is openness (e.g., the open exploration of "a tradeoff between the precisions" of the position and momentum of a sub-atomic particle, according to the uncertainty principle), there is hiddenness (e.g., the hidden bias in the uncertainty principle, insofar as there are criticisms against it, like the view by Einstein that the loss of certainty in quantum mechanical measurements can be due to our ignorance of some hidden variables in the real world, together with other problems as described earlier). And the reverse direction also holds true.

In relation to the denseness-emptiness principle, if there is denseness (e.g., the relatively denser concentration of concerns with indeterminism by those like Heiseberg in favor of the uncertainty principle), there is emptiness (e.g., the relatively less dense, or more empty, concentration of concerns with indeterminism by those like Einstein in favor of the argument about "some hidden variables in the real world"). And the reverse direction also holds true.

In relation to the change-constancy principle, if there is change (e.g., according to Niels Bohr's view on quantum mechanics, "the probability distributions…depend on which measurements we choose to perform," so different measurements can yield different probability distributions), there is constancy (e.g., according to to the Copenhagen interpretation of quantum mechanics, there is the constancy that "there is no way to say what the state of a system fundamentally is, only what the result of observations might be"). And the reverse direction also holds true.

In relation to the order-chaos principle, if there is order (e.g., the more or less orderly view on a particle, if looked from the vantage point of the "wave mechanics," when "σ_x and σ_p have an inverse relationship or are at least bounded from below"), there is chaos (e.g., the more or less chaotic view on a particle, if looked from the vantage point of the "wave mechanics," when, "in the case of the single-moded plane wave,…the particle position is extremely uncertain in the sense that it could be essentially anywhere along the wave packet"). And the reverse direction also holds true.

In relation to the slowness-quickness principle, if there is slowness (e.g., the relatively slower speed of a particle like an electron, if "the photon has a long wavelength and low momentum," so "the collision does not disturb the electron's momentum very much," according to the experiment with "an experimenter trying to measure the position and momentum of an electron by shooting a photon at it"), there is quickness (e.g., the relatively quicker speed of a particle like an electron, if "the photon has a short wavelength, and therefore, a large momentum,…transferring a large and uncertain amount of momentum to the electron," according to the experiment with "an experimenter trying to measure the position and momentum of an electron by shooting a photon at it"). And the reverse direction also holds true.

In relation to the expansion-contraction principle, if there is expansion (e.g., the relative expansion of the influence of quantum mechanics in the

20th ceutury), there is contraction (e.g., the relative decline of the influence of classical mechanics in the 20th ceutury). And the reverse direction also holds true.

In relation to the theory-praxis principle, if there is theory (e.g., the theoretical construction of the uncertainty principle in physics), there is praxis (e.g., the practical application of the uncertainty principle in physics to the field of philosophy, like the debate on the existence of "free will"). And the reverse direction also holds true.

In relation to the convention-novelty principle, if there is convention (e.g., the conventional wisdom about determinism in classical mechanics), there is novelty (e.g., the alternative novel challenge to the conventional wisdom about determinism in classical mechanics, by the new idea of indeterminism in quantum mechanics in the 20th century). And the reverse direction also holds true.

In relation to the evolution-transformation principle, if there is evolution (e.g., the natural evolution of humans to observe things in the state of nature, albeit in a primitive way), there is transformation (e.g., the technical transformation of human ability to observe things in nature by the invention of quantum mechanics). And the reverse direction also holds true.

In relation to the symmetry-asymmetry principle, if there is symmetry (e.g., the existence of different ways to understand reality, like classical mechanics and quantum mechanics), there is asymmetry (e.g., the popularity of quantum mechanics among 20th century physicists, especially in regard to the micro-world of particles—but the popularity of classical mechanics among 19th century physicists, especially in regard to the macro-world of everyday life). And the reverse direction also holds true.

In relation to the softness-hardness principle, if there is softness (e.g., the defense of the uncertainty principle by Heisenberg), there is hardness (e.g., the critique of the uncertainty principle by Einstein). And the reverse direction also holds true.

In relation to the seriousness-playfulness principle, if there is seriousness (e.g., the serious business of those physicists who try to understand quantum mechanics), there is playfulness (e.g., the playful part of those physicists who try to understand quantum mechanics, when they play around with different versions and interpretations over the decades, like "matrix mechanics," "wave mechanics," "statistics of subatomic particles," "the Copenhagen interpretation," "many-worlds interpretation," "philosophical speculation about the role of the observer," "quantum chemistry," "quantum electronics," "quantum optics," "quantum information science," etc.). And the reverse direction also holds true.

In relation to the activeness-inactivess principle, if there is activeness (e.g., the relative activeness of those physicists who actively try to apply quantum mechanics to the micro-world of particles), there is inactiveness (e.g., the relative inactiveness of those physicists who have difficutly to apply quantum mechanics to the macro-world of gravity, where the theory of relativity is more relevant instead).

In relation to the selfness-otherness principle, if there is selfness (e.g., the tendency to be mostly centered on one's self or group, as shown in the wishful attempt by some people to use the uncertainty principle for the understanding of whether they have free will or not, even though the critics had showed that "nontrivial biological mechanisms requiring quantum mechanics are unlikely, due to the rapid decoherence time of quantum systems at room temperature"), there is otherness (e.g., the tendency to be mostly centered on others, as shown in the work on the uncertainty principle by Heisenberg, as an advance in the knowledge of physics for humanity). And the reverse direction also holds true.

In relation to the regression-progression principle, if there is regression (e.g., the regression made by the uncertainty principle, as shown in the problems pointed out by the critics), there is progression (e.g., the progress made by the uncertainty principle, as shown in its contribution to the understanding of the "tradeoff between the precisions" of the position and momentum of a sub-atomic particle). And the reverse direction also holds true.

In relation to the same-difference principle, if there is similarity in outcome (e.g., the contribution to the molding and control of beliefs, values, and behaviors, regardless of whether this be done by way of quantum mechanics, or by way of classical mechanics), there is difference in outcome (e.g., the contribution to the molding and control of beliefs, values, and behaviors by way of quantum mechanics for a relatively more indetermistic worldview—but the contribution to the molding and control of beliefs, values, and behaviors by way of classical mechanics for a relatively more deterministic worldview). And the reverse direction also holds true.

In relation to the stability-reaction principle, if there is stability (e.g., the relatively more stable condition of classical mechanics in the older days), there is reaction (e.g., classical mechanics in the older days then led to further changes later on, when it was challenged by quantum mechanics in the 20th century for a more indeterministic view of the world). And the reverse direction also holds true.

In relation to the functionality-nonfunctionality principle, if there is functionality (e.g., the presence of action for which something or someone is fitted, as shown in the influence of classical mechanics in the older days, when it served the function of holding a deterministic worldview at that time), there is nonfunctionality (e.g., the relative lack of action for which something or someone is fitted, as shown in the rise of quantum mechanics in the 20th century, such that the older deterministic worldview of classical mechanics is no longer function in our relatively more scientific age). And the reverse direction also holds true.

In relation to the intentionality-nonintentionality principle, if there is intentionality (e.g., the planning or design of something for a certain outcome, as shown in the original intention of Heisenberg to use the uncertainty principle to shed light on the "tradeoff between the precisions" of the position and momentum of a sub-atomic particle), there is nonintentionality (e.g., the relative absence of planning or design of something for a certain outcome, such that the outcome can even be counter-intentional in being the opposite of what was originally planned, as shown in the use of the uncertainty principle by those people nowadays for the understanding of free will in philosophy, contrary to the original intention of Heisenberg). And the reverse direction also holds true.

And in relation to the survivability-nonsurvivability principle, if there is survivability (e.g., the relative survival of quantum mechanics in the 21st century, after extensive debate about its validity in much of the 20th century), there is non-survivability (e.g., the relative non-survival of classical mechanics in the 21st century, after it was replaced by quantum mechanics and the theory of relativity in the 20th century, although it retains some "limited validity" as an approximation to special relativity and quantum mechanics). (WK 2012eee) And the reverse direction also holds true.

Subjectivity and Society

Subjectivity, when examined from the perspective of society, can reveal to us some powerful features in relation to its possibility (or impossibility) and desirability (or undesirability), and this can be shown by way of two case studies, namely, (3.4.1) subjectivity, communicative action, and power, and (3.4.2) subjectivity, mathematical pratice, and sociology—to be addressed hereafter.

Subjectivity, Communicative Action, and Power

A good way to evaluate the possibility (or impossibility) and desirability (or undesirability) of subjectivity from the perspective of society is to explore subjectivity, commuicative action, and power.

An argument in the literature is that human reasoning as understood in probability study (e.g., Bayesian probability) is not much useful, because it does not take into account the social context of power, so a better way to construct rationality can be in the form of "communicative action," especially in relation to the interactions among different subjectivities for the "intersubjective mutality" of understanding. (WK 2012fff)

History of Universal Pragmatics

The issue of "communicative action" is fundamental in "universal pragmatics" (or "UP" in short), which refers to the "study of the necessary conditions for reaching an understanding through communication. The philosopher Jürgen Habermas coined the term in his essay 'What is Universal Pragmatics?'" in 1979, "where he suggests that human competition, conflict, and strategic action are attempts to achieve understanding that have failed because of modal confusions. The implication is that coming to terms with how people understand or misunderstand one another could lead to a reduction of social conflict." (WK 2012fff)

So "the goal of coming to an understanding is 'intersubjective mutuality…shared knowledge, mutual trust, and accord with one another.' In other words, the underlying goal of coming to an understanding would help to foster the enlightenment, consensus, and good will necessary for establishing socially beneficial norms. Habermas' goal is not primarily for subjective feeling alone, but for development of shared (intersubjective) norms which in turn establish the social coordination needed for practical action in pursuit of shared and individual objectives." (WK 2012fff; J. Habermas 1979)

In the end, "universal pragmatics (UP) seeks to overcome three dichotomies: the dichotomy between body and mind, between theory and practice, and between analytic and continental philosophy. It is part of a larger project to rethink the relationship between philosophy and the individual sciences during a period of social crisis. The project is within the tradition of Critical Theory, a program that traces back to the work of Max Horkheimer." (WK 2012fff)

The UP project in fact "includes two different traditions that Habermas and his collaborator, colleague, and friend Karl-Otto Apel have attempted to reconcile. On the one hand, ideas are drawn from the tradition of Plato, Aristotle, and Kant, wherein words and concepts are regarded as universally valid idealizations of shared meanings. And, on the other hand, inspiration is drawn from the American Pragmatist tradition (feat. Charles Sanders Peirce, George Herbert Mead, Charles W. Morris), for whom words are arbitrary signs devoid of intrinsic meaning, and whose function is to denote the things and processes in the objective world that surrounds the speakers." (WK 2012fff)

In addition, "UP shares with speech act theory, semiotics, and linguistics an interest in the details of language use and communicative action. However, unlike those fields, it insists on a difference between the linguistic data that we observe in the 'analytic' mode, and the rational reconstruction of the rules of symbol systems that each reader/listener possesses intuitively when interpreting strings of words. In this sense, it is an examination of the two ways that language usage can be analyzed: as an object of scientific investigation, and as a 'rational reconstruction' of intuitive linguistic 'know-how.'" (WK 2012fff)

Goals and Methods of Universal Pragmatics

Universal pragmatics is mainly concerned with "the philosophical method of rational reconstruction." (WK 2012fff)

To this end, "the basic concern in universal pragmatics is utterances (or speech acts) in general. This is in contrast to most other fields of linguistics, which tend to be more specialized, focusing exclusively on very specific sorts of utterances such as sentences (which in turn are made up of words, morphemes, and phonemes)." (WK 2012fff)

In UP, "the most significant difference between a sentence and an utterance is in that sentences are judged according to how well they make sense grammatically, while utterances are judged according to their communicative validity." (WK 2012fff; J. Habermas 1979)

So, it is no wonder that "universal pragmatics is also distinct from the field of sociolinguistics. This is because UP is only interested in the meanings of utterances if they have to do with claims about truth or rightness, while sociolinguistics is interested in all utterances in their social contexts." (WK 2012fff; J. Habermas 1979)

Aspects of Universal Pragmatics

With this focus on utterances, UP proposes three ways to deal with them, namely, (a) theories of "elementary propositions," (b) theories of "first-person sentences," and (c) theories of "speech acts" (as summarized in *Table 3.5*). (WK 2012fff)

(a) Theories of Elementary Propositions

By definition, "a theory of elementary propositions investigates those things in the real world that are being referenced by an utterance, and the things that are implied by an utterance, or predicate it. For example, the utterance 'The first Prime Minister of Canada' refers to a man who went by the name of Sir John A. Macdonald. And when a speaker delivers the utterance, 'My husband is a lawyer,' it implies that the speaker is married to a man." (WK 2012fff)

(b) Theories of First-Person Sentences

By definition, "a theory of first-person sentences examines the expression of the intentions of the actor(s) through language and in the first-person." (WK 2012fff)

(c) Theories of Speech Acts

Finally, "a theory of speech acts examines the setting of standards for interpersonal relations through language. The basic goal for speech act theory is to explain how and when utterances in general are *performative*. Central to the notion of speech acts are the ideas of 'illocutionary force' and perlocutionary force, both terms coined by philosopher J. L. Austin. *Illocutionary force* describes the intent of the speaker, while *perlocutionary force* means *the effect an utterance has in the world*, or more specifically, the effect on others." (WK 2012fff; J. Habermas 1979)

Thus, "a performative utterance is a sentence where an action being performed is done by the utterance itself. For example: 'I inform you that you have a moustache,' or 'I promise you I will not burn down the house.' In these cases, the words are also taken as significant actions: the act of informing and promising (respectively)." (WK 2012fff)

What is interesting here is that "Habermas adds to this the observation that speech acts can either succeed or fail, depending on whether or not they succeed on influencing another person in the intended way. This last method of evaluation—the theory of speech acts—is the domain that Habermas is most interested in developing as a theory of communicative action." (WK 2012fff; J. Habermas 1979)

Communicative Action in Universal Pragmatics

This theory of communicative action can be summarized in terms of its three major domains, namely, (a) modes of social integration, (b) communicative competence, and (c) validity of speech, as explained below (and summarized in *Table 3.5*).

(a) Modes of Social Integration

First, the theory of communicative action provides different modes of social integration.

Of course, "the coordination and integration of social action has been explained in many ways by many theories. Rational choice theory and game theory are two examples, which describe the integration of individuals into social groups by detailing the complex manner in which individuals motivated only by self-interest will form mutually beneficial and cooperative social arrangements." (WK 2012fff)

But Habermas (1987) made a distinction between two social realms, namely, "the system" and "the lifeworld," as "two distinct modes of social integration" as explained below: (WK 2012fff)

- The System—refers to "the functional integration of the consequences of actions. It bypasses the consciousness of individuals and does not depend upon their being oriented towards acting collectively. Economic and industrial systems are great examples, often producing complex forms of social integration and interdependence despite the openly competitive orientations of individuals."
- The Lifeworld—refers to "the coordination of action plans and the conscious action-orientations of individuals. It relies on processes of human interaction involving symbolic and cultural forms of meaning. More specifically, as Habermas maintains, the coordination of the lifeworld is accomplished through communicative action."

In this way, "communicative action is an indispensable facet of society. It is at the heart of the lifeworld and is, Habermas claims, responsible for accomplishing several fundamental social functions: reaching understanding, cultural reproduction, coordinating action-plans, and socializing individuals." (WK 2012fff)

But Habermas qualified that "different modes of interaction can (in some ways) facilitate these social functions and achieve integration within the lifeworld. This points towards the second key distinction Habermas makes, which differentiates communicative action from strategic action. The coordination of action plans, which constitutes the social integration of the lifeworld, can be accomplished either through consensus or influence," as elaborated below: (WK 2012fff)

- Communicative Action—in which "actors…raw on the uniquely impelling force of mutual understanding to align the orientation of their action plans. It is this subtle but insistent binding force of communicative interactions that opens the door to an understanding of their meanings."
- Strategic Action—in which "actors…relate to others with no intention of reaching consensus or mutual understanding, but only the intention of accomplishing pre-determined ends unrelated to reaching an understanding. Strategic action often involves the use of communicative actions to achieve the isolated intentions of individuals, manipulating shared understanding in the service of private interests. Thus, Habermas claims, strategic action is parasitic on communicative action, which means communicative action is the primary mode of linguistic interaction. Reaching a reciprocally defined understanding is communication's basic function."

This then is "the formal piece of formal pragmatics: a rational reconstruction of the deep generative structures that are the universal conditions for the possibility of a binding and compelling mutual understanding." (WK 2012fff)

(b) Communicative Competence

Second, the theory of communicative action identifies some "universal conditions of possible mutual understanding," in that "when speakers are communicating successfully, they will have to defend their meaning by using these four claims," as shown below: (WK 2012fff)

- "That they have uttered something understandably—or their statements are intelligible"
- "That they have given other people something to understand—or are speaking something true"
- "That the speaker is therefore understandable—or their intentions are recognized and appreciated for what they are"
- "That they have come to an understanding with another person—or, they have used words that both actors can agree upon"

And "Habermas is emphatic that these claims are universal—no human communication oriented at achieving mutual understanding could possibly fail to raise all of these validity claims. Additionally, to illustrate that all other forms of communication are derived from that which is oriented toward mutual understanding, he argues that there are no other kinds of validity claims whatsoever. This is important, because it is the basis of Habermas' critique of postmodernism." (WK 2012fff)

Thus, "any meaning that meets the above criteria, and is recognized by another as meeting the criteria, is considered 'vindicated' or communicatively competent." (WK 2012fff)

(c) Validity of Speech

Third, the theory of communicative action spells out the conditions for validity claims, that is, for the assessment of the validity of speech.

This is important for Habermas, because "communication rests upon a non-egoistic understanding of the world, which is an idea he borrowed from thinkers like Jean Piaget. A subject capable of a de-centered understanding can take up three fundamentally different attitudes to the world," as M. Cooke (1994) clarified them in what follows: (WK 2012fff)

- "Claims to Truth (IT)"—have to do with "the objective world of facts and existing states of affairs"
- "Claims to Truthfulness (I)"—have to do with "the subjective world of inner experience"
- "Claims to Rightness (WE)"—have to do with "the social world of normatively regulated interactions"

In this way, "a hearer may reject the offering of a speech act on the grounds that it is invalid" because of the following three reasons: (WK 2012fff)

- Untruth—in that the speech act "presupposes or explicates states of affairs which are not the case (IT)"
- Untruthfulness—in that the speech act "raises doubts about the intentions or sincerity of the speaker (I)"
- Wrongness—in that the speech act "does not conform to accepted normative expectations (WE)"

For Habermas, "the three dimensions of validity pointed out above are implicated in any attempt at communication," and "this begins to point towards the idea of communicative rationality, which is the potential for rationality that is implicit in the validity basis of everyday communication, the shape of reason that can be extracted from Habermas's formal-pragmatic analyses." (WK 2012fff)

But Habermas also made a distinction between "communicative action" and "communicative discourse," in that the latter refers to the more vigorous form of communication "that come sufficiently close to actually satisfying these presuppositions" or conditions for validity claims in the "idealized" form of communication (which is "counterfactual"). (WK 2012fff)

Problems with Communication Action

However, for the critics, the theory of communicative action has some major problems. Consider, for illustration, a few examples below (as summarized in *Table 3.5*).

Firstly, one main criticism of the theory of communicative action is that it is unwarrantedly biased for "the lifeworld" and against "the system."

For example, the critic like Hans-Georg Gadamer argued that the theroy of communicative action is not immune from "the hermeneutic circle" and raised four specific criticisms, namely, "(i) a speaker's reflection is always partial, limited, and prejudiced, (ii) a speaker cannot 'get "behind" language to the "real" conditions which it historically develops,' as it serves as a medium in all domains of social life, (iii) only through an intersubjective understanding on the different structures of inevitable prejudices can reflection be possible, and (iv) understanding presupposes both the rejection of unjustifiable prejudices and the acceptance of justifiable ones." (P. Baofu 2000: 431-433; T. McCarthy 1978)

Secondly, another main criticism of the theory of communicative action is that it falsely presupposes a powerful role of reason that it does not have, as it is inherently embedded within the ideology of Western Enlightenment.

For instance, "Barry Smith's criticism (1988) is instructive, in that 'to imagine speech-situations in which all such differences were eradicated or equalized and thus "free" of all so-called "distortions" of communication is to imagine a superlunary universe.' And 'even there, it seems, the conditions of perfection will always call forth someone, an archangel perhaps, who will introduce difference into the company.'" (P. Baofu 2000: 432) Thus, for the critics, the theory of communicative action "constitutes, for his critics, a 'technocratic fantasy'....It is just 'another form of that logocentric desire for perfect presence which dominated Western metaphysics for millennia.'" (P. Baofu 2000: 433; M. Jay 1982; D. LaCapra 1977)

Thirdly, still another main criticism of the theory of communicative action is that there is a serious tension between "communicative action" and "communicative discourse."

For instance, for the critics, "the Habermasian distinction between communicative *action* (where background knowledge is allowed) and communicative *discourse* (where only problematized claims are relevant) creates a great dilemma which his critics do not fail to notice....On the one hand, if under a communicative *action* speakers are allowed to hold their general views of the world and self (background knowledge), they cannot expect their substantive claims (as they are different from each other), to reach any consensus. But if in a communicative *discourse* the speakers are required to detach from any general view so understood, it is no longer clear what sense there would be in a rational conversation so understood, or what would occur, if anything." (P. Baofu 2000: 432-433; C. Larmore 1987)

Fourthly, an additional main criticism of the theory of communicative action is that it does not give a serious attention to the lasting social spectre of power.

For instance, "historian Ian McNeeley [2003], for instance, contrasts Habermas' view with Michel Foucault's notion of communication as embodying pre-existing power relationships: 'Jürgen Habermas subscribes to an unrealistic ideal of power-free communication...Michel Foucault remedies this idealism by treating knowledge as power; his work is in fact suffused with applications of knowledge for the control of human bodies.' In a like manner, the discursive fiction of consensus achieved through rational argumentation might be used as a legitimating prop for social action to the detriment of marginalized members—this is the basis of much feminist critique of Habermas' notions." (WK 2012ggg)

And fifthly, another additional main criticism of the theory of communicative action is that its appeal to the claim to rightness is itself highly problematic.

For instance, for the critics, when different subjectivities (or social actors) are in dispute and appeal to "rightness" (or "the social world of normatively regulated interactions") for adjudication, "such a notion of truth, however, seems dangerously circular enough, since it takes whatever the participants consent upon as truth—which, however, might well be conventional in nature." (P. Baofu 2000: 430; T. McCarthy 1973; C. Larmore 1987)

Beyond Commuicative Action

These criticisms of the theroy of communicative action, however, do not imply a total rejection of its usefulness but serve to show us the two opposing sides of the debate, such that the possibility (or impossibility) and desirability (or undesirability) of subjectivity (from the perspective of society with subjectivity, commuicative action, and power as a case study here) are not to the extent that the respective defenders would like us to believe.

Moreover, the analysis of subjectivity, commuicative action, and power reveals something insightful about the ontological principles in existential dialectics, and good examples include the formalness-informalness principle, the absoluteness-relativeness principle, the partiality-totality principle, the predictability-unpredictability principle, the explicability-inexplicability principle, the fiction-reality principle, the cognitiveness-noncognitiveness principle, the finiteness-transfiniteness principle, the preciseness-vagueness principle, the simpleness-complicatedness principle, the openness-hiddenness principle, the denseness-emptiness principle, the change-constancy principle, the order-chaos principle, the slowness-quickness principle, the expansion-contraction principle, the theory-praxis principle, the convention-novelty principle, the evolution-transformation principle, the symmetry-asymmetry principle, the softness-hardness principle, the seriousness-playfulness principle, the activeness-inactivess principle, the selfness-otherness principle, the regression-progression principle, the same-difference principle, the stability-reaction principle, the functionality-nonfunctionality principle, and the intentionality-nonintentionality principle.

For instance, in relation to the formalness-informalness principle, if there is formalness (e.g., the formal logical requirement of soundness, completeness, and consistency in a system of ideas, as in the theroy of communicative action), there is informalness (e.g., the non-formal existence of unsoundness, incompleteness, and inconsistency in the theroy of communicative action, as shown in the criticisms that it is unwarrantedly biased for "the lifeworld" and against "the system," that it falsely presupposes a powerful role of reason that it does not have, that there is a serious tension between "communicative action" and "communicative discourse," that it does not give a serious attention to the lasting social spectre of power, and that its appeal to the claim to rightness is itself highly problematic). And the reverse direction also holds true.

In relation to the absoluteness-relativeness principle, if there is absoluteness (e.g., the absolute view on communicative action by

Habermas, in that "communicative action is an indispensable facet of society"), there is relativeness (e.g., what is right for Habermas in relation to his claim that "communicative action is an indispensable facet of society" is not necessarily so for those who favor "strategic action" instead, not "communicative action"). And the reverse direction also holds true.

In relation to the partiality-totality principle, if there is partiality (e.g., the partial view on communication by Habermas in regard to his "communicative action"), there is totality (e.g., the more holistic view on communication, such that the whole is not the sum of its parts, that is, it cannot be reduced to the sum of opposing views of Habermas, Gadamer, Foucault, Smith, etc., since there will emerge new views in the future not yet known today, just as there are already alternative views nowadays which do not side exclusively with any of them, as is the analysis here). And the reverse direction also holds true.

In relation to the predictability-unpredictability principle, if there is predictability (e.g., the predictable tendency by the followers of Habermas to assess communication on the basis of the claims to truth, truthfulness, and rightness), there is unpredictability (e.g., the more difficult task to predict exactly the extent to which communicative action can ever be realized in its ideal form on a particular future occasion, since it is "counterfactual"). And the reverse direction also holds true.

In relation to the explicability-inexplicability principle, if there is explicability (e.g., the explanation by Habermas of the desirability of communicative action on the basis of the "intersubjective mutuality" of understanding), there is inexplicability (e.g., the lack of sufficient explanation by Habermas of why communicative action is necessarily desirable, since, for the critics, it is unwarrantedly biased for "the lifeworld" and against "the system," that it falsely presupposes a powerful role of reason that it does not have, that it does not give a serious attention to the lasting social spectre of power, etc.). And the reverse direction also holds true.

In relation to the fiction-reality principle, if there is fiction (e.g., the fictional aspect of the theory of communicative action, insofar as its ideal form is counterfactual), there is reality (e.g., the realistic aspect of the theory of communicative action, insofar as it strives for mutual understanding in communication, if some specific conditions are fulfilled). And the reverse direction also holds true.

In relation to the cognitiveness-noncognitiveness principle, if there is cognitiveness (e.g., a view of things on the basis of reason and evidences, as shown in the requirements of "truth," "truthfulness," and "rightness" in communicative action), there is noncognitiveness (e.g., a view of things on the basis of non-cognitive factors like envy, jealousy, power, nationality, race, gender, age, class, greed, lust, status, faith, anger, sadness, joy, fear, wish, etc., as shown in strategic action, which is based on "the isolated intentions of individuals...in the service of private interests," with "no intention of reaching consensus or mutual understanding" for "truth," for example). And the reverse direction also holds true.

In relation to the finiteness-transfiniteness principle, if there is finiteness (e.g., the finite number of the categories of "the social integration of the lifeworld," namely, through "consensus" or through "influence," for Habermas), there is transfiniteness (e.g., the transfinite number of all the individual cases of achieving social integration through "consensus" or "influence" that have ever existed in history). And the reverse direction also holds true.

In relation to the preciseness-vagueness principle, if there is preciseness (e.g., the precise identification of 3 claims in the assessment of the validity of speech in the theory of communicative action, namely, truth, truthfulness, and rightness), there is vagueness (e.g., the vagueness in the identification of the 3 claims in the assessment of the validity of speech in the theory of communicative action, since it is not clear why there must be only 3, not 4, 5, 6, and so on). And the reverse direction also holds true.

In relation to the simpleness-complicatedness principle, if there is simpleness (e.g., the relatively simple analysis of communicative action by the adcocates), there is complicatedness (e.g., the relatively more complicated analysis of communicative action, by challenging the claims and assumptions, as shown in the problems pointed out by the critics as cited). And the reverse direction also holds true.

In relation to the openness-hiddenness principle, if there is openness (e.g., the open exploration of using communicative action for mutual understanding), there is hiddenness (e.g., the hidden bias in communicative action for mutual understanding, since, for the critics, it is unwarrantedly biased for "the lifeworld" and against "the system," falsely presupposes a powerful role of reason that it does not have, does not give a serious attention to the lasting social spectre of power, etc.). And the reverse direction also holds true.

In relation to the denseness-emptiness principle, if there is denseness (e.g., the relatively denser concentration of concerns with "mutual understadning" by Habermas), there is emptiness (e.g., the relatively less dense, or more empty, concentration of concerns with "mutual understanding" by Foucault). And the reverse direction also holds true.

In relation to the change-constancy principle, if there is change (e.g., the appeal to "rightness" is conventional in nature, so what is accepted as "truth" can change over time, according to the critics of communicative action), there is constancy (e.g., the constant existence of the hermeneutic circle, according to the critics of communicative action like Gadamer). And the reverse direction also holds true.

In relation to the order-chaos principle, if there is order (e.g., the more or less orderly view on communication, if looked from the vantage point of communicative action, such that the participants seriously follows the three requirements for the validity of speech for genuine mutual understanding, so that there will be no more, or few, social conflicts), there is chaos (e.g., the more or less chaotic view on communication, if looked from the vantage point of strategic action, such that each participant only follows his "private interests," with "no intention of reaching consensus or mutual understanding" for "truth," so that conflicts abound among them). And the reverse direction also holds true.

In relation to the slowness-quickness principle, if there is slowness (e.g., the relatively slower readiness of someone to be "untruthful" in the "idealized" communicative discourse), there is quickness (e.g., the relatively quicker readiness of someone to be "untruthful" in the realistic strategic action). And the reverse direction also holds true.

In relation to the expansion-contraction principle, if there is expansion (e.g., the relatively more developed ability of those in the idealized communicative discourse to engage in "mutual understanding"), there is contraction (e.g., the relatively less developed ability of of those in the idealized communicative discourse to engage in manipulation for private interests). And the reverse direction also holds true.

In relation to the theory-praxis principle, if there is theory (e.g., the theoretical construction of "speech acts" in linguistics), there is praxis (e.g., the practical application of "speech acts" in linguistics to the field of communication, as shown in the theory of communicative action). And the reverse direction also holds true.

In relation to the convention-novelty principle, if there is convention (e.g., the conventional concern with "strategic action"), there is novelty (e.g., the alternative novel challenge to the conventional concern with "strategic action," by the new idea of "commuicative action"). And the reverse direction also holds true.

In relation to the evolution-transformation principle, if there is evolution (e.g., the natural evolution of humans to communicate with each other in the state of nature, albeit in a primitive way), there is transformation (e.g., the technical transformation of human ability to communicate with each other by the invention of the theory of "communicative action"). And the reverse direction also holds true.

In relation to the symmetry-asymmetry principle, if there is symmetry (e.g., the symmetric status among participants in the idealized world of communicative discourse for consensus and mutual understanding), there is asymmetry (e.g., the asymmetric status among participants in the realistic world of strategic action for power and influence). And the reverse direction also holds true.

In relation to the softness-hardness principle, if there is softness (e.g., consensus and mutual understanding), there is hardness (e.g., power and influence). And the reverse direction also holds true.

In relation to the seriousness-playfulness principle, if there is seriousness (e.g., the serious business of those who try to understand human communication), there is playfulness (e.g., the playful part of those who try to understand human communication, when they play around with different options over time, like "communicative action," "strategic action," "communicative discourse," etc.). And the reverse direction also holds true.

In relation to the activeness-inactivess principle, if there is activeness (e.g.,the relative activeness of those who seek "mutual understanding" in a situation of "communicative action"), there is inactiveness (e.g., the relative inactiveness of those who seek "mutual understanding" in a situation of "strategic action," since they will be frustrated, discouraged, and even ignored).

In relation to the selfness-otherness principle, if there is selfness (e.g., the concern with "influence" for "private interests"), there is otherness (e.g., the concern with "consensus" for "mutual understanding"). And the reverse direction also holds true.

In relation to the regression-progression principle, if there is regression (e.g., the regression made by the theory of communicative action, as shown in the problems pointed out by the critics aforecited), there is progression (e.g., the progress made by the theory of communicative action, as shown in its contribution to the understanding of "mutual understanding" in communication aforecited). And the reverse direction also holds true.

In relation to the same-difference principle, if there is similarity in outcome (e.g., the contribution to the molding and control of beliefs, values, and behaviors, regardless of whether this be done by way of communicative action, or by way of strategic action), there is difference in outcome (e.g., the contribution to the molding and control of beliefs, values, and behaviors by way of communicative action for a relatively more cooperative lifeworld—but the contribution to the molding and control of beliefs, values, and behaviors by way of strategic action for a relatively more competitive lifeworld). And the reverse direction also holds true.

In relation to the stability-reaction principle, if there is stability (e.g., the relatively more stable condition of the understanding of "social action" in the older days, like "rational choice theory and game theory"), there is reaction (e.g., the understanding of "social action" in the older days, like "rational choice theory and game theory," then led to further changes later on, when they are challenged by the theory of communicative action). And the reverse direction also holds true.

In relation to the functionality-nonfunctionality principle, if there is functionality (e.g., the presence of action for which something or someone is fitted, as shown in the popularity of the idea of the "survival of the fittest" for the pursuit of "private interests" in strategic action in the 19th century, which served the function of perpetuting the ideology of "laissez-faire capitalism" at the time), there is nonfunctionality (e.g., the relative lack of action for which something or someone is fitted, as shown in the popularity, among some people nowadays, of the idea of "consensus" and "mutual understanding," even if to be understood as an ideal type only, such that the idea of the "survival of the fittest" for the pursuit of "private interests" in strategic action is no longer much influential in our democratic age nowadays, when capitalism is becoming more and more "regulated" for the broader interests of society). And the reverse direction also holds true.

In relation to the intentionality-nonintentionality principle, if there is intentionality (e.g., the planning or design of something for a certain outcome, as shown in the original intention of Habermas to address the social issue of power by providing a power-free communication like communicative action as a solution), there is nonintentionality (e.g., the relative absence of planning or design of something for a certain outcome, such that the outcome can even be counter-intentional in being the opposite of what was originally planned, as shown in the critique of the historian Ian McNeeley [2003], who argued "Jürgen Habermas subscribes to an unrealistic ideal of power-free communication," and "the discursive fiction of consensus achieved through rational argumentation might be used as a legitimating prop for social action to the detriment of marginalized members—this is the basis of much feminist critique of Habermas' notions," contrary to the original intention of Habermas to resolve the issue of power in social action). And the reverse direction also holds true.

And in relation to the survivability-nonsurvivability principle, if there is survivability (e.g., the survival of those who win in the battle of "influence" and "power" in strategic action, such that they become dominant), there is non-survivability (e.g., the non-survival of those who lose in the battle of "influence" and "power" in strategic action, such that they become marginalized or even eliminated). And the reverse direction also holds true.

Subjectivity, Mathematical Practice, and Sociology

Another good way to evaluate the possibility (or impossibility) and desirability (or undesirability) of subjectivity from the perspective of society is to explore subjectivity, mathematical practice, and sociology.

An argument in the literature is that mathematics (including probability study) is not as objective as conventionally assumed, since it is socially constructed, in accordance to what is known as "the sociology of mathematical knowledge."

Sociology of Mathematical Knowledge

The term "sociology of mathematical knowledge" is part of "the sociology of scientific knowledge (SSK)," which refers to "the study of science as a social activity, especially dealing with 'the social conditions and effects of science, and with the social structures and processes of scientific activity.' The sociology of knowledge, by contrast, focuses on the production of non-scientific ideas....Crucially, the field does not set out to promote relativism or to attack the scientific project; the aim of the researcher is to explain why one interpretation rather than another succeeds due to external social and historical circumstances. The field emerged in the late 1960s and early 1970s...." (WK 2012hhh; J. Ben-David 1975)

In the specific context of the sociology of mathematical knowledge, the focus here is "on the community of those who practice mathematics and their common assumptions." (WK 2012hhh)

For instance, "Eugene Wigner raised the issue in 1960 and Hilary Putnam made it more rigorous in 1975," and "the question of why fields such as physics and mathematics should agree so well has been debated. Proposed solutions point out that the fundamental constituents of mathematical thought, space, form-structure, and number-proportion are also the fundamental constituents of physics....Another approach is to suggest that there is no deep problem, that the division of human scientific thinking through using words such as 'mathematics' and 'physics' is only useful in their practical everyday function to categorize and distinguish." (WK 2012hhh)

Others like David Bloor and Ludwig Wittgenstein "claim that mathematical knowledge is socially constructed and has irreducible contingent and historical factors woven into it. More recently Paul Ernest has proposed a social constructivist account of mathematical knowledge, drawing on the works of both of these sociologists." (WK 2012hhh)

History of Mathematical Practice

The history of "mathematical practice" can be a good way to examine the social construction of mathematical knowledge in action, since "mathematical practice" is used "to distinguish the working practices of professional mathematicians (e.g. selecting theorems to prove, using informal notations to persuade themselves and others that various steps in the final proof can be formalised, and seeking peer review and publication) from the end result of proven and published theorems." (WK 2012iii)

Historically, "the evolution of mathematical practice was slow, and some contributors to modern mathematics did not follow even the practice of their time, e.g. Pierre de Fermat who was infamous for withholding his proofs, but nonetheless had a vast reputation for correct assertions of results. Likewise there is contrast between the practices of Pythagoras and Euclid. While Euclid was the originator of what we now understand as the published geometric proof, Pythagoras created a closed community and suppressed results; he is even said to have drowned a student in a barrel for revealing the existence of irrational numbers. Modern mathematicians admire Euclid's practices, and usually frown on those of both Fermat and Pythagoras. Nonetheless, all three are considered important contributors to mathematics, despite the variance in method." (WK 2012iii)

In fact, an important reason "to study mathematical practice is that, despite much work in the 20th century, some still feel that the foundations of mathematics remain unclear and ambiguous. One proposed remedy is to shift focus to some degree onto 'what is meant by a proof', and other such questions of method." (WK 2012iii)

For illustration, consider two specific developments in the history of mathematical practice, namely, (a) quasi-empiricism in mathematics and (b) folk mathematics, as described below (and summarized in *Table 3.6*), in that order.

Quasi-Empiricism in Mathematics

The first illustration of the history of mathematical practice has to do with "quasi-empiricism in mathematics," which "denies the possibility of foundations of mathematics and attempts to refocus attention on the ways in which mathematicians arrive at mathematical statements." (WK 2012iii)

Hereafter are three primary arguments for "quasi-empiricism in mathematics," namely, (a) mathematical bafflement, (b) manipulative effectiveness, and (c) informal proofs and proof by authority. (WK 2012jjj)

(a) Mathematical Bafflement

The first primary argument is by Eugene Wigner (1960), who argued that "the miracle of the appropriateness of the language of mathematics…is a wonderful gift which we neither understand nor deserve. We should be grateful for it and hope that it will remain valid in future research and that it will extend, for better or for worse, to our pleasure, even though perhaps also to our bafflement, to wide branches of learning." (WK 2012jjj)

Specifically, "Wigner used several examples to demonstrate why 'bafflement' is an appropriate description, such as showing how mathematics adds to situational knowledge in ways that are either not possible otherwise or are so outside normal thought to be of little notice." (WK 2012jjj)

(b) Manipulative Effectiveness

The second primary argument is by Richard Hamming (1980), who "wrote about applications of mathematics as a central theme to this topic and suggested that successful use can trump, sometimes, proof, in the following sense: where a theorem has evident veracity through applicability, later evidence that shows the theorem's proof to be problematic would result more in trying to firm up the theorem rather than in trying to redo the applications or to deny results obtained to date. Hamming had four explanations for the 'effectiveness' that we see with mathematics and definitely saw this topic as worthy of discussion and study," as shown below: (WK 2012jjj)

- "We see what we look for"—"Why 'quasi' is apropos in reference to this discussion."
- "We select the kind of mathematics to use"—"Our use and modification of mathematics is essentially situational and goal driven."

- "Science in fact answers comparatively few problems"—"What still needs to be looked at is a larger set."
- "The evolution of man provided the model"—"There may be limits attributable to the human element."

Of course, there can be other explanations, but Hamming only offered these four examples.

(c) Informal Proofs and Proof by Authority

And the third primary argument is by Hilary Putnam (1975), who "stated that mathematics had accepted informal proofs and proof by authority, and had made and corrected errors all through its history. Also, he stated that Euclid's system of proving geometry theorems was unique to the classical Greeks and did not evolve similarly in other mathematical cultures in China, India, and Arabia. This and other evidence led many mathematicians to reject the label of Platonists, along with Plato's ontology—which, along with the methods and epistemology of Aristotle, had served as a foundation ontology for the Western world since its beginnings. A truly international culture of mathematics would, Putnam and others (1983) argued, necessarily be at least 'quasi'-empirical (embracing 'the scientific method' for consensus if not experiment)." (WK 2012jjj)

Folk Mathematics

And the second illustration of the history of mathematical practice has to do with "folk mathematics," which refers to those mathematical practices which "are generally without the accompanying proof discipline." (WK 2012iii)

Hereafter are two versions of "folk mathematics" in action, namely, (a) mathematical folklore and (b) informal mathematics. (WK 2012kkk)

(a) Mathematical Folklore

The first version of "folk mathematics" concerns "mathematical folklore," which refers to "theorems, definitions, proofs, or mathematical facts or techniques that are found by investigation and may circulate among mathematicians by word-of-mouth but have not appeared in print, either in books or in scholarly journals." (WK 2012lll)

In fact, "quite important at times for researchers are folk theorems, which are results known, at least to experts in a field, and considered to have established status, but not published in complete form. Sometimes these are only alluded to in the public literature. For example, in tidying up loose ends of the classification of finite simple groups around 2004 (a result which had been claimed, somewhat prematurely, to be proved around 1980), Michael Aschbacher [2004] devoted an entire volume to proving various infrastructural results, some of which had not previously been proved in print." (WK 2012lll)

Another example is "wellknowable mathematics, a term introduced by John Conway. This consists of matters that are known and factual, but not in active circulation in relation with current research." (WK 2012lll)

In addition, "mathematical folklore may also refer to unusual (and possibly apochryphal) stories or jokes involving mathematicians or mathematics that are told verbally in mathematics departments. Compilations include tales collected in G. H. Hardy's *A Mathematician's* Apology" and the work of S. Krantz (2005); concrete examples include, for illustration, the following: (WK 2012lll)

- "Galileo dropping weights from the Leaning Tower of Pisa."
- "An apple falling on Isaac Newton's head to inspire his theory of gravitation."
- "Richard Feynman cracking safes in the Manhattan Project."
- "The 'turtles all the way down' story told by Stephen Hawking."
- "If anyone were to successfully divide by zero, a massive paradox would form, destroying the known universe."

Of course, there can be other examples, so the ones above are illustrative.

(b) Informal Mathematics

And the second version of "folk mathematics" concerns "informal mathematics," which refers to "the predominant form of mathematics at most times and in most cultures, and is the subject of modern ethno-cultural studies of mathematics." (WK 2012mmm)

In particular, "informal mathematics means any informal mathematical practices, as used in everyday life, or by aboriginal or ancient peoples, without historical or geographical limitation. Modern mathematics, exceptionally from that point of view, emphasizes formal and strict proofs of all statements from given axioms. This can usefully be called therefore *formal mathematics.* Informal practices are usually understood intuitively and justified with examples—there are no axioms. This is of direct interest in anthropology and psychology: it casts light on the perceptions and agreements of other cultures. It is also of interest in developmental psychology as it reflects a naïve understanding of the relationships between numbers and things." (WK 2012mmm)

For instance, "there has long been a standard account of the development of geometry in ancient Egypt, followed by Greek mathematics and the emergence of deductive logic. The modern sense of the term *mathematics*, as meaning only those systems justified with reference to axioms, is however an anachronism if read back into history. Several ancient societies built impressive mathematical systems and carried out complex calculations based on proofless heuristics and practical approaches. Mathematical facts were accepted on a pragmatic basis....Commerce, engineering, calendar creation and the prediction of eclipses and stellar progression were practiced by ancient cultures on at least three continents." (WK 2012mmm)

Problems with the Sociology of Mathematical Practice

However, for the critics, the sociology of mathematical practice has some major problems. Consider, for illustration, a few examples below (as summarized in *Table 3.6*).

Firstly, one main criticism of the sociology of mathematical practice is that it is too obsessed with social construction in a reductionistic way.

For example, for the critics, the sociology of scientific knowledge (including the sociology of mathematical practice) "received criticism from theorists of the French school of science and technology studies, Actor-network theory (ANT). These theorists criticise SSK for sociological reductionism and a human centered universe. SSK, they say, relies too heavily on human actors and social rules and conventions settling scientific controversies. The debate is discussed in an article *Epistemological Chicken*." (WK 2012hhh; H. Collins 1992)

Secondly, another main criticism of the sociology of mathematical practice is that it is not critical enough to folklore mathematics.

For instance, for the critics, many "stories or jokes involving mathematicians or mathematics" in mathematical folklore can be spurious, questionable, or simply "apochryphal." (WK 2012lll) And many "informal mathematical practices...by aboriginal or ancient peoples" are often treated as "naïve," because their mathematical practices were "usually understood intuitively" without any proofs (that is, "based on proofless heuristics")—although "this is of direct interest in anthropology and psychology," "casts light on the perceptions and agreements of other cultures," and "reflects a naïve understanding of the relationships between numbers and things." (WK 2012mmm)

Thirdly, still another main criticism of the sociology of mathematical practice is that there are necessary arguments for the foundation of mathematics, contrary to the main view in the sociology of mathematical knowledge.

For instance, Willard Quine and Hilary Putnam (1985) provided "one of the most challenging arguments in favor of the acceptance of the existence of abstract mathematical entities, such as numbers and sets....Both Putnam and Quine invoke naturalism to justify the exclusion of all non-scientific entities....The assertion that 'all' entities postulated in scientific theories, including numbers, should be accepted as real is justified by confirmation holism. Since theories are not confirmed in a piecemeal fashion, but as a whole, there is no justification for excluding any of the entities referred to in well-confirmed theories. This puts the nominalist who wishes to exclude the existence of sets and non-Euclidean geometry, but to include the existence of quarks and other undetectable entities of physics, for example, in a difficult position." (WK 2012nnn Yablo, S. 1998)

And fourthly, an additional main criticism of the sociology of mathematical practice is that its social constructionism is incoherent.

For instance, the critic like Paul Boghossian (2006) pointed out the inconsistency in the argument for social constructionism, as "he follows Ian Hackings argument that many adapt social constructionism because of its potentially liberating fact: if things are the way that they are only because of our social conventions, as opposed to being so naturally, then we can possible change manipulate them into being how we would rather have them be. He then states that social constructionists argue that we should refrain from making absolute judgements about what is true, and instead state that something is true in the light of theory x," but Boghossian countered with his argument in this way: "But it is hard to see how we might coherently follow this advice. Given that the propositions which make up epistemic systems are just very general propositions about what absolutely justifies what, it makes no sense to insist that we abandon making absolute particular judgements about what justifies what while allowing us to accept absolute general judgements about what justifies what. But in effect this is what the epistemic relativist is recommending." (WK 2012ooo)

Beyond the Sociology of Mathematical Practice

These criticisms of the sociology of mathematical practice, however, do not entail a total rejection of its usefulness but serve to show us the two opposing sides of the debate, such that the possibility (or impossibility) and desirability (or undesirability) of subjectivity (from the perspective of society with subjectivity, mathematical practice, and sociology as a case study here) are not to the extent that the respective defenders would like us to believe.

Furthermore, the analysis of subjectivity, mathematical practice, and sociology is didactic in regard to the ontological principles in existential dialectics, and good examples include the formalness-informalness principle, the absoluteness-relativeness principle, the partiality-totality principle, the predictability-unpredictability principle, the explicability-inexplicability principle, the fiction-reality principle, the cognitiveness-noncognitiveness principle, the finiteness-transfiniteness principle, the preciseness-vagueness principle, the simpleness-complicatedness principle, the openness-hiddenness principle, the denseness-emptiness principle, the change-constancy principle, the order-chaos principle, the slowness-quickness principle, the expansion-contraction principle, the theory-praxis principle, the convention-novelty principle, the evolution-transformation principle, the symmetry-asymmetry principle, the softness-hardness principle, the seriousness-playfulness principle, the activeness-inactivess principle, the selfness-otherness principle, the regression-progression principle, the same-difference principle, the stability-reaction principle, the functionality-nonfunctionality principle, and the intentionality-nonintentionality principle.

For instance, in relation to the formalness-informalness principle, if there is formalness (e.g., the formal logical requirement of soundness, completeness, and consistency in a system of ideas, as in the sociology of mathematical knowledge), there is informalness (e.g., the non-formal existence of unsoundness, incompleteness, and inconsistency in the sociology of mathematical knowledge, as shown in the criticisms that it is too obsessed with social construction in a reductionistic way, that it is not critical enough to folklore mathematics, that there are necessary arguments for the foundation of mathematics, and that its social constructionism is incoherent). And the reverse direction also holds true.

In relation to the absoluteness-relativeness principle, if there is absoluteness (e.g., the absolute view by Ian Hackings when he argued that "we should refrain from making absolute judgements about what is true, and instead state that something is true in the light of theory x"), there is relativeness (e.g., what is right for Ian Hackings when he argued that "we should refrain from making absolute judgements about what is true, and instead state that something is true in the light of theory x" is not necessarily so for Paul Boghossian, who countered with his argument in this way: "But it is hard to see how we might coherently follow this advice. Given that the propositions which make up epistemic systems are just very general propositions about what absolutely justifies what, it makes no sense to insist that we abandon making absolute particular judgements about what justifies what while allowing us to accept absolute general judgements about what justifies what. But in effect this is what the epistemic relativist is recommending"). And the reverse direction also holds true.

In relation to the partiality-totality principle, if there is partiality (e.g., the partial view by Ian Hackings on social constructionism), there is totality (e.g., the more holistic view on social constructionism, such that the whole is not the sum of its parts, that is, it cannot be reduced to the sum of opposing views of Hackings, Boghossian, etc., since there will emerge new views in the future not yet known today, just as there are already alternative views nowadays which do not side exclusively with any of them, as is the analysis here). And the reverse direction also holds true.

In relation to the predictability-unpredictability principle, if there is predictability (e.g., the predictable tendency of many "informal mathematical practices…by aboriginal or ancient peoples" to be "usually understood intuitively" without any proofs [that is, "based on proofless heuristics"]), there is unpredictability (e.g., the more difficult task to predict exactly the extent to which any particular "informal mathematical practices…by aboriginal or ancient peoples" will one day be proved by formal mathematicians, just as Aschbacher once proved "various infrastructural results" of "the classification of finite simple groups"). And the reverse direction also holds true.

In relation to the explicability-inexplicability principle, if there is explicability (e.g., the explanation, by those in formal mathematics, of the foundation of modern mathematics on the basis of "formal and strict proofs of all statements from given axioms"), there is inexplicability (e.g., the lack of sufficient explanation by those in formal mathematics of why the foundation of modern mathematics is necessarily robust, since, for the critics, there is "quasi-empiricism in mathematics," with examples like mathematical bafflement, manipulative effectiveness, and informal proofs and proof by authority). And the reverse direction also holds true.

In relation to the fiction-reality principle, if there is fiction (e.g., the fictional aspect of mathematics, insofar as there exists "quasi-empiricism in mathematics," with examples like mathematical bafflement, manipulative effectiveness, and informal proofs and proof by authority), there is reality (e.g., the realistic aspect of mathematics, insofar as there are necessary arguments for the foundation of mathematics as suggested by Willard Quine and Hilary Putnam). And the reverse direction also holds true.

In relation to the cognitiveness-noncognitiveness principle, if there is cognitiveness (e.g., a view of things on the basis of reason and evidences, as shown in modern mathematics on the basis of "formal and strict proofs of all statements from given axioms"), there is noncognitiveness (e.g., a view of things on the basis of non-cognitive factors like envy, jealousy, power, nationality, race, gender, age, class, greed, lust, status, faith, anger, sadness, joy, fear, wish, etc., as shown in the "hope" or faith in mathematics as pointed out by Eugene Wigner, who argued that "the miracle of the appropriateness of the language of mathematics…is a wonderful gift which we neither understand nor deserve. We should be grateful for it and hope that it will remain valid in future research and that it will extend, for better or for worse, to our pleasure, even though perhaps also to our bafflement, to wide branches of learning"). And the reverse direction also holds true.

In relation to the finiteness-transfiniteness principle, if there is finiteness (e.g., the finite number of the categories of mathematical practice in its history, namely, quasi-empiricism and folk mathematics), there is transfiniteness (e.g., the transfinite number of all the individual instances of quasi-empiricism and folk mathematics that have ever existed in history). And the reverse direction also holds true.

In relation to the preciseness-vagueness principle, if there is preciseness (e.g., the precise identification of 3 arguments for "quasi-empiricism in mathematics," namely, mathematical bafflement, manipulative effectiveness, and informal proofs and proof by authority, in the article aforecited), there is vagueness (e.g., the vagueness in the identification of the 3 arguments for "quasi-empiricism in mathematics," namely, mathematical bafflement, manipulative effectiveness, and informal proofs and proof by authority, in the article aforecited, since it is not clear why there must be only 3, not 4, 5, 6, and so on). And the reverse direction also holds true.

In relation to the simpleness-complicatedness principle, if there is simpleness (e.g., the relatively simple analysis of the sociology of mathematical knowledge by its advocates), there is complicatedness (e.g., the relatively more complicated analysis of the sociology of mathematical knowledge, by challenging the claims and assumptions, as shown in the problems pointed out by the critics as cited). And the reverse direction also holds true.

In relation to the openness-hiddenness principle, if there is openness (e.g., the open exploration of "Euclid's system of proving geometry theorems…along with Plato's ontology" and "the methods and epistemology of Aristotle" to "serve…as a foundation ontology for the Western world since its beginnings"), there is hiddenness (e.g., the hidden bias in "Euclid's system of proving geometry theorems…along with

Plato's ontology" and "the methods and epistemology of Aristotle," since, for the critic like Hilary Putnam, "Euclid's system of proving geometry theorems was unique to the classical Greeks and did not evolve similarly in other mathematical cultures in China, India, and Arabia"). And the reverse direction also holds true.

In relation to the denseness-emptiness principle, if there is denseness (e.g., the relatively denser concentration of concerns with "formal and strict proofs of all statements from given axioms" in formal mathematics), there is emptiness (e.g., the relatively less dense, or more empty, concentration of concerns with "formal and strict proofs of all statements from given axioms" in informal mathematics, since many "informal mathematical practices…by aboriginal or ancient peoples" were "usually understood intuitively" without any proofs, that is, "based on proofless heuristics"). And the reverse direction also holds true.

In relation to the change-constancy principle, if there is change (e.g., the ever new ways to understand mathematics, as shown in the sociology of mathematical knowledge), there is constancy (e.g., the ever constant existence of different problems with the ever new ways to understand mathematics, as shown in the problems with the sociology of mathematical knowledge pointed out by the critics). And the reverse direction also holds true.

In relation to the order-chaos principle, if there is order (e.g., the more or less orderly view about mathematics, if looked only from the vantage point of "formal and strict proofs of all statements from given axioms" in formal mathematics), there is chaos (e.g., the more or less chaotic view about mathematics, if looked only from the vantage point of folklore mathematics, where many "stories or jokes involving mathematicians or mathematics" can be spurious, questionable, or simply "apochryphal," and many "informal mathematical practices…by aboriginal or ancient peoples" were "usually understood intuitively" without any proofs, that is, "based on proofless heuristics"—such that almost anything goes, for practical purposes). And the reverse direction also holds true.

In relation to the slowness-quickness principle, if there is slowness (e.g., the relatively slower speed of mathematics to be accepted if it is to be judged on the basis of "formal and strict proofs of all statements from given axioms," which are not easily available), there is quickness (e.g., the relatively quicker speed of mathematics to be accepted if it is to be judged "intuitively" without any proofs, that is, "based on proofless heuristics"). And the reverse direction also holds true.

In relation to the expansion-contraction principle, if there is expansion (e.g., the relatively more developed ability of judging mathematical practices "intuitively" without any proofs, that is, "based on proofless heuristics," among those "aboriginal or ancient peoples" in informal mathematics), there is contraction (e.g., the relatively less developed ability of doing "formal and strict proofs of all statements from given axioms" among those "aboriginal or ancient peoples" in informal mathematics). And the reverse direction also holds true.

In relation to the theory-praxis principle, if there is theory (e.g., the theoretical construction of formal mathematics), there is praxis (e.g., the practical application of formal mathematics to the field of physics, as pointed out by Eugene Wigner in 1960). And the reverse direction also holds true.

In relation to the convention-novelty principle, if there is convention (e.g., the conventional wisdom about the foundation of mathematics in formal mathematics), there is novelty (e.g., the alternative novel challenge to the conventional wisdom about the foundation of mathematics in formal mathematics, by the new idea of "quasi-empiricism in mathematics"). And the reverse direction also holds true.

In relation to the evolution-transformation principle, if there is evolution (e.g., the natural evolution of humans to understand "space, form-structure, and number-proportion" in the state of nature, albeit in a primitive way), there is transformation (e.g., the technical transformation of human ability to understand "space, form-structure, and number-proportion" by the invention of formal mathematics, especially when applied to physics). And the reverse direction also holds true.

In relation to the symmetry-asymmetry principle, if there is symmetry (e.g., the existence of different ways of doing mathematics), there is asymmetry (e.g., the popularity of formal mathematics among those professional mathematicians in contemporary mainstream Western society—but the popularity of folklore mathematics among those "aboriginal or ancient peoples" in underdeveloped countries). And the reverse direction also holds true.

In relation to the softness-hardness principle, if there is softness (e.g., the defense of social constructionism by Ian Hackings), there is hardness (e.g., the critique of social constructionism by Boghossian). And the reverse direction also holds true.

In relation to the seriousness-playfulness principle, if there is seriousness (e.g., the serious business of those who try to understand mathematics), there is playfulness (e.g., the playful part of those who try to understand mathematics, when they play around with different interpretations over time, like formal mathematics, quasi-empiricism in mathematics, mathematical folklore, informal mathematics, etc.). And the reverse direction also holds true.

In relation to the activeness-inactivess principle, if there is activeness (e.g., the relative activeness of those in formal mathematics who actively try to do "formal and strict proofs of all statements from given axioms"), there is inactiveness (e.g., the relative inactiveness of those in formal mathematics when their "formal and strict proofs of all statements from given axioms" are not appropriate among those "aboriginal or ancient peoples," so they may well get frustrated and cannot do much there). And the reverse direction also holds true.

In relation to the selfness-otherness principle, if there is selfness (e.g., the tendency to be mostly centered on one's self or group, as shown in the use of informal mathematics for "commerce, engineering, calendar creation," etc. among those "aboriginal or ancient peoples"), there is otherness (e.g., the tendency to be mostly centered on others, as shown in the work of Eugene Wigner on mathematics and physics in 1960, as a contribution to the advance in mathematical knowledge for humanity). And the reverse direction also holds true.

In relation to the regression-progression principle, if there is regression (e.g., the regression made by the sociology of mathematical knowledge, as shown in the problems pointed out by the critics), there is progression (e.g., the progress made by the sociology of mathematical knowledge, as shown in its contribution to the understanding of the social constructionism of mathematics aforecited). And the reverse direction also holds true.

In relation to the same-difference principle, if there is similarity in outcome (e.g., the contribution to the molding and control of beliefs, values, and behaviors, regardless of whether this be done by way of formal mathematics, or by way of mathematical folklore), there is difference in outcome (e.g., the contribution to the molding and control of beliefs, values, and behaviors by way of formal mathematics for a relatively more scientific worldview—but the contribution to the molding and control of beliefs, values, and behaviors by way of mathematical folklore for a relatively more pragmatic worldview). And the reverse direction also holds true.

In relation to the stability-reaction principle, if there is stability (e.g., the relatively more stable condition of the foundation of formal mathematics in the older days), there is reaction (e.g., the foundation of formal mathematics in the older days then led to further changes later on, when it was questioned by the sociology of mathematical knowledge). And the reverse direction also holds true.

In relation to the functionality-nonfunctionality principle, if there is functionality (e.g., the presence of action for which something or someone is fitted, as shown in the use of mathematical folklore among those "aboriginal or ancient peoples," which served the function of maintaining the pre-scientific lifeworld at that time), there is nonfunctionality (e.g., the relative lack of action for which something or someone is fitted, as shown in the use of formal mathematics nowadays, such that the use of mathematical folklore is no longer as functional nowadays in our relatively more scientifically advanced society). And the reverse direction also holds true.

In relation to the intentionality-nonintentionality principle, if there is intentionality (e.g., the planning or design of something for a certain outcome, as shown in the original intention or hope of Euclid to use his "system of proving geometry theorems" as a foundation of mathematics), there is nonintentionality (e.g., the relative absence of planning or design of something for a certain outcome, such that the outcome can even be counter-intentional in being the opposite of what was originally planned, as shown in the work of Putnam, which showed that "Euclid's system of proving geometry theorems was unique to the classical Greeks and did not evolve similarly in other mathematical cultures in China, India, and Arabia," so "a truly international culture of mathematics would...necessarily be at least 'quasi'-empirical," contrary to the original intention or hope of Euclid). And the reverse direction also holds true.

And in relation to the survivability-nonsurvivability principle, if there is survivability (e.g., the survival of the sociology of mathematical practice in our time), there is non-survivability (e.g., the non-survival, in our time, of the old-fashioned "standard account of the development of geometry in ancient Egypt, followed by Greek mathematics and the emergence of deductive logic," since mathematicians now realized that "several ancient societies built impressive mathematical systems and carried out complex calculations based on proofless heuristics and practical approaches," different from the "standard" account of the development of geometry"). And the reverse direction also holds true.

Subjectivity and Culture

Subjectivity, when examined from the perspective of culture, can cast an illuminating light on its possibility (or impossibility) and desirability (or undesirability), and this can be shown by way of two case studies, namely, (3.5.1) subjectivity, experience, and phenomenology, and (3.5.2) subjectivity, the subject, and philosophical traditions—to be addressed hereafter, in that order.

Subjectivity, Experience, and Phenomenology

An excellent case study here to understand subjectivity from the perspective of culture has to do with subjectivity, experience, and phenomenology.

An argument here is that the study of probability (say, about an expectation that an event will likely occur) is inadequate because it does not give sufficient attention to the role of a subject who experiences events in the world in the first place. The Western philosophical tradition of phenomenology casts some light on this issue.

Subjective Experience and Phenomenology

The relation among subjectivity, experience and phenomenology is close enough, since the term "phenomenology (from Greek: *phainómenon* 'that which appears'; and *lógos* 'study')" refers to "the philosophical study of the structures of subjective experience and consciousness. As a philosophical movement it was founded in the early years of the 20th century by Edmund Husserl and was later expanded upon by a circle of his followers at the universities of Göttingen and Munich in Germany. It then spread to France, the United States, and elsewhere…." (WK 2012ppp; D. Zahavi 2003)

According to Husserl, "phenomenology…is primarily concerned with the systematic reflection on and study of the structures of consciousness and the phenomena that appear in acts of consciousness. This phenomenological ontology can be clearly differentiated from the Cartesian method of analysis which sees the world as objects, sets of objects, and objects acting and reacting upon one another." (WK 2012ppp)

History of Phenomenology

From the historical perspective, Stephen Hicks (2004) argued that "to understand phenomenology, one must identify its roots in the philosophy of Immanuel Kant (1724–1804). In his *Critique of Pure Reason*, Kant distinguished between 'phenomena' (objects as interpreted by human sensibility and understanding), and 'noumena' (objects as things-in-themselves, which humans cannot directly experience). According to Hicks, 19th-century Kantianism operated in two broad camps," namely: (WK 2012ppp)

- "structural linguistics"
- "phenomenology"

For Hicks (2004), "the Structuralists were seeking subjective noumenal categories, and the Phenomenologists were content with describing the phenomena without asking what connection to an external reality those experiences might have." (WK 2012ppp)

Originally, "phenomenology thus attempts to create conditions for the...study of topics usually regarded as subjective: consciousness and the content of conscious experiences such as judgments, perceptions, and emotions. Although phenomenology seeks to be scientific, it does not attempt to study consciousness from the perspective of clinical psychology or neurology. Instead, it seeks through systematic reflection to determine the essential properties and structures of experience." (WK 2012ppp)

Husserl was influenced by "the works and lectures of his teachers, the philosophers and psychologists Franz Brentano and Carl Stumpf. An important element of phenomenology that Husserl borrowed from Brentano is intentionality (often described as 'aboutness'), the notion that consciousness is always consciousness of something. The object of consciousness is called the intentional object, and this object is constituted for consciousness in many different ways, through, for instance, perception, memory, retention and protention, signification, etc. Throughout these different intentionalities, though they have different structures and different ways of being 'about' the object, an object is still constituted as the identical object; consciousness is directed at the same intentional object in direct perception as it is in the immediately following retention of this object and the eventual remembering of it." (WK 2012ppp: R. Rollinger 1999)

The term "phenomenology" was "previously employed by Hegel in his *Phenomenology of Spirit*," but "it was Husserl's adoption of this term (circa 1900) that propelled it into becoming the designation of a philosophical school. As a philosophical perspective, phenomenology is its method, though the specific meaning of the term varies according to how it is conceived by a given philosopher. As envisioned by Husserl, phenomenology is a method of philosophical inquiry that rejects the rationalist bias that has dominated Western thought since Plato in favor of a method of reflective attentiveness that discloses the individual's 'lived experience.' Loosely rooted in an epistemological device, with Sceptic roots, called epoché, Husserl's method entails the suspension of judgment while relying on the intuitive grasp of knowledge, free of presuppositions and intellectualizing." (WK 2012ppp; E. Husserl 1970)

Phenomenology is therefore called at times as the "science of experience," because "the phenomenological method is rooted in intentionality, Husserl's theory of consciousness (developed from Brentano). Intentionality represents an alternative to the representational theory of consciousness, which holds that reality cannot be grasped directly because it is available only through perceptions of reality that are representations of it in the mind. Husserl countered that consciousness is not 'in' the mind but rather conscious of something other than itself (the intentional object), whether the object is a substance or a figment of imagination (i.e., the real processes associated with and underlying the figment). Hence the phenomenological method relies on the description of phenomena as they are given to consciousness, in their immediacy." (WK 2012ppp)

In its practice, phenomenology "entails an unusual combination of discipline and detachment to suspend, or bracket, theoretical explanations and second-hand information while determining one's 'naive' experience of the matter. The phenomenological method serves to momentarily erase the world of speculation by returning the subject to his or her primordial experience of the matter, whether the object of inquiry is a feeling, an idea, or a perception. According to Husserl the suspension of belief in what we ordinarily take for granted or infer by conjecture diminishes the power of what we customarily embrace as objective reality. According to Rüdiger Safranski (1998), '[Husserl and his followers'] great ambition was to disregard anything that had until then been thought or said about consciousness or the world [while] on the lookout for a new way of letting the things [they investigated] approach them, without covering them up with what they already knew.'" (WK 2012ppp)

Main Ideas of Phenomenology

Phenomenology can be explained in terms of the following six main ideas, namely, (a) intentionality, (b) intuition, (c) evidence, (d) noesis and noema, (e) empathy and intersubjectivty, and (f) lifeworld (as summarized in *Table 3.7*).

(a) Intentionality

The first main idea of phenomenology concerns "intentionality."

The idea of "intentionality" refers to "the notion that consciousness is always the consciousness of something. The word itself should not be confused with the 'ordinary' use of the word intentional, but should rather be taken as playing on the etymological roots of the word. Originally, intention referred to a 'stretching out' ('in tension,' lat. *intendere*), and in this context it refers to consciousness *'stretching out'* towards its object (although one should be careful with this image, seeing as there is not some consciousness first that, subsequently, stretches out to its object. Rather, consciousness occurs as the simultaneity of a conscious act and its object.) Intentionality is often summed up as 'aboutness.'" (WK 2012ppp)

However, "whether this something that consciousness is about is in direct perception or in fantasy is inconsequential to the concept of intentionality itself; whatever consciousness is directed at, that is what consciousness is conscious of. This means that the object of consciousness doesn't have to be a physical object apprehended in perception: it can just as well be a fantasy or a memory. Consequently, these 'structures' of consciousness, i.e., perception, memory, fantasy, etc., are called intentionalities." (WK 2012ppp)

Historically, "the term intentionality originated with the Scholastics in the medieval period and was resurrected by Brentano who in turn influenced Husserl's conception of phenomenology, who refined the term and made it the cornerstone of his theory of consciousness. The meaning of the term is complex and depends entirely on how it is conceived by a given philosopher. The term should not be confused with 'intention' or the psychoanalytic conception of unconscious 'motive' or 'gain.'" (WK 2012ppp)

(b) Intuition

The second main idea of phenomenology concerns "intuition."

The idea of "intuition" refers to "those cases where the intentional object is directly present to the intentionality at play; if the intention is 'filled' by the direct apprehension of the object, you have an intuited object. Having a cup of coffee in front of you, for instance, seeing it, feeling it, or even imagining it—these are all filled intentions, and the object is then intuited. The same goes for the apprehension of mathematical formulae or a number. If you do not have the object as referred to directly, the object is not intuited, but still intended, but then emptily. Examples of empty intentions can be signitive intentions—intentions that only imply or refer to their objects." (WK 2012ppp)

(c) Evidence

The third main idea of phenomenology concerns "evidence."

In everyday life," we use the word evidence to signify a special sort of relation between a state of affairs and a proposition: State A is evidence for the proposition 'A is true.' In phenomenology, however, the concept of evidence is meant to signify the 'subjective achievement of truth.' This is not an attempt to reduce the objective sort of evidence to subjective 'opinion,' but rather an attempt to describe the structure of having something present in intuition with the addition of having it present as intelligible: 'Evidence is the successful presentation of an intelligible object, the successful presentation of something whose truth becomes manifest in the evidencing itself.'" (WK 2012ppp; R. Sokolowski 2000)

(d) Noesis and Noema

The fourth main idea of phenomenology concerns "noesis and noema."

These two terms are "from the Greek *nous* (mind), designate respectively the real content, noesis, and the ideal content, noema, of an intentional act (an act of consciousness). The Noesis is the part of the act that gives it a particular sense or character (as in judging or perceiving something, loving or hating it, accepting or rejecting it, and so on). This is real in the sense that it is actually part of what takes place in the consciousness (or psyche) of the subject of the act. The Noesis is always correlated with a Noema; for Husserl, the full Noema is a complex ideal structure comprising at least a noematic sense and a noematic core....[And] the noematic sense is generally understood as the ideal meaning of the act and the noematic core as the act's referent or object as *it is meant in the act*." (WK 2012ppp)

(e) Empathy and Intersubjectivity

The fifth main idea of phenomenology concerns "empathy and intersubjectivity."

The idea of "empathy" refers to "the experience of one's own body as another. While we often identify others with their physical bodies, this type of phenomenology requires that we focus on the subjectivity of the other, as well as our intersubjective engagement with them. In Husserl's original account, this was done by a sort of apperception built on the experiences of your own lived-body. The lived body is your own body as experienced by yourself, as yourself. Your own body manifests itself to you mainly as your possibilities of acting in the world. It is what lets you reach out and grab something, for instance, but it also, and more importantly, allows for the possibility of changing your point of view. This helps you differentiate one thing from another by the experience of moving around it, seeing new aspects of it (often referred to as making the absent present and the present absent), and still retaining the notion that this is the same thing that you saw other aspects of just a moment ago (it is identical). Your body is also experienced as a duality, both as object (you can touch your own hand) and as your own subjectivity (you experience being touched)." (WK 2012ppp)

And the idea of "intersubjectivity" refers to a subjective form of "objectivity (i.e., what you experience as objective is experienced as being intersubjectively available—available to all other subjects. This does not imply that objectivity is reduced to subjectivity nor does it imply a relativist position, cf. for instance intersubjective verifiability). In the experience of intersubjectivity, one also experiences oneself as being a subject among other subjects, and one experiences oneself as existing objectively for these Others; one experiences oneself as the noema of Others' noeses, or as a subject in another's empathic experience. As such, one experiences oneself as objectively existing subjectivity." (WK 2012ppp)

(f) Lifeworld

And the sixth main idea of phenomenology concerns "lifeworld."

The idea of "lifeworld (German: *Lebenswelt*)" refers to "the 'world' each one of us lives in. One could call it the 'background' or 'horizon' of all experience, and it is that on which each object stands out as itself (as different) and with the meaning it can only hold for us. The lifeworld is both personal and intersubjective (it is then called a 'homeworld'), and, as such, it does not enclose each one of us in a solus ipse" (that is, only oneself). (WK 2012ppp)

The Transcendental Turn in Phenomenology

But Husserl had a change of heart in later years, when he took a transcendental turn in phenomenology, since "some years after the publication of the *Logical Investigations*, Husserl made some key elaborations that led him to the distinction between the act of consciousness (*noesis*) and the phenomena at which it is directed (the *noemata*)," as shown below (and summarized in *Table 3.7*): (WK 2012ppp)

- "noetic"—refers to "the intentional act of consciousness (believing, willing, etc.)"
- "noematic"—refers to "the object or content (noema), which appears in the noetic acts (the believed, wanted, hated, and loved ...)."

This means that "what we observe is not the object as it is in itself, but how and inasmuch it is given in the intentional acts. Knowledge of essences would only be possible by 'bracketing' all assumptions about the existence of an external world and the inessential (subjective) aspects of how the object is concretely given to us. This procedure Husserl called *epoché*." (WK 2012ppp)

With this change in mind, "Husserl in a later period concentrated more on the ideal, essential structures of consciousness. As he wanted to exclude any hypothesis on the existence of external objects, he introduced the method of phenomenological reduction to eliminate them. What was left over was the pure transcendental ego, as opposed to the concrete empirical ego. Now Transcendental Phenomenology is the study of the essential structures that are left in pure consciousness: This amounts in practice to the study of the noemata and the relations among them." (WK 2012ppp)

Problems with Phenomenology

However, for the critics, Husserl's phenomenology has some major problems. Consider, for illustration, a few examples below (as summarized in *Table 3.7*).

Firstly, one main criticism of Husserl's phenomenology is that it is too subjectivist.

For example, "Martin Heidegger modified Husserl's conception of phenomenology because of (what Heidegger perceived as) Husserl's subjectivist tendencies. Whereas Husserl conceived humans as having been constituted by states of consciousness, Heidegger countered that consciousness is peripheral to the primacy of one's existence (i.e., the mode of being of Dasein), which cannot be reduced to one's consciousness of it. From this angle, one's state of mind is an 'effect' rather than a determinant of existence, including those aspects of existence that one is not conscious of. By shifting the center of gravity from consciousness (psychology) to existence (ontology), Heidegger altered the subsequent direction of phenomenology, making it at once both personal and mysterious. As one consequence of Heidegger's modification of Husserl's conception, phenomenology became increasingly relevant to psychoanalysis. Whereas Husserl gave priority to a depiction of consciousness that was fundamentally alien to the psychoanalytic conception of the unconscious, Heidegger offered a way to conceptualize experience that could accommodate those aspects of one's existence that lie on the periphery of sentient awareness." (WK 2012ppp; M. Natanson 1973)

Secondly, another main criticism of Husserl's phenomenology is that its ideas are often used in an obscure (controversial) way.

For instance, for the critics, "the correct interpretation of what Husserl meant by the Noema has long been controversial....One element of controversy is whether this noematic object is the same as the actual object of the act (assuming it exists) or is some kind of ideal object." (WK 2012ppp: D. Smith 2007)

Thirdly, still another main criticism of Husserl's phenomenology is that its foundational stance in regard to the transcendental ego is illusory.

For instance, the critic "Merleau-Ponty objects to the ego's transcendence of the world" and proposed "existential phenomenology," which "differs from transcendental phenomenology by its rejection of the transcendental ego." (WK 2012ppp) Similarly, another critic "Theodor Adorno criticised Husserl's concept of phenomenological epistemology in his metacritique *Against Epistemology*, which is anti-foundationalist in its stance." (WK 2012ppp)

Fourthly, an additional main criticism of Husserl's phenomenology is that it evolves over time into different conflicting versions.

For instance, for the critics, "Husserl's conception of phenomenology has been criticized and developed not only by himself but also," later on, "by students such as Edith Stein, by existentialists, such as Max Scheler, Nicolai Hartmann, Maurice Merleau-Ponty, Jean-Paul Sartre, and by other philosophers, such as Paul Ricoeur, Emmanuel Levinas, and sociologists Alfred Schütz and Eric Voegelin." (WK 2012ppp)

And fifthly, another additional main criticism of Husserl's phenomenology is that there are alternative approaches competing with it.

For instance, "in *Ethnomethodology's Program* (2002), Garfinkel speaks of phenomenological texts and findings as being, 'appropriated' and intentionally 'misread,' for the purposes of exploring topics in the study of social order. These appropriations and methodical 'misread[ings]' of phenomenological texts and findings are clearly made for the purposes of furthering ethnomethodological analyses, and should not be mistaken for logical extensions of these phenomenological texts and findings." (WK 2012qqq)

Beyond Phenomenology

These criticisms of Husser's phenomenology, however, do not mean a total rejection of its usefulness but serve to show us the two opposing sides of the debate, such that the possibility (or impossibility) and desirability (or undesirability) of subjectivity (from the perspective of culture with subjectivity, experience, and phenomenology as a case study here) are not to the extent that the respective defenders would like us to believe.

Furthermore, the analysis of subjectivity, experience, and phenomenology can reveal the workings of the ontological principles in existential dialectics, and good examples include the formalness-informalness principle, the absoluteness-relativeness principle, the partiality-totality principle, the predictability-unpredictability principle, the explicability-inexplicability principle, the fiction-reality principle, the cognitiveness-noncognitiveness principle, the finiteness-transfiniteness principle, the preciseness-vagueness principle, the simpleness-complicatedness principle, the openness-hiddenness principle, the denseness-emptiness principle, the change-constancy principle, the order-chaos principle, the slowness-quickness principle, the expansion-contraction principle, the theory-praxis principle, the convention-novelty principle, the evolution-transformation principle, the symmetry-asymmetry principle, the softness-hardness principle, the seriousness-playfulness principle, the activeness-inactivess principle, the selfness-otherness principle, the regression-progression principle, the same-difference principle, the stability-reaction principle, the functionality-nonfunctionality principle, and the intentionality-nonintentionality principle.

For instance, in relation to the formalness-informalness principle, if there is formalness (e.g., the formal logical requirement of soundness, consistency, and completeness in a system of ideas, as in Husserl's phenomenology), there is informalness (e.g., the non-formal existence of unsoundness, inconsistency, and incompleteness in Husserl's phenomenology, as shown in the criticisms that it is too subjectivist, that its ideas are often used in an obscure [controversial] way, that its foundational stance in regard to the transcendental ego is illusory, that it evolves over time into different conflicting versions, and that there are alternative approaches competing with it). And the reverse direction also holds true.

In relation to the absoluteness-relativeness principle, if there is absoluteness (e.g., the absolute view by Husserl on the transcendental ego), there is relativeness (e.g., what is right for Husserl on the transcendental ego is not necessarily so for Merleau-Ponty, who "objects to the ego's transcendence of the world" and proposes "existential phenomenology," which "differs from transcendental phenomenology by its rejection of the transcendental ego"). And the reverse direction also holds true.

In relation to the partiality-totality principle, if there is partiality (e.g., the partial view on subjective experience from Husserl's phenomenological vantage point), there is totality (e.g., the more holistic view on subjective experience, such that the whole is not the sum of its parts, that is, it cannot be reduced to the sum of opposing views of Husserl, Heidegger, Merleau-Ponty, Garfinkel, etc., since there will emerge new views in the future not yet known today, just as there are already alternative views nowadays which do not side exclusively with any of them, as is the analysis here). And the reverse direction also holds true.

In relation to the predictability-unpredictability principle, if there is predictability (e.g., the predictable tendency of a subject to be conscious of something, with the help of intuition and evidence, in the world of Husserl's phenomenology), there is unpredictability (e.g., the more difficult task to predict exactly the extent to which a subject can ever achieve "pure consciousness" on a particular occasion, in the world of Husserl's phenomenology). And the reverse direction also holds true.

In relation to the explicability-inexplicability principle, if there is explicability (e.g., the explanation by Husserl of the desirability of phenomenology on the basis of "the method of phenomenological reduction to eliminate" or exclude "any hypothesis on the existence of external objects" or any conventional prejudice, so as to reach "the pure transcendental ego, as opposed to the concrete empirical ego"), there is inexplicability (e.g., the lack of sufficient explanation by Husserl of why transcendental phenomenology is necessarily desirable or even possible, since, for the critic like Heidegger, "consciousness is peripheral to the primacy of one's existence [i.e., the mode of being of Dasein], which cannot be reduced to one's consciousness of it," so, in this way, Heidegger shifted "the center of gravity from consciousness [psychology] to existence [ontology]"). And the reverse direction also holds true.

In relation to the fiction-reality principle, if there is fiction (e.g., the fictional aspect of Husserl's phenomenology, insofar as the critics reject "the transcendental ego" as illusory, as "Merleau-Ponty objects to the ego's transcendence of the world" and proposed "existential phenomenology," which "differs from transcendental phenomenology by its rejection of the transcendental ego"), there is reality (e.g., the realistic aspect of Husserl's phenomenology, insofar as it focuses on "intuition," and "evidence" to back up one's subjective experience). And the reverse direction also holds true.

In relation to the cognitiveness-noncognitiveness principle, if there is cognitiveness (e.g., a view of things on the basis of reason and evidences, as shown in the focus on "intuition," and "evidence" in Husserl's phenomenology), there is noncognitiveness (e.g., a view of things on the basis of non-cognitive factors like envy, jealousy, power, nationality, race, gender, age, class, greed, lust, status, faith, anger, sadness, joy, fear, wish, etc., as shown in the hope by those in Husserl's phenomenology to eventually reach "the pure transcendental ego, as opposed to the concrete empirical ego"). And the reverse direction also holds true.

In relation to the finiteness-transfiniteness principle, if there is finiteness (e.g., the finiteness of "the empirical ego," because of its being bounded), there is transfiniteness (e.g., the transfiniteness of "the transcendental ego," because of its being unbounded, or largely so). And the reverse direction also holds true.

In relation to the preciseness-vagueness principle, if there is preciseness (e.g., the precise identification of 6 main ideas of Husserl's phenomenology in the article aforecited), there is vagueness (e.g., the vagueness in the identification of the 6 main ideas of Husserl's phenomenology in the article aforecited, since it is not clear why there must be only 6, not 7, 8, 9, and so on). And the reverse direction also holds true.

In relation to the simpleness-complicatedness principle, if there is simpleness (e.g., the relatively simple analysis of Husserl's phenomenology by its advocates), there is complicatedness (e.g., the relatively more complicated analysis of Husserl's phenomenology, by challenging the claims and assumptions, as shown in the problems pointed out by the critics as cited). And the reverse direction also holds true.

In relation to the openness-hiddenness principle, if there is openness (e.g., the open exploration of using "bracketing" to question "all assumptions about the existence of an external world and the inessential [subjective] aspects of how the object is concretely given to us," so as to eventually reach "the transcendental ego" in transcendental phenomenology), there is hiddenness (e.g., the hidden bias in transcendental phenomenology, because, for the critics like Merleau-Ponty, the idea of "the transcendental ego" is illusory). And the reverse direction also holds true.

In relation to the denseness-emptiness principle, if there is denseness (e.g., the relatively denser concentration of concerns with "transcendental reduction" by those in Husserl's transcentental phenomenology), there is emptiness (e.g., the relatively less dense, or more empty, concentration of concerns with "the production of social order" by those in Husserl's phenomenology, although it is central to ethnomethodology). And the reverse direction also holds true.

In relation to the change-constancy principle, if there is change (e.g., the ever new ways to understand subjective experience, as shown in Husserl's phenomenology), there is constancy (e.g., the ever constant existence of different problems with the ever new ways to understand subjective experience, as shown in the problems with Husserl's phenomenology pointed out by the critics). And the reverse direction also holds true.

In relation to the order-chaos principle, if there is order (e.g., the more or less orderly view on reality, if looked from the vantage point of ethnomethdology, where people find ways to reproduce "social order"), there is chaos (e.g., the more or less chaotic view on reality, if looked from the vantage point of Husserl's transcendental phenomenology, where a subject is to question or "bracket…all assumptions about the existence of an external world"). (WK 2012qqq & 2012ppp) And the reverse direction also holds true.

In relation to the slowness-quickness principle, if there is slowness (e.g., the relatively slower speed of a subject to understand his experience of an object, if "intuition" and "evidence" are absent, in accordance to Husserl's phenomenology), there is quickness (e.g., the relatively quicker speed of a subject to understand his experience of an object, if "intuition" and "evidence" are used, in accordance to Husserl's phenomenology). And the reverse direction also holds true.

In relation to the expansion-contraction principle, if there is expansion (e.g., the relatively more developed ability of a subject in Husserl's phenomenology to question or "bracket…all assumptions about the existence of an external world"), there is contraction (e.g., the relatively less developed ability of a subject in Husserl's phenomenology to reproduce "social order"). And the reverse direction also holds true.

In relation to the theory-praxis principle, if there is theory (e.g., the theoretical construction of Husserl's phenomenology), there is praxis (e.g., the practical application of Husserl's phenomenology to the field of ethnomethodology, when "Garfinkel speaks of phenomenological texts and findings as being, 'appropriated' and intentionally 'misread,' for the purposes of exploring topics in the study of social order"). And the reverse direction also holds true.

In relation to the convention-novelty principle, if there is convention (e.g., the conventional wisdom about consciousness as understood by Husserl when he "conceived humans as having been constituted by states of consciousness"), there is novelty (e.g., the alternative novel challenge to the conventional wisdom about consciousness as understood by Husserl, by the new idea of Heidegger, who "countered that consciousness is peripheral to the primacy of one's existence [i.e., the mode of being of Dasein], which cannot be reduced to one's consciousness of it"). And the reverse direction also holds true.

In relation to the evolution-transformation principle, if there is evolution (e.g., the natural evolution of humans to experience consciousness in the state of nature), there is transformation (e.g., the technical transformation of human ability to understand consciousness by the invention of phenemonology by Husserl and, later on, others). And the reverse direction also holds true.

In relation to the symmetry-asymmetry principle, if there is symmetry (e.g., the existence of different ways to do phenomenology, like Husserl's and Heidegger's), there is asymmetry (e.g., the popularity of Husserl's phenomenology among those who are more interested in psychology—but the popularity of Heidegger's phenomenology among those who are more interested in ontology). And the reverse direction also holds true.

In relation to the softness-hardness principle, if there is softness (e.g., the defense of phenomenology by Husserl), there is hardness (e.g., the critique of phenomenology by Garfinkel). And the reverse direction also holds true.

In relation to the seriousness-playfulness principle, if there is seriousness (e.g., the serious business of those who try to understand consciousness), there is playfulness (e.g., the playful part of those who try to understand consciousness, when they play around with different versions over the decades, like the ones by Husserl, Heidegger, Merleau-Ponty, Edith Stein, Max Scheler, Nicolai Hartmann, Jean-Paul Sartre, Paul Ricoeur, Emmanuel Levinas, Alfred Schütz and Eric Voegelin). And the reverse direction also holds true.

In relation to the activeness-inactivess principle, if there is activeness (e.g., the relative activeness of those followers of Husserl's phenomenology to try to reach "pure consciousness"), there is inactiveness (e.g., the relative inactiveness of those followers of Husserl's phenomenology to try to reach "pure consciousness" in the community of ethnomethodology, since they do not quite fit in the latter's project for the understanding of the reproduction of social order).

In relation to the selfness-otherness principle, if there is selfness (e.g., the tendency to be mostly centered on one's self or group, as shown in the confession by Garfinkel who spoke of "phenomenological texts and findings as being, 'appropriated' and intentionally 'misread,' for the purposes of exploring topics in the study of social order"), there is otherness (e.g., the tendency to be mostly centered on others, as shown in the philosophical work of Husserl for the study of "the structures of subjective experience and consciousness," as an advance in knowledge for humanity). And the reverse direction also holds true.

In relation to the regression-progression principle, if there is regression (e.g., the regression made by Husserl's phenomenology, as shown in the problems pointed out by the critics aforecited), there is progression (e.g., the progress made by Husserl's phenomenology, as shown in the contribution to the study of "the structures of subjective experience and consciousness" aforecited). And the reverse direction also holds true.

In relation to the same-difference principle, if there is similarity in outcome (e.g., the contribution to the molding and control of beliefs, values, and behaviors, regardless of whether this be done by way of Husserl's phenomenology or by way of Garfinkel's phenomenology), there is difference in outcome (e.g., the contribution to the molding and control of beliefs, values, and behaviors by way of Husserl's phenomenology for a relatively more psychological worldview—but the contribution to the molding and control of beliefs, values, and behaviors by way of Garfinkel's phenomenolog for a relatively more sociological worldview). And the reverse direction also holds true.

In relation to the stability-reaction principle, if there is stability (e.g., the relatively more stable condition of transcendental phenomenology by Husserl in the older days), there is reaction (e.g., transcendental phenomenology by Husserl in the older days then led to further changes later on, when Merleau-Ponty proposed the alternative approach known as "existential phenomenology"). And the reverse direction also holds true.

In relation to the functionality-nonfunctionality principle, if there is functionality (e.g., the presence of action for which something or someone is fitted, as shown in the silence of Heidegger on the influence of Okakura Kakuzo's concept of *das-in-der-Welt-sein* [being in the world] expressed in *The Book of Tea* to describe Zhuangzi's philosophy, which Imamichi's teacher had offered to Heidegger in 1919, after having studied with him the year before"—which served the function of Eurocentrism in the Western world at the time, without giving much credit to Eastern philosophy), there is nonfunctionality (e.g., the relative lack of action for which something or someone is fitted, as shown by "some researchers in phenomenology" nowadays, who "see possibilities of establishing dialogues with traditions of thought outside of the so-called Western philosophy, particularly with respect to East-Asian thinking," such that the old-fashioned Eurocentric prejudice in the old days is no longer as influential in our age of globalization). (WK 2012ppp; M. Heidegger 1971; T. Imamichi 2006) And the reverse direction also holds true.

In relation to the intentionality-nonintentionality principle, if there is intentionality (e.g., the planning or design of something for a certain outcome, as shown in the original intention of Husserl to use phenomenology for the study of "the structures of subjective experience and consciousness"), there is nonintentionality (e.g., the relative absence of planning or design of something for a certain outcome, such that the outcome can even be counter-intentional in being the opposite of what was originally planned, as shown by Garfinkel who spoke of "phenomenological texts and findings as being, 'appropriated' and intentionally 'misread,' for the purposes of exploring topics in the study of social order"—contrary to the original intention of Husserl for phenomenology). And the reverse direction also holds true.

And in relation to the survivability-nonsurvivability principle, if there is survivability (e.g., the relative survival of Husserl's earlier phenomenology known as "realist phenomenology," as many "members of the Munich group," for example, "preferred the earlier realist phenomenology of the first edition of the *Logical Investigations*"), there is non-survivability (e.g., the relative non-survival of Husserl's later phenomenology known as "transcendental phenomenology," since "many phenomenologists took a critical stance towards his new theories"). (WK 2012ppp) And the reverse direction also holds true.

Subjectivity, the Subject, and Philosophical Traditions

Another way to evaluate the possibility (or impossibility) and desirability (or undesirability) of subjectivity from the perspective of culture concerns subjectivity, the subject, and philosophical traditions.

An argument here is that mathematics (including probability study) has a lot to say about objects (for quantitative analysis) but little to say about subjects (for qualitative analysis)—so, with this weakness in mind, different philosophical traditions are to illuminate what a "sujbect" is, as opposed to an "object."

Subject and Object

This weakness in mathematics (including probability study) can be magnified by examining the distinction between a "subject" and an "object" in Western philosophy, where "a subject is a being who has unique experiences and a unique consciousness, or an entity that has a relationship with another entity that exists outside of itself (called an 'object'). A *subject* is an observer and an *object* is a thing observed." (WK 2012rrr)

This concept of "the subject" (or in a different wording, the "personhood" in Anglo-American analytical philosophy, and the "I" in critical theory and psychology) is "especially important in [European] continental philosophy, where 'the Subject' is a central term in debates over human autonomy and the nature of the self." (WK 2012rrr)

In fact, this "sharp distinction between subject and object corresponds to the distinction, in the philosophy of René Descartes, between thought and extension. Descartes believed that thought (subjectivity) was the essence of the mind, and that extension (the occupation of space) was the essence of matter." (WK 2012rrr)

Later on, "in the modern continental tradition, which may plausibly be said to date from Descartes, debates over the nature of the Subject play a role comparable to debates over personhood within the distinct Anglo-American tradition of analytical philosophy." (WK 2012rrr)

And "in critical theory and psychology, subjectivity is also the actions or discourses that produce individuals or 'I'—the 'I' is the subject." (WK 2012rrr)

In any event, the idea of the "subject" in continental philosophy (aka the "I" in critical theory and psychology, or the "personhood" in Anglo-American analytic philosophy) can be understood in different ways in different philosophical traditions.

Consider, for illustration, the following three main Western philosophical traditions, namely, (a) German idealism, (b) postmodernism, and (c) analytic philosophy, in that order.

Subject in German Idealism

The first philosophical tradition to address the concept of the "subject" has to do with "German idealism."

Historically, "*subject* as a key-term in thinking about human consciousness began its career with the German Idealists, in response to David Hume's radical skepticism. The idealists' starting point was Hume's conclusion that there is nothing to the self over and above a big, fleeting bundle of perceptions. The next step was to ask how this undifferentiated bundle comes to be experienced as a unity—as a single subject. Hume had offered the following proposal: '*...the imagination must by long custom acquire the same method of thinking, and run along the parts of space and time in conceiving its objects*.'" (WK 2012rrr)

Later, "Kant, Hegel and their successors sought to flesh out the process by which the subject is constituted out of the flow of sense impressions. Hegel, for example, stated in his *Preface to the Phenomenology of Spirit* that a subject is constituted by 'the process of reflectively mediating itself with itself.'" (WK 2012rrr)

More specifically, "Hegel begins his definition of the subject at a standpoint derived from Aristotelian physics: 'the unmoved which is also *self-moving*'....That is, what is not moved by an outside force, but which propels itself, has a *prima facie* case for subjectivity. Hegel's next step, however, is to identify this power to move, this unrest that is the subject, as *pure negativity*. Subjective self-motion, for Hegel, comes not from any pure or simple kernel of authentic individuality, but rather, it is '...the bifurcation of the simple; it is the doubling which sets up opposition, and then again the negation of this indifferent diversity and of its anti-thesis.'" (WK 2012rrr)

Hegel then addressed "the…subject's *modus operandi*" as "cutting, splitting and introducing distinctions by injecting negation into the flow of sense-perceptions. Subjectivity is thus a kind of structural effect—what happens when Nature is diffused, refracted around a field of negativity and the 'unity of the subject' for Hegel is in fact a second-order effect, a 'negation of negation.' The subject experiences itself as a unity only by purposively negating the very diversity it itself had produced. The Hegelian subject may therefore be characterized either as 'self-restoring sameness' or else as 'reflection in otherness within itself.' In short, a subject in the Hegelian sense is subjected to subjection." (WK 2012rrr)

Subject in Postmodernism

The second philosophical tradition to address the concept of the "subject" has to do with "postmodernism."

As a critique of German idealism and others, "the thinking of Marx and Freud provided a point of departure for questioning the notion of a unitary, autonomous Subject, which for many thinkers in the Continental tradition is seen as the foundation of the liberal theory of the social contract. These thinkers opened up the way for the deconstruction of the subject as a core-concept of metaphysics." (WK 2012rrr)

Then, "Sigmund Freud's explorations of the unconscious mind added up to a wholesale indictment of Enlightenment notions of subjectivity. Among the most radical re-thinkers of human self-consciousness was Heidegger, whose concept of *Dasein* or 'Being-there' displaces traditional notions of the personal subject altogether. Jacques Lacan, inspired by Heidegger and Saussure, built on Freud's psychoanalytic model of the subject, in which the 'split subject' is constituted by a double bind: alienated from jouissance when he or she leaves the Real, enters into the Imaginary (during the mirror stage), and separates from the Other when he or she comes into the realm of language, difference, and demand in the Symbolic or the Name of the Father." (WK 2012rrr)

Later on, "thinkers such as Althusser, Foucault or Bourdieu theorize the subject as a social construction. According to Althusser, the 'subject' is an ideological construction (more exactly, constructed by the 'Ideological State Apparatuses'). One's subjectivity exists, 'always already' and is discovered through the process of interpellation. Ideology inaugurates one into being a subject. According to Foucault, it is the 'effect' of power and 'disciplines'" like the "construction of the subject as student, soldier, 'criminal,' etc." in his book *Discipline and Punish.* (WK 2012rrr)

Subject in Analytic Philosophy

And the third philosophical tradition to address the concept of the "subject" has to do with "analytic philosophy."

In Anglo-American analytic philosophy, "the issue of subject—and more specifically the 'point of view' of the subject, or 'subjectivity'—has received attention as one of the major intractable problems in philosophy of mind (a related issue being the mind-body problem)." (WK 2012rrr)

It is interesting to note that, in the essay "What is it like to be a bat?," Thomas Nagel "famously argued that explaining subjective experience—the 'what it is like' to be something—is currently beyond the reach of scientific inquiry, because scientific understanding by definition requires an objective perspective, which, according to Nagel, is diametrically opposed to the subjective first-person point of view. These additional features of subjective experience are often referred to as qualia" (as in Frank Cameron Jackson's "Mary's room"). (WK 2012rrr)

Problems with Philosophical Traditions on Subjectivity

However, for the critics, these different philosophical traditions have some major problems. Consider, for illustration, a few examples below.

Firstly, one main criticism of these different philosophical traditions is that they often offer opposing viewpoints.

For example, the opposing views on the subject by those in German idealism and by others in postmodernism are often incompatible; for instance, "Sigmund Freud's explorations of the unconscious mind added up to a wholesale indictment of Enlightenment notions of subjectivity. Among the most radical re-thinkers of human self-consciousness was Heidegger, whose concept of *Dasein* or 'Being-there' displaces traditional notions of the personal subject altogether." (WK 2012rrr)

Secondly, another main criticism of these different philosophical traditions is that they often ignore alternative philosophical traditions in the Non-West.

For instance, none of the philosophical traditions aforediscussed has much to say about non-Western philosophical traditions on the subject. For instance, one of the most notorious cases involves Heidegger himself, about his "silence" on the influence of Eastern philosophical thought on his work, as "it has been claimed that a number of elements within phenomenology (mainly Heidegger's thought) have some resonance with Eastern philosophical ideas, particularly with Zen Buddhism and Taoism. According to Tomonubu Imamichi, the concept of *Dasein* was inspired—although Heidegger remains silent on this—by Okakura Kakuzo's concept of *das-in-der-Welt-sein* (being in the world) expressed in *The Book of Tea* to describe Zhuangzi's philosophy, which Imamichi's teacher had offered to Heidegger in 1919, after having studied with him the year before." (WK 2012ppp; G. Parkes 1987; T. Imamichi 2006)

Thirdly, still another main criticism of these different philosophical traditions is that the tension between subjectivity and objectivity may not be resolvable.

For instance, for the critic like Nagel, "explaining subjective experience—the 'what it is like' to be something—is currently beyond the reach of scientific inquiry, because scientific understanding by definition requires an objective perspective, which, according to Nagel, is diametrically opposed to the subjective first-person point of view." (WK 2012rrr)

And fourthly, an additional main criticism of these different philosophical traditions is that its views can be nonsensual.

For instance, for the critics, an excellent example involves "postmodernism" in the so-called "Sokal affair, also known as the Sokal hoax," which "was a publishing hoax perpetrated by Alan Sokal, a physics professor at New York University. In 1996, Sokal submitted an article to *Social Text*, an academic journal of postmodern cultural studies. In subsequent publications, Sokal claimed that the submission was an experiment to test the journal's intellectual rigor and, specifically, to investigate whether such a journal would 'publish an article liberally salted with nonsense if it (a) sounded good and (b) flattered the editors' ideological preconceptions.' The article 'Transgressing the Boundaries: Towards a Transformative Hermeneutics of Quantum Gravity,' published in the *Social Text* Spring/Summer 1996 'Science Wars' issue, proposed that quantum gravity is a social and linguistic construct....On its date of publication (May 1996), Sokal revealed in *Lingua Franca* that the article was a hoax, identifying it as 'a pastiche of Left-wing cant, fawning references, grandiose quotations, and outright nonsense...structured around the silliest quotations [by postmodernist academics] he could find about mathematics and physics.'" (WK 2012hhh; A. Sokai 1996 & 1996a)

Beyond Philosophical Traditions on Subjectivity

These criticisms of different philosophical traditions, however, should not be misconstrued as a total rejection of its usefulness but serve to show us the two opposing sides of the debate, such that the possibility (or impossibility) and desirability (or undesirability) of subjectivity (from the perspective of culture with subjectivity, the subject, and philosophical traditions as a case study here) are not to the extent that the respective defenders would like us to believe.

Furthermore, the analysis of subjectivity, the subject, and philosophical traditions can teach us something important about the ontological principles in existential dialectics, and good examples include the formalness-informalness principle, the absoluteness-relativeness principle, the partiality-totality principle, the predictability-unpredictability principle, the explicability-inexplicability principle, the fiction-reality principle, the cognitiveness-noncognitiveness principle, the finiteness-transfiniteness principle, the preciseness-vagueness principle, the simpleness-complicatedness principle, the openness-hiddenness principle, the denseness-emptiness principle, the change-constancy principle, the order-chaos principle, the slowness-quickness principle, the expansion-contraction principle, the theory-praxis principle, the convention-novelty principle, the evolution-transformation principle, the symmetry-asymmetry principle, the softness-hardness principle, the seriousness-playfulness principle, the activeness-inactivess principle, the selfness-otherness principle, the regression-progression principle, the same-difference principle, the stability-reaction principle, the functionality-nonfunctionality principle, and the intentionality-nonintentionality principle.

For instance, in relation to the formalness-informalness principle, if there is formalness (e.g., the formal logical requirement of soundness, consistency, and completeness in a system of ideas, as in the philosophical traditions on the "subject"), there is informalness (e.g., the non-formal existence of unsoundness, inconsistency, and incompleteness in the philosophical traditions on the "subject," as shown in the criticisms that they often offer opposing viewpoints, that they often ignore alternative philosophical traditions in the Non-West, that the tension between subjectivity and objectivity may not be resolvable, and that its views can be nonsensual). And the reverse direction also holds true.

In relation to the absoluteness-relativeness principle, if there is absoluteness (e.g., the absolute view on "a unitary, autonomous Subject, which for many thinkers…is seen as the foundation of the liberal theory of the social contract"), there is relativeness (e.g., what is right for those who defended "a unitary, autonomous Subject…as the foundation of the liberal theory of the social contract" is not necessarily so for the critics like Marx and Freud, who "provided a point of departure for questioning the notion of a unitary, autonomous Subject" and "opened up the way for the deconstruction of the subject as a core-concept of metaphysics"). And the reverse direction also holds true.

In relation to the partiality-totality principle, if there is partiality (e.g., the partial view on the "subject" by those in German idealism), there is totality (e.g., the more holistic view on the "subject," such that the whole is not the sum of its parts, that is, it cannot be reduced to the sum of opposing views by those in German idealism, postmodernism, analytic philosophy, the liberal theory of the social contract, etc., since there will emerge new views in the future not yet known today, just as there are already alternative views nowadays which do not side exclusively with any of them, as is the analysis here). And the reverse direction also holds true.

In relation to the predictability-unpredictability principle, if there is predictability (e.g., the predictable tendency of different philosophical traditions in the West to debate on the nature of the "subject"), there is unpredictability (e.g., the more difficult task to predict exactly the extent to which the debate on the "subject" will one day be resolved by the scientific method, since for the critic like Nagel, "scientific understanding…requires an objective perspective, which, according to Nagel, is diametrically opposed to the subjective first-person point of view"). And the reverse direction also holds true.

In relation to the explicability-inexplicability principle, if there is explicability (e.g., the explanation, by those in defense of the older version of the liberal tradition, of the desirability of "a unitary, autonomous Subject" on the basis of its being "the foundation" of "social contract" theory), there is inexplicability (e.g., the lack of sufficient explanation by those in defense of the older version of the liberal tradition, of why "a unitary, autonomous Subject" is necessarily desirable, since the critics like Marx and Freud "opened up the way for the deconstruction of the subject as a core-concept of metaphysics"). And the reverse direction also holds true.

In relation to the fiction-reality principle, if there is fiction (e.g., the fictional aspect of the "subject," insofar as "thinkers such as Althusser, Foucault or Bourdieu theorize the subject as a social construction," or, "as "an ideological construction" for Althusser), there is reality (e.g., the realistic aspect of the "subject," insofar as it involves "consciousness" or "experience"). And the reverse direction also holds true.

In relation to the cognitiveness-noncognitiveness principle, if there is cognitiveness (e.g., a view of things on the basis of reason and evidences, as shown in "scientific understanding," which, for Nagel, "requires an objective perspective" based on evidences to be verified), there is noncognitiveness (e.g., a view of things on the basis of non-cognitive factors like envy, jealousy, power, nationality, race, gender, age, class, greed, lust, status, faith, anger, sadness, joy, fear, wish, etc., as shown in the "politics" of "flatter[ing] the editors' ideological preconceptions," so as to have a nonsensical article published, as revealed in the "Sokal affair"). And the reverse direction also holds true.

In relation to the finiteness-transfiniteness principle, if there is finiteness (e.g., the finite number of ways to refer to subjectivity, namely, the idea of the "subject" in continental philosophy, the "I" in critical theory and psychology, and the "personhood" in Anglo-American analytic philosophy), there is transfiniteness (e.g., the transfinite number of all the possible individual subjects, be they known as the "subject," the "I," or the "personhood," that have ever existed in history). And the reverse direction also holds true.

In relation to the preciseness-vagueness principle, if there is preciseness (e.g., the precise identification of 3 philosophical traditions on the "subject" in the article aforecited), there is vagueness (e.g., the vagueness in the identification of the 3 philosophical traditions on the "subject" in the article aforecited, since it is not clear why there must be only 3, not 4, 5, 6, and so on). And the reverse direction also holds true.

In relation to the simpleness-complicatedness principle, if there is simpleness (e.g., the relatively simple analysis of "a unitary, autonomous Subject" in "the liberal theory of the social contract" by its advocates), there is complicatedness (e.g., the relatively more complicated analysis of "a unitary, autonomous Subject" in "the liberal theory of the social contract," by challenging the claims and assumptions, as shown in the problems pointed out by the critics as cited). And the reverse direction also holds true.

In relation to the openness-hiddenness principle, if there is openness (e.g., the open exploration of the "subject" as "a unitary, autonomous Subject" in "the liberal theory of the social contract"), there is hiddenness (e.g., the hidden bias in the analysis of the "subject" as "a unitary, autonomous Subject" in "the liberal theory of the social contract," since the critics like "Althusser, Foucault or Bourdieu theorize the subject as a social construction," or, "as "an ideological construction" for Althusser). And the reverse direction also holds true.

In relation to the denseness-emptiness principle, if there is denseness (e.g., the relatively denser concentration of concerns with "a unitary, autonomous Subject" by those in "the liberal theory of the social contract"), there is emptiness (e.g., the relatively less dense, or more empty, concentration of concerns with "a unitary, autonomous Subject" by those in "postmodernism"). And the reverse direction also holds true.

In relation to the change-constancy principle, if there is change (e.g., the ever new ways to study the "subject," as shown in the different philosophical traditions), there is constancy (e.g., the ever constant existence of different problems with the ever new ways to study the "subject," as shown in the problems with the different philosophical traditions pointed out by the critics aforecited). And the reverse direction also holds true.

In relation to the order-chaos principle, if there is order (e.g., the more or less orderly view on the "subject," if looked from the vantage point of "a unitary, autonomous Subject" in "the liberal theory of the social contract"), there is chaos (e.g., the more or less chaotic view on the "subject," if looked from the vantage point of "Freud's psychoanalytic model of the subject," as further developed by Lacan, in which "the 'split subject' is constituted by a double bind: alienated from jouissance when he or she leaves the Real, enters into the Imaginary [during the mirror stage], and separates from the Other when he or she comes into the realm of language, difference, and demand in the Symbolic or the Name of the Father"). And the reverse direction also holds true.

In relation to the slowness-quickness principle, if there is slowness (e.g., the relatively slower speed of a subject to decide what to do, when confronted with "a double bind," which "is an emotionally distressing dilemma in communication in which an individual…receives two or more conflicting messages, in which one message negates the other"), there is quickness (e.g., the relatively quicker speed of of a subject to decide what to do, when he does not suffer from "a double bind" but benefits from being "a unitary, autonomous Subject"). (WK 2012sss) And the reverse direction also holds true.

In relation to the expansion-contraction principle, if there is expansion (e.g., the relative spread, in the 20th century, of the influence of "Freud's psychoanalytic model of the subject"), there is contraction (e.g., the relative decline, in the 20th century, of the influence of "the liberal theory of the social contract" in terms of "a unitary, autonomous Subject"). And the reverse direction also holds true.

In relation to the theory-praxis principle, if there is theory (e.g., the theoretical construction of "Freud's psychoanalytic model of the subject"), there is praxis (e.g., the practical application of "Freud's psychoanalytic model of the subject" to the field of communication, when Lacan used it to discuss the "split subject," which "is constituted by a double bind: alienated from jouissance when he or she leaves the Real, enters into the Imaginary [during the mirror stage], and separates from the Other when he or she comes into the realm of language, difference, and demand in the Symbolic or the Name of the Father"). And the reverse direction also holds true.

In relation to the convention-novelty principle, if there is convention (e.g., the conventional idea about "a unitary, autonomous Subject" in "the liberal theory of the social contract"), there is novelty (e.g., the alternative novel challenge to the conventional idea about "a unitary, autonomous Subject" in "the liberal theory of the social contract," by the new idea of the "split subject" in "Freud's psychoanalytic model of the subject," as further developed by Lacan). And the reverse direction also). And the reverse direction also holds true.

In relation to the evolution-transformation principle, if there is evolution (e.g., the natural evolution of humans to have consciousness in the state of nature), there is transformation (e.g., the technical transformation of human ability to understand consciousness by the invention of different philosophical traditions on the "subject"). And the reverse direction also holds true.

In relation to the symmetry-asymmetry principle, if there is symmetry (e.g., the existence of different ways to understand the "subject," like German idealism in the older days and postmodernism in the 20th century unto today), there is asymmetry (e.g., the popularity of German idealism in the older days of the 18th and 19th centuries—but the popularity of postmodernism in the second half of the 20th century unto today). And the reverse direction also holds true.

In relation to the softness-hardness principle, if there is softness (e.g., the defense of "a unitary, autonomous Subject" by those in "the liberal theory of the social contract"), there is hardness (e.g., the critique of "a unitary, autonomous Subject" by those in contemporary postmodernism). And the reverse direction also holds true.

In relation to the seriousness-playfulness principle, if there is seriousness (e.g., the serious business of those who try to understand the nature of the "subject"), there is playfulness (e.g., the playful part of those who try to understand the nature of the "subject," when they play around with different versions in different philosophical traditions over time, like German idealism, postmodernism, analytic philosophy, the liberal theory of the social contract, etc.). And the reverse direction also holds true.

In relation to the activeness-inactivess principle, if there is activeness (e.g., the relative activeness of those who defend "a unitary, autonomous Subject" in the community of "the liberal theory of the social contract"), there is inactiveness (e.g., the relative inactiveness of those in the community of "the liberal theory of the social contract," when they are not welcome in the postmodernist circle, as the latter camp is more concerned with "the subject as a social construction" or as "an ideological construction" instead). And the reverse direction also holds true.

In relation to the selfness-otherness principle, if there is selfness (e.g., the tendency to be mostly centered on one's self or group, as shown in the editors of *Social Text*, who accepted a "nonsensical" article by Alan Sokal, who "flattered the editors' ideological preconceptions" in the so-called "Sokal hoax"), there is otherness (e.g., the tendency to be mostly centered on others, as shown in "the resultant academic and public quarrels" after the "Sokal hoax," in regard to "the scholarly merit of humanistic commentary about the physical sciences; the influence of postmodern philosophy on social disciplines in general; academic ethics, including whether Sokal was right or wrong to deceive the editors and readers of Social Text; and whether the journal had exercised the appropriate intellectual rigor before publishing the pseudoscientific article"—as a way to improve the "intellectual rigor" in fuure publications, as an advance in knowledge for humanity). (WK 2012hhh) And the reverse direction also holds true.

In relation to the regression-progression principle, if there is regression (e.g., the regression made by the philosophical traditions, as shown in the problems pointed out by the critics aforecited), there is progression (e.g., the progress made by the philosophical traditions, as shown in the contribution to a better understanding of the evolution of the idea of the "subject" aforecited). And the reverse direction also holds true.

In relation to the same-difference principle, if there is similarity in outcome (e.g., the contribution to the molding and control of beliefs, values, and behaviors, regardless of whether this be done by way of "a unitary, autonomous Subject" in "the liberal theory of the social contract," or by way of the "split subject"' in "Freud's psychoanalytic model of the subject" and its further development by Lacan), there is difference in outcome (e.g., the contribution to the molding and control of beliefs, values, and behaviors by way of "a unitary, autonomous Subject" for a relatively more conservative worldview—but the contribution to the molding and control of beliefs, values, and behaviors by way of the "split subject"' for a relatively more subversive worldview). And the reverse direction also holds true.

In relation to the stability-reaction principle, if there is stability (e.g., the relatively more stable condition of the idea of "a unitary, autonomous Subject" in the older days of "the liberal theory of the social contract"), there is reaction (e.g., "a unitary, autonomous Subject" in the older days of "the liberal theory of the social contract" then led to further changes later on, when it was challenged by different philosophical traditions in the 20th century aforediscussed). And the reverse direction also holds true.

In relation to the functionality-nonfunctionality principle, if there is functionality (e.g., the presence of action for which something or someone is fitted, as shown in the belief in "a unitary, autonomous Subject" in the older days of "social contract" theory, which served the function of perpetuating the "liberal" tradition at the time), there is nonfunctionality (e.g., the relative lack of action for which something or someone is fitted, as shown in the rise of postmodernism in our time, with its "deconstruction of the subject as a core-concept of metaphysics," such that the old-fashioned belief in "a unitary, autonomous Subject" in the older days of "social contract" theory is no longer much functional in our contemporary postmodernist era). And the reverse direction also holds true.

In relation to the intentionality-nonintentionality principle, if there is intentionality (e.g., the planning or design of something for a certain outcome, as shown in the original intention of those in the "liberal theory of the social contract" to use the idea of "a unitary, autonomous Subject" as its "foundation"), there is nonintentionality (e.g., the relative absence of planning or design of something for a certain outcome, such that the outcome can even be counter-intentional in being the opposite of what was originally planned, as shown in the works of Marx and Freud, who "provided a point of departure for questioning the notion of a unitary, autonomous Subject" and "opened up the way for the deconstruction of the subject as a core-concept of metaphysics"—contrary to the original intention of those in the "liberal theory of the social contract" to use the idea of "a unitary, autonomous Subject" as its "foundation"). And the reverse direction also holds true.

And in relation to the survivability-nonsurvivability principle, if there is survivability (e.g., the relative survival of "the issue of subject" as "one of the major intractable problems in philosophy of mind" in our time, especially by those in contemporary analytic philosophy), there is non-survivability (e.g., the relative non-survival of "the notion of a unitary, autonomous Subject" in our time, especially among those in postmodernism—although it was "the foundation of the liberal theory of the social contract" in the older days). And the reverse direction also holds true.

The Opinionatedness of Subjectivity

This comprehensive analysis of the possibility (or impossibility) and desirability (or undesirability) of subjectivity in the context of probability study, from the perspectives of the mind, nature, society, and culture, is to show us the different ways in which subjectivity is both possible and desirable, but not to the extent that the spokespersons from each side would like us to believe.

Subjectivity thus has its opinionatedness, just as it has its flexibility too, as the other side of the same coin.

Of course, the other side of the story concerns objectivity, the opposite of subjectivity, but the issues concerning objectivity were already addressed in Chapter Two.

These dual analyses should not be underestimated, because those on the side of objectivity often dismiss subjectivity merely as the other side without really appreciating it from the vantage point of subjectivity too—and vice versa.

Now that we have reached the end of this intellectual journey, what then can be concluded in regard to the future of probability, in the context of both subjectivity and objectivity?

This is the question which will be answered in the following concluding chapter on the future of probability—for which we now turn to Chapter Four.

Table 3.1. Subjectivity and Its Duality

- **Subjectivity and the Mind**
  —Ex: subjectivity, Bayesian probability, and personal belief
  —Ex: subjectivity, Q analysis, and the association in thinking

- **Subjectivity and Nature**
  —Ex: subjectivity, measurement, and natural condition
  —Ex: subjectivity, quantum mechanics, and uncertainty

- **Subjectivity and Society**
  —Ex: subjectivity, comunicative action, and power
  —Ex: subjectivity, mathematical practice, and sociology

- **Subjectivity and Culture**
  —Ex: subjectivity, experience, and phenomenology
  —Ex: subjectivity, the subject, and philosophical traditions

Notes: The examples in the categories are solely illustrative (not exhaustive), and the comparison is relative (not absolute), nor are they necessarily mutually exclusive. And some can be easily re-classified elsewhere. As generalities, they allow exceptions.
Source: A summary of Ch.3 of *FPHPROB*

Table 3.2. Subjectivity, Bayesian Probability, and Personal Belief

- **Justifications of Bayesian Probability**
  —Ex: axiomatic approach
  —Ex: Dutch book approach
  —Ex: decision theory approach

- **Characteristics of Bayesian Probability**
  —Ex: "the sequential use of the Bayes' formula: when more data become available after calculating a posterior distribution, the posterior becomes the next prior."
  —Ex: "for the frequentist, a hypothesis is a proposition (which must be either true or false), so that the frequentist probability of a hypothesis is either one or zero. In Bayesian statistics, a probability can be assigned to a hypothesis that can differ from 0 or 1 if the truth value is uncertain."
  —Ex: "the use of hierarchical models and marginalization over the values of nuisance parameters [which are not of immediate interest]. In most cases, the computation is intractable, but good approximations can be obtained using Markov chain Monte Carlo methods."

- **Problems with Bayesian Probability**
  —Firstly, there is an alternative interpretation of Bayesian probability.
  —Secondly, the probability of a random event can be quantified without the need of Bayesian statistics.
  —Thirdly, there can be an eclectic approach which is neither Bayesian (subjectivist) nor objectivist.
  —Fourthly, different individual Bayesians disagree with each other about a particular probability.
  —Fifthly, that its influence is still limited.
  —Sixthly, its justifications are problematic.

Notes: The examples in the categories are solely illustrative (not exhaustive), and the comparison is relative (not absolute), nor are they necessarily mutually exclusive. And some can be easily re-classified elsewhere. As generalities, they allow exceptions.
Source: From *Sec. 3.2.1* of *FPHPROB*. See book for citations.

Table 3.3. Subjectivity, Measurement, and Natural Condition

- **Types of Measurement Uncertainty**
  —Ex: systematic error
  —Ex: random Error

- **Evaluation of Measurement Uncertainty**
  —Ex: Type A evaluation
  —Ex: Type B evaluation

- **Evaluation of Measurement Uncertainty**
  —Ex: formulation stage
  —Ex: calculation stage

- **Problems with the the Study of Measurement Uncertainty**
  —Firstly, one main criticism of the study of measurement uncertainty is that there are alternative perspectives to understand uncertainty.
  —Secondly, another main criticism of the study of measurement uncertainty is that its assumptions can be questioned.
  —Thirdly, still another main criticism of the study of measurement uncertainty is that there are more robust ways to do measurement uncertainty.
  —Fourthly, an additional main criticism of the study of measurement uncertainty is that it has its own inherent limits.
  —And fifthly, another additional main criticism of the study of measurement uncertainty is that there are conflicting interpretations of probability in measurement uncertainty.

Notes: The examples in the categories are solely illustrative (not exhaustive), and the comparison is relative (not absolute), nor are they necessarily mutually exclusive. And some can be easily re-classified elsewhere. As generalities, they allow exceptions.
Source: From *Sec. 3.3.1* of *FPHPROB*. See book for citations.

Table 3.4. Subjectivity, Quantum Mechanics, and Uncertainty

- **Uncertainty and Related Concepts**
  —Ex: "uncertainty"—"the lack of certainty, A state of having limited knowledge where it is impossible to exactly describe the existing state, a future outcome, or more than one possible outcome."
  —Ex: "risk"—"a state of uncertainty where some possible outcomes have an undesired effect or significant loss"

- **Formulations of the Uncertainty Principle**
  —Ex: "wave mechanics" formulation
  —Ex: "matrix mechanics" formulation

- **Problems with the Uncertainty Principle**
  —Firstly, the loss of certainty in quantum mechanical measurements can be due to our ignorance of some hidden variables in the real world.
  —Secondly, quantum mechanics itself has gone through different revisions and interpretations over time.
  —Thirdly, it can be falsified.
  —And fourthly, there can be an alternative in-between perspective.

Notes: The examples in the categories are solely illustrative (not exhaustive), and the comparison is relative (not absolute), nor are they necessarily mutually exclusive. And some can be easily re-classified elsewhere. As generalities, they allow exceptions.
Source: From *Sec. 3.3.2* of *FPHPROB.* See book for citations.

Table 3.5. Subjectivity, Communicative Action, and Power

- **Aspects of Universal Pragmatics**
  —Ex: theories of "elementary propositions"
  —Ex: theories of "first-person sentences"
  —Ex: theories of "speech acts"

- **Communicative Action in Universal Pragmatics**
  —Ex: modes of social integration
  —Ex: communicative competence
  —Ex: validity of speech

- **Problems with Communication Action**
  —Firstly, one main criticism of the theory of communicative action is that it is unwarrantedly biased for "the lifeworld" and against "the system."
  —Secondly, another main criticism of the theory of communicative action is that it falsely presupposes a powerful role of reason that it does not have, as it is inherently embedded within the ideology of Western Enlightenment.
  —Thirdly, still another main criticism of the theory of communicative action is that there is a serious tension between "communicative action" and "communicative discourse."
  —Fourthly, an additional main criticism of the theory of communicative action is that it does not give a serious attention to the lasting social spectre of power.
  —And fifthly, another additional main criticism of the theory of communicative action is that its appeal to the claim to rightness is itself highly problematic.

Notes: The examples in the categories are solely illustrative (not exhaustive), and the comparison is relative (not absolute), nor are they necessarily mutually exclusive. And some can be easily re-classified elsewhere. As generalities, they allow exceptions.
Source: From *Sec. 3.4.1* of *FPHPROB*. See book for citations.

Table 3.6. Subjectivity, Mathematical Practice, and Sociology

- **Quasi-Empiricism in Mathematics**
  - —Ex: mathematical bafflement
  - —Ex: manipulative effectiveness
  - —Ex: informal proofs and proof by authority

- **Folk Mathematics**
  - —Ex: mathematical folklore
  - —Ex: informal mathematics

- **Problems with the Sociology of Mathematical Practice**
  - —Firstly, one main criticism of the sociology of mathematical practice is that it is too obsessed with social construction in a reductionistic way.
  - —Secondly, another main criticism of the sociology of mathematical practice is that it is not critical enough to folklore mathematics.
  - —Thirdly, still another main criticism of the sociology of mathematical practice is that there are necessary arguments for the foundation of mathematics, contrary to the main view in the sociology of mathematical knowledge.
  - —And fourthly, an additional main criticism of the sociology of mathematical practice is that its social constructionism is incoherent.

Notes: The examples in the categories are solely illustrative (not exhaustive), and the comparison is relative (not absolute), nor are they necessarily mutually exclusive. And some can be easily re-classified elsewhere. As generalities, they allow exceptions.
Source: From *Sec. 3.4.2* of *FPHPROB*. See book for citations.

Table 3.7. Subjectivity, Experience, and Phenomenology

- **Main Ideas of Phenomenology**
  —Ex: intentionality
  —Ex: intuition
  —Ex: evidence
  —Ex: noesis and noema
  —Ex: empathy and intersubjectivty
  —Ex: lifeworld

- **The Transcendental Turn in Phenomenology**
  —Ex: "what we observe is not the object as it is in itself, but how and inasmuch it is given in the intentional acts. Knowledge of essences would only be possible by 'bracketing' all assumptions about the existence of an external world and the inessential (subjective) aspects of how the object is concretely given to us. This procedure Husserl called *epoché*." (WK 2012ppp)
  —Ex: as Husserl "wanted to exclude any hypothesis on the existence of external objects, he introduced the method of phenomenological reduction to eliminate them. What was left over was the pure transcendental ego, as opposed to the concrete empirical ego. Now Transcendental Phenomenology is the study of the essential structures that are left in pure consciousness: This amounts in practice to the study of the noemata and the relations among them." (WK 2012ppp)

- **Problems with Phenomenology**
  —Firstly, it is too subjectivist.
  —Secondly, its ideas are often used in an obscure (controversial) way.
  —Thirdly, its foundational stance in regard to the transcendental ego is illusory.
  —Fourthly, it evolves over time into different conflicting versions.
  —And fifthly, there are alternative approaches competing with it.

Notes: The examples in the categories are solely illustrative (not exhaustive), and the comparison is relative (not absolute), nor are they necessarily mutually exclusive. And some can be easily re-classified elsewhere. As generalities, they allow exceptions.
Source: From *Sec. 3.5.1* of *FPHPROB*. See book for citations.

• PART FOUR •

Conclusion

CHAPTER 4
CONCLUSION—THE FUTURE OF PROBABILITY

A creation needs not only subjectivity,
but also objectivity.
__Stephen Chow (FQ 2012)

Beyond Objectivity and Subjectivity

The analyses of objectivity (in Chapter Two) and subjectivity (in Chapter Three) show the extent of their possibility (or impossibility) and desirability (or undesirability) in the study of probability.

These two analyses are valuable for the evaluation of the opposing views on probability (and other views as already introduced in *Sec. 1.5* and the rest of the book), such that probability (in relation to both objectivity and subjectivity) are neither possible (or impossible) nor desirable (or undesirable) to the extent that the respective ideologues (on different sides) would like us to believe.

As already indicated in *Sec.1.1*, this questioning of the opposing interpretations on probability does not entail that probability is useless, or that those fields (related to probability) like "statistics, finance, gambling, science, artificial intelligence/machine learning and philosophy" are not worth studying. Needless to say, neither of these extreme views is reasonable.

Instead, this book offers an alternative (better) way to understand the future of probability, especially in the dialectic context of objectivity and subjectivity—while learning from different approaches in the literature but without favoring any one of them (nor integrating them, since they are not necessarily compatible with each other).

More specifically, this book offers a new theory (that is, *the interpretivist theory of probability*) to go beyond the existing approaches in a novel way.

My *interpretivist theory of probability* has four major distinct features.

Firstly, my theory makes good use of all theoretical approaches to probability study in relation to objectivity and subjectivity, be they about the mental argument, the natural argument, the societal argument, or the cultural argument—especially from the most comprehensive combined perspectives of the mind, nature, society, and culture (as already dealt with in Chapter Two and Chapter Three).

Secondly, just like many other theories of mine in my previous books, my theory here does not heavily favor any specific theory over others in the literature, nor trying to integrate them (as they are not necessarily compatible with each other).

Thirdly, my theory treats the issue of probability in the distinctive dialectic context of no objectivity without subjectivity, and vice versa.

And fourthly, it contains thirty-one major theses, namely, (a) the first thesis: the formalness-informalness principle, (b) the second thesis: the absoluteness-relativeness principle, (c) the third thesis: the partiality-totality principle, (d) the fourth thesis: the predictability-unpredictability principle, (e) the fifth thesis: the explicability-inexplicability principle, (f) the sixth thesis: the fiction-reality principle, (g) the seventh thesis: the cognitiveness-noncognitiveness principle, (h) the eighth thesis: the finiteness-transfiniteness principle, (i) the ninth thesis: the preciseness-vagueness principle, (j) the tenth thesis: the simpleness-complicatedness principle, (k) the eleventh thesis: the openness-hiddenness principle, (l) the twelfth thesis: the denseness-emptiness principle, (m) the thirteenth thesis: the change-constancy principle, (n) the fourteenth thesis: the order-chaos principle, (o) the fifteenth thesis: the slowness-quickness principle, (p) the sixteenth thesis: the expansion-contraction principle, (q) the seventeenth thesis: the theory-praxis principle, (r) the eighteenth thesis: the convention-novelty principle, (s) the nineteenth thesis: the evolution-transformation principle, (t) the twentieth thesis: the symmetry-asymmetry principle, (u) the twenty-first thesis: the softness-hardness principle, (v) the twenty-second thesis: the seriousness-playfulness principle, (w) the twenty-third thesis: the activeness-inactiveness principle, (x) the twenty-fourth thesis: the selfness-otherness principle, (y) the twenty-fifth thesis: the regression-progression principle, (z) the twenty-sixth thesis: the sameness-difference principle, (aa) the twenty-seventh thesis: the stability-reaction principle, (bb) the twenty-eighth thesis: the functionality-nonfunctionality principle, (cc) the twenty-ninth thesis: the intentionality-nonintentionality principle, (dd) the thirtieth thesis: the survivability-nonsurvivability principle, and (ee) the thirty-first thesis: the post-human interpretiveness—to be elaborated in this chapter.

Because of the last thesis on "the post-human interpretiveness," the word "interpretivist" in the title of my theory is suggestive, because it refers to the different interpretations of probability in the face of changing circumstances over time (like the opposing objectivist and subjectivist interpretations)—which then require a multi-faceted inquiry of the dialectic relations between objectivity and subjectivity, as well as an understanding of the different views regarding these dialectic relations in probability study (together with other debates as already discussed in the rest of the book)—and in the end it calls for a need to transcend them by future advanced humans and later post-humans in the context of the great transformations of probability in the future world (as will be discussed in this chapter).

With this clarification in mind—the seminal project will fundamentally change the way that we think about objectivity and subjectivity in probability study (together with other debates as already discussed in the book), from the combined perspectives of the mind, nature, society, and culture, with enormous implications for the human future and what I originally called its "post-human" fate.

In Relation to Method

The first seven theses are in relation to method in existential dialectics, as shown below.

1st Thesis: The Formalness-Informalness Principle

The first thesis refers to the formalness-informalness principle (on the formal requirements of logical systems) in the category of "method" in existential dialectics, which was first proposed in *FPHML* (and later, *FPHH*, *FPHGAM*, *FPHSEM*, and other books of mine), in that the formal requirements of a logical system (e.g., consistency, soundness, and completeness) have both usefulness and non-usefulness, to the extent that, if there are formal systems requiring them, there are alternative ones which do not.

The formalness-informalness principle reveals something interesting in existential dialectics, in that it goes beyond both classical and non-classical logics (while learning something from both).

Thus, it does *not* exclude classical logics but simply goes beyond both classical and non-classical logics, while learning something from each. There is no formalness without informalness—and vice versa.

For illustration, existential dialectics can make use of both classical logics under certain conditions (e.g., especially, though not exclusively, when they are clear-cut, etc.) and non-classical logics under alternative conditions (especially, though not exclusively, when they are "unknown," "irrelevant," "ambiguous," "possible," with "different degrees of truth," empirically inconsistent in a desirable way, etc.). (WK 2008u)

For instance, on the one hand, there is formalness (e.g., the formal logical requirement of soundness, completeness, and consistency in a system of ideas, as in the argument on imagination and mental models, in *Sec. 2.2.1*; the formal logical requirement of soundness, completeness, and consistency in a system of ideas, as in the argument on emotion and rationality, in *Sec. 2.2.2*; the formal logical requirement of soundness and completeness in a system of ideas, as in the argument on the physical world and discrete probability, in *Sec. 2.3.1*; the formal logical requirement of soundness and completeness in a system of ideas, as in the argument on complexity and emergence, in *Sec. 2.3.2*; the formal logical requirement of soundness, completeness, and consistency in a system of ideas, as in the study of reliability engineering, in *Sec. 2.4.1*; the formal logical requirement of soundness and completeness in a system of ideas, as in the argument for optimal decision, in *Sec. 2.4.2*; the formal logical requirement of soundness and completeness in a system of ideas, as in propositional calculus, in *Sec. 2.5.1*; the formal logical requirement of soundness and completeness in a system of ideas, as in the argument on history and common causes, in *Sec. 2.5.2*; the formal logical requirement of soundness, completeness, and consistency in a system of ideas, as in the argument for Bayesian probability, in *Sec. 3.2.1*; the formal logical requirement of soundness, completeness, and consistency in a system of ideas, as in Q methodology, in *Sec. 3.2.2*; the formal logical requirement of soundness, completeness, and consistency in a system of ideas, as in the study of measurement uncertainty, in *Sec. 3.3.1*; the formal logical requirement of soundness and consistency in a system of ideas, as in the uncertainty principle, in *Sec. 3.3.2*; the formal logical requirement of soundness, completeness, and consistency in a system of ideas, as in the theroy of communicative action, in *Sec. 3.4.1*; the formal logical requirement of soundness, completeness, and consistency in a system of ideas, as in the sociology of mathematical knowledge, in *Sec. 3.4.2*; the formal logical requirement of soundness, consistency, and completeness in a system of ideas, as in Husserl's phenomenology, in *Sec. 3.5.1*; and the formal logical requirement of soundness, consistency, and completeness in a system of ideas, as in the philosophical traditions on the "subject," in *Sec. 3.5.2*).

On the other hand, there is also informalness (e.g., the non-formal existence of unsoundness, incompleteness, and inconsistency in the argument on imagination and mental models, as shown in the criticisms that there are alternative explanations of human reasoning, that there is no general consensus of whether mental models are better than other models in explaining human reasoning, that mental models are simplistic, that they have their own limits, that a counter-example does not necessarily refute an inductive conclusion totally, and that there are complex situations which require optimal decision making [not on the basis of the simplification of reality as in mental models], in *Sec. 2.2.1*; the non-formal existence of unsoundness, incompleteness, and inconsistency in the argument on emotion and rationality, as shown in the criticisms that there is no general consensus on the usage of "rationality" in different fields, that the embedded assumptions in each rationality make it relative in relation to others, that the conventional views about rationality do not give sufficient weight to the issue of power, that the theories of rationality are mixed up in classification, that it is limited in practice, that it assumes a false conception of human nature, and that there are alternative explanations of rationality, in *Sec. 2.2.2*; the non-formal existence of unsoundness and incompleteness in the argument on the physical world and discrete probability, as shown in the criticisms that mathematics assumes certain forms of perfection which are not true in the real world, that there are alternative probability distributions which are not discrete, that mathematics use certain constructs of infinity which are not possible to apply in the real world, that discreteness is very difficult to be computed in certain situations, and that probability theory does not explain how probability mass functions come into existence, in *Sec. 2.3.1*; the non-formal existence of unsoundness in the argument on complexity and emergence, as shown in the criticisms that there is a tradeoff between solvability and size [e.g., the larger the size of a problem is, the less solvable the problem becomes], that solving complex problems is not practical in certain cases, that there is the problem of Bremermann's fundamental limit on computation, and that some complex problems cannot be predicted, etc., in *Sec. 2.3.2*; the non-formal existence of unsoundness, incompleteness, and inconsistency in the study of reliability engineering, as shown in the criticisms that there is a trade-off between "reliability" and "safety," that reliability testing can be infeasible in practice, that there is the problem of definition in testing, that testing does not always need to be formal, that the bias of perception can screw up the understanding of reliability, that the validity of testing can be problematic due to questionable assumptions, and that test errors are hard to avoid, in *Sec. 2.4.1*; the non-formal existence of unsoundness and incompleteness in the argument for optimal decision, as shown in the criticisms that optimization is seldom achieved in real life because decision making can be more intuitive and less formal, that optimal and non-optimal decision-making strategies are relative to different personality traits, that its scientific objectivity is questionable, that it is often taken out of context, that it has many questionable assumptions, that it does not give sufficient attention to non-rational factors, and that it ignores power and corruption, in *Sec. 2.4.2*; the non-formal existence of unsoundness and incompleteness in propositional calculus, as shown in the criticisms that the quest for completeness in mathematical logic is problematic, that it does not include modal logic, that it does not include alternative many-valued logics, that it does not pay attention to context-specific rules of inference, that it does not give weight to the factors of power and social action, that it is too formal and thus ignores less formal and more intuitive approaches to human thinking, and that it is too simplistic because there are are higher-order logics, in *Sec. 2.5.1*; the non-formal existence of unsoundness and incompleteness in the argument on history and common causes, as shown in the criticisms that there are "special causes" outside the historical experience base which are not predictable, that there are causes which can be within the historical experience base but are subjective in orientation, and that common causes are not necessarily desirable and any attempt to correct them could lead to more undesirable outcomes, in *Sec. 2.5.2*; the non-formal existence of unsoundness, incompleteness, and inconsistency in the argument for Bayesian probability, as shown in the criticisms that there is an alternative interpretation of Bayesian probability, that the probability of a random event can be quantified without the need of Bayesian statistics, that there can be an eclectic approach which is neither Bayesian [subjectivist] nor objectivist, that different individual Bayesians disagree with each other about a particular probability, that its influence is still limited, and that its justifications are problematic, in *Sec. 3.2.1*; the non-formal existence of unsoundness, incompleteness, and inconsistency in Q methodology, as shown in the criticisms that its analysis is theory-dependent, that it does not explain why the association of ideas in someone's mind exists in the way that it does, that there are alternative approaches, that the validity of its analysis is relative to the data collected about an individual, and that its overall evaluation as a method has yet to be fully accessed, in *Sec. 3.2.2*; the non-formal existence of unsoundness, incompleteness, and inconsistency in the study of measurement uncertainty, as shown in the criticisms that there are alternative perspectives to understand uncertainty, that its assumptions can be questioned, that there are more robust ways to do measurement uncertainty, that it has its own inherent limits, and that there are conflicting interpretations of probability in measurement uncertainty, in *Sec. 3.3.1*; the non-formal existence of unsoundness and inconsistency in the uncertainty principle, as shown in the criticisms that the loss of certainty in quantum mechanical measurements can be due to our ignorance of some hidden variables in the real world, that quantum mechanics itself has gone through different formulations over time, that it can be falsified, and that there can be an alternative in-between perspective, in *Sec. 3.3.2*; the non-formal existence of unsoundness, incompleteness, and inconsistency in the theroy of communicative action, as shown in the criticisms that it is unwarrantedly biased for "the lifeworld" and against "the system," that it falsely presupposes a powerful role of reason that it does not have, that there is a serious tension between "communicative action" and "communicative discourse," that it does not give a serious attention to the lasting social spectre of power, and that its appeal to the claim to rightness is itself highly problematic, in *Sec. 3.4.1*; the non-formal existence of unsoundness, incompleteness, and inconsistency in the sociology of mathematical knowledge, as shown in the criticisms that it is too obsessed with social construction in a reductionistic way, that it is not critical enough to folklore mathematics, that there are necessary arguments for the foundation of mathematics, and that its social constructionism is incoherent, in *Sec. 3.4.2*; the non-formal existence of unsoundness, inconsistency, and incompleteness in Husserl's phenomenology, as shown in the criticisms that it is too subjectivist, that its ideas are often used in an obscure [controversial] way, that its foundational stance in regard to the transcendental ego is illusory, that it evolves over time into different conflicting versions, and that there are alternative approaches competing with it, in *Sec. 3.5.1*; and the non-formal existence of unsoundness, inconsistency, and incompleteness in the philosophical traditions on the "subject," as shown in the criticisms that they often offer opposing viewpoints, that they often ignore alternative philosophical traditions in the Non-West, that the tension between subjectivity and objectivity may not be resolvable, and that its views can be nonsensual, in *Sec. 3.5.2*).

And the reverse direction also holds true.

However, it should be stressed (as this is something that I regularly did in my previous books, *almost verbatim*) that there are different shades of gray (or different degrees of truth) in the two opposites, and the classification is not necessarily mutually exclusive either. In fact, even when some combinations of the two occur, they only end up sharing the same dialectic relationship, but in a different degree.

Also, there are some other relationships (as a kind of family resemblance) which have something in common with the principle, but they are not exactly the same but only more or less comparable, which varies from case to case.

Good examples of family resemblance in relation to the principle are logicalness-nonlogicalness, rationality-nonrationality, etc.

2nd Thesis: The Absoluteness-Relativeness Principle

The second thesis refers to the absoluteness-relativeness principle (on the multiplicity of things) in the category of "method" in existential dialectics, which was first proposed in *FPHK*, in that there is the multiplicity of things in reality, be they about entities, qualities (or properties), and relationships. If there is something absolute, there is likewise something relative. And there is no absoluteness without relativeness—and vice versa.

Both absoluteness and relativeness here are also relevant to different modalities often cited in the literature on ontology, such as possibility (e.g., something "can" happen) and its opposite (e.g., impossibility) in alethic modalities, probability (e.g., something "will" happen) and its opposite (e.g., improbability) in temporal modalities, and necessity (e.g., something "should" happen) and its opposite (e.g., contingency) in alethic modalities.

For instance, on the one hand, there is absoluteness (e.g., the absolute view by those who favor mental models, in *Sec. 2.2.1*; the absolute view within a model of rationality, like "a model in which benefiting oneself is optimal," in *Sec. 2.2.2*; the absolute focus on discrete events by the advocates, in *Sec. 2.3.1*; system operates within its own "initial conditions," in *Sec. 2.3.2*; the absolute view of those who focus on "reliability" in reliability engineering, in *Sec. 2.4.1*; the absolute view on optimal decision making by its advocates, in *Sec. 2.4.2*; the absolute "assignment to each propositional symbol of $\mathcal{P}$ of one or the other [but not both] of the truth values truth [T] and falsity [F]" in propositional calculus, in *Sec. 2.5.1*; the absolute view by those who are tempted "to react to an extreme outcome and to see it as significant," in *Sec. 2.5.2*; the absolute view by those for the objectivist interpretation of probability, like the "frequentist probability…claim…that the probability of a random event denotes the *relative frequency of occurrence* of an experiment's outcome, when repeating the experiment," such that the probability of a random event can be quantified without the need of Bayesian statistics, in *Sec. 3.2.1*; the absolute view on analysis by those for R analysis, with its focus on "correlations between variables [say, height and age] across a sample of subjects," in *Sec. 3.2.2*; the absolute view by those in mainstream metrology who claimed that "random variables are proper mathematical models for uncertain quantities and simple probability distributions are sufficient for representing all forms of measurement uncertainties," in *Sec. 3.3.1*; the absolute view by Einstein on the loss of certainty in quantum mechanical measurements due to our ignorance of some hidden variables in the real world, in *Sec. 3.3.2*; the absolute view on communicative action by Habermas, in that "communicative action is an indispensable facet of society," in *Sec. 3.4.1*; the absolute view by Ian Hackings when he argued that "we should refrain from making absolute judgements about what is true, and instead state that something is true in the light of theory x," in *Sec. 3.4.2*; the absolute view by Husserl on the transcendental ego, in *Sec. 3.5.1*; and the absolute view on "a unitary, autonomous Subject, which for many thinkers…is seen as the foundation of the liberal theory of the social contract," in *Sec. 3.5.2*).

On the other hand, there is relativeness (e.g., what is right for those who favor mental models is not necessarily so for those who argue that there are alternative views on human reasoning, which are not based on mental models, and good examples are those based on "formal rules of inference" and "domain-specific rules of inference," for example, in *Sec. 2.2.1*; what is right for a model of rationality, like "a model in which benefiting oneself is optimal," is not necessarily so for a different rationality "in which benefiting the group is optimal," since "purely selfish behavior is deemed irrational" in the latter, in *Sec. 2.2.2*; what is appropriate for the advocates of discrete events is not necessarily so for the advocates of continuous events, in *Sec. 2.3.1*; different systems have different "initial conditions," so each of these systems is relative to others, since each "is highly sensitive" to its "initial conditions," in *Sec. 2.3.2*; what is right for those who focus on "reliability" in reliability engineering is not necessarily so for those who focus on "safety" instead, in *Sec. 2.4.1*; what is right for the advocates of optimal decision making is not necessarily so for those critics who prefer satisficing decision making, in *Sec. 2.4.2*; what is right for propositional calculus to assign "one or the other [but not both] of the truth values truth [T] and falsity [F]" is not necessarily so for "many-valued logics" which "allow…values other than *true* and *false*," in *Sec. 2.5.1*; what is right for those who are tempted "to react to an extreme outcome and to see it as significant" is not necessarily so for the critics, who pointed out that "its causes are common to many situations and the distinctive circumstances surrounding its occurrence," which are "the results of mere chance," in *Sec. 2.5.2*; what is right for those in favor of the objectivist interpretation of probability is not necessarily so for those in favor of the subjectivist version of Bayesian probability, in *Sec. 3.2.1*; what is right for those for R analysis, with its focus on "correlations between variables…across a sample of subjects," is not necessarily so for those in favor of Q analysis, which looks "for correlations between subjects across a sample of variables," since "for mathematical reasons no one data matrix would be suitable for analysis with both Q and R," in *Sec. 3.2.2*; what is right for those in mainstream metrology who claimed that "random variables are proper mathematical models for uncertain quantities and simple probability distributions are sufficient for representing all forms of measurement uncertainties" is not necessarily so for the critics who offered alternative perspectives to understand uncertainty, like the idea that "a mathematical interval rather than a probability distribution might be a better model of uncertainty," in *Sec. 3.3.1*; what is right for Einstein in regard to the loss of certainty in quantum mechanical measurements due to our ignorance of some hidden variables in the real world is not necessarily so for Heisenberg because of the uncertainty principle, in *Sec. 3.3.2*; what is right for Habermas in relation to his claim that "communicative action is an indispensable facet of society" is not necessarily so for those who favor "strategic action" instead, not "communicative action," in *Sec. 3.4.1*; what is right for Ian Hackings when he argued that "we should refrain from making absolute judgements about what is true, and instead state that something is true in the light of theory x" is not necessarily so for Paul Boghossian, who countered with his argument in this way: "But it is hard to see how we might coherently follow this advice. Given that the propositions which make up epistemic systems are just very general propositions about what absolutely justifies what, it makes no sense to insist that we abandon making absolute particular judgements about what justifies what while allowing us to accept absolute general judgements about what justifies what. But in effect this is what the epistemic relativist is recommending," in *Sec. 3.4.2*; what is right for Husserl on the transcendental ego is not necessarily so for Merleau-Ponty, who "objects to the ego's transcendence of the world" and proposes "existential phenomenology," which "differs from transcendental phenomenology by its rejection of the transcendental ego," in *Sec. 3.5.1*; and what is right for those who defended "a unitary, autonomous Subject…as the foundation of the liberal theory of the social contract" is not necessarily so for the critics like Marx and Freud, who "provided a point of departure for questioning the notion of a unitary, autonomous Subject" and "opened up the way for the deconstruction of the subject as a core-concept of metaphysics," in *Sec. 3.5.2*).

And the reverse direction also holds true.

However, it should be stressed (as this is something that I regularly did in my previous books, *almost verbatim*) that there are different shades of gray (or different degrees of truth) in the two opposites, and the classification is not necessarily mutually exclusive either. In fact, even when some combinations of the two occur, they only end up sharing the same dialectic relationship, but in a different degree.

Also, there are some other relationships (as a kind of family resemblance) which have something in common with the principle, but they are not exactly the same but only more or less comparable, which varies from case to case.

Good examples of family resemblance in relation to the principle are internalness-externalness, immanence-transcendence, independence-dependence, isolatedness-interconnectiveness, internality-externality, autonomy-interdependency, universalness-relativeness, and so forth.

3rd Thesis: The Partiality-Totality Principle

The third thesis refers to the partiality-totality principle (on the relationships between whole and parts) in the category of "method" in existential dialectics, which was first proposed in *FC* and later *FPHGEOL*, based on earlier works like *FHC* and *FPHC*), in that any analysis of a phenomenon in the world requires both the examination of the individual parts and the whole of them (with the whole as being not the sum of its parts), without privileging the whole as more important than the parts, or vice-versa (that is, the parts as more important than the whole).

In fact, this principle targets two kinds of misleading methodologies as used over the ages, that is, again in the absence of better words, what I called (in *FHC*, *FPHC* and *FC*) "reductionism" and its opposite, "reverse-reductionism," which can take four major forms in relation to concept, theory, methodology, and ontology (as summarized in *Sec. 1.8*).

Thus, reductionism and reverse-reductionism can be (i) conceptual, (ii) theoretical, (iii) methodological, and (iv) ontological. And the whole is not the sum of its parts, and there is no partiality without totality—and vice versa.

Works on emergence (as in *FC*), gestalt psychology (as in *FPHGEOL*), and the hermeneutic circle (as in *FPHDA*), just to name a few of them, have contributed to the understanding of whole and parts.

For instance, on the one hand, there is partiality (e.g., the partial view on event anticipation by those who favor mental models, in *Sec. 2.2.1*; in *Sec. 2.2.1*; the partial view about rationality, if looked from the vantage point of a model of rationality, like "a model in which benefiting oneself is optimal," in *Sec. 2.2.2*; the partial view by those who focus on discreteness, in *Sec. 2.3.1*; the partial view on complexity from the vantage point of a particular meaning of complexity in a particular field, in *Sec. 2.3.2*; the partial view of "failure" if looked from the vantage point of "reliability," in *Sec. 2.4.1*; the partial view about decision making, if looked from the vantage point of optimal decision, in *Sec. 2.4.2*; the partial view about inference rules in propositional calculus, in *Sec. 2.5.1*; the partial view about a prison riot in history by those who are tempted "to react to an extreme outcome and to see it as significant," in *Sec. 2.5.2*; the partial view on probability by those in favor of the subjectivist version of Bayesian probability, in *Sec. 3.2.1*; the partial view on analysis by those for R analysis, in *Sec. 3.2.2*; the partial view on measurement uncertainty by those in the Joint Committee for Guides in Metrology, when they proposed a "subjectivist" interpretation in treating "measurement uncertainty…as a measure of how well one believes one knows the true value of the measurand," and "the notion of 'belief' is an important one, since it moves metrology into a realm where results of measurement need to be considered and quantified in terms of probabilities that express degrees of belief," in *Sec. 3.3.1*; the partial view on quantum mechanics by Heisenberg, in *Sec. 3.3.2*; the partial view on communication by Habermas in regard to his "communicative action," in *Sec. 3.4.1*; the partial view by Ian Hackings on social constructionism, in *Sec. 3.4.2*; the partial view on subjective experience from Husserl's phenomenological vantage point, in *Sec. 3.5.1*; and the partial view on the "subject" by those in German idealism, in *Sec. 3.5.2*).

On the other hand, there is also totality (e.g., the more holistic view on event anticipation, such that the whole is not the sum of its parts, that is, it cannot be reduced to the sum of opposing views by the advocates of "mental models," "formal rules of inference," "domain-specific rules of inference," and "probabilities," etc., since there will emerge new views in the future not yet known today, just as there are already alternative views nowadays which do not side exclusively with any of them, as is the analysis here, in *Sec. 2.2.1*; the more holistic view of about rationality, such that the whole is not the sum of its parts, that is, it cannot be reduced to the sum of opposing views like the four rational constructs by Weber, since there will emerge new views in the future not yet known today, just as there are already alternative views nowadays which do not side exclusively with any of them, as is the analysis here, in *Sec. 2.2.2*; the more holistic view of probability distributions, such that the whole is not the sum of its parts, that is, it cannot be reduced to the sum of opposing views about discrete probability distributions, continuous probability distributions, etc., since there will emerge new views in the future not yet known today, just as there are already alternative views nowadays which do not side exclusively with any of them, as is the analysis here, in *Sec. 2.3.1*; the more holistic view of complexity, such that the whole is not the sum of its parts, that is, it cannot be reduced to the sum of opposing views about the meanings of complexity in different fields, since there will emerge new views in the future not yet known today, just as there are already alternative views nowadays which do not side exclusively with any of them, as is the analysis here, in *Sec. 2.3.2*; the more holistic view of "failure," such that the whole is not the sum of its parts, that is, it cannot be reduced to the sum of opposing views like "reliability," "safety," "tradeoff," "definition," etc., since there will emerge new views in the future not yet known today, just as there are already alternative views nowadays which do not side exclusively with any of them, as is the analysis here, in *Sec. 2.4.1*; the more holistic view about decision making, such that the whole is not the sum of its parts, that is, it cannot be reduced to the sum of opposing views by those for optimal decision, satisficing decision, and so on, since there will emerge new views in the future not yet known today, just as there are already alternative views nowadays which do not side exclusively with any of them, as is the analysis here, in *Sec. 2.4.2*; the more holistic view about inference rules, such that the whole is not the sum of its parts, that is, it cannot be reduced to the sum of opposing views in propositional calculus, in higher-order logics, in modal logic, in many-valued logics, in incompleteness theorem, in everyday decision techniques, etc., since there will emerge new views in the future not yet known today, just as there are already alternative views nowadays which do not side exclusively with any of them, as is the analysis here, in *Sec. 2.5.1*; the more holistic view about a prison riot in history, such that the whole is not the sum of its parts, that is, it cannot be reduced to the sum of opposing views by the officials, the critics, etc., since there will emerge new views in the future not yet known today, just as there are already alternative views nowadays which do not side exclusively with any of them, as is the analysis here, in *Sec. 2.5.2*; the more holistic view on probability, such that the whole is not the sum of its parts, that is, it cannot be reduced to the sum of opposing views like the objectivist interpretation of probability, the two opposing versions of Bayesian probability, etc., since there will emerge new views in the future not yet known today, just as there are already alternative views nowadays which do not side exclusively with any of them, as shown in the argument by some that there can be an eclectic approach which is neither Bayesian [subjectivist] nor objectivist, in *Sec. 3.2.1*; the more holistic view on analysis, such that the whole is not the sum of its parts, that is, it cannot be reduced to the sum of opposing views like R analysis, Q analysis, Cultural Consensus Theory, etc., since there will emerge new views in the future not yet known today, just as there are already alternative views nowadays which do not side exclusively with any of them, as is the analysis here, in *Sec. 3.2.2*; the more holistic view on measurement uncertainty, such that the whole is not the sum of its parts, that is, it cannot be reduced to the sum of opposing views like the "subjectivist" interpretation, the "objectivist" interpretation, etc., since there will emerge new views in the future not yet known today, just as there are already alternative views nowadays which do not side exclusively with any of them, as is the analysis here, in *Sec. 3.3.1*; the more holistic view on quantum mechanics, such that the whole is not the sum of its parts, that is, it cannot be reduced to the sum of opposing views by Heisenberg, Einstein, Everett, etc., since there will emerge new views in the future not yet known today, just as there are already alternative views nowadays which do not side exclusively with any of them, as is the analysis here, in *Sec. 3.3.2*; the more holistic view on communication, such that the whole is not the sum of its parts, that is, it cannot be reduced to the sum of opposing views of Habermas, Gadamer, Foucault, Smith, etc., since there will emerge new views in the future not yet known today, just as there are already alternative views nowadays which do not side exclusively with any of them, as is the analysis here, in *Sec. 3.4.1*; the more holistic view on social constructionism, such that the whole is not the sum of its parts, that is, it cannot be reduced to the sum of opposing views of Hackings, Boghossian, etc., since there will emerge new views in the future not yet known today, just as there are already alternative views nowadays which do not side exclusively with any of them, as is the analysis here, in *Sec. 3.4.2*; the more holistic view on subjective experience, such that the whole is not the sum of its parts, that is, it cannot be reduced to the sum of opposing views of Husserl, Heidegger, Merleau-Ponty, Garfinkel, etc., since there will emerge new views in the future not yet known today, just as there are already alternative views nowadays which do not side exclusively with any of them, as is the analysis here, in *Sec. 3.5.1*; and the more holistic view on the "subject," such that the whole is not the sum of its parts, that is, it cannot be reduced to the sum of opposing views by those in German idealism, postmodernism, analytic philosophy, the liberal theory of the social contract, etc., since there will emerge new views in the future not yet known today, just as there are already alternative views nowadays which do not side exclusively with any of them, as is the analysis here, in *Sec. 3.5.2*;

And the reverse direction also holds true.

However, it should be stressed (as this is something that I regularly did in my previous books, *almost verbatim*) that there are different shades of gray (or different degrees of truth) in the two opposites, and the classification is not necessarily mutually exclusive either. In fact, even when some combinations of the two occur, they only end up sharing the same dialectic relationship, but in a different degree.

Also, there are some other relationships (as a kind of family resemblance) which have something in common with the principle, but they are not exactly the same but only more or less comparable, which varies from case to case.

Good examples of family resemblance in relation to the principle are individualisticness-holisticness, partness-wholeness, analysis-synthesis, and the like.

4th Thesis: The Predictability-Unpredictability Principle

The fourth thesis is called the predictability-unpredictability principle (on the occurrence of events) in the category of "method" in existential dialectics (which was first proposed in *FC* and, later, other books of mine), in that both predictability and unpredictability have a major role to play in the occurrence of things, so that neither determinism nor indeterminism wins the centuries-old fight. There is no predictability without unpredictability—and vice versa.

There are events which are predictable, just as there are those which are not. Or what is regarded as unpredictable at one point in time may turn out to be predictable later, and, conversely, what is deemed as predictable may turn out to not be so predictable. Even in predictability, outcomes are subject to uncertainty, the degree of which varies from case to case.

Works on ecological uncertainty (as in *BNR*), complexity theory (as in *FC*), quantum mechanics (as in *FPHST* and *FPHG*), frequency probability, subjective (or Bayesian) probability (as in *FPHPROB*), "special causes" (as in *FPHGAM*), and measurement uncertainty (as in

FPHPROB), just to name a few of them, have contributed to the understanding of uncertainity.

The predictability-unpredictability principle is not to be confused with the absoluteness-relativeness principle; although both can address the issue of probability, the former is concerned with the extent of certainty in prediction, whereas the latter focuses on the relativity of viewpoints.

For instance, on the one hand, there is predictability (e.g., the predictable tendency of "reasoners" with mental models to reject a conclusion if they find a counterexample, in *Sec. 2.2.1*; the predictable tendency of "humans" to be "rational in principle, according to some scholars like Philip Johnson-Laird and Ruth M.J. Byrne, in *Sec. 2.2.2*; the predictable tendency of a dice to have a probability of rolling each number in terms of 1/*s* in principle, in *Sec. 2.3.1*; the predictable tendency of "the properties of the system as a whole" in organized complexity, since they "can be understood by using probability and statistical methods," in *Sec. 2.3.2*; the predictable tendency of those in reliability engineering to make certain predictions aforecited, in *Sec. 2.4.1*; the predictable tendency of an optimal decision in situation of uncertainty to make use of a probabilistic approach, with "the probability distribution for the possible outcomes described by the conditional probability density *p [o|d]*, in *Sec. 2.4.2*; the predictable tendency of propositional calculus to be refined by Predicate Logic, Natural Deduction, Truth-Trees, Truth-Tables, and so on, in *Sec. 2.5.1*; the predictable tendency of the existence of special causes in history, in *Sec. 2.5.2*; the predictable tendency by those for Bayesian probability to argue that one hypothesis can be "more probable than the other," in *Sec. 3.2.1*; the predictable tendency of those in Q analysis to argue that "the individual who gains the highest factor loading on an Operant factor is the person most able to conceive the norm for the factor," in *Sec. 3.2.2*; the predictable tendency of a "random error" in measurement uncertainty to be "associated with the fact that when a measurement is repeated, it will generally provide a measured value that is different from the previous value," in *Sec. 3.3.1*; the predictable tendency by the advocates of the uncertainty principle that "the more precisely the position of some particle is determined, the less precisely its momentum can be known," in *Sec. 3.3.2*; the predictable tendency by the followers of Habermas to assess communication on the basis of the claims to truth, truthfulness, and rightness, in *Sec. 3.4.1*; the predictable tendency of many "informal mathematical practices…by aboriginal or ancient peoples" to be "usually understood intuitively" without any proofs [that is, "based on proofless heuristics"], in *Sec. 3.4.2*; the predictable tendency of a subject to be conscious of something, with the help of intuition and evidence, in the world of Husserl's phenomenology, in *Sec. 3.5.1*; and the predictable tendency of different philosophical traditions in the West to debate on the nature of the "subject," in *Sec. 3.5.2*).

On the other hand, there is unpredictability (e.g., the more difficult task to predict exactly the extent to which "reasoners" with mental models who focus on a subset of a single model will end up rejecting a particular conclusion on a particular occasion, in *Sec. 2.2.1*; the more difficult task to predict exactly the extent to which humans are rational on a particular occasion in practice, since "humans…err in practice, that is,…their performance is limited by various factors," according to some scholars like Philip Johnson-Laird and Ruth M.J. Byrne, in *Sec. 2.2.2*; the more difficult task to predict exactly the extent to which rolling a dice on a particular occasion will have a particular number as its outcome in practice, in *Sec. 2.3.1*; the more difficult task to predict exactly the extent to which "the interactions of the parts in a 'disorganized complexity' situation" will behave on a particular future occasion, since they "can be seen as largely random" and there is "the lack of correlation between elements in the system," in *Sec. 2.3.2*; the more difficult task to predict exactly the extent to which a particular reliability model will succeed in making correct predictions on a particular occasion, because of the problems pointed out by the critics aforecited, in *Sec. 2.4.1*; the more difficult task to predict exactly the extent to which "the outcome of a particular decision" will be on a particular occasion, in a situation of uncertainty, in *Sec. 2.4.2*; the more difficult task to predict exactly the extent to which propositional calculus will be refined in a particular way on a particular future occasion, since there will like be new logics in the future not yet knonw today, in *Sec. 2.5.1*; the more difficult task to predict exactly the extent to which a particular special cause will occur on a particular future occasion, since special causes are "new, unanticipated, emergent or previously neglected phenomena within the system" and their "variation" is "inherently unpredictable, even probabilistically," in *Sec. 2.5.2*; the more difficult task to predict exactly which "individual Bayesians might differ about which was the more probable and by how much, by virtue of having used different priors," in *Sec. 3.2.1*; the more difficult task to predict exactly the extent to which, in Q analysis, "the norm for the factor" is to mean, since "what the norm means is a matter, always, for conjecture and refutation [Popper],…so "these are all matters for future determination," in *Sec. 3.2.2*; the more difficult task to predict exactly what the next measured value is on a particular occasion, since a "random error" in measurement uncertainty is "random in that the next measured value cannot be predicted exactly from previous such values," in *Sec. 3.3.1*; the more difficult task to predict exactly "which certain pairs of physical properties of a particle, such as position x and momentum p, can be known simultaneously," which is not possible in accordance to the uncertainty principle, in *Sec. 3.3.2*; the more difficult task to predict exactly the extent to which communicative action can ever be realized in its ideal form on a particular future occasion, since it is "counterfactual," in *Sec. 3.4.1*; the more difficult task to predict exactly the extent to which any particular "informal mathematical practices…by aboriginal or ancient peoples" will one day be proved by formal mathematicians, just as Aschbacher once proved "various infrastructural results" of "the classification of finite simple groups," in *Sec. 3.4.2*; the more difficult task to predict exactly the extent to which a subject can ever achieve "pure consciousness" on a particular occasion, in the world of Husserl's phenomenology, in *Sec. 3.5.1*; and the more difficult task to predict exactly the extent to which the debate on the "subject" will one day be resolved by the scientific method, since for the critic like Nagel, "scientific understanding…requires an objective perspective, which, according to Nagel, is diametrically opposed to the subjective first-person point of view," in *Sec. 3.5.2*).

And the reverse direction also holds true.

Again, it should be stressed, however, that there are different shades of gray (or different degrees of truth) in the two opposites, and the classification is not necessarily mutually exclusive either. In fact, even when some combinations of the two occur, they only end up sharing the same dialectic relationship, but in a different degree.

Also, there are some other relationships (as a kind of family resemblance) which have something in common with the principle, but they are not exactly the same but only more or less comparable, which varies from case to case.

Good examples of family resemblance in relation to the principle are sureness-arbitrariness, deterministicness-randomness, likeliness-riskiness, sureness-doubtfulness, deterministicness-indeterministicness, certainty-uncertainty, reliability-unreliability, and so forth.

5th Thesis: The Explicability-Inexplicability Principle

The fifth thesis is called the explicability-inexplicability principle (on the underlying mechanisms of things) in the category of "method" in existential dialectics (which was proposed in *FPHU* and other books of mine), in that both explicability and inexplicability are part of the understanding of things. There is no explicability without inexplicability—and vice versa.

This principle tells us the dual nature of the research dilemma, in that, if reality can be explained in some ways, it also has its other ways which are not quite explainable, at a given point in time.

For instance, on the one hand, there is explicability (e.g., the explanation by "reasoners" with mental models that these models are desirable on the basis of their provision of a sense of reality about a given possibility, that is, about "what is common to all the different ways in which the possibility may occur," in *Sec. 2.2.1*; the explanation by those of benefiting oneself as rational on the basis of "a model in which benefiting oneself is optimal," in *Sec. 2.2.2*; the explanation by those in probability theory that "the function *f(x)* mapping a point in the sample space to the 'probability' value" is "a probability mass function," in *Sec. 2.3.1*; the explanation, by the advocates of cybernetics in the early history of complexity study, of the research on "the interaction of parts in systems and the structure of systems" on the basis of its contribution to the understanding of the constraints and possibilities of systems, in *Sec. 2.3.2*; the explanation by those in reliability engineering of why it is desirable on the basis of being concerned with "cost," etc., in *Sec. 2.4.1*; the explanation by the advocates that an optimal decision is desirable because it gives "the best option available," in *Sec. 2.4.2*; the explanation by those in propositional calculus of the importance of "assignment to each propositional symbol of $\mathcal{P}$ of one or the other [but not both] of the truth values truth [T] and falsity [F]," on the basis of classical laws of thought for two-valued logic, in *Sec. 2.5.1*; the explanation by those officials of the importance to intervene in a "prison riot" on the basis of its being "an extreme outcome" which requires urgent action, in *Sec. 2.5.2*; the explanation by those like Cox in favor of Bayesian probability on the basis of the "axiomatic approach," in that Bayesian updating follows from several axioms," in *Sec. 3.2.1*; the explanation, by those for Q analysis, of its desirability on the basis of its focus on "correlations between subjects across a sample of variables," so as to show that "people think about ideas in relation to other ideas, rather than in isolation," in *Sec. 3.2.2*; the explanation, by those in mainstream metrology, of the desirability of using "random variables" and "simple probability distributions" for understanding measurement uncertainties," in *Sec. 3.3.1*; the explanation, by those in favor of the Copenhagen interpretation of quantum mechanics, that "there is no way to say what the state of a system fundamentally is, only what the result of observations might be," in *Sec. 3.3.2*; the explanation by Habermas of the desirability of communicative action on the basis of the "intersubjective mutuality" of understanding, in *Sec. 3.4.1*; the explanation, by those in formal mathematics, of the foundation of modern mathematics on the basis of "formal and strict proofs of all statements from given axioms," in *Sec. 3.4.2*; the explanation by Husserl of the desirability of phenomenology on the basis of "the method of phenomenological reduction to eliminate" or exclude "any hypothesis on the existence of external objects" or any conventional prejudice, so as to reach "the pure transcendental ego, as opposed to the concrete empirical ego," in *Sec. 3.5.1*; and the explanation, by those in defense of the older version of the liberal tradition, of the desirability of "a unitary, autonomous Subject" on the basis of its being "the foundation" of "social contract" theory in *Sec. 3.5.2*).

On the other hand, there is inexplicability (e.g., the lack of sufficient explanation by "reasoners" with mental models of why these models are necessarily desirable, since, for the critics, there is no general consensus of whether mental models are better than other models in explaining human reasoning, in *Sec. 2.2.1*; the lack of sufficient explanation by those accepting "a model in which benefiting oneself is optimal" of why benefiting oneself is necessrily desirable, since, for the critics, there are alternative models "in which benefiting the group is optimal" and "purely selfish behavior is deemed irrational," in *Sec. 2.2.2*; the lack of sufficient explanation by those in probability theory of "how probability mass functions are obtained," since the theory "instead…assumes their existence," in *Sec. 2.3.1*; the lack of sufficient explanation by the advocates of cybernetics in the early history of complexity study, of why this way of studying systems is necessarily desirable, since, for the critics in the later history of complexity study, there are such unresolvable problems as the problem of Bremermann's fundamental limit on computation and the problem of intractability, among others, in *Sec. 2.3.2*; the lack of sufficient explanation by those in "reliability engineering of why it is necessarily desirable to focus on "cost," since, for the critics, there are other important concerns like "safety," and so on, in *Sec. 2.4.1*; the lack of sufficient explanation by the advocates of why an optimal decision is necessarily desirable, because, for the critics, optimization is seldom achieved in real life because decision making can be more intuitive and less formal, and non-optimal decision-making strategies are relative to different personality traits, its scientific objectivity is questionable, and it has many questionable assumptions, in *Sec. 2.4.2*; the lack of sufficient explanation by those in propositional calculus of why the two-value logic is necessarily desirable, since, for the critics, there are alternative "many-valued logics" which "allow…values other than *true* and *false*," in *Sec. 2.5.1*; the lack of sufficient explanation by those officials of why it is necessarily desirable to intervene a "prison riot" on the basis of its being "an extreme outcome," since, for the critics, there is a temptation "to react to an extreme outcome and to see it as significant, even where its causes are common to many situations and the distinctive circumstances surrounding its occurrence," which are "the results of mere chance," so this "temptation to react to an extreme outcome...often lead...to ad-hoc interventions that merely increase the level of variation and frequency of undesirable outcomes," in *Sec. 2.5.2*; the lack of sufficient explanation, by those in favor of Bayesian probability on the basis of the "axiomatic approach," of why the "axiomatic approach" is necessarily desirable, since, for the critics, "Cox's 1961 development [mainly copied by Jaynes] is non-rigorous, and in fact a counterexample has been found by Halpern [1999]," in *Sec. 3.2.1*; the lack of sufficient explanation, by those for Q analysis, of why it is necessarily desirable, since, for the critics, there are alternatives like R methodology, Cultural Consensus Theory, etc., in *Sec. 3.2.2*; the lack of sufficient explanation, by those in mainstream metrology, of why using "random variables" and "simple probability distributions" are necessarily sufficient for understanding measurement uncertainties, since, for the critics, "in some situations, however, a mathematical interval rather than a probability distribution might be a better model of uncertainty," in *Sec. 3.3.1*; the lack of sufficient explanation, by those in favor of the Copenhagen interpretation of quantum mechanics, of why the uncertainty principle is necessarily true, since the critic like Einstein argued that "randomness is a reflection of our ignorance of some fundamental property of reality," so "the present theory [of quantum mechanics] was incomplete," in *Sec. 3.3.2*; the lack of sufficient explanation by Habermas of why communicative action is necessarily desirable, since, for the critics, it is unwarrantedly biased for "the lifeworld" and against "the system," that it falsely presupposes a powerful role of reason that it does not have, that it does not give a serious attention to the lasting social spectre of power, etc., in *Sec. 3.4.1*; the lack of sufficient explanation by those in formal mathematics of why the foundation of modern mathematics is necessarily robust, since, for the critics, there is "quasi-empiricism in mathematics," with examples like mathematical bafflement, manipulative effectiveness, and informal proofs and proof by authority, in *Sec. 3.4.2*; the lack of sufficient explanation by Husserl of why transcendental phenomenology is necessarily desirable or even possible, since, for the critic like Heidegger, "consciousness is peripheral to the primacy of one's existence [i.e., the mode of being of Dasein], which cannot be reduced to one's consciousness of it," so, in this way, Heidegger shifted "the center of gravity from consciousness [psychology] to existence [ontology]," in *Sec. 3.5.1*; and the lack of sufficient explanation by those in defense of the older version of the liberal tradition, of why "a unitary, autonomous Subject" is necessarily desirable, since the critics like Marx and Freud "opened up the way for the deconstruction of the subject as a core-concept of metaphysics," in *Sec. 3.5.2*).

And the reverse direction also holds true.

Once more, it should be stressed, however, that there are different shades of gray (or different degrees of truth) in the two opposites, and the classification is not necessarily mutually exclusive either. In fact, even when some combinations of the two occur, they only end up sharing the same dialectic relationship, but in a different degree.

Also, there are some other relationships (as a kind of family resemblance) which have something in common with the principle, but they are not exactly the same but only more or less comparable, which varies from case to case.

Good examples of family resemblance in relation to the principle are underlyingness-regularness, causation-regularity, causation-correlation, explanation-description, explanation-narration, explanation-interpretation, and so forth.

6th Thesis: The Fiction-Reality Principle

The sixth thesis is called the fiction-reality principle (on the different faces of reality) in the category of "method" in existential dialectics (as first proposed in *FPHLIT*), in that both fiction and reality co-exist. There is no reality without fiction—and vice versa.

This principle tells us the different categories of reality, in that, if there is reality, there is also some fiction in it—just as, if there is fiction, there is also some reality in it.

For instance, on the one hand, there is fiction (e.g., the fictional aspect of mental models, insofar as "mental models can represent what is false, temporarily assumed to be true, for example, in the case of counterfactual conditionals and counterfactual thinking," in *Sec. 2.2.1*; the fictional aspect of a rationality in individuals or organizatins which "are called rational if they make optimal decisions in pursuit of their goals," insofar as, "for such 'rationality,' the decision maker's goals are taken as part of the model, and not made subject to criticism, ethical or otherwise, in *Sec. 2.2.2*; the fictional aspect of the central limit theorem about a normal distribution "as *n* approaches infinity," insofar as "[i]n actual practice..., we cannot flip a coin an infinite number of times," in *Sec. 2.3.1*; the fictional aspect of complexity study, insofar as, for the critics in the later history of complexity study, there are such unresolvable problems such as the problem of Bremermann's fundamental limit on computation and the problem of intractability, in *Sec. 2.3.2*; the fictional aspect of reliability about air travel, insofar as "one bad experience can be magnified in the mind of the customer, inflating the perceived unreliability of the product," in *Sec. 2.4.1*; the fictional aspect of optimal decision, insofar as its scientific objectivity is questionable and it has many questionable assumptions, in *Sec. 2.4.2*; the fictional aspect of completeness in propositional calculus, insofar as "Gödel's incompleteness theorems are two theorems of mathematical logic that establish inherent limitations of all but the most trivial axiomatic systems capable of doing arithmetic," in *Sec. 2.5.1*; the fictional aspect of a response to a prison riot, insofar as there is a temptation by officials "to react to an extreme outcome and to see it as significant, even where its causes are common to many situations and the distinctive circumstances surrounding its occurrence," which are "the results of mere chance," in *Sec. 2.5.2*; the fictional aspect of the Dutch book approach, insofar as the critic like Ian Hacking argued that "traditional Dutch book arguments did not specify Bayesian updating: they left open the possibility that non-Bayesian updating rules could avoid Dutch books," in *Sec. 3.2.1*; the fictional aspect of Q analysis, insofar as "the data collected" are not valid or reliable since its analysis "can be only as good as the data allows," in *Sec. 3.2.2*; the fictional aspect of Type A evaluation, insofar as the assumption that "the distribution best describing an input quantity X given repeated measured values of it [obtained independently] is a Gaussian distribution" is problematic in other cases, especially "when the measured values are not obtained independently," in *Sec. 3.3.1*; the fictional aspect of the uncertainty principle, insofar as Popper pointed out that, "in…his statistical interpretation, a particular measurement may be made to arbitrary precision without invalidating the quantum theory" and "this directly contrasts the Copenhagen interpretation of quantum mechanics, which is non-deterministic but lacks local hidden variables," in *Sec. 3.3.2*; the fictional aspect of the theory of communicative action, insofar as its ideal form is counterfactual, in *Sec. 3.4.1*; the fictional aspect of mathematics, insofar as there exists "quasi-empiricism in mathematics," with examples like mathematical bafflement, manipulative effectiveness, and informal proofs and proof by authority, in *Sec. 3.4.2*; the fictional aspect of Husserl's phenomenology, insofar as the critics reject "the transcendental ego" as illusory, as "Merleau-Ponty objects to the ego's transcendence of the world" and proposed "existential phenomenology," which "differs from transcendental phenomenology by its rejection of the transcendental ego," in *Sec. 3.5.1*; and the fictional aspect of the "subject," insofar as "thinkers such as Althusser, Foucault or Bourdieu theorize the subject as a social construction," or, "as "an ideological construction" for Althusser, in *Sec. 3.5.2*).

On the other hand, there is reality (e.g., the realistic aspect of mental mdoels, insofar as "mental models are based on a principle of truth: they typically represent only those situations that are possible, and each model of a possibility represents only what is true in that possibility according to the proposition," in *Sec. 2.2.1*; the realistic aspect of a rationality in individuals or organizatins, insofar as they are allowed to "make optimal decisions in pursuit of their goals" or to be selfish as rational in this case, in *Sec. 2.2.2*; the realistic aspect of the central limit theorem, insofar as a normal distribution can be approximated, although it cannot be exact, as n approaches "a sufficiently large number," in *Sec. 2.3.1*; the realistic aspect of complexity study, insofar as it contributes to the resolution of some problems which can be solved, in *Sec. 2.3.2*; the realistic aspect of reliability about air travel, insofar there are actual cases of "plane crash where hundreds of passengers died," in *Sec. 2.4.1*; the realistic aspect of optimal decision, insofar as it "attempts to find the best option available," at least as an ideal point of reference, in *Sec. 2.4.2*; the realistic aspect of completeness in propositional calculus, insofar as the 3 rules aforecited are complete, in *Sec. 2.5.1*; the realistic aspect of a response to a prison riot, insofar as there is a need for officials to deal with a prison riot, whenever it happens, in *Sec. 2.5.2*; the realistic aspect of the Dutch book approach, insofar as, under some specified conditions, "a Dutch book is made when a clever gambler places a set of bets that guarantee a profit, no matter what the outcome is of the bets," but "if a bookmaker follows the rules of the Bayesian calculus in the construction of his odds, a Dutch book cannot be made," in *Sec. 3.2.1*; the realistic aspect of Q analysis, insofar as "the data collected" are indeed valid or reliable, in *Sec. 3.2.2*; the realistic aspect of Type A evaluation, insofar as the assumption that "the distribution best describing an input quantity X given repeated measured values of it (obtained independently) is a Gaussian distribution" is valid in some situations, especially when the measured values are obtained independently, in *Sec. 3.3.1*; the realistic aspect of the uncertainty principle, insofar as Heisenberg pointed out that there is "a tradeoff between the precisions" of the position and momentum of a sub-atomic particle under certain conditions, in *Sec. 3.3.2*; the realistic aspect of the theory of communicative action, insofar as it strives for mutual understanding in communication, if some specific conditions are fulfilled, in *Sec. 3.4.1*; the realistic aspect of mathematics, insofar as there are necessary arguments for the foundation of mathematics as suggested by

Willard Quine and Hilary Putnam, in *Sec. 3.4.2*; the realistic aspect of Husserl's phenomenology, insofar as it focuses on "intuition," and "evidence" to back up one's subjective experience, in *Sec. 3.5.1*; and the realistic aspect of the "subject," insofar as it involves "consciousness" or "experience," in *Sec. 3.5.1*).

And the reverse direction also holds true.

Once more, it should be stressed, however, that there are different shades of gray (or different degrees of truth) in the two opposites, and the classification is not necessarily mutually exclusive either. In fact, even when some combinations of the two occur, they only end up sharing the same dialectic relationship, but in a different degree.

Also, there are some other relationships (as a kind of family resemblance) which have something in common with the principle, but they are not exactly the same but only more or less comparable, which varies from case to case.

Good examples of family resemblance in relation to the principle are fiction-nonfiction, fiction-fact, imagination-fact, imagination-nonfiction, subjectivity-objectivity, inaccuracy-accuracy, and so forth.

7th Thesis: The Cognitiveness-Noncognitiveness Principle

The seventh thesis is called the cognitiveness-noncognitiveness principle (on the understanding of meaning) in the category of "method" in existential dialectics (as first proposed in *BNR*), in that, if a view of things is based on reason and evidences, it is also affected by non-cognitive factors (such that it is neither true nor false in the latter case)—and vice versa. In other words, there is no cognitiveness without noncognitiveness—and vice versa.

Noncognitive factors can come in all shapes and sizes, and good examples include envy, jealousy, power, nationality, race, gender, age, class, greed, lust, status, faith, anger, sadness, joy, fear, wish, etc.

Communication, even in its ideal type, is seldom based on reason and evidences only, as falsely assumed during the Age of the Enlightenment and its continued supporters nowadays. What is understood about something is often twisted on the basis of something else too. In these cases, what is at stake is often neither true nor false (which is something that classical logic has sidelined and ignored).

For instance, on the one hand, there is cognitiveness (e.g., a view of things on the basis of reason and evidences, as shown in the reliance of mental models on "deeply held images of thinking and acting" which are "so basic to understanding of the world," as well as "on a principle of truth," in *Sec. 2.2.1*; a view of things on the basis of reason and evidences, as shown in the use of information to make a decision in a model of rationality, depending on "how much information is available" for the purpose at hand, in *Sec. 2.2.2*; a view of things on the basis of reason and evidences, as shown in the reasoning of the central limit theorem about a normal distribution "as *n* approaches infinity," in *Sec. 2.3.1*; a view of things on the basis of reason and evidences, as shown inWeaver's research on the distinction between "disorganized complexity" and "organized complexity," in *Sec. 2.3.2*; a view of things on the basis of reason and evidences, as shown in the facts about some accidents of "plane crash" which did occur in the history of air travel, in *Sec. 2.4.1*; a view of things on the basis of reason and evidences, as shown in the different steps of a rational decision making model to follow, aforecited, in *Sec. 2.4.2*; a view of things on the basis of reason and evidences, as shown in the use of axioms and inference rules in propositional calculus, in *Sec. 2.5.1*; a view of things on the basis of reason and evidences, as shown in the empirical research by Keynes that "economic systems did not automatically tend to the equilibrium of full employment owing to their agents' inability to predict the future," in *Sec. 2.5.2*; a view of things on the basis of reason and evidences, as shown in the "Bayesian methods," in that "one hypothesis" is "more probable than the other… by virtue of having used different priors, in *Sec. 3.2.1*; a view of things on the basis of reason and evidences, as shown in Q analysis, since its analysis "can be only as good as the data allows," in *Sec. 3.2.2*; a view of things on the basis of reason and evidences, as shown in the 2 stages of evaluating measurement uncertainty, in regard to output quantity Y and input quantities $X_1, \ldots, X_N$, in *Sec. 3.3.1*; a view of things on the basis of reason and evidences, as shown in the 2 mathematical formulations of the uncertainty principle aforecited, in *Sec. 3.3.2*; a view of things on the basis of reason and evidences, as shown in the requirements of "truth," "truthfulness," and "rightness" in communicative action, in *Sec. 3.4.1*; a view of things on the basis of reason and evidences, as shown in modern mathematics on the basis of "formal and strict proofs of all statements from given axioms," in *Sec. 3.4.2*; a view of things on the basis of reason and evidences, as shown in the focus on "intuition," and "evidence" in Husserl's phenomenology, in *Sec. 3.5.1*; and a view of things on the basis of reason and evidences, as shown in "scientific understanding," which, for Nagel, "requires an objective perspective" based on evidences to be verified, in *Sec. 3.5.1*).

On the other hand, there is noncognitiveness (e.g., a view of things on the basis of non-cognitive factors like envy, jealousy, power, nationality, race, gender, age, class, greed, lust, status, faith, anger, sadness, joy, fear, wish, etc., as shown in the "hope" that mental models are better than other models in explaining human reasoning, even though "scientific debate continues" about this in the field, with no consensus, in *Sec. 2.2.1*; a view of things on the basis of non-cognitive factors like envy, jealousy, power, nationality, race, gender, age, class, greed, lust, status, faith, anger, sadness, joy, fear, wish, etc., as shown in "the background model assumptions describing how the problem is framed and formulated" in a model of rationality, so that they are to be "taken as part of the model, and not made subject to criticism, ethical or otherwise, in *Sec. 2.2.2*; a view of things on the basis of non-cognitive factors like envy, jealousy, power, nationality, race, gender, age, class, greed, lust, status, faith, anger, sadness, joy, fear, wish, etc., as shown in the hopeful assumptions in regard to a "fair" coin and the existence of infinity, since "we cannot flip a coin an infinite number of times; so in general, this formula most accurately applies to situations in which we have already assigned an a priori probability to a particular outcome [in this case, our assumption that the coin was a 'fair' coin]," among other assumptions, which are not proved and are therefore neither true nor false within the formula, in *Sec. 2.3.1*; a view of things on the basis of non-cognitive factors like envy, jealousy, power, nationality, race, gender, age, class, greed, lust, status, faith, anger, sadness, joy, fear, wish, etc., as shown in the hope of those in complexity study to bridge the gap between "disorganized complexity" and "organized complexity" one day, as Weaver had addressed the problem only in "a preliminary way, in *Sec. 2.3.2*; a view of things on the basis of non-cognitive factors like envy, jealousy, power, nationality, race, gender, age, class, greed, lust, status, faith, anger, sadness, joy, fear, wish, etc., as shown in the "fear" in "a large percentage of the flying consumer population," when they heard about a "plane crash where hundreds of passengers died,... regardless of actual reliability data about the safety of air travel," in *Sec. 2.4.1*; a view of things on the basis of non-cognitive factors like envy, jealousy, power, nationality, race, gender, age, class, greed, lust, status, faith, anger, sadness, joy, fear, wish, etc., as shown in the dependence on "authority" by those who use it as an everyday decision technique and "acquiesce to a person in authority or an 'expert,' just following orders," not on the basis of reason and evidences, in *Sec. 2.4.2*; a view of things on the basis of non-cognitive factors like envy, jealousy, power, nationality, race, gender, age, class, greed, lust, status, faith, anger, sadness, joy, fear, wish, etc., as shown in the assumption of two-valued logic in propositional calculus, even though there are alternative many-valued logics, in *Sec. 2.5.1*; a view of things on the basis of non-cognitive factors like envy, jealousy, power, nationality, race, gender, age, class, greed, lust, status, faith, anger, sadness, joy, fear, wish, etc., as shown in the ideological politics against Keynes' finding by those in "the classical liberalism of the Austrian school of economics" at the time, who did not appreciate it, in *Sec. 2.5.2*; a view of things on the basis of non-cognitive factors like envy, jealousy, power, nationality, race, gender, age, class, greed, lust, status, faith, anger, sadness, joy, fear, wish, etc., as shown in the "Bayesian methods," in which "individual Bayesians might differ about which was the more probable and by how much," due to differences in "personal belief," in *Sec. 3.2.1*; a view of things on the basis of non-cognitive factors like envy, jealousy, power, nationality, race, gender, age, class, greed, lust, status, faith, anger, sadness, joy, fear, wish, etc., as shown in the good faith in Q analysis, insofar as "little information on validation of the methods is available currently," in *Sec. 3.2.2*; a view of things on the basis of non-cognitive factors like envy, jealousy, power, nationality, race, gender, age, class, greed, lust, status, faith, anger, sadness, joy, fear, wish, etc., as shown in the confession by the Joint Committee for Guides in Metrology in 1997 when it treated "measurement uncertainty…as a measure of how well one believes one knows the true value of the measurand," on the basis of "degrees of belief" or faith, in *Sec. 3.3.1*; a view of things on the basis of non-cognitive factors like envy, jealousy, power, nationality, race, gender, age, class, greed, lust, status, faith, anger, sadness, joy, fear, wish, etc., as shown in the wishful belief by some people that the uncertainty principle can be applied for the understanding of free will, even though the critics had showed that "nontrivial biological mechanisms requiring quantum mechanics are unlikely, due to the rapid decoherence time of quantum systems at room temperature," in *Sec. 3.3.2*; a view of things on the basis of non-cognitive factors like envy, jealousy, power, nationality, race, gender, age, class, greed, lust, status, faith, anger, sadness, joy, fear, wish, etc., as shown in strategic action, which is based on "the isolated intentions of individuals…in the service of private interests," with "no intention of reaching consensus or mutual understanding" for "truth," for example, in *Sec. 3.4.1*; a view of things on the basis of non-cognitive factors like envy, jealousy, power, nationality, race, gender, age, class, greed, lust, status, faith, anger, sadness, joy, fear, wish, etc., as shown in the "hope" or faith in mathematics as pointed out by Eugene Wigner, who argued that "the miracle of the appropriateness of the language of mathematics…is a wonderful gift which we neither understand nor deserve. We should be grateful for it and hope that it will remain valid in future research and that it will extend, for better or for worse, to our pleasure, even though perhaps also to our bafflement, to wide branches of learning," in *Sec. 3.4.2*; a view of things on the basis of non-cognitive factors like envy, jealousy, power, nationality, race, gender, age, class, greed, lust, status, faith, anger, sadness, joy, fear, wish, etc., as shown in the hope by those in Husserl's phenomenology to eventually reach "the pure transcendental ego, as opposed to the concrete empirical ego," in *Sec. 3.5.1*; and a view of things on the basis of non-cognitive factors like envy, jealousy, power, nationality, race, gender, age, class, greed, lust, status, faith, anger, sadness, joy, fear, wish, etc., as shown in the "politics" of "flatter[ing] the editors' ideological preconceptions," so as to have a nonsensical article published, as revealed in the "Sokal affair," in *Sec. 3.5.1*).

And the reverse direction also holds true.

Once more, it should be stressed, however, that there are different shades of gray (or different degrees of truth) in the two opposites, and the classification is not necessarily mutually exclusive either. In fact, even when some combinations of the two occur, they only end up sharing the same dialectic relationship, but in a different degree.

Also, there are some other relationships (as a kind of family resemblance) which have something in common with the principle, but they are not exactly the same but only more or less comparable, which varies from case to case.

Good examples of family resemblance in relation to the principle are proposition-attitude, cognition-emotion, cognition-prescription, theory-extratheory, justification-nonjustification, and so forth.

In Relation to Structure

The next five theses are in relation to structure in existential dialectics, as shown below.

8th Thesis: The Finiteness-Transfiniteness Principle

The eighth thesis is called the finiteness-transfiniteness principle (on the nature of numbers) in the category of "structure" in existential dialectics (as first proposed in *FPHG*), in that, if there are finite things, there are likewise transfinite ones—and vice versa.

To avoid confusion, my usage of the word "transfinite" here differs radically from the one used by Cantor (and other mathematicians) for "relative" infinity—and is more limited, in light of the problems confronting any attempt to understand the idea of infinity, be it by intuition, imagination, and conception (as detailedly analyzed in *Sec. 2.2.3* of *FPHG*).

Instead, by "transfinity," I allow numbers which can be many times larger—or smaller, for that matter—than the finite things that we encounter in daily life, but they do not have to be related to the idea of infinity at all (which may not exist).

Of course, there may be some *borderline* cases, in which it is not clear whether the number in question is transfinite (in my usage) or simply a mathematical convenience. A good example of a borderline case is the Planck unit of length for "the smallest space possibly measured in nature," which is "less than billionths of trillionths of trillionths of an inch" (or something like 1.6×10^{-35} meters). (P. Baofu 2006a; N. McAleer 1987: 219; WKV 2008; D. Corbett 2008a)

That said—my usage of "transfinity" can also resolve (or better, dissolve) an age-old problem in philosophy known as "Zeno's paradoxes" (as already explained in *Sec. 2.2.3* of *FPHG*).

In any event, in relation to the finiteness-transfiniteness principle, on the other hand, there is finiteness (e.g., the finite bits of information used in mental models for the simplication of reality, in *Sec. 2.2.1*; the finite amount of time and information needed to find a satisfying rational decision for simple problems, as in bounded rationality, in *Sec. 2.2.2*; the finite number of times in flipping a coin or rolling a dice as *n* approaches "a sufficiently large number," so that a normal distribution can be approximated, though not exact, in *Sec. 2.3.1*; the finite number of parts in a small system of "organized complexity," in *Sec. 2.3.2*; the finite number of "test levels" like "component, circuit board, unit, assembly, subsystem and system levels," in *Sec. 2.4.1*; the finite number of bits of information needed for a satisficing decision, in *Sec. 2.4.2*; the finite number of possible interpretations, when $2^1=2$, in *Sec. 2.5.1*; the finite number of the categories of causes in history as identified by Keynes, namely, "frequency probability," "subjective or Bayesian probability," and "special causes," in *Sec. 2.5.2*; the finite number of procedures which characterize "Bayesian methods," as described earlier in regard to the 3 examples, in *Sec. 3.2.1*; the finite number of subjects used in Q analysis, since "one salient difference between Q and other social science research methodologies, such as surveys, is that it typically uses many fewer subjects," in *Sec. 3.2.2*; the finite number of the categories of situation, in which "a mathematical interval rather than a probability distribution might be a better model of uncertainty," like the situations involving "periodic measurements, binned data values, censoring, detection limits, or plus-minus ranges of measurements where no particular probability distribution seems justified or where one cannot assume that the errors among individual measuresments are completely independent," in *Sec. 3.3.1*; the finite number of the categories of worlds, like "histories," "futures," etc. in the "many-worlds interpretation" of quantum mechanics, in *Sec. 3.3.2*; the finite number of the categories of "the social integration of the lifeworld," namely, through "consensus" or through "influence," for Habermas, in *Sec. 3.4.1*; the finite number of the categories of mathematical practice in its history, namely, quasi-empiricism and folk mathematics, in *Sec. 3.4.2*; the finiteness of "the empirical ego," because of its being bounded, in *Sec. 3.5.1*; and the finite number of ways to refer to subjectivity, namely, the idea of the "subject" in continental philosophy, the "I" in critical theory and psychology, and the "personhood" in Anglo-American analytic philosophy, in *Sec. 3.5.2*).

On the other hand, there is transfinitty (e.g., the transfinite bits of information used in optimal decision-making models when the situation in question is extremely complex, in *Sec. 2.2.1*; the transfinite amount of time and information needed to "find…the rational decision for arbitrarily complex problems," in *Sec. 2.2.2*; the transfinite idea of "*n*" as its approaches "infinity" in the central limit theorem, in *Sec. 2.3.1*; the transfinite number of parts in an extremely large system of "disorganized complexity," in *Sec. 2.3.2*; the transfinite number of all the individual items in history which can be included for testing at any of the levels in reliability engineering, in principle, in *Sec. 2.4.1*; the transfinite number of bits of information needed for an optimal decision, by those who follow a strict interpretation of the rational decision making model, in *Sec. 2.4.2*; the transfinite number of possible interpretations, when $2^{\aleph_0} = \mathfrak{c}$), in *Sec. 2.5.1*; the transfinite number of all the individual phenomena in history that can be classified under "frequency probability," "subjective or Bayesian probability," or "special causes," in *Sec. 2.5.2*; the transfinite amount of time in the computation of "intractable" problems in some Bayesian procedures, like "the use of hierarchical models and marginalization over the values of nuisance parameters [which are not of immediate interest]," in *Sec. 3.2.1*; the transfinite number of all the indiviudal subjects in history that have ever existed and can be used in Q analysis, at least in principle, in *Sec. 3.2.2*; the transfinite number of all the differentiable units in a normal or Gaussian distribution in a Type A evaluation of measurement uncertainty, since "the function is supersmooth…, implying that it is infinitely differentiable," in *Sec. 3.3.1*; the transfinite number of all the possible individual "alternative histories and futures," according to the "many-worlds interpretation" of quantum mechanics, in *Sec. 3.3.2*; the transfinite number of all the individual cases of achieving social integration through "consensus" or "influence" that have ever existed in history, in *Sec. 3.4.1*; the transfinite number of all the individual instances of quasi-empiricism and folk mathematics that have ever existed in history, in *Sec. 3.4.2*; the transfiniteness of "the transcendental ego," because of its being unbounded, or largely so, in *Sec. 3.5.1*; and the transfinite number of all the possible individual subjects, be they known as the "subject," the "I," or the "personhood," that have ever existed in history, in *Sec. 3.5.2*).

And the reverse direction also holds true.

Once more, it should be stressed, however, that there are different shades of gray (or different degrees of truth) in the two opposites, and the classification is not necessarily mutually exclusive either. In fact, even when some combinations of the two occur, they only end up sharing the same dialectic relationship, but in a different degree.

Also, there are some other relationships (as a kind of family resemblance) which have something in common with the principle, but they are not exactly the same but only more or less comparable, which varies from case to case.

Good examples of family resemblance in relation to the principle are boundedness-quasiunboundedness, smallness-largeness, microness-macroness, boundedness-unboundedness, and the like.

9th Thesis: The Preciseness-Vagueness Principle

The ninth thesis is called the preciseness-vagueness principle (on the refinement of things) in the category of "structure" in existential dialectics (which was first out worked out in *FIA*), in that both preciseness and vagueness are important, not that one is better than the other, but that both are used, in different degrees of preference, in accordance to the contextual application from the perspectives of nature, the mind, culture, and society. There is no preciseness without vagueness—and vice versa.

For illustration, in relation to taxonomy, preciseness has its taxonomic clarity, just as vagueness has its classificatory flexibility. And in relation to network, vagueness has its explorative liberty, just as preciseness has its conceptual definitiveness.

For instance, on the one hand, there is preciseness (e.g., the precise identification of 4 principles of mental models in the article aforecited, in *Sec. 2.2.1*; the precise identification of 4 types of rationality by Weber in the article aforecited, in *Sec. 2.2.2*; the precise identification of 6 properties of discreteness in the article aforecited, in *Sec. 2.3.1*; the precise identification of 6 attributes of complexity in the article aforecited, in *Sec.*

2.3.2; the precise identification of 2 main approaches to reliability modeling in the article aforecited, namely, "the physics of failure approach" and "the parts stress modeling approach," in *Sec. 2.4.1*; the precise identification of 7 decision making steps in the article aforecited, in *Sec. 2.4.2*; the precise identification of 3 inference rules in the article aforecited, in *Sec. 2.5.1*; the precise identification of 4 features of common causes in the article aforecited, in *Sec. 2.5.2*; the precise identification of 3 justifications of Bayesian probability, namely, "axiomatic approach," "Dutch book approach," and "decision theory approach," in *Sec. 3.2.1*; the precise identification of 8 applications of Q analysis in the article aforecited, namely, "nursing, veterinary medicine, public health, transportation, education, rural sociology, hydrology and mobile communication," in *Sec. 3.2.2*; the precise classification of 2 stages of evaluating measurement uncertainty in the article aforecited, in *Sec. 3.3.1*; the precise identification of 2 formulations of the uncertainty principle in the article aforecited, namely, "wave mechanics" and "matrix mechanics," in *Sec. 3.3.2*; the precise identification of 3 claims in the assessment of the validity of speech in the theory of communicative action, namely, truth, truthfulness, and rightness, in *Sec. 3.4.1*; the precise identification of 3 arguments for "quasi-empiricism in mathematics," namely, mathematical bafflement, manipulative effectiveness, and informal proofs and proof by authority, in the article aforecited, in *Sec. 3.4.2*; the precise identification of 6 main ideas of Husserl's phenomenology in the article aforecited, in *Sec. 3.5.1*; and the precise identification of 3 philosophical traditions on the "subject" in the article aforecited, in *Sec. 3.5.2*).

On the other hand, there is vagueness (e.g., the vagueness in the identification of the 4 principles of mental models in the article aforecited, since it is not clear why there must be only 4, not 6, 7, 8, and so on, in *Sec. 2.2.1*; the vagueness in the identification of the 4 types of rationality by Weber in the article aforeicted, since it is not clear why there must be only 4, not 2, 3, 5, 6, 7, and so on, in *Sec. 2.2.2*; the vagueness in the identification of the 6 properties of discreteness in the article aforecited, since it is not clear why there must be only 6, not 7, 8, 9, 10, 11, 13, 14, 17, and so on, in *Sec. 2.3.1*; the vagueness in the identification of the 6 attributes of complexity in the article aforecited, since it is not clear why there must be only 6, not 7, 8, 9, 10, and so on, in *Sec. 2.3.2*; the vagueness in the identification of the 2 main approaches to reliability modeling in the article aforecited, since it is not clear why there must be only 2, not 4, 7, 8, 12, 13, and so on, in *Sec. 2.4.1*; the vagueness in the identification of the 7 decision making steps in the article aforecited, since it is not clear why there must be only 7, not 8, 9, 10, and so on, in *Sec.*

2.4.2; the vagueness in the identification of the 3 inference rules in the article aforecited, since it is not clear why there must be only 3, not 4, 6, 7, 9, 13, and so on, in *Sec. 2.5.1*; the vagueness in the identification of the 4 features of common causes in the article aforecited, since it is not clear why there must be only 4, not 5, 6, 7, 8, and so on, in *Sec. 2.5.2*; the vagueness in the identification of the `3 justifications of Bayesian probability, since it is not clear why there must be only 3, not 4, 5, 6, 7, and so on, in *Sec. 3.2.1*; the vagueness in the identification of the 8 applications of Q analysis in the article aforecited, since it is not clear why there must be only 8, not, 9, 13, 15, and so on, in *Sec. 3.2.2*; the vagueness in the classification of 2 stages of evaluating measurement uncertainty in the article aforecited, since it is not clear why there must be only 2, not 3, 4, 5, and so on, in *Sec. 3.3.1*; the vagueness in the identification of the 2 formulations of the uncertainty principle in the article aforecited, namely, "wave mechanics" and "matrix mechanics," since it is not clear why there must be only 2, not 3, 4, 5, and so on, in *Sec. 3.3.2*; the vagueness in the identification of the 3 claims in the assessment of the validity of speech in the theory of communicative action, since it is not clear why there must be only 3, not 4, 5, 6, and so on, in *Sec. 3.4.1*; the vagueness in the identification of the 3 arguments for "quasi-empiricism in mathematics," namely, mathematical bafflement, manipulative effectiveness, and informal proofs and proof by authority, in the article aforecited, since it is not clear why there must be only 3, not 4, 5, 6, and so on, in *Sec. 3.4.2*; the vagueness in the identification of the 6 main ideas of Husserl's phenomenology in the article aforecited, since it is not clear why there must be only 6, not 7, 8, 9, and so on, in *Sec. 3.5.1*; and the vagueness in the identification of the 3 philosophical traditions on the "subject" in the article aforecited, since it is not clear why there must be only 3, not 4, 5, 6, and so on, in *Sec. 3.5.2*).

And the reverse direction also holds true.

Once more, it should be stressed, however, that there are different shades of gray (or different degrees of truth) in the two opposites, and the classification is not necessarily mutually exclusive either. In fact, even when some combinations of the two occur, they only end up sharing the same dialectic relationship, but in a different degree.

Also, there are some other relationships (as a kind of family resemblance) which have something in common with the principle, but they are not exactly the same but only more or less comparable, which varies from case to case.

Good examples of family resemblance in relation to the principle are clarity-ambiguity, directness-indirectness, quantitativeness-qualitativeness, specificity-obscurity, describability-nondescribability, concretness-abstractness, translatability-untranslatability, specificity-generality, clearcutness-transitionalness, sharpness-roughness, and the like.

10th Thesis: The Simpleness-Complicatedness Principle

The tenth thesis refers to the simpleness-complicatedness principle (on the interconnection among things) in the category of "structure" in existential dialectics (which was first out worked out in *FIA*, *FPHPROB*, nd other books), in that both simpleness and complicatedness are vital, without favoring one over the other, and each is utilized, depending on the basis of the perspectives of nature, the mind, culture, and society. There is no simpleness without complicatedness—and vice versa.

As an illustration, in relation to taxonomy, simpleness has its heuristic usefulness, just as complicatedness has its realistic representation. In relation to network, simpleness has its economical attractiveness, just as complicatedness has its practical reliability. And in relation to logic, there is the simple zeroth-order logic, just as there is the complicated higher-order logic.

For instance, on the one hand, there is simpleness (e.g., the simplification of reality in mental models, in *Sec. 2.2.1*; the relatively simple analysis of the argument on emotion and rationality by the advocates, in *Sec. 2.2.2*; the relatively simple analysis of discrete probability and the physical world by the advocates, in *Sec. 2.3.1*; the relatively simple analysis of complexity study by the advocates, in *Sec. 2.3.2*; the relatively simple analysis of reliability engineering by the practitioners, in *Sec. 2.4.1*; the relatively simple analysis of optimal decision by the advocates, in *Sec. 2.4.2*; the relatively simple analysis of logic from the vantage point of propositional logic as "zeroth-order logic," in *Sec. 2.5.1*; the relatively simple analysis of history and common causes by the advocates, in *Sec. 2.5.2*; the relatively simple analysis of Bayesian probability by its advocates, in *Sec. 3.2.1*; the relatively simple analysis of Q methodology by its advocates, in *Sec. 3.2.2*; the relatively simple analysis of measurement uncertainty by its advocates, in *Sec. 3.3.1*; the relatively simple analysis of the uncertainty principle, by its advocates, in *Sec. 3.3.2*; the relatively simple analysis of communicative action by the adcocates, in *Sec. 3.4.1*; the relatively simple analysis of the sociology of mathematical knowledge by its advocates, in *Sec. 3.4.2*; the relatively simple analysis of Husserl's phenomenology by its advocates, in *Sec. 3.5.1*; and the relatively simple analysis of "a unitary, autonomous

Subject" in "the liberal theory of the social contract" by its advocates, in *Sec. 3.5.2*).

On the other hand, there is complicatedness (e.g., the complexity of reality in optimal decision-making models, in *Sec. 2.2.1*; the relatively more complicated analysis of the argument on emotion and rationality, by challenging the claims and assumptions, as shown in the problems pointed out by the critics as cited, in *Sec. 2.2.2*; the relatively more complicated analysis of discrete probability and the physical world, by challenging the claims and assumptions, as shown in the problems pointed out by the critics as cited, in *Sec. 2.3.1*; the relatively more complicated analysis of complexity study, by challenging the claims and assumptions, as shown in the problems pointed out by the critics as cited, in *Sec. 2.3.2*; the relatively more complicated analysis of reliability engineering, by challenging the claims and assumptions, as shown in the problems pointed out by the critics as cited, in *Sec. 2.4.1*; the relatively more complicated analysis of optimal decision, by challenging the claims and assumptions, as shown in the problems pointed out by the critics as cited, in *Sec. 2.4.2*; the relatively more complicated analysis of logic from the vantage point of higher-order logics, in *Sec. 2.5.1*; the relatively more complicated analysis of history and common causes, by challenging the claims and assumptions, as shown in the problems pointed out by the critics as cited, in *Sec. 2.5.2*; the relatively more complicated analysis of Bayesian probability, by challenging the claims and assumptions, as shown in the problems pointed out by the critics as cited, in *Sec. 3.2.1*; the relatively more complicated analysis of Q methodology, by challenging the claims and assumptions, as shown in the problems pointed out by the critics as cited, in *Sec. 3.2.2*; the relatively more complicated analysis of measurement uncertainty, by challenging the claims and assumptions, as shown in the problems pointed out by the critics as cited, in *Sec. 3.3.1*; the relatively more complicated analysis of the uncertainty principle, by challenging the claims and assumptions, as shown in the problems pointed out by the critics as cited, in *Sec. 3.3.2*; the relatively more complicated analysis of communicative action, by challenging the claims and assumptions, as shown in the problems pointed out by the critics as cited, in *Sec. 3.4.1*; the relatively more complicated analysis of the sociology of mathematical knowledge, by challenging the claims and assumptions, as shown in the problems pointed out by the critics as cited, in *Sec. 3.4.2*; the relatively more complicated analysis of Husserl's phenomenology, by challenging the claims and assumptions, as shown in the problems pointed out by the critics as cited, in *Sec. 3.5.1*; and the relatively more complicated analysis of "a unitary, autonomous Subject" in "the liberal theory of the social contract," by challenging the claims and assumptions, as shown in the problems pointed out by the critics as cited, in *Sec. 3.5.1*).

And the reverse direction also holds true.

Again, it should be stressed, however, that there are different shades of gray (or different degrees of truth) in the two opposites, and the classification is not necessarily mutually exclusive either. In fact, even when some combinations of the two occur, they only end up sharing the same dialectic relationship, but in a different degree.

Also, there are some other relationships (as a kind of family resemblance) which have something in common with the principle, but they are not exactly the same but only more or less comparable, which varies from case to case.

Good examples of family resemblance in relation to the principle are inflexibility-flexibility, standardization-specialization, imperfectness-perfectness, superficiality-depth, shallowness-deepness, economicalness-elaboratedness, plainness-circumspection, onesidedness-multisidedness, succinctness-duplicatedness, simpleness-sophisticatedness, crudeness-subtleness, and the like.

11th Thesis: The Openness-Hiddenness Principle

The eleventh thesis refers to the openness-hiddenenss principle (on the detection of things) in the category of "structure" in existential dialectics (which was already worked out in my previous books, especially in *FPHU*), in that reality has its hidden face, just as it is open to outside view in some other ways. There is no openness without hiddenness—and vice versa.

For instance, on the one hand, there is openness (e.g., the open exploration of reality by being "flexible" in mental models, in *Sec. 2.2.1*; the open exploration of 4 types of rationality by Weber, in *Sec. 2.2.2*; the open exploration of a normal distribution "as *n* approaches infinity" in the central limit theorem, in *Sec. 2.3.1*; the open exploration of the relationships among different parts in a system by the advocates of complexity study, in *Sec. 2.3.2*; the open exploration of "reliability" by those in reliability engineering, in *Sec. 2.4.1*; the open exploration of achieving optimality in decision making within organizational behavior, in *Sec. 2.4.2*; the open exploration of two truth values in propositional calculus, in *Sec. 2.5.1*; the open exploration of common causes in history, in *Sec. 2.5.2*; the open exploration of calculating probability on the basis of using "different priors" in the Bayesian methods, in *Sec. 3.2.1*; the open exploration of how "people think about ideas in relation to other ideas, rather than in isolation" by Q analysis, in *Sec. 3.2.2*; the open exploration of evaluating measurement uncertainty by Type A and Type B, in *Sec. 3.3.1*; the open exploration of "a tradeoff between the precisions" of the position and momentum of a sub-atomic particle, according to the uncertainty principle, in *Sec. 3.3.2*; the open exploration of using communicative action for mutual understanding, in *Sec. 3.4.1*; the open exploration of "Euclid's system of proving geometry theorems...along with Plato's ontology" and "the methods and epistemology of Aristotle" to "serve...as a foundation ontology for the Western world since its beginnings," in *Sec. 3.4.2*; the open exploration of using "bracketing" to question "all assumptions about the existence of an external world and the inessential [subjective] aspects of how the object is concretely given to us," so as to eventually reach "the transcendental ego" in transcendental phenomenology, in *Sec. 3.5.1*; and the open exploration of the "subject" as "a unitary, autonomous Subject" in "the liberal theory of the social contract," in *Sec. 3.5.2*).

On the other hand, there is hiddenness (e.g., the hidden bias in mental models, because of its "effects as information filter" so as to "cause...selective perception, perception of only selected parts of information," in *Sec. 2.2.1*; the hidden bias in the 4 types of rationality by Weber, since, for the critics like the Habermasians and feminists, they do not give sufficient weight to the issue of power, in *Sec. 2.2.2*; the hidden bias in the central limit theorem, since, in practice, "we cannot flip a coin an infinite number of times," so "this formula most accurately applies to situations in which we have already assigned an a priori probability to a particular outcome [in this case, our assumption that the coin was a 'fair' coin]," among other assumptions, in *Sec. 2.3.1*; the hidden bias in complexity study, since, for the critics, there are different unresolvable problems such as the problem of Bremermann's fundamental limit on computation, the problem of intractability, and so on, in *Sec. 2.3.2*; the hidden bias in the focus on "reliability" by those in reliability engineering, since there are other important issues not emphasized, like "safety" as stressed by those in safety engineering, in *Sec. 2.4.1*; the hidden bias in the quest for achieving optimality in decision making within organizational behavior, because its scientific objectivity is questionable, it is often taken out of context, it has many questionable assumptions, it does not give sufficient attention to non-rational factors, and it ignores power and corruption, in *Sec. 2.4.2*; the hidden bias in the two truth values in propositional calculus, since there are "values other than *true* and *false*" in many-valued logics, in *Sec. 2.5.1*; the hidden bias in the analysis of common causes in history, because there are alternative causes like "special causes," in *Sec. 2.5.2*; the hidden bias in the Bayesian methods, since "individual Bayesians might differ about which was the more probable and by how much, by virtue of having used different priors," in *Sec. 3.2.1*; the hidden bias in Q analysis, as shown in the problems that its analysis is theory-dependent, that the validity of its analysis is relative to the data collected about an individual, and that its overall evaluation as a method has yet to be fully accessed, in *Sec. 3.2.2*; the hidden bias in Type A and Type B evaluations of measurement uncertainty, because of its questionable assumptions afore-discussed, in *Sec. 3.3.1*; the hidden bias in the uncertainty principle, insofar as there are criticisms against it, like the view by Einstein that the loss of certainty in quantum mechanical measurements can be due to our ignorance of some hidden variables in the real world, together with other problems as described earlier, in *Sec. 3.3.2*; the hidden bias in communicative action for mutual understanding, since, for the critics, it is unwarrantedly biased for "the lifeworld" and against "the system," falsely presupposes a powerful role of reason that it does not have, does not give a serious attention to the lasting social spectre of power, etc., in *Sec. 3.4.1*; the hidden bias in "Euclid's system of proving geometry theorems...along with Plato's ontology" and "the methods and epistemology of Aristotle," since, for the critic like Hilary Putnam, "Euclid's system of proving geometry theorems was unique to the classical Greeks and did not evolve similarly in other mathematical cultures in China, India, and Arabia," in *Sec. 3.4.2*; the hidden bias in transcendental phenomenology, because, for the critics like Merleau-Ponty, the idea of "the transcendental ego" is illusory, in *Sec. 3.5.1*; and the hidden bias in the analysis of the "subject" as "a unitary, autonomous Subject" in "the liberal theory of the social contract," since the critics like "Althusser, Foucault or Bourdieu theorize the subject as a social construction," or, "as "an ideological construction" for Althusser, in *Sec. 3.5.2*).

And the reverse direction also holds true.

Again, it should be stressed, however, that there are different shades of gray (or different degrees of truth) in the two opposites, and the classification is not necessarily mutually exclusive either. In fact, even when some combinations of the two occur, they only end up sharing the same dialectic relationship, but in a different degree.

Also, there are some other relationships (as a kind of family resemblance) which have something in common with the principle, but they are not exactly the same but only more or less comparable, which varies from case to case.

Good examples of family resemblance in relation to the principle are overtness-covertness, participation-obstruction, publicness-privateness, openness-closedness, surface-depth, transparency-secrecy, openness-biasedness, , exposition-illusion, and so on.

12th Thesis: The Denseness-Emptiness Principle

The twelfth thesis concerns the denseness-emptiness principle (on the distribution of entities in space) in the category of "structure" in existential dialectics, which is first proposed in *FPHUP*, in that both density and void are needed, in relation to the mind, nature, culture, and society, albeit in different ways. There is no denseness without emptiness—and vice versa.

Lest any misunderstanding occurs, the term "void" is used here only as an approximation of emptiness (depending on the degree of the lack of density), since, in physics, it is well known that "empty" space is not really empty all the way, because it can be full of energy (e.g., random quantum fluctuations at the sub-atomic level, and, for that matter, dark energy in the universe) and matter (e.g., different versions of sub-atomic particles, and, for that matter, dark matter in the universe). (F. Wilczek 2008)

For instance, on the one hand, there is denseness (e.g., the relatively denser concentration of info used in optimal decisin-making models, in *Sec. 2.2.1*; the relatively denser concentration of concerns with selfishness by those accepting "a model in which benefiting oneself is optimal," in *Sec. 2.2.2*; the relatively denser concentration of concerns with discreteness in the Poisson distribution, in *Sec. 2.3.1*; the relatively denser concentration of correlations among different parts in an organized complex system, in *Sec. 2.3.2*; the relatively denser concentration of concerns with "reliability" by those in reliability engineerin, in *Sec. 2.4.1*; the relatively denser concentration of concerns with optimality for those in the rational decision making model, in *Sec. 2.4.2*; the relatively denser concentration of concerns with "zeroth-order logic" in propositional logic, in *Sec. 2.5.1*; the relatively denser concentration of concerns with "the equilibrium of full employment" in the free market by those in the Austrial School of Economics in *Sec. 2.5.2*; the relatively denser concentration of concerns with "personal belief" in the Bayesian probability interpretation, in *Sec. 3.2.1*; the relatively denser concentration of concerns with "correlations between variables [say, height and age] across a sample of subjects" by those in R analysis, in *Sec. 3.2.2*; the relatively denser concentration of concerns with "random variables" and "simple probability distributions" for evaluating measurement uncertainties by those in mainstream metrology, in *Sec. 3.3.1*; the relatively denser concentration of concerns with indeterminism by those like Heiseberg in favor of the uncertainty principle, in *Sec. 3.3.2*; the relatively denser concentration of concerns with "mutual understadning" by Habermas, in *Sec. 3.4.1*; the relatively denser concentration of concerns with "formal and strict proofs of all statements from given axioms" in formal mathematics, in *Sec. 3.4.2*; the relatively denser concentration of concerns with "transcendental reduction" by those in Husserl's transcentental phenomenology, in *Sec. 3.5.1*; and the relatively denser concentration of concerns with "a unitary, autonomous Subject" by those in "the liberal theory of the social contract," in *Sec. 3.5.2*).

On the other hand, there is emptiness (e.g., the relatively less dense, or more empty, concentration of info used in mental models, in *Sec. 2.2.1*; the relatively less dense, or more empty, concentration of concerns with selfishness by those accepting a model "in which benefiting the group is optimal" and "purely selfish behavior is deemed irrational," in *Sec. 2.2.2*; the relatively less dense, or more empty, concentration of concerns with discreteness in the chi-squared distribution, in *Sec. 2.3.1*; the relatively less dense, or more empty, concentration of correlations among different parts in a disorganized complex system, in *Sec. 2.3.2*; the relatively less dense, or more empty, concentration of concerns with "reliability" by those in safety engineering, in *Sec. 2.4.1*; the relatively less dense, or more empty, concentration of concerns with optimality for those in the satisficing decision making model, in *Sec. 2.4.2*; the relatively less dense, or more empty, concentration of concerns with "zeroth-order logic" in higher-order logics, in *Sec. 2.5.1*; the relatively less dense, or more empty, concentration of concerns with "the equilibrium of full employment" in the free market by those critics like Keynes, who challenged the idea, in *Sec. 2.5.2*; the relatively less dense, or more empty, concentration of concerns with "personal belief" in the objectivist probability interpretation, in *Sec. 3.2.1*; the relatively less dense, or more empty, concentration of concerns with "correlations between variables [say, height and age] across a sample of subjects" by those in Q analysis, since they focus on "correlations between subjects across a sample of variables" instead, in *Sec. 3.2.2*; the relatively less dense, or more empty, concentration of concerns with "random variables" and "simple probability distributions" for evaluating measurement uncertainties by the critics who offered an alternative perspective, that is, the idea that "a mathematical interval rather than a probability distribution might be a better model of uncertainty," in *Sec. 3.3.1*; the relatively less dense, or more empty, concentration of concerns with indeterminism by those like Einstein in favor of the argument about "some hidden variables in the real world," in *Sec. 3.3.2*; the relatively less dense, or more empty, concentration of concerns with "mutual understanding" by Foucault, in *Sec. 3.4.1*; the relatively less dense, or more empty, concentration of concerns with "formal and strict proofs of all statements from given axioms" in informal mathematics, since many "informal mathematical practices…by aboriginal or ancient peoples" were "usually understood intuitively" without any proofs, that is, "based on proofless heuristics," in *Sec. 3.4.2*; the relatively less dense, or more empty, concentration of concerns with "the production of social order" by those in Husserl's phenomenology, although it is central to ethnomethodology, in *Sec. 3.5.1*; and the relatively less dense, or more empty, concentration of concerns with "a unitary, autonomous Subject" by those in "postmodernism", in *Sec. 3.5.2*).

And the reverse direction also holds true.

Again, it should be stressed, however, that there are different shades of gray (or different degrees of truth) in the two opposites, and the classification is not necessarily mutually exclusive either. In fact, even when some combinations of the two occur, they only end up sharing the same dialectic relationship, but in a different degree.

Also, there are some other relationships (as a kind of family resemblance) which have something in common with the principle, but they are not exactly the same but only more or less comparable, which varies from case to case.

Good examples of family resemblance in relation to the principle are fullness-voidness, nearness-farness, concentration-dispersion, denseness-nothingness, something-nothing, heaviness-lightness, composition-emptiness, solidness-void, presence-absence, focus-unfocus, emphasis-disemphasis, and the like.

In Relation to Process

The next four theses are in relation to process in existential dialectics, as shown below.

13th Thesis: The Change-Constancy Principle

The thirteenth thesis concerns the change-constancy principle (on the alteration of things) in the category of "process" in existential dialectics, which is first named in *BCPC* (and were worked out in my books like *FHC*, *FCD*, and *FPHC*), in that change occurs over time, although constancy is also allowed.

Old players fade away, and new ones emerges, with ever new causes and ever new forms—and asymmetry undergoes changes over time too, so does symmetry. There is no change without constancy—and vice versa.

For instance, on the one hand, there is change (e.g., the ever new ways to deal with info, as shown in mental models, in *Sec. 2.2.1*; the ever new ways to deal with reasoning, as shown in the argument on emotion and rationality, in *Sec. 2.2.2*; the ever new ways to understand mathematical structures, as shown in the construction of discreteness, in *Sec. 2.3.1*; the ever new ways to understand the linkages among different parts, as shown in complexity study, in *Sec. 2.3.2*; the ever new ways to cope with failure, as shown in reliability engineering, in *Sec. 2.4.1*; the ever new ways to deal with decision making, as shown by optimal decision making in organizational behavior, in *Sec. 2.4.2*; the ever new ways to understand logic, as shown in propositional logic, in *Sec. 2.5.1*; the ever new ways to study history, as shown in the argument on history and common causes, in *Sec. 2.5.2*; the ever new ways to study probability, as shown in the construction of the Bayesian methods, in *Sec. 3.2.1*; the ever new ways to study analysis, as shown in Q analysis, in *Sec. 3.2.2*; in a random error, "when a measurement is repeated, it will generally provide a measured value that is different from the previous value," in *Sec. 3.3.1*; according to Niels Bohr's view on quantum mechanics, "the probability distributions...depend on which measurements we choose to perform," so different measurements can yield different probability distributions, in *Sec. 3.3.2*; the appeal to "rightness" is conventional in nature, so what is accepted as "truth" can change over time, according to the critics of communicative action, in *Sec. 3.4.1*; the ever new ways to understand mathematics, as shown in the sociology of mathematical knowledge, in *Sec. 3.4.2*; the ever new ways to understand subjective experience, as shown in Husserl's phenomenology, in *Sec. 3.5.1*; and the ever new ways to study the "subject," as shown in the different philosophical traditions, in *Sec. 3.5.2*).

On the other hand, there is constancy (e.g., the ever constant existence of different problems with the ever new ways to deal with info, as shown in the problems with mental models pointed out by the critics, in *Sec. 2.2.1*; the ever constant existence of different problems with the ever new ways to deal with reasoning, as shown in the problems with the argument on emotion and rationality pointed out by the critcs, in *Sec. 2.2.2*; the ever constant existence of different problems with the ever new ways to understand mathematical structures, as shown in the problems with discreteness pointed out by the critics aforecited, in *Sec. 2.3.1*; the ever constant existence of different problems with the ever new ways to understand the linkages among different parts, as shown in the problems with complexity study pointed out by the critics, in *Sec. 2.3.2*; the ever constant existence of different problems with the ever new ways to cope with failure, as shown in the problems with reliability engineering pointed out by the critics, in *Sec. 2.4.1*; the ever constant existence of different problems with the ever new ways to deal with decision making, as shown in the problems with optimal decision making in organizational behavior pointed out by the critics aforecited, in *Sec. 2.4.2*; the ever constant existence of different problems with the ever new ways to understand logic, as shown in the problems with propositional logic pointed out by the critics, in *Sec. 2.5.1*; the ever constant existence of different problems with the ever new ways to study history, as shown in the problems with the argument on history and common causes pointed out by the critics, in *Sec. 2.5.2*; the ever constant existence of different problems with the ever new ways to study probability, as shown in the problems with the Bayesian methods pointed out by the critics, in *Sec. 3.2.1*; the ever constant existence of different problems with the ever new ways to study analysis, as shown in the problems with Q analysis pointed out by the critics, in *Sec. 3.2.2*; the existence of "quantities representing physical constants," like "modulus of elasticity and specific heat" in measurement uncertainty, in *Sec. 3.3.1*; according to to the Copenhagen interpretation of quantum mechanics, there is the constancy that "there is no way to say what the state of a system fundamentally is, only what the result of observations might be," in *Sec. 3.3.2*; the constant existence of the hermeneutic circle, according to the critics of communicative action like Gadamer, in *Sec. 3.4.1*; the ever constant existence of different problems with the ever new ways to understand mathematics, as shown in the problems with the sociology of mathematical knowledge pointed out by the critics, in *Sec. 3.4.2*; the ever constant existence of different problems with the ever new ways to understand subjective experience, as shown in the problems with Husserl's phenomenology pointed out by the critics, in *Sec. 3.5.1*; and the ever constant existence of different problems with the ever new ways to study the "subject," as shown in the problems with the different philosophical traditions pointed out by the critics aforecited, in *Sec. 3.5.2*).

And the reverse direction also holds true.

Again, it should be stressed, however, that there are different shades of gray (or different degrees of truth) in the two opposites, and the classification is not necessarily mutually exclusive either. In fact, even when some combinations of the two occur, they only end up sharing the same dialectic relationship, but in a different degree.

Also, there are some other relationships (as a kind of family resemblance) which have something in common with the principle, but they are not exactly the same but only more or less comparable, which varies from case to case.

Good examples of family resemblance in relation to the principle are dynamicness-staticness, instability-stability, and the like.

14th Thesis: The Order-Chaos Principle

The fourteenth thesis is called the the order-chaos principle (on the pattern of things) in the category of "process" in existential dialectics (as first worked out in *FC*), in that both order and chaos are vital in the process of change in the world.

The preference for order is biased, since it does not give sufficient attention to the vital role of chaos in the transformation of the world (without somehow reducing it for the understanding of order). There is no order without chaos—and vice versa.

The scientific search for order in the world is often a hidden bias in its ontological obsession with order, since chaos is often treated as the "bad" guy, with order as the "good" guy (for the end goal of science).

Neither order nor chaos is the final end of the world, and one is not to be treated as the means for the other in the transformation of things. Both are fundamental in their recurrent dialectical interactions with each other over time, without reducing one for the other.

For instance, on the one hand, there is order (e.g., the more or less orderly view of reality, if looked from the vantage point of the "effects as information filter" in mental models, so as to "cause…selective perception, perception of only selected parts of information" for a simplified sense of reality, in *Sec. 2.2.1*; the more or less orderly view about rationality, if looked only from the vantage point of a particular type of rationality for a particular goal, like "a model in which benefiting oneself is optimal," in *Sec. 2.2.2*; the more or less orderly view about discreteness, if looked from the vantage point of a dice with a probability of obtaining each number in terms of 1/*s* in principle, in *Sec. 2.3.1*; the more or less orderly view of reality, if looked from the vantage point of highly correlated parts in a small organized complex system, in *Sec. 2.3.2*; the more or less orderly view about products and services, if looked from the vantage point of their "successes" like good performances, in *Sec. 2.4.1*; the more or less orderly view about decision making, if looked from the vantage point of following the mathematical equations for optimal decision aforecited, in *Sec. 2.4.2*; the more or less orderly view about logic, if looked only from the vantage point of propositional logic, in *Sec. 2.5.1*; the more or less orderly view about the free market, if looked from the vantage point of those in the Austrian School of Economics, who have strong faith on the ability of the free market to reach "the equilibrium of full employment," in *Sec. 2.5.2*; the more or less orderly view on probability, if looked from the vantage point of "the frequentist probability of a hypothesis," which must be "either one or zero," in an orderly way, in *Sec. 3.2.1*; the more or less orderly view on the ideas of an individual, if looked from the vantage point of Q analysis, which tries to show that "people think about ideas in relation to other ideas, rather than in isolation," in *Sec. 3.2.2*; the more or less orderly view on random error, if looked from the vantage point that, as is well expected, "when a measurement is repeated it will generally provide a measured value that is different from the previous value," in *Sec. 3.3.1*; the more or less orderly view on a particle, if looked from the vantage point of the "wave mechanics," when "σ_x and σ_p have an inverse relationship or are at least bounded from below," in *Sec. 3.3.2*; the more or less orderly view on communication, if looked from the vantage point of communicative action, such that the participants seriously follows the three requirements for the validity of speech for genuine mutual understanding, so that there will be no more, or few, social conflicts, in *Sec. 3.4.1*; the more or less orderly view about mathematics, if looked only from the vantage point of "formal and strict proofs of all statements from given axioms" in formal mathematics, in *Sec. 3.4.2*; the more or less orderly view on reality, if looked from the vantage point of ethnomethdology, where people find ways to reproduce "social order," in *Sec. 3.5.1*; and the more or less orderly view on the "subject," if looked from the vantage point of "a unitary, autonomous Subject" in "the liberal theory of the social contract," in *Sec. 3.5.2*).

On the other hand, there is chaos (e.g., the more or less chaotic view of reality, if lookd from "hardly qualifiable, impugnable, obscure, or incomplete facts" in mental models, well before they are even "selected" and "filtered," in *Sec. 2.2.1*; the more or less chaotic view abour rationality, if looked from the conflicting views that there is no general consensus on the usage of "rationality" in different fields, that the embedded assumptions in each rationality make it relative in relation to others, that the theories of rationality are mixed up in classification, etc., in *Sec. 2.2.2*; the more or less chaotic view about discreteness, if looked from the vantage point of the difficult case when "there exist groups for which computing discrete logarithms is apparently difficult," like cases concerning "large prime order subgroups of groups $(Z_p)^{\times}$ when "there is not only no efficient algorithm known for the worst case, but also the average-case complexity can be shown to be about as hard as the worst case using random self-reducibility," in *Sec. 2.3.1*; the more or less chaotic view of reality, if looked from the vantage point of many uncorrelated parts in an extremely large disorganized complex system, in *Sec. 2.3.2*; the more or less chaotic view about products and services, if looked from the vantage point of their "failures" or unpredicted accidents, in *Sec. 2.4.1*; the more or less chaotic view about decision making, if looked from the vantage point of "flipism," "divination," or other chanceful techniques, in *Sec. 2.4.2*; the more or less chaotic view about logic, if looked from the multiple, often conflicting, vantage points like propositional calculus, higher-order logics, modal logic, many-valued logics, incompleteness theorem, everyday decision techniques, mental models, etc., in *Sec. 2.5.1*; the more or less chaotic view about the free market, if looked from the vantage point of Keynes, who argued that "economic systems did not automatically tend to the equilibrium of full employment owing to their agents' inability to predict the future" and "the inherent unpredictability of economic systems," in *Sec. 2.5.2*; the more or less chaotic view on probability, if looked from the vantage point of Bayesian statistics, in which "a probability can be assigned to a hypothesis that can differ from 0 or 1," in that "the truth value is uncertain" or unknown, such that it can go in any direction, in *Sec. 3.2.1*; the more or less chaotic view on the ideas of an individual, if looked from the vantage point of understanding ideas in isolation in the mindset of an individual, such that they are not closely linked with each other, in *Sec. 3.2.2*; the more or less chaotic view on random error, if looked from the vantage point that "the next measured value cannot be predicted exactly from previous such values," such that it can go in any direction, in *Sec. 3.3.1*; the more or less chaotic view on a particle, if looked from the vantage point of the "wave mechanics," when, "in the case of the single-moded plane wave,...the particle position is extremely uncertain in the sense that it could be essentially anywhere along the wave packet," in *Sec. 3.3.2*; the more or less chaotic view on communication, if looked from the vantage point of strategic action, such that each participant only follows his "private interests," with "no intention of reaching consensus or mutual understanding" for "truth," so that conflicts abound among them, in *Sec. 3.4.1*; the more or less chaotic view about mathematics, if looked only from the vantage point of folklore mathematics, where many "stories or jokes involving mathematicians or mathematics" can be spurious, questionable, or simply "apochryphal," and many "informal mathematical practices...by aboriginal or ancient peoples" were "usually understood intuitively" without any proofs, that is, "based on proofless heuristics"—such that almost anything goes, for practical purposes, in *Sec. 3.4.2*; the more or less chaotic view on reality, if looked from the vantage point of Husserl's transcendental phenomenology, where a subject is to question or "bracket...all assumptions about the existence of an external world," in *Sec. 3.5.1*; and the more or less chaotic view on the "subject," if looked from the vantage point of "Freud's psychoanalytic model of the subject," as further developed by Lacan, in which "the 'split subject' is constituted by a double bind: alienated from jouissance when he or she leaves the Real, enters into the Imaginary [during the mirror stage], and separates from the Other when he or she comes into the realm of language, difference, and demand in the Symbolic or the Name of the Father, in *Sec. 3.5.2*).

And the reverse direction also holds true.

Once more, it should be stressed, however, that there are different shades of gray (or different degrees of truth) in the two opposites, and the classification is not necessarily mutually exclusive either. In fact, even when some combinations of the two occur, they only end up sharing the same dialectic relationship, but in a different degree.

Also, there are some other relationships (as a kind of family resemblance) which have something in common with the principle, but they are not exactly the same but only more or less comparable, which varies from case to case.

Good examples of family resemblance in relation to the principle are lawfulness-disorder, order-disorder, pattern-chaos, linearity-nonlinearity, unity-disunity, linearity-irreversalness, etc.

15th Thesis: The Slowness-Quickness Principle

The fifteenth thesis concerns the slowness-quickness principle (on the speed of change) in the category of "process" in existential dialectics (which was first worked out in *FIA*), in that both slowness and quickness co-exist, with their own internal tension, to the extent that each fights for its own relevance with the other, in accordance to the perspectives of nature, the mind, culture, and society, without one being the victor and the other being the vanquished in the long haul. There is no slowness without quickness—and vice versa.

For instance, on the one hand, there is slowness (e.g., the relatively slower speed of understadning reality if "hardly qualifiable, impugnable, obscure, or incomplete facts" in mental models are not "selected" and "filtered," in *Sec. 2.2.1*; the relatively slower speed of decision making, if the goal is to "find...the rational decision for arbitrarily complex problems" in an optimal way, in *Sec. 2.2.2*; the relatively slower speed of finishing the task of rolling a dice many times before the probability of each number will approach $1/s$ in practice, not just in principle, in *Sec. 2.3.1*; the relatively slower speed of a problem to be solved, if the size is extremely large, in *Sec. 2.3.2*; the relatively slower speed of a product to fail in the laboratory if a normal life testing, not an "accelerated life testing," is used for testing instead, in *Sec. 2.4.1*; the relatively slower speed of making a decision if an optimal decision making model is used, in *Sec. 2.4.2*; the relatively slower speed to calculate possible interpretations if when $2^{\aleph_0} = \mathfrak{c}$, in *Sec. 2.5.1*; the relatively slower speed of the free market to ever reach "the equilibrium of full employment," for Keynes, in *Sec. 2.5.2*; the relatively slower speed of computation to be completed, if it is "intractable," in cases involving "the use of hierarchical models and marginalization over the values of nuisance parameters" in Bayesian methods, in *Sec. 3.2.1*; the relatively slower speed of doing Q sorts in Q analysis, if the traditional way of "a paper template and the sample of statements or other stimuli printed on individual cards" is used instead, in *Sec. 3.2.2*; the relatively slower speed of evaluating measurement uncertainty, if the task is complicated by including difficult situations like a huge number of output quantities and input quantities, "periodic measurements, binned data values, censoring, detection limits,…plus-minus ranges of measurements," etc., in *Sec. 3.3.1*; the relatively slower speed of a particle like an electron, if "the photon has a long wavelength and low momentum," so "the collision does not disturb the electron's momentum very much," according to the experiment with "an experimenter trying to measure the position and momentum of an electron by shooting a photon at it," in *Sec. 3.3.2*; the relatively slower readiness of someone to be "untruthful" in the "idealized" communicative discourse, in *Sec. 3.4.1*; the relatively slower speed of mathematics to be accepted if it is to be judged on the basis of "formal and strict proofs of all statements from given axioms," which are not easily available, in *Sec. 3.4.2*; the relatively slower speed of a subject to understand his experience of an object, if "intuition" and "evidence" are absent, in accordance to Husserl's phenomenology, in *Sec. 3.5.1*; and the relatively slower speed of a subject to decide what to do, when confronted with "a double bind," which "is an emotionally distressing dilemma in communication in which an individual…receives two or more conflicting messages, in which one message negates the other," in *Sec. 3.5.2*).

On the other hand, there is quickness (e.g., the relatively quicker speed of understadning reality if "hardly qualifiable, impugnable, obscure, or incomplete facts" in mental models are carefully "selected" and "filtered" to make sense of reality, in *Sec. 2.2.1*; the relatively quicker speed of decision making, if the goal is merely to find a satisfying rational decision for simple problems, as in bounded rationality, in *Sec. 2.2.2*; the relatively quicker speed of calculating the probability of obtaining each number of a dice in terms of 1/*s* in principle, in *Sec. 2.3.1*; the relatively quicker speed of of a problem to be solved, if the size is extremely small, in *Sec. 2.3.2*; the relatively quicker speed of a product to fail in the laboratory if "accelerated life testing" is used for testing, in *Sec. 2.4.1*; the relatively quicker speed of making a decision if a satisficing decision making model is used, in *Sec. 2.4.2*; the relatively quicker speed to calculate possible interpretations if 2^1=2, in *Sec. 2.5.1*; the relatively quicker speed of the free market to ever reach "the equilibrium of full employment," if government intervention is undertaken, for Keynes, in *Sec. 2.5.2*; the relatively quicker speed of computation to be completed, if "good approximations can be obtained using Markov chain Monte Carlo methods," in cases involving "the use of hierarchical models and marginalization over the values of nuisance parameters" in Bayesian methods, in *Sec. 3.2.1*; the relatively quicker speed of doing Q sorts in Q analysis, if sophisticated computer software applications are used, in *Sec. 3.2.2*; the relatively quicker speed of evaluating measurement uncertainty, if the task is simplified by focusing only on situations involving "random variables" and "simple probability distributions," together with the inherent restrictive assumptions, in *Sec. 3.3.1*; the relatively quicker speed of a particle like an electron, if "the photon has a short wavelength, and therefore, a large momentum,…transferring a large and uncertain amount of momentum to the electron," according to the experiment with "an experimenter trying to measure the position and momentum of an electron by shooting a photon at it," in *Sec. 3.3.2*; the relatively quicker readiness of someone to be "untruthful" in the realistic strategic action, in *Sec. 3.4.1*; the relatively quicker speed of mathematics to be accepted if it is to be judged "intuitively" without any proofs, that is, "based on proofless heuristics," in *Sec. 3.4.2*; the relatively quicker speed of a subject to understand his experience of an object, if "intuition" and "evidence" are used, in accordance to Husserl's phenomenology, in *Sec. 3.5.1*; and the relatively quicker speed of of a subject to decide what to do, when he does not suffer from "a double bind" but benefits from being "a unitary, autonomous Subject," in *Sec. 3.5.2*).

And the reverse direction also holds true.

Again, it should be stressed, however, that there are different shades of gray (or different degrees of truth) in the two opposites, and the classification is not necessarily mutually exclusive either. In fact, even when some combinations of the two occur, they only end up sharing the same dialectic relationship, but in a different degree.

Also, there are some other relationships (as a kind of family resemblance) which have something in common with the principle, but they are not exactly the same but only more or less comparable, which varies from case to case.

Good examples of family resemblance in relation to the principle are inconvenience-convenience, passiveness-activenessness, gradualness-abruptness, deceleration-acceleration, difficulty-ease, nonresilience-resilience, patience-impatience, and the like.

16th Thesis: The Expansion-Contraction Principle

The sixteenth thesis is called the expansion-contraction principle in the category of "process" in existential dialectics, in that entities in the world can both expand in some ways and contract in other ones, as part of their nature. There is no expansion without contraction—and vice versa.

This principle, although not so explicitly called, was already used in my previous works on different topics (e.g., the theory of floating consciousness in *FCD* and *FPHC*, the union of the unions in *BWT*, and the cyclical progression of system fragmentation and integration in *FCD*).

For instance, on the one hand, there is expansion (e.g., the relatively more developed ability of those in mental models to deal with the simplification of reality, in *Sec. 2.2.1*; the relative expansion of the use of the argument for non-emotional rationality in the field of artificial intelligence, in *Sec. 2.2.2*; the relative spread of the "modern definition" of probability nowadays, in *Sec. 2.3.1*; the relative expansion of "web science" in the research on complexity study nowadays, in *Sec. 2.3.2*; the relatively more developed ability to deal with "stress" in the physics of failure approach to reliability engineering, in *Sec. 2.4.1*; the relatively more developed ability to deal with optimality for the practitioners of the rational decision making model, in *Sec. 2.4.2*; the relative spread of propositional calculus in the 19th and 20th centuries, in *Sec. 2.5.1*; the probability of the joint failure of two" components in a system increase to "one in one million per year, provided that the two events are statistically independent," in the example about the principle of redundancy in the article aforecited, in *Sec. 2.5.1*; the relative "dramatic growth in research and applications of Bayesian methods, mostly attributed to the discovery of Markov chain Monte Carlo methods," in the 1980s, in *Sec. 2.5.2*; in *Sec. 3.2.1*; the relative expansion of using computer software applications for doing Q sorts nowadays, in *Sec. 3.2.2*; the expansion of the size of a mathematical model to evaluate measurement uncertainty, if it includes a huge number of output quantities and input quantities, "periodic measurements, binned data values, censoring, detection limits,...plus-minus ranges of measurements," etc., in *Sec. 3.3.1*; the relative expansion of the influence of quantum mechanics in the 20th ceutury, in *Sec. 3.3.2*; the relatively more developed ability of those in the idealized communicative discourse to engage in "mutual understanding," in *Sec. 3.4.1*; the relatively more developed ability of judging mathematical practices "intuitively" without any proofs, that is, "based on proofless heuristics," among those "aboriginal or ancient peoples" in informal mathematics, in *Sec. 3.4.2*; the relatively more developed ability of a subject in Husserl's phenomenology to question or "bracket...all assumptions about the existence of an external world," in *Sec. 3.5.1*; and the relative spread, in the 20th century, of the influence of "Freud's psychoanalytic model of the subject," in *Sec. 3.5.2*).

On the other hand, there is contraction (e.g., the relatively less developed ability of those in mental models to deal with the complexity of reality as in optimal decision-making models, in *Sec. 2.2.1*; the relative decline of the use of the argument for non-emotional rationality in the field of feminism, in *Sec. 2.2.2*; the relative decline of the "classical definition" of probability nowadays, in *Sec. 2.3.1*; the relative decline of "cybernetics" in the research on complexity study nowadays, in *Sec. 2.3.2*; the relatively less developed ability to deal with "redundancy" in the physics of failure approach to reliability engineering, in *Sec. 2.4.1*; the relatively less developed ability to deal with authority as an everyday decision making for the practitioners of the rational decision making model, in *Sec. 2.4.2*; the relative decline of traditional syllogistic logic in the 19th and 20th centuries, in *Sec. 2.5.1*; "the probability of failure of a component of a system" decreases to "one in one thousand per year," in the example about the principle of redundancy in the article aforecited, in *Sec. 2.5.2*; the relative decline of Bayesian statistics in "most undergraduate teaching," which "is still based on frequentist statistics," in *Sec. 3.2.1*; the relative decline of using "a paper template and the sample of statements or other stimuli printed on individual cards" for doing Q sorts nowadays, in *Sec. 3.2.2*; the relative shrinkage of the size of a mathematical model to evaluate measurement uncertainty, if it makes things easier by focusing only on simple situations involving only a few "random variables" and "simple probability distributions," together with the inherent restrictive assumptions, in *Sec. 3.3.1*; the relative decline of the influence of classical mechanics in the 20th ceutury, in *Sec. 3.3.2*; the relatively less developed ability of of those in the idealized communicative discourse to engage in manipulation for private interests, in *Sec. 3.4.1*; the relatively less developed ability of doing "formal and strict proofs of all statements from given axioms" among those "aboriginal or ancient peoples" in informal mathematics, in *Sec. 3.4.2*; the relatively less developed ability of a subject in Husserl's phenomenology to reproduce "social order," in *Sec. 3.5.1*; and the relative decline, in the 20th century, of the influence of "the liberal theory of the social contract" in terms of "a unitary, autonomous Subject," in *Sec. 3.5.2*).

And the reverse direction also holds true.

Again, it should be stressed, however, that there are different shades of gray (or different degrees of truth) in the two opposites, and the classification is not necessarily mutually exclusive either. In fact, even when some combinations of the two occur, they only end up sharing the same dialectic relationship, but in a different degree.

Also, there are some other relationships (as a kind of family resemblance) which have something in common with the principle, but they are not exactly the same but only more or less comparable, which varies from case to case.

Good examples of family resemblance in relation to the principle are conquest-autarky, rise-fall, spread-shrink, extendingness-shorteningness, widening-narrowing, construction-deconstruction, expansion-destruction, construction-demolition, creation-destruction, expansion-contraction, expansion-constriction, expansion-decline, and so forth.

In Relation to Agency

The next eight theses are in relation to agency in existential dialectics, as shown below.

17th Thesis: The Theory-Praxis Principle

The seventeenth thesis is the theory-praxis principle (on the duality of knowledge) in the category of "agency" in existential dialectics (which was first worked out in *FPHE*), in that, if there is theoretical construction, there is likewise its practical application, both technical and normative. There is no theory without praxis—and vice versa.

For instance, on the one hand, there is theory (e.g., the theoretical construction of the "principles" of mental models, in *Sec. 2.2.1*; the theoretical construction of an "objective" account of rationality without the distortion of emotion in philosophy, in *Sec. 2.2.2*; the theoretical construction of discrete probability in mathematics, in *Sec. 2.3.1*; the theoretical construction of "emergence" in complexity study, in *Sec. 2.3.2*; the theoretical construction of "probability" in mathematics like mainstream probability study and statistics, in *Sec. 2.4.1*; the theoretical construction of "probability distribution" in mathematics, in *Sec. 2.4.2*; the theoretical construction of "graphs" as a subfield of mathematics, in *Sec. 2.5.1*; the theoretical construction of "common cause" and "special cause" in mathematics, in *Sec. 2.5.2*; the theoretical construction of "Bayesian methods" in the study of probability, in *Sec. 3.2.1*; the theoretical construction of Q methodology, in *Sec. 3.2.2*; the theoretical construction of the models of evaluating measurement uncertainty in metrology, in *Sec. 3.3.1*; the theoretical construction of the uncertainty principle in physics, in *Sec. 3.3.2*; the theoretical construction of "speech acts" in linguistics, in *Sec. 3.4.1*; the theoretical construction of formal mathematics, in *Sec. 3.4.2*; the theoretical construction of Husserl's phenomenology, in *Sec. 3.5.1*; and the theoretical construction of "Freud's psychoanalytic model of the subject," in *Sec. 3.5.2*).

On the other hand, there is praxis (e.g., the practical application of the "principles" of mental models to the field of "system dynamics," in *Sec. 2.2.1*; the practical application of an "objective" account of rationality without the distortion of emotion in philosophy to the field of artificial intelligence, in *Sec. 2.2.2*; the practical application of discrete probability in mathematics to the field of games, like dice games and the game of cleromancy, in *Sec. 2.3.1*; the practical application of "emergence" in complexity study to the field of software engineering, like the issue of "programming complexity," in *Sec. 2.3.2*; , the practical application of probability" in mathematics like mainstream probability study and statistics to the field of engineering, like "reliability engineering," in *Sec. 2.4.1*; the practical application of "probability distribution" in mathematics to the field of organizational behavior for optimal decision making, in *Sec. 2.4.2*; the practical application of "graphs" as a subfield of mathematics to another subfield like propositional logic for "graphical calculi," in *Sec. 2.5.1*; the practical application of "common cause" and "special cause" in mathematics to the field of engineering in the article aforecited, in *Sec. 2.5.2*; the practical application of "Bayesian methods" in the study of probability to the fields of "machine learning and talent analytics," in *Sec. 3.2.1*; the practical application of Q methodology to the applied fields like "nursing, veterinary medicine, public health, transportation, education, rural sociology, hydrology and mobile communication", in *Sec. 3.2.2*; the practical application of the models of evaluating measurement uncertainty by such applied organizations like the International Electrotechnical Commission, the International Federation of Clinical Chemistry and

Laboratory Medicine, the International Laboratory Accreditation Cooperation, the International Organization for Standardization, the International Union of Pure and Applied Chemistry, etc., in *Sec. 3.3.1*; the practical application of the uncertainty principle in physics to the field of philosophy, like the debate on the existence of "free will," in *Sec. 3.3.2*; the practical application of "speech acts" in linguistics to the field of communication, as shown in the theory of communicative action, in *Sec. 3.4.1*; the practical application of formal mathematics to the field of physics, as pointed out by Eugene Wigner in 1960, in *Sec. 3.4.2*; , in *Sec. 3.4.2*; the practical application of Husserl's phenomenology to the field of ethnomethodology, when "Garfinkel speaks of phenomenological texts and findings as being, 'appropriated' and intentionally 'misread,' for the purposes of exploring topics in the study of social order," in *Sec. 3.5.1*; and the practical application of "Freud's psychoanalytic model of the subject" to the field of communication, when Lacan used it to discuss the "split subject," which "is constituted by a double bind: alienated from jouissance when he or she leaves the Real, enters into the Imaginary [during the mirror stage], and separates from the Other when he or she comes into the realm of language, difference, and demand in the Symbolic or the Name of the Father, in *Sec. 3.5.2*).

And the reverse direction also holds true.

Again, it should be stressed, however, that there are different shades of gray (or different degrees of truth) in the two opposites, and the classification is not necessarily mutually exclusive either. In fact, even when some combinations of the two occur, they only end up sharing the same dialectic relationship, but in a different degree.

Also, there are some other relationships (as a kind of family resemblance) which have something in common with the principle, but they are not exactly the same but only more or less comparable, which varies from case to case.

Good examples of family resemblance in relation to the principle are discovery-application, knowledge-action, invention-innovation, theory-intentionality, discourse-action, knowledge-interest, knowledge-motivation, universality-particularity, potentiality-actuality, theory-practice, and so forth.

18th Thesis: The Convention-Novelty Principle

The eighteenth thesis is the convention-novelty principle (on the nature of creative thinking) in the category of "agency" in existential dialectics (which was first worked out in *FPHCT*), in that, if there is conventional wisdom, there is likewise novel challenge, to the extent that both convergent and divergent thinking are part of life. There is no convention without novelty—and vice versa.

As srummarized from *FPHCT* (almost verbatim here), there are (a) "creative techniques" and (b) "creative traits," which, when satisfied—in relation to the larger context of the mind, nature, society, and culture—can be used to enhance creative works.

In addition, creative thinking has its own possibilities and limits (in relation to invention), just as it has its own promises and pitfalls (in relation to innovation)— as already analyzed in *FPHCT*).

In the end, creative thinking has its own desirability and dark sides (as also already analyzed in *FPHCT*).

With these clarifications in mind—there are good empirical examples for the convention-novelty princple.

For instance, on the one hand, there is convention (e.g., the conventional wisdom about the formal rules of inference in Aristotle's logic, in *Sec. 2.2.1*; the conventional wisdom about an "objective" account of rationality without the distortion of emotion in philosophy, in *Sec. 2.2.2*; the "classical definition" of probability by James Bernoulli and Pierre-Simon Laplace in the older days, in *Sec. 2.3.1*; the conventional wisdom about complexity study to understand the linkages among different parts in a system, in *Sec. 2.3.2*; the mainstream of probability study and statistics, in *Sec. 2.4.1*; the conventional wisdom about optimal decision making in organizational behavior in the middle of the 20th century, in *Sec. 2.4.2*; the tradition of syllogistic logic in ancient Greece, like the Aristotelian syllogisms, in *Sec. 2.5.1*; the conventional wisdom about "common cause" by Jacob Bernoulli in 1703, in *Sec. 2.5.2*; the conventional understanding of Bayesian probability as worked out by Bayes and Laplace in the older days of the 18th and 19th centuries, in *Sec. 3.2.1*; the conventional wisdom about using R analysis in the older days, in *Sec. 3.2.2*; the conventional wisdom about using only "random variables" and "simple probability distributions" for evaluating measurement uncertainties by those in mainstream metrology, in *Sec. 3.3.1*; the conventional wisdom about determinism in classical mechanics, in *Sec. 3.3.2*; the conventional concern with "strategic action," in *Sec. 3.4.1*; the conventional wisdom about the foundation of mathematics in formal mathematics, in *Sec. 3.4.2*; the conventional wisdom about consciousness as understood by Husserl when he "conceived humans as having been constituted by states of consciousness," in *Sec. 3.5.1*; and the conventional idea about "a unitary, autonomous Subject" in "the liberal theory of the social contract," in *Sec. 3.5.2*).

On the other hand, there is novelty (e.g., the alternative novel challenge to the conventional wisdom about the formal rules of inference, by the new idea of mental models by Kenneth Craik in 1943, in *Sec. 2.2.1*; the alternative novel challenge to the conventional wisdom about an "objective" account of rationality without the distortion of emotion in philosophy, by the new idea of the Habermasians and feminists that the conventional views about rationality do not give sufficient weight to the issue of power, in *Sec. 2.2.2*; the alternative novel challenge to the "classical definition" of probability, by the "modern definition" of probability by John Venn and George Boole, in *Sec. 2.3.1*; the alternative novel challenge to the conventional wisdom about complexity study to understand the linkages among different parts in a system, by the new idea that there are different unresolvable problems such as the problem of Bremermann's fundamental limit on computation, the problem of intractability, and so on, in *Sec. 2.3.2*; the alternative novel challenge to the mainstream of probability study and statistics, by the new idea of "reliability theory," which "developed apart from the mainstream of probability and statistics," in *Sec. 2.4.1*; the alternative novel challenge to the conventional wisdom about optimal decision making in organizational behavior in the middle of the 20th century, by the new idea of "bounded rationality" in the 1960s and 1970s, in *Sec. 2.4.2*; the alternative novel challenge to the tradition of syllogistic logic in ancient Greece, by the new idea of propositional logic in modern times, in *Sec. 2.5.1*; the alternative novel challenge to the conventional wisdom about "common cause" by Jacob Bernoulli, by the new idea of "special cause" by Keynes and others in the 20th century, in *Sec. 2.5.2*; the alternative novel challenge to the conventional understanding of Bayesian probability as worked out by Bayes and Laplace in the older days of the 18th and 19th centuries, by the new ideas in the 20th century, when "the ideas of Laplace were further developed in two different directions, giving rise to objective and subjective currents in Bayesian practice," in *Sec. 3.2.1*; the alternative novel challenge to the conventional wisdom about using R analysis in the older days, by the new idea of using Q analysis nowadays, in *Sec. 3.2.2*; the alternative novel challenge to the conventional wisdom about using only "random variables" and "simple probability distributions" for evaluating measurement uncertainties by those in mainstream metrology, by the new idea of the critics, in that "a mathematical interval rather than a probability distribution might be a better model of uncertainty," in *Sec. 3.3.1*; the alternative novel challenge to the conventional wisdom about determinism in classical mechanics, by the new idea of indeterminism in quantum mechanics in the 20th century, in *Sec. 3.3.2*; the alternative novel challenge to the conventional concern with "strategic action," by the new idea of "commuicative action," in *Sec. 3.4.1*; the alternative novel challenge to the conventional wisdom about the foundation of mathematics in formal mathematics, by the new idea of "quasi-empiricism in mathematics," in *Sec. 3.4.2*; the alternative novel challenge to the conventional wisdom about consciousness as understood by Husserl, by the new idea of Heidegger, who "countered that consciousness is peripheral to the primacy of one's existence [i.e., the mode of being of Dasein], which cannot be reduced to one's consciousness of it," in *Sec. 3.5.1*; and the alternative novel challenge to the conventional idea about "a unitary, autonomous Subject" in "the liberal theory of the social contract," by the new idea of the "split subject" in "Freud's psychoanalytic model of the subject," as further developed by Lacan, in *Sec. 3.5.2*).

And the reverse direction also holds true.

Again, it should be stressed, however, that there are different shades of gray (or different degrees of truth) in the two opposites, and the classification is not necessarily mutually exclusive either. In fact, even when some combinations of the two occur, they only end up sharing the same dialectic relationship, but in a different degree.

Also, there are some other relationships (as a kind of family resemblance) which have something in common with the principle, but they are not exactly the same but only more or less comparable, which varies from case to case.

Good examples of family resemblance in relation to the principle are convergence-divergence (or convergent thinking vs. divergent thinking), normalness-nonnormalness, conformity-nonconformity, familiarity-freshness, inertia-newness, equilibrium-disturbance, and so on.

19th Thesis: The Evolution-Transformation Principle

The nineteenth thesis refers to the evolution-transformation principle (on the multiple kinds of agency). in the category of "agency" in existential dialectics (which was first worked out in *FAE* and then in other books of mine).

This principle—and the symmetry-asymmetry principle, for instance—are both about agency. More precisely, the word "agency," in a formal definition, refers to "a person or thing through which power is exerted or an end is achieved." (MWD 2007) It therefore does not have to necessarily involve an intelligent lifeform.

Because of this dual meaning in agency, the evolution-transformation principle is more concerned with *the kind of agency*, that is, both about the *evolution* in the state of nature (e.g., an object of natural beauty) and the *transformation* in the world of intelligent lifeforms (e.g., a work of art). There is no evolution without transformation—and vice versa.

In classical Darwinian evolutionary theory (as more detailedly analyzed in *BNN*), evolution is "blind." But in the human world, change often takes place because of the conscious intervention of humans in transforming society and culture, just to cite two instances.

And the transformative part of the principle precisely refers to the other dimension in the dual meaning of agency, in giving technology (as invented by intelligent lifeforms like humans and, soon, post-humans) a major role to play in the change of the world, which is something that I extensively analyzed in *FHC* in the context of the technophilic lifeworld, espeically though not exclusively since modern times. The same theme is also elaborated in many other books of mine.

For instance, on the one hand, there is evolution (e.g., the natural evolution of humans to process information in the state of nature, in *Sec. 2.2.1*; the natural evolution of humans to have reasoning and emotion in the state of nature, in *Sec. 2.2.2*; the natural evolution of humans to distinguish "separate" or "distinct" events in the state of nature, in *Sec. 2.3.1*; the natural evolution of humans to observe different parts of things in the state of nature, in *Sec. 2.3.2*; the natural evolution of humans to experience failures in the state of nature, in *Sec. 2.4.1*; the natural evolution of humans to make decision in the state of nature, in *Sec. 2.4.2*; the natural evolution of humans to think in the state of nature in a primitive way, in *Sec. 2.5.1*; the natural evolution of humans to experience variation of events in the state of nature in a primitive way, in *Sec. 2.5.2*; the natural evolution of humans to hold personal beliefs in the state of nature, in *Sec. 3.2.1*; the natural evolution of humans to observe and analyze things in the state of nature, albeit in a primitive way, in *Sec. 3.2.2*; the natural evolution of humans to measure things in the state of nature in a primitive way, in *Sec. 3.3.1*; the natural evolution of humans to observe things in the state of nature, albeit in a primitive way, in *Sec. 3.3.2*; the natural evolution of humans to communicate with each other in the state of nature, albeit in a primitive way, in *Sec. 3.4.1*; the natural evolution of humans to understand "space, form-structure, and number-proportion" in the state of nature, albeit in a primitive way, in *Sec. 3.4.2*; the natural evolution of humans to experience consciousness in the state of nature, in *Sec. 3.5.1*; and the natural evolution of humans to have consciousness in the state of nature, in *Sec. 3.5.2*).

On the other hand, there is transformation (e.g., the technical transformation of human ability to process information by the invention of mental models, in *Sec. 2.2.1*; the technical transformation of human ability to reason without emotion by the invention of an "objective" account of rationality without the distortion of emotion in philosophy, in *Sec. 2.2.2*; the technical transformation of human ability to distinguish "separate" or "distinct" events by the invention of discrete probability theory, in *Sec. 2.3.1*; the technical transformation of human ability to understand different parts of things by the invention of complexity theory, in *Sec. 2.3.2*; the technical transformation of human ability to cope with failures by the invention of reliability engineering, in *Sec. 2.4.1*; the technical transformation of human ability to make decision by the invention of the mathematical formulation of optimal decision, in *Sec. 2.4.2*; the technical transformation of human ability to think by the invention of propositional calculus, in *Sec. 2.5.1*; the technical transformation of human ability to understand variation of events by the invention of "common cause" and "special cause" in mathematics and statistics, in *Sec. 2.5.2*; the technical transformation of human ability to understand personal beliefs by the invention of Bayesian methods in the study of probability, in *Sec. 3.2.1*; the technical transformation of human ability to observe and analyze things by the invention of Q methodology, in *Sec. 3.2.2*; the technical transformation of human ability to understand measurement by the invention of "metrology" for studying measurement uncertainty, in *Sec. 3.3.1*; the technical transformation of human ability to observe things in nature by the invention of quantum mechanics, in *Sec. 3.3.2*; the technical transformation of human ability to communicate with each other by the invention of the theory of "communicative action," in *Sec. 3.4.1*; the technical transformation of human ability to understand "space, form-structure, and number-proportion" by the invention of formal mathematics, especially when applied to physics, in *Sec. 3.4.2*; the technical transformation of human ability to understand consciousness by the invention of phenemonology by Husserl and, later on, others, in *Sec. 3.5.1*; and the technical transformation of human ability to understand consciousness by the invention of different philosophical traditions on the "subject," in *Sec. 3.5.2*).

And the reverse direction also holds true.

Again, it should be stressed, however, that there are different shades of gray (or different degrees of truth) in the two opposites, and the classification is not necessarily mutually exclusive either. In fact, even when some combinations of the two occur, they only end up sharing the same dialectic relationship, but in a different degree.

Also, there are some other relationships (as a kind of family resemblance) which have something in common with the principle, but they are not exactly the same but only more or less comparable, which varies from case to case.

Good examples of family resemblance in relation to the principle are inorganicness-volition, naturalness-technologicalness, naturalness-nonnaturalness, nonwillingness-willingness, inorganicness-motivation, evolution-creation, naturalness-artificialness, and so on.

20th Thesis: The Symmetry-Asymmetry Principle

The twentieth thesis is labeled as the symmetry-asymmetry principle (on the relationships among existents) in the category of "agency" in existential dialectics (which was also already worked out in my previous books), in that there is no asymmetry without symmetry—and vice versa.

For instance, the Same can be symmetric and asymmetric towards the Others. But in case of asymmetry (as analyzed in *BDPD*), oppression and self-oppression can occur. So, when the Same is asymmetric towards the Others, the Same can also be relatively asymmetric towards itself in self-oppression, just as the Others can be likewise towards themselves. The subsequent oppressiveness is dualistic, as much by the Same against the Others and itself, as by the Others against the Same and themselves. Both oppression and self-oppression can be achieved by way of downgrading differences between the Same and the Others and of accentuating them.

This is true, even though not all forms of asymmetry have to be about oppression and self-oppression.

In addition, from Chapter Three of *FPHG*, symmetry is not perfect, to be understood in an approximate sense under many life circumstances. With this clarification in mind, hereafter are some empirical examples.

For instance, on the one hand, there is symmetry (e.g., the co-existence of animals and humans, in *Sec. 2.2.1*; the co-existence of animals and humans, in *Sec. 2.2.2*; both animals and humans can perceive separate or distinct things in nature, in *Sec. 2.3.1*; the co-existence of different problems studied by those in complexity study, in *Sec. 2.3.2*; the co-existence of different risks in reliability testing, in *Sec. 2.4.1*; the co-existence of different everyday decision techniques, like "authority" and "mistrusted authorites" in the article aforecited, in *Sec. 2.4.2*; the co-existence of humans and animals, in *Sec. 2.5.1*; the co-existence of different views on uncertainty and prediction in economics, in *Sec. 2.5.2*; the existence of different ways to study probability, like Bayesian statistics and frequentist statistics, in *Sec. 3.2.1*; the co-existence of different analyses, like Q analysis, Cultural Consensus Theory, etc., in *Sec. 3.2.2*; the coexistence of different ways for evaluating measurement uncertainties, in *Sec. 3.3.1*; existence of different ways to understand reality, like classical mechanics and quantum mechanics, in *Sec. 3.3.2*; the symmetric status among participants in the idealized world of communicative discourse for consensus and mutual understanding, in *Sec. 3.4.1*; the existence of different ways of doing mathematics, in *Sec. 3.4.2*; the existence of different ways to do phenomenology, like Husserl's and Heidegger's, in *Sec. 3.5.1*; and the existence of different ways to understand the "subject," like German idealism in the older days and postmodernism in the 20th century unto today, in *Sec. 3.5.2*).

On the other hand, there is asymmetry (e.g., humans can create systematic rules of inference, princples of mental models, etc.—but animals cannot, in *Sec. 2.2.1*; humans can propose different systematic models of rationality—but animals cannot, in *Sec. 2.2.2*; human can invent discrete probability theory to understand discrete probability distributions—but animals cannot, in *Sec. 2.3.1*; some problems in complexity study are "tractable"—but some others are "intractable," in *Sec. 2.3.2*; "the risk of incorrectly accepting a bad design" in "type 2 error"—but "the risk of incorrectly rejecting a good design" in type 1 error," in *Sec. 2.4.1*; some individuals follow "authority"—but others "mistrust authorities," in *Sec. 2.4.2*; humans have the intellectual power to create propositional logic—but animals do not, in *Sec. 2.5.1*; the view of the free market to reach "the equilibrium of full employment" is more influential among those in the Austrian School —but the view about the "inherent unpredictability of economic systems" is more influential among those like Keynes, in *Sec. 2.5.2*; the popularity of Bayesian statistics in the fields of "machine learning and talent analytics"—but the popularity of frequentist statistics in "most undergraduate teaching," in *Sec. 3.2.1*; the popularity of Q analysis among those who research on how "people think about ideas in relation to other ideas, rather than in isolation"—but the popularity of Cultural Consensus Theory among those who research on "the cultural 'truth' of the statements used in the test," in *Sec. 3.2.2*; the popularity of the simple way of using only "random variables" and "simple probability distributions" for evaluating measurement uncertainties among those in mainstream metrology—but the popularity of the more complicated way of using "a mathematical interval rather than a probability distribution," together with multivariate mathematical modeling, for evaluating measurement uncertainties among the critics, in *Sec. 3.3.1*; the popularity of quantum mechanics among 20th century physicists, especially in regard to the micro-world of particles—but the popularity of classical mechanics among 19th century physicists, especially in regard to the macro-world of everyday life, in *Sec. 3.3.2*; the asymmetric status among participants in the realistic world of strategic action for power and influence, in *Sec. 3.4.1*; the popularity of formal mathematics among those professional mathematicians in contemporary mainstream Western society—but the popularity of folklore mathematics among those "aboriginal or ancient peoples" in underdeveloped countries, in *Sec. 3.4.2*; the popularity of Husserl's phenomenology among those who are more interested in psychology—but the popularity of Heidegger's phenomenology among those who are more interested in ontology, in *Sec. 3.5.1*; and the popularity of German idealism in the older days of the 18th and 19th centuries—but the popularity of postmodernism in the second half of the 20th century unto today, in *Sec. 3.5.2*).

And the reverse direction also holds true.

Again, for the last time, it should be stressed, however, that there are different shades of gray (or different degrees of truth) in the two opposites, and the classification is not necessarily mutually exclusive either. In fact, even when some combinations of the two occur, they only end up sharing the same dialectic relationship, but in a different degree.

Also, there are some other relationships (as a kind of family resemblance) which have something in common with the principle, but they are not exactly the same but only more or less comparable, which varies from case to case.

Good examples of family resemblance in relation to the principle are balance-extremity, reflexiveness-unreflexiveness, equality-inequality, harmoniousness-unharrmoniousness, center-periphery, compatibility-incompatibility, proportion-disproportion, balance-unbalance, and something like that.

21st Thesis: The Softness-Hardness Principle

The twenty-first thesis refers to the softness-hardness principle (on the force of change) in the category of "agency" in existential dialectics (which was first worked out in *ALD*), in that any change by an agent, be it organic (like humans) or non-organic (like natural objects), can occur in a forceful (aggressive) or gentle (pacific) way, which can come in all shapes and sizes, of course. There is no softness without hardness—and vice versa.

For instance, on the one hand, there is softness (e.g., the inclusion of certain bits of informaton in mental models, when they are selected, in *Sec. 2.2.1*; the inclusion of emotion in the understanding of rationality by the feminists, in *Sec. 2.2.2*; the defense of the "classical definition" of probability by James Bernoulli and Pierre-Simon Laplace, in *Sec. 2.3.1*;

the defense of complexity study by the advocates, in *Sec. 2.3.2*; the defense of focusing on "reliability" by the advocates, in *Sec. 2.4.1*; some individuals "acquiesce to a person in authority or an 'expert,' just following orders," in *Sec. 2.4.2*; the defense of syllogistic logic by Aristotle in ancient Greece, in *Sec. 2.5.1*; the defense of the free market to reach "the equilibrium of full employment" by those in the Austrian School, in *Sec. 2.5.2*; the defense of the Dutch book approach by de Finetti, in *Sec. 3.2.1*; the defense of Q methodology by William Stephenson, in *Sec. 3.2.2*; the defense of using only "random variables" and "simple probability distributions" for evaluating measurement uncertainties by those in mainstream metrology aforecited, in *Sec. 3.3.1*; the defense of the uncertainty principle by Heisenberg, in *Sec. 3.3.2*; consensus and mutual understanding, in *Sec. 3.4.1*; the defense of social constructionism by Ian Hackings, in *Sec. 3.4.2*; the defense of phenomenology by Husserl, in *Sec. 3.5.1*; and the defense of "a unitary, autonomous Subject" by those in "the liberal theory of the social contract," in *Sec. 3.5.2*).

On the other hand, there is hardness (e.g., the exclusion of certain bits of information in mental models, when they are ignored in the process of "filtering" info, in *Sec. 2.2.1*; the exclusion of emotion in the understanding of rationality by philosophers like A. C. Grayling, in *Sec. 2.2.2*; the critique of the "classical definition" of probability by John Venn and George Boole, in *Sec. 2.3.1*; the critique of complexity study by the critics, in *Sec. 2.3.2*; the critique of focusing on reliability by the critics, like those who focus on "safety" instead, in *Sec. 2.4.1*; some individuals take "the most opposite action compared to the advice of mistrusted authorities [parents, police officers, partners]," in *Sec. 2.4.2*; the critique of syllogistic logic by Gottlob Frege for propositional calculus in modern times, in *Sec. 2.5.1*; the critique of the free market to reach "the equilibrium of full employment" by those in the Keynesian School, in *Sec. 2.5.2*; the critique of the Dutch book approach by Ian Hacking, in *Sec. 3.2.1*; the critique of Q methodology by the critics, as shown in the problems aforecited, in *Sec. 3.2.2*; the critique of using only "random variables" and "simple probability distributions" for evaluating measurement uncertainties by the critics aforecited, in *Sec. 3.3.1*; the critique of the uncertainty principle by Einstein, in *Sec. 3.3.2*; power and influence, in *Sec. 3.4.1*; the critique of social constructionism by Boghossian, in *Sec. 3.4.2*; the critique of phenomenology by Garfinkel, in *Sec. 3.5.1*; and the critique of "a unitary, autonomous Subject" by those in contemporary postmodernism, in *Sec. 3.5.2*).

And the reverse direction also holds true.

Again, it should be stressed, however, that there are different shades of gray (or different degrees of truth) in the two opposites, and the classification is not necessarily mutually exclusive either. In fact, even when some combinations of the two occur, they only end up sharing the same dialectic relationship, but in a different degree.

Also, there are some other relationships (as a kind of family resemblance) which have something in common with the principle, but they are not exactly the same but only more or less comparable, which varies from case to case.

Good examples of family resemblance in relation to the principle are peacefulness-violence, cooperation-competition, pacificity-aggression, dovishness-hawkishness, reward-punishment, peace-war, friendliness-unfriendliness, kindness-hostility, love-hatred, acceptance-discrimination, admiration-belittlement, inclusion-exclusion, like-dislike, and the like.

22nd Thesis: The Seriousness-Playfulness Principle

The twenty-second thesis refers to the seriousness-playfulness principle (on the extent of seriousness) in the category of "agency" in existential dialectics (which is first worked out in *FPHH*), in that, if there is seriousness, there is also playfulness, especially for any agent with some kind of intelligence life, be it about humans, animals, or, later, post-humans. There is no seriousness without playfulness—and vice versa.

Lest any misunderstanding occurs, the word "playfulness" here should not be confused with other terms like "laughter," "smile," "joke," "tickleness," and "laugh-like vocalization" (in some animals), for instance.

Being playful can come in all shapes and sizes and therefore is not necessarily tied up with any of these terms, alhough it can result from any of them, or a combination of them, or something else altogether.

This principle, together with some other principles (especially though not exclusively, like the formalness-informalness principle), have important implications for what constitutes rationality, be it about "rationality of application" at the practical level or "rationality of knowledge" at the meta-theoretical level (as summarized in *Sec. 4.20* of *FPHH*).

For instance, on the one hand, there is seriousness (e.g., the serious business of those who try to figure out how humans anticipate events, in *Sec. 2.2.1*; the serious business of those who try to understand the nature of rationality, in *Sec. 2.2.2*; the serious business of those who try to understand probability, in *Sec. 2.3.1*; the serious business of those who try to understand complexity, in *Sec. 2.3.2*; the serious business of those who work on reliability models, in *Sec. 2.4.1*; the serious business of those who try to make good decisions, in *Sec. 2.4.2*; the serious business of those who try to understand inference rules, in *Sec. 2.5.1*; the serious business of those who try to understand the causes of history, in *Sec. 2.5.2*; the serious business of those who try to understand the nature of probability, in *Sec. 3.2.1*; the serious business of those who research on the nature of analysis, in *Sec. 3.2.2*; the serious business of those who try to understand the business of evaluating uncertainty, in *Sec. 3.3.1*; the serious business of those physicists who try to understand quantum mechanics, in *Sec. 3.3.2*; the serious business of those who try to understand human communication, in *Sec. 3.4.1*; the serious business of those who try to understand mathematics, in *Sec. 3.4.2*; the serious business of those who try to understand consciousness, in *Sec. 3.5.1*; and the serious business of those who try to understand the nature of the "subject," in *Sec. 3.5.2*).

On the other hand, there is playfulness (e.g., the playful part of those who try to figure out how humans anticipate events, when they play around with different options over time, like "formal rules of inference," "domain-specific rules of inference," "mental models," "probabilities," etc., in *Sec. 2.2.1*; the playful part of those who try to understand the nature of rationality, when they play around with different views about rationality, in relation to the theories, levels, interpretations, etc., in *Sec. 2.2.2*; the playful part of those who try to understand probability, when they play around with different "interpretations" over time, like the "classical" and "modern" definitions aforecited, in *Sec. 2.3.1*; the playful part of those who try to understand complexity, when they play around with different topics and/or different fields over time, like "complex behaviour," "complex mechanisms," "complex simulations," "complex systems," "complexity in data," "complexity in molecular recognition," "complexity management," etc., in *Sec. 2.3.2*; the playful part of those who work on reliability models, when they play around with different approaches to modeling over time, like "the physics of failure approach," "the parts stress modeling approach," etc., in *Sec. 2.4.1*; the playful part of those who try to make good decisions, when they play around with different "everyday decision techniques" over time as cited earlier, in *Sec. 2.4.2*; the playful part of those who try to understand inference rules, when they play around with different options over time, like propositional calculus, higher-order logics, modal logic, many-valued logics, incompleteness theorem, everyday decision techniques, mental models, etc., in *Sec. 2.5.1*; the playful part of those who try to understand the causes of history, when they play around with different interpretations over time, like "frequency probability," "subjective or Bayesian probability," and "special causes," in *Sec. 2.5.2*; the playful part of those who try to understand the nature of probability, when they play around with different interpretations and versions over time, like the "objectivist" interpretation, the "subjectivist" interpretation, the two different versions of Bayesian methods, the "eclectic" alternative, etc., in *Sec. 3.2.1*; the playful part of those who research on the nature of analysis, when they play around with different approaches over time, like R analysis, Q analysis, Cultural Consensus Theory, etc., in *Sec. 3.2.2*; ,the playful part of those who try to understand the business of evaluating uncertainty, when they play around with different proposals and interpretations over time, like the conventional way of using only "random variables" and "simple probability distributions," the more robust way of using "interval [*a*,*b*]," the "subjectivist" interpretation, the "objectivist" interpretation, etc., in *Sec. 3.3.1*; the playful part of those physicists who try to understand quantum mechanics, when they play around with different versions and interpretations over the decades, like "matrix mechanics," "wave mechanics," "statistics of subatomic particles," "the Copenhagen interpretation," "many-worlds interpretation," "philosophical speculation about the role of the observer," "quantum chemistry," "quantum electronics," "quantum optics," "quantum information science," etc., in *Sec. 3.3.2*; the playful part of those who try to understand human communication, when they play around with different options over time, like "communicative action," "strategic action," "communicative discourse," etc., in *Sec. 3.4.1*; the playful part of those who try to understand mathematics, when they play around with different interpretations over time, like formal mathematics, quasi-empiricism in mathematics, mathematical folklore, informal mathematics, etc., in *Sec. 3.4.2*; the playful part of those who try to understand consciousness, when they play around with different versions over the decades, like the ones by Husserl, Heidegger, Merleau-Ponty, Edith Stein, Max Scheler, Nicolai Hartmann, Jean-Paul Sartre, Paul Ricoeur, Emmanuel Levinas, Alfred Schütz and Eric Voegelin, in *Sec. 3.5.1*; and the playful part of those who try to understand the nature of the "subject," when they play around with different versions in different philosophical traditions over time, like German idealism, postmodernism, analytic philosophy, the liberal theory of the social contract, etc., in *Sec. 3.5.2*).

And the reverse direction also holds true.

Again, it should be stressed, however, that there are different shades of gray (or different degrees of truth) in the two opposites, and the classification is not necessarily mutually exclusive either. In fact, even when some combinations of the two occur, they only end up sharing the same dialectic relationship, but in a different degree.

Also, there are some other relationships (as a kind of family resemblance) which have something in common with the principle, but they are not exactly the same but only more or less comparable, which varies from case to case.

Good examples of family resemblance in relation to the principle are seriousness-jokingness, seriousness-humorousness, seriousness-wittiness, seriousness-nonseriousness, Apollo-Dionysus, seriousness-restingness, seriousness-relaxedness, seriousness-clownishness, and so forth.

23rd Thesis: The Activeness-Inactiveness Principle

The twenty-third thesis refers to the activeness-inactiveness principle (on the extent of activeness) in the category of "agency" in existential dialectics (which was first worked out in *FPHTRAN*), in that, if there is activeness, there is also inactiveness, especially (though not exclusively) for any agent with some kind of intelligence life, be it about humans, animals, or, later, post-humans. There is no activeness without inactiveness—and vice versa.

Being active can come in all shapes and sizes, just as a person (or an animal) can be inactive because he (or it) is tired, old, injured, sick, sleepy (or hibernating), uninterested, in a state of passive inertia (or disposition), etc.—and an object can be inactive because it is at rest, out of work (or order), etc.

This principle, together with some other principles (especially though not exclusively, like the formalness-informalness principle and seriousness-playfulness principle), have important implications for what constitutes rationality, be it about "rationality of application" at the practical level or "rationality of knowledge" at the meta-theoretical level (as summarized in *Sec. 4.20* of *FPHH*).

For instance, on the one hand, there is activeness (e.g., the relative activeness of those in mental models when they filter info to make sense of reality, in *Sec. 2.2.1*; the relative activeness of those experts who excel in making good use of rationality in practice, in *Sec. 2.2.2*; the relative activeness of a dice when it is still new and can be used by players for role-playing games, for example, in *Sec. 2.3.1*; the relative activeness of complexity study when it can be used to solve some problems which are tractable, in *Sec. 2.3.2*; the relative activeness of those products which perform well in reliability testing, in *Sec. 2.4.1*; the relative activeness of those who are able to make a lot of decisions based on satisficing techniques, in *Sec. 2.4.2*; the relative activeness of those followers of propositional calculus in the community of its advocates, in *Sec. 2.5.1*; the relative activeness of those who are lucky to be able to work, even when the free market does not reach "the equilibrium of full employment" due to "special causes" contributing to the "inherent unpredictability of economic systems," in *Sec. 2.5.2*; the relative activeness of those for Bayesian methods, when they actively practice them in the fields of "machine learning and talent analytics," in *Sec. 3.2.1*; the relative activeness of those in favor of Q methodology when it is widely used in fields like "nursing, veterinary medicine, public health, transportation, education, rural sociology, hydrology and mobile communication," in *Sec. 3.2.2*; the relative activeness of those in favor of using "random variables" and "simple probability distributions" for evaluating measurement uncertainties, when they try to apply them in simple situations, in *Sec. 3.3.1*; the relative activeness of those physicists who actively try to apply quantum mechanics to the micro-world of particles, in *Sec. 3.3.2*; the relative activeness of those who seek "mutual understanding" in a situation of "communicative action," in *Sec. 3.4.1*; the relative activeness of those in formal mathematics who actively try to do "formal and strict proofs of all statements from given axioms," in *Sec. 3.4.2*; the relative activeness of those followers of Husserl's phenomenology to try to reach "pure consciousness," in *Sec. 3.5.1*; and the relative activeness of those who defend "a unitary, autonomous Subject" in the community of "the liberal theory of the social contract," in *Sec. 3.5.1*).

On the other hand, there is inactiveness (e.g., the relative inactiveness of those who suffer from such "restrictions" like old age and failing working memory, so that they cannot effectively use mental models, in *Sec. 2.2.1*; the relative inactiveness of those humans who cannot be rational "in practice, in that, although they "have the competence to be rational but their performance is limited by various factors," in *Sec. 2.2.2*; the relative inactiveness of a dice for role-playing games when it wears so much that it renders "the die unusable," in *Sec. 2.3.1*; the relative inactiveness of complexity study when it cannot be used to solve some problems which are intractable, in *Sec. 2.3.2*; the relative inactiveness of those products which fail in reliability testing, in *Sec. 2.4.1*; the relative inactiveness of those who are not able to make a lot of decisions, because they follow the optimal decision making approach which requires "a great deal of time" and a "great deal of information," etc., in *Sec. 2.4.2*; the relative inactiveness of those followers of propositional calculus, when they are not welcome in the community of those who follow the alternative theory of "domain-specific or content-sensitive rules of inference," in *Sec. 2.5.1*; the relative inactiveness of those who are not lucky and cannot find work when the free market does not reach "the equilibrium of full employment" due to "special causes" contributing to the "inherent unpredictability of economic systems," in *Sec. 2.5.2*; the relative inactiveness of those for Bayesian methods, when they are not welcome in "most undergraduate teaching," which "is still based on frequentist statistics," in *Sec. 3.2.1*; the relative inactiveness of those in favor of Q methodology, when it is not used in other fields like "marketing, product management, operations research, and other applied sciences that deal with large quantities of data," where R analysis is more widely used instead, in *Sec. 3.2.2*; the relative inactiveness of those in favor of using "random variables" and "simple probability distributions" for evaluating measurement uncertainties, when they confront difficult situations which require different practices, like a huge number of output quantities and input quantities, "periodic measurements, binned data values, censoring, detection limits,...plus-minus ranges of measurements," etc., which then require alternative ways of evaluating measurement uncertainty, in *Sec. 3.3.1*; the relative inactiveness of those physicists who have difficutly to apply quantum mechanics to the macro-world of gravity, where the theory of relativity is more relevant instead, in *Sec. 3.3.2*; the relative inactiveness of those who seek "mutual understanding" in a situation of "strategic action," since they will be frustrated, discouraged, and even ignored, in *Sec. 3.4.1*; the relative inactiveness of those in formal mathematics when their "formal and strict proofs of all statements from given axioms" are not appropriate among those "aboriginal or ancient peoples," so they may well get frustrated and cannot do much there, in *Sec. 3.4.2*; the relative inactiveness of those followers of Husserl's phenomenology to try to reach "pure consciousness" in the community of ethnomethodology, since they do not quite fit in the latter's project for the understanding of the reproduction of social order, in *Sec. 3.5.1*; and the relative inactiveness of those in the community of "the liberal theory of the social contract," when they are not welcome in the postmodernist circle, as the latter camp is more concerned with "the subject as a social construction" or as "an ideological construction" instead, in *Sec. 3.5.2*).

And the reverse direction also holds true.

Again, it should be stressed, however, that there are different shades of gray (or different degrees of truth) in the two opposites, and the classification is not necessarily mutually exclusive either. In fact, even when some combinations of the two occur, they only end up sharing the same dialectic relationship, but in a different degree.

Also, there are some other relationships (as a kind of family resemblance) which have something in common with the principle, but they are not exactly the same but only more or less comparable, which varies from case to case.

Good examples of family resemblance in relation to the principle are alertness-unalertness, activeness-decrepidness, alertness-sleepiness, activeness-tiredness, alertness-unwareness, alertness-uninterestedness, soberness-inactiveness, alertness-inactiveness, soberness-unawareness, awakenness-hibernatedness, and so on.

24th Thesis: The Selfness-Otherness Principle

The twenty-fourth thesis refers to the selfness-otherness principle (on the magnitude of self-centeredness) in the category of "agency" in existential dialectics (which is first worked out in *FPHCA*), in that, if there is the tendency to be mostly centered on one's self or group, there is likewise the tendency to be mostly centered on others, especially (though not exclusively) for any agent with some kind of intelligence life, be it about humans, animals, or, later, post-humans. There is no selfness without otherness—and vice versa.

Being self-centered and being other-centered do not have to be mutually exclusive, as the two can come in all shapes and sizes, or in different degrees. And other-centeredness does not entail desirability, any more than self-centeredness implies non-desirability, as both can go either way.

In addition, whether self-centeredness constitutes a moral alternative or, conversely, an alternative to morality is of course debatable (as addressed in my other book titled *Beyond Ethics to Post-Ethics*).

For instance, on the one hand, there is selfness (e.g., the tendency to be mostly centered on one's self or group, as shown by those in mental models to simplify reality so as to make sense of it for themselves, in *Sec. 2.2.1*; the tendency to be mostly centered on one's self or group, as shown by those who follow a model of rationality in which benefiting oneself is optimal," in *Sec. 2.2.2*; the tendency to be mostly centered on one's self or group, as shown by those who play dice games like *Dungeons & Dragons* for their personal enjoyment, in *Sec. 2.3.1*; the tendency to be mostly centered on one's self or group, as shown by those in business who uses "complexity management" so as to "efficiently control value-adding complexity" for profits, in *Sec. 2.3.2*; the tendency to be mostly centered on one's self or group, as shown in the focus on reliability by those in a company which is "only concerned with cost" to minmize the "loss of revenue for the company," in *Sec. 2.4.1*; the tendency to be mostly centered on one's self or group, as shown by those who follow the "political" everyday technique to negotiate choices for their own interest groups, in *Sec. 2.4.2*; the tendency to be mostly centered on one's self or group, as shown in the "controversy" surrounding the credit for "the invention of truth-tables," as there has been disputes as to who should receive the honor to invent it first, in *Sec. 2.5.1*; the tendency to be mostly centered on one's self or group, as shown in the emphasis by those in the Austrian School on the individual choices of economic agents to pursue their own self-interests in the free market, in *Sec. 2.5.2*; the tendency to be mostly centered on one's self or group, as shown in "a clever gambler," who "places a set of bets that guarantee a profit, no matter what the outcome is of the bets," such that "a Dutch book is made," in *Sec. 3.2.1*; the tendency to be mostly centered on one's self or group, as shown in Q analysis, when the "norm" for a factor by an individual "may be indicative of the wisest solution, or the most responsible, the most important, or an optimized-balanced solution," which he most desires for his own well-being, in *Sec. 3.2.2*; the tendency to be mostly centered on one's self or group, as shown in those manufacturing firms, which use metrology for their products, like "the bathroom scale" which "convert[s] a measured extension of a spring into an estimate of the measurand, the mass of the person on the scale," in *Sec. 3.3.1*; the tendency to be mostly centered on one's self or group, as shown in the wishful attempt by some people to use the uncertainty principle for the understanding of whether they have free will or not, even though the critics had showed that "nontrivial biological mechanisms requiring quantum mechanics are unlikely, due to the rapid decoherence time of quantum systems at room temperature," in *Sec. 3.3.2*; the concern with "influence" for "private interests," in *Sec. 3.4.1*; the tendency to be mostly centered on one's self or group, as shown in the use of informal mathematics for "commerce, engineering, calendar creation," etc. among those "aboriginal or ancient peoples," in *Sec. 3.4.2*; the tendency to be mostly centered on one's self or group, as shown in the confession by Garfinkel who spoke of "phenomenological texts and findings as being, 'appropriated' and intentionally 'misread,' for the purposes of exploring topics in the study of social order," in *Sec. 3.5.1*; and the tendency to be mostly centered on one's self or group, as shown in the editors of *Social Text*, who accepted a "nonsensical" article by Alan Sokal, who "flattered the editors' ideological preconceptions" in the so-called "Sokal hoax," in *Sec. 3.5.2*).

On the other hand, there is otherness (e.g., the tendency to be mostly centered on others, as shown by those scholars of optimal decision-making models to take the big trouble of going through complex information for a better understanding of the world, as an advance in human knowledge, in *Sec. 2.2.1*; the tendency to be mostly centered on others, as shown by those who follow a different model rationality "in which benefiting the group is optimal," in *Sec. 2.2.2*; the tendency to be mostly centered on others, as shown by the Pythagoreans who "used the Platonic solids as dice" in their quest for understanding "the universe through an understanding of geometry," as an advancement in knowledge for humanity, in *Sec. 2.3.1*; the tendency to be mostly centered on others, as shown in the work by Weaver on the distinction between "disorganized complexity" and "organized complexity," as an advancement in knowledge for humanity, in *Sec. 2.3.2*; the tendency to be mostly centered on others, as shown in the focus on "safety" by those in government for the welfare of the larger "society," in *Sec. 2.4.1*; the tendency to be mostly centered on others, as shown by those scholars who contribute to the rational decision making model, as an advancement in the knowledge of organizational behavior for humanity, in *Sec. 2.4.2*; the tendency to be mostly centered on others, as shown in the efforts of Gottfried Leibniz, who "has been credited with being the founder of symbolic logic for his work with the calculus ratiocinator" but "it was unknown to the larger logical community" at the time and was not much appreciated then, in *Sec. 2.5.1*; the tendency to be mostly centered on others, as shown in the call for government intervention for the benefits of larger society by those in the Keynesian School to fix the problems of the free market, when it cannot reach "the equilibrium of full employment," in *Sec. 2.5.2*; the tendency to be mostly centered on others, as shown in the Dutch book argument proposed by de Finetti, as a contribution to the advance in the scientific knowledge of probability for humanity, in *Sec. 3.2.1*; the tendency to be mostly centered on others, as shown in Q analysis, when it contributes to the understanding of how "people think about ideas in relation to other ideas, rather than in isolation," as an advance in the knowledge of psychology for humanity, in *Sec. 3.2.2*; the tendency to be mostly centered on others, as shown in the Joint Committee for Guides in Metrology to provide guidelines for the advance in the knowledge of measurement uncertainty for humanity, in *Sec. 3.3.1*; the tendency to be mostly centered on others, as shown in the work on the uncertainty principle by Heisenberg, as an advance in the knowledge of physics for humanity, in *Sec. 3.3.2*; the concern with "consensus" for "mutual understanding," in *Sec. 3.4.1*; the tendency to be mostly centered on others, as shown in the work of Eugene Wigner on mathematics and physics in 1960, as a contribution to the advance in mathematical knowledge for humanity, in *Sec. 3.4.2*; the tendency to be mostly centered on others, as shown in the philosophical work of Husserl for the study of "the structures of subjective experience and consciousness," as an advance in knowledge for humanity, in *Sec. 3.5.1*; and the tendency to be mostly centered on others, as shown in "the resultant academic and public quarrels" after the "Sokal hoax," in regard to "the scholarly merit of humanistic commentary about the physical sciences; the influence of postmodern philosophy on social disciplines in general; academic ethics, including whether Sokal was right or wrong to deceive the editors and readers of Social Text; and whether the journal had exercised the appropriate intellectual rigor before publishing the pseudoscientific article"—as a way to improve the "intellectual rigor" in fuure publications, as an advance in knowledge for humanity, in *Sec. 3.5.2*).

And the reverse direction also holds true.

Again, it should be stressed, however, that there are different shades of gray (or different degrees of truth) in the two opposites, and the classification is not necessarily mutually exclusive either. In fact, even when some combinations of the two occur, they only end up sharing the same dialectic relationship, but in a different degree.

Also, there are some other relationships (as a kind of family resemblance) which have something in common with the principle, but they are not exactly the same but only more or less comparable, which varies from case to case.

Good examples of family resemblance in relation to the principle are egoisticness-altruisticness, selfishness-unselfishness, selfness-selflessness, egocentricness-othercentricness, and so on.

In Relation to Outcome

The next six theses are in relation to outcome in existential dialectics, as shown below.

25th Thesis: The Regression-Progression Principle

The twenty-fifth thesis is called the regression-progression princple (on the direction of history) in the category of "outcome" in existential dialectics (which was also already worked out in my previous books), in that neither the cyclical nor the linear views are adequate for explaining many phenomena at all levels. There is no regression without progression—and vice versa.

History progresses to more advanced forms, but with a regressive touch. Examples include no freedom without unfreedom, no equality without inequality, and no civilization without barbarity. This is not an inevitable law, but merely a highly likely empirical trend.

For instance, on the one hand, there is regression (e.g., the regression made by mental models, as shown in the problems pointed out by the critics, in *Sec. 2.2.1*; the regression made by the argument on emotion and rationality, as shown in the problems pointed out by the critics aforecited, in *Sec. 2.2.2*; the regression made by discrete probability theory, as shown in the problems pointed out by the critics, in *Sec. 2.3.1*; the regression made by complexity study, as shown in the problems pointed out by the critics aforecited, in *Sec. 2.3.2*; the regression made by reliability engineering, as shown in the problems pointed out by the critics, in *Sec. 2.4.1*; the regression made by optimal decision, as shown in the problems pointed out by the critics, in *Sec. 2.4.2*; the regression made by propositional calculus, as shown in the problems pointed out by the critics, in *Sec. 2.5.1*; the regression made by the study on history and common causes, as shown in the problems pointed out by the critics, in *Sec. 2.5.2*; the regression made by Bayesian methods, as shown in the problems pointed out by the critics, in *Sec. 3.2.1*; the regression made by Q methodology, as shown in the problems pointed out by the critics, in *Sec. 3.2.2*; the regression made by the study of measurement uncertainty, as shown in the problems pointed out by the critics, in *Sec. 3.3.1*; the regression made by the uncertainty principle, as shown in the problems pointed out by the critics, in *Sec. 3.3.2*; the regression made by the theory of communicative action, as shown in the problems pointed out by the critics aforecited, in *Sec. 3.4.1*; the regression made by the sociology of mathematical knowledge, as shown in the problems pointed out by the critics, in *Sec. 3.4.2*; the regression made by Husserl's phenomenology, as shown in the problems pointed out by the critics aforecited, in *Sec. 3.5.1*; and the regression made by the philosophical traditions, as shown in the problems pointed out by the critics aforecited, in *Sec. 3.5.2*).

On the other hand, there is progress (e.g., the progress made by mental models, as shown in its contribution to the effort of making sense of reality, in *Sec. 2.2.1*; the progress made by the argument on emotion and rationality, as shown in its contribution to the understanding of human reasoning aforecited, in *Sec. 2.2.2*; the progress made by discrete probability theory, as shown in its contribution to the understanding of discreteness aforecited, in *Sec. 2.3.1*; the progress made by complexity study, as shown in its contribution to the understanding of the linkages among different parts in a system aforecited, in *Sec. 2.3.2*; the progress made by reliability engineering, as shown in its contribution to the enhancement of reliability aforecited, in *Sec. 2.4.1*; the progress made by optimal decision, as shown in its contribution to the understanding of finding the best option available in decision making aforecited, in *Sec. 2.4.2*; the progress made by propositional calculus, as shown in its contribution to the understanding of inference rules aforecited, in *Sec. 2.5.1*; the progress made by the study on history and common causes, as shown in its contribution to the understanding of "the usual, historical, quantifiable variation in a system" aforecited, in *Sec. 2.5.2*; the progress made by Bayesian methods, as shown in their contribution to the fields of "machine learning and talent analytics" aforecited, in *Sec. 3.2.1*; the progress made by Q methodology, as shown in its contribution to the understanding of how "people think about ideas in relation to other ideas, rather than in isolation," in *Sec. 3.2.2*; the progress made by the study of measurement uncertainty, as shown in the contribution to the understanding of systematic and random errors aforecited, in *Sec. 3.3.1*; the progress made by the uncertainty principle, as shown in its contribution to the understanding of the "tradeoff between the precisions" of the position and momentum of a sub-atomic particle, in *Sec. 3.3.2*; the progress made by the theory of communicative action, as shown in its contribution to the understanding of "mutual understanding" in communication aforecited, in *Sec. 3.4.1*; the progress made by the sociology of mathematical knowledge, as shown in its contribution to the understanding of the social constructionism of mathematics aforecited, in *Sec. 3.4.2*; the progress made by Husserl's phenomenology, as shown in the contribution to the study of "the structures of subjective experience and consciousness" aforecited, in *Sec. 3.5.1*; and the progress made by the philosophical traditions, as shown in the contribution to a better understanding of the evolution of the idea of the "subject" aforecited, in *Sec. 3.5.2*).

And the reverse direction also holds true.

Again, it should be stressed, however, that there are different shades of gray (or different degrees of truth) in the two opposites, and the classification is not necessarily mutually exclusive either. In fact, even when some combinations of the two occur, they only end up sharing the same dialectic relationship, but in a different degree.

Also, there are some other relationships (as a kind of family resemblance) which have something in common with the principle, but they are not exactly the same but only more or less comparable, which varies from case to case.

Good examples of family resemblance in relation to the principle are undesirability-desirability, risk-opportunity, badness-goodness, cost-benefit, dysfunctionality-functionality, cyclicity-linearity, and so on.

26th Thesis: The Same-Difference Principle

The twenty-sixth thesis refers to the same-difference principle (on the metamorphosis of change) in the category of "outcome" in existential dialectics (which was worked out in *ALD* and other books of mine), in that an entity, as it evolves over time, can be both different from and similar to its opposing alternatives and does not have to be solely more different from them over time. There is no similarity without difference—and vice versa.

Opposites are not absolute in a black-or-white fashion; so, an entity can become relatively more similar to (or more different from) its opposite over time.

This is further constrained by another principle, that is, the symmetry-asymmetry principle about the relationships among existents under the category about agency in existential dialectics, in that if there is symmetry (equality) between two entities, there is likewise asymmetry (inequality) emerging in a different way.

For instance, on the one hand, there is similarity in outcome (e.g., the contribution to the molding and control of beliefs, values, and behaviors, regardless of whether this be done by way of mental models, or by way of optimal decision-making models, in *Sec. 2.2.1*; the contribution to the molding and control of beliefs, values, and behaviors, regardless of whether this be done by way of Weberian rationality, or by way of Habermasian rationality, in *Sec. 2.2.2*; the contribution to the molding and control of beliefs, values, and behaviors, regardless of whether this be done by way of playing dice games like *Dungeons & Dragons*, or by way of playing dice games like cleromancy, in *Sec. 2.3.1*; the contribution to the molding and control of beliefs, values, and behaviors, regardless of whether this be done by way of the use of complexity study in mathematics, or by way of the use of complexity study in business, in *Sec. 2.3.2*; the contribution to the molding and control of beliefs, values, and behaviors, regardless of whether this be done by way of focusing on reliability, or by way of focusing on safety, in *Sec. 2.4.1*; the contribution to the molding and control of beliefs, values, and behaviors, regardless of whether this be done by way of rational decision-making model, or by way of bounded rationality, in *Sec. 2.4.2*; the contribution to the molding and control of beliefs, values, and behaviors, regardless of whether this be done by way of two-valued logic, or by way of many-valued logics, in *Sec. 2.5.1*; the contribution to the molding and control of beliefs, values, and behaviors, regardless of whether this be done by way of the Keynesian School, or by way of the Austrial School, in *Sec. 2.5.2*; the contribution to the molding and control of beliefs, values, and behaviors, regardless of whether this be done by way of the objectivist interpretation of probability, or by way of the Bayesian interpretation of probability, in *Sec. 3.2.1*; the contribution to the molding and control of beliefs, values, and behaviors, regardless of whether this be done by way of Q methodology, or by way of Cultural Consensus Theory, in *Sec. 3.2.2*; the contribution to the molding and control of beliefs, values, and behaviors, regardless of whether this be done by way of "modern standards" for measurement, or by way of "arbitrary standards set up by regional or local authorities" in antiquity, in *Sec. 3.3.1*; the contribution to the molding and control of beliefs, values, and behaviors, regardless of whether this be done by way of quantum mechanics, or by way of classical mechanics, in *Sec. 3.3.2*; the contribution to the molding and control of beliefs, values, and behaviors, regardless of whether this be done by way of communicative action, or by way of strategic action, in *Sec. 3.4.1*; the contribution to the molding and control of beliefs, values, and behaviors, regardless of whether this be done by way of formal mathematics, or by way of mathematical folklore, in *Sec. 3.4.2*; the contribution to the molding and control of beliefs, values, and behaviors, regardless of whether this be done by way of Husserl's phenomenology or by way of Garfinkel's phenomenology, in *Sec. 3.5.1*; and the contribution to the molding and control of beliefs, values, and behaviors, regardless of whether this be done by way of "a unitary, autonomous Subject" in "the liberal theory of the social contract," or by way of the "split subject"' in "Freud's psychoanalytic model of the subject" and its further development by Lacan, in *Sec. 3.5.2*).

On the other hand, there is difference in outcome (e.g., the contribution to the molding and control of beliefs, values, and behaviors by way of mental models for a relatively more satisfying view of the world—but the contribution to the molding and control of beliefs, values, and behaviors by way of optimal decision-making models for a relatively more maximizing view of the world, in *Sec. 2.2.1*; the contribution to the molding and control of beliefs, values, and behaviors by way of Weberian rationality for a relatively more value-neutral worldview, as in the relativity of rationality—but the contribution to the molding and control of beliefs, values, and behaviors by way of Habermasian rationality for a relatively more context-specific worldview, as in the critique of social power, in *Sec. 2.2.2*; the contribution to the molding and control of beliefs, values, and behaviors by way of playing dice games like *Dungeons & Dragons* for a relatively more entertaining lifeworld—but the contribution to the molding and control of beliefs, values, and behaviors by way of playing dice games like cleromancy for a relatively more spiritual lifeworld, in *Sec. 2.3.1*; the contribution to the molding and control of beliefs, values, and behaviors by way of the use of complexity study in mathematics for a relatively more logical-formalistic worldview—but the contribution to the molding and control of beliefs, values, and behaviors by way of the use of complexity study in business for a relatively more materialstic worldview, in *Sec. 2.3.2*; the contribution to the molding and control of beliefs, values, and behaviors by way of focusing on reliability for a relatively more efficient lifeworld—but the contribution to the molding and control of beliefs, values, and behaviors by way of focusing on safety for a relatively more effective lifeworld, in *Sec. 2.4.1*; the contribution to the molding and control of beliefs, values, and behaviors by way of rational decision-making model for a relatively more maximizing personalitiy type—but the contribution to the molding and control of beliefs, values, and behaviors by way of bounded rationality for a relatively more satisficing personality type, in *Sec. 2.4.2*; the contribution to the molding and control of beliefs, values, and behaviors by way of two-valued logic for a relatively more simple worldview—but the contribution to the molding and control of beliefs, values, and behaviors by way of many-valued logics for a relatively more complex worldview, in *Sec. 2.5.1*; the contribution to the molding and control of beliefs, values, and behaviors by way of the Keynesian School for a relatively more government-interventionist lifeworld—but the contribution to the molding and control of beliefs, values, and behaviors by way of the Austrial School for a relatively more free-market lifeworld, in *Sec. 2.5.2*; the contribution to the molding and control of beliefs, values, and behaviors by way of the objectivist interpretation of probability for a relatively more scientific worldview—but the contribution to the molding and control of beliefs, values, and behaviors by way of of the Bayesian interpretation of probability for a relatively more psychological perosnal worldview, in *Sec. 3.2.1*; the contribution to the molding and control of beliefs, values, and behaviors by way of Q methodology for a relatively more psychological worldview—but the contribution to the molding and control of beliefs, values, and behaviors by way of Cultural Consensus Theory for a relatively more cultural worldview, in *Sec. 3.2.2*; the contribution to the molding and control of beliefs, values, and behaviors by way of "modern standards" for measurement for a relatively more scientific worldview—but the contribution to the molding and control of beliefs, values, and behaviors by way of "arbitrary standards set up by regional or local authorities" in antiquity for a relatively more pre-scientific worldview, in *Sec. 3.3.1*; the contribution to the molding and control of beliefs, values, and behaviors by way of quantum mechanics for a relatively more indeterministic worldview—but the contribution to the molding and control of beliefs, values, and behaviors by way of classical mechanics for a relatively more deterministic worldview, in *Sec. 3.3.2*; the contribution to the molding and control of beliefs, values, and behaviors by way of communicative action for a relatively more cooperative lifeworld—but the contribution to the molding and control of beliefs, values, and behaviors by way of strategic action for a relatively more competitive lifeworld, in *Sec. 3.4.1*; the contribution to the molding and control of beliefs, values, and behaviors by way of formal mathematics for a relatively more scientific worldview—but the contribution to the molding and control of beliefs, values, and behaviors by way of mathematical folklore for a relatively more pragmatic worldview, in *Sec. 3.4.2*; the contribution to the molding and control of beliefs, values, and behaviors by way of Husserl's phenomenology for a relatively more psychological worldview—but the contribution to the molding and control of beliefs, values, and behaviors by way of Garfinkel's phenomenolog for a relatively more sociological worldview, in *Sec. 3.5.1*; and the contribution to the molding and control of beliefs, values, and behaviors by way of "a unitary, autonomous Subject" for a relatively more conservative worldview—but the contribution to the molding and control of beliefs, values, and behaviors by way of the "split subject'" for a relatively more subversive worldview, in *Sec. 3.5.2*).

And the reverse direction also holds true.

Again, it should be stressed, however, that there are different shades of gray (or different degrees of truth) in the two opposites, and the classification is not necessarily mutually exclusive either. In fact, even when some combinations of the two occur, they only end up sharing the same dialectic relationship, but in a different degree.

Also, there are some other relationships (as a kind of family resemblance) which have something in common with the principle, but they are not exactly the same but only more or less comparable, which varies from case to case.

Good examples of family resemblance in relation to the principle are homogeneity-heterogeneity, we-they, similarity-contrast, and so forth.

27th Thesis: The Stability-Reaction Principle

The twenty-seventh thesis is called the stability-reaction princple (on the feedback mechanism for further changes) in the category of "outcome" in existential dialectics (which was first worked out in *FPHCHEM*), in that once an outcome becomes stabilized over time (which varies from case to case), it will undergo further changes in its multifaceted interactions with the outside world (e.g., with other entities, energy, space, and so on).

There is no stability without reaction for further changes—and vice versa.

Works on chemical reactions (as in *FPHCHEM*) and feedback mechanisms for learning in adaptive management (as in *BNR*), just to name a few of them, have contributed to the understanding of stability and reaction.

For instance, on the one hand, there is stability in outcome (e.g., the relatively more stable condition of formal rules of inference in the older days, like Aristotle's logic, in *Sec. 2.2.1*; the relatively more stable condition of the fashionable view about rationality, as shown in the influential Weberian account of 4 types of rationality, in *Sec. 2.2.2*; the relatively more stable condition of the "classical" interpretation of probability in the older days, in *Sec. 2.3.1*; the relatively more stable condition of cybernetics in the early history of complexity study, when it was widely studied, in *Sec. 2.3.2*; the relatively more stable condition of companies which focused more on cost in the early days of industrialization, in *Sec. 2.4.1*; the relatively more stable condition of the "scientific management" approach to organizational behavior in the early part of the 20th century, in *Sec. 2.4.2*; the relatively more stable condition of syllogistic logic in the older days, when it was influential in ancient Greece, in *Sec. 2.5.1*; the relatively more stable condition of the free market ideology in the older days, when people believed in the free market to eventually reach "the equilibrium of full employment," in *Sec. 2.5.2*; the relatively more stable condition of the ideas of probability as worked out by Bayes and Laplace in the older days of the 18th and 19th centuries, in *Sec. 3.2.1*; the relatively more stable condition of factor analysis like R analysis, when it was pioneered by Charles Spearman and others in the older days, in *Sec. 3.2.2*; the relatively more stable condition of "arbitrary standards set up by regional or local authorities" for measurement in antiquity, in *Sec. 3.3.1*; the relatively more stable condition of classical mechanics in the older days, in *Sec. 3.3.2*; the relatively more stable condition of the understanding of "social action" in the older days, like "rational choice theory and game theory," in *Sec. 3.4.1*; the relatively more stable condition of the foundation of formal mathematics in the older days, in *Sec. 3.4.2*; the relatively more stable condition of transcendental phenomenology by Husserl in the older days, in *Sec. 3.5.1*; and the relatively more stable condition of the idea of "a unitary, autonomous Subject" in the older days of "the liberal theory of the social contract," in *Sec. 3.5.1*).

On the other hand, there is reaction in outcome (e.g., formal rules of inference in the older days then led to further changes later on, when alternative ways of reasoning like mental models are in fashion nowadays, in *Sec. 2.2.1*; the influential Weberian account of 4 types of rationality then led to further changes later on, when nowadays, the feminist perspective rejects it as "imbued with masculine values and oriented toward the maintenance of male power," in *Sec. 2.2.1*; the "classical" interpretation of probability in the older days then led to further changes later on, when nowadays it is replaced by the "modern" interpretation, in *Sec. 2.3.1*; cybernetics in the early history of complexity study led to further changes later on, when nowadays it is superseded by different new sub-fields, like "web science" and "e-science" in the current history of complexity study, in *Sec. 2.3.2*; the stable condition of companies which focused more on cost in the early days of industrialization then led to further changes later on, when in our times, there are laws governing "consumer protection" for safety, like the establishment of the Federal Trade Commission in 1914, in *Sec. 2.4.1*; the "scientific management" approach to organizational behavior in the early part of the 20th century then led to further changes later on, when nowadays, the satisficing approach and the cultural approach gain more attention, in *Sec. 2.4.2*; syllogistic logic in the older days then led to further changes later on, when nowadays it is superseded by propositional calculus and other logics, in *Sec. 2.5.1*; the free market ideology in the older days led to further changes later on, when nowadays, there are those in the Keynesian School who advocate "the inherent unpredictability of economic systems," which requires some government intervention, in *Sec. 2.5.2*; the ideas of probability as worked out by Bayes and Laplace in the older days of the 18th and 19th centuries led to further changes later on, when "the ideas of Laplace were further developed in two different directions, giving rise to objective and subjective currents in Bayesian practice" in the 20th century, in *Sec. 3.2.1*; the development of factor analysis led to further changes later on, when it further evolved into Q analysis by William Stephenson and others, in *Sec. 3.2.2*; "arbitrary standards set up by regional or local authorities" for measurement in antiquity led to further changes later on, when "modern standards" for measurement were set up since the scientific revolution, in *Sec. 3.3.1*; classical mechanics in the older days then led to further changes later on, when it was challenged by quantum mechanics in the 20th century for a more indeterministic view of the world, in *Sec. 3.3.2*; the understanding of "social action" in the older days, like "rational choice theory and game theory," then led to further changes later on, when they are challenged by the theory of communicative action, in *Sec. 3.4.1*; the foundation of formal mathematics in the older days then led to further changes later on, when it was questioned by the sociology of mathematical knowledge, in *Sec. 3.4.2*; transcendental phenomenology by Husserl in the older days then led to further changes later on, when Merleau-Ponty proposed the alternative approach known as "existential phenomenology," in *Sec. 3.5.1*; and "a unitary, autonomous Subject" in the older days of "the liberal theory of the social contract" then led to further changes later on, when it was challenged by different philosophical traditions in the 20th century aforediscussed, in *Sec. 3.5.2*).

And the reverse direction also holds true.

Again, it should be stressed, however, that there are different shades of gray (or different degrees of truth) in the two opposites, and the classification is not necessarily mutually exclusive either. In fact, even when some combinations of the two occur, they only end up sharing the same dialectic relationship, but in a different degree.

Also, there are some other relationships (as a kind of family resemblance) which have something in common with the principle, but they are not exactly the same but only more or less comparable, which varies from case to case.

Good examples of family resemblance in relation to the principle are stability-forwardness, normalcy-feedback, balance-forwardness, balance-reaction, consolidation-modification, normalcy-feedback, stability-breakup, rest-motion, and the like.

28th Thesis: The Functionality-Nonfunctionality Principle

The twenty-eighth thesis is called the functionality-nonfunctionality princple (on the presence of function) in the category of "outcome" in existential dialectics (which is first worked out in *FPHCA*), in that, if there is the presence of action for which something or someone is fitted, there is likewise the relative lack of action for which something or someone is fitted. There is no functionality without nonfunctionallity—and vice versa.

A *function* should not be confused with an *alternative function,* which is just another kind of function, just as *nonfunction* should not be confused with *dysfunction* (which has more to do with malfunction and is thus related more to the regression-progression principle).

In addition, being functional (or useful) and being nonfunctional (or useless) do not have to be mutually exclusive, as the two can come in all shapes and sizes, or in different degrees. And what is functional at one time may become nonfunctional at another time, and vice-versa.

For instance, on the one hand, there is functionality in outcome (e.g., the presence of action for which something is fitted, as shown in the creation of formal rules of inference in Aristotle's logic in the older days, when they served the function of providing a relatively more deterministic, logical worldview, in *Sec. 2.2.1*; the presence of action for which something is fitted, as shown in the influential Weberian account of 4 types of rationality in the older days, when they served the function of propounding "masculine values" and being "oriented toward the maintenance of male power," in *Sec. 2.2.2*; the presence of action for which something is fitted, as shown in the use of dice by the Tibetans, which serve the important function of being a form of divination to make decisions about "health, work" and the like, in *Sec. 2.3.1*; the presence of action for which something is fitted, as shown in the use of creation theory in the older days, when it served the function of explaining the complexity of things in the world in terms of the "guiding hand" of God, in *Sec. 2.3.2*; the presence of action for which something is fitted, as shown in the focus on cost in the older days, which served the function of laissez-faire capitalism in the early days of industrialization, in *Sec. 2.4.1*; the presence of action for which something is fitted, as shown in the rise of the "scientific management" approach to organizational behavior in the early part of the 20th century, which served the function of scientific optimism at the time, like the belief that "rationalizing the organization with precise sets of instructions and time-motion studies would lead to increased productivity," in *Sec. 2.4.2*; the presence of action for which something is fitted, as shown in the efforts by mathematicians in "the logicist period from the Begriffsschrift of Frege to the *Principia Mathematica* of Russell and Whitehead," which served the function to "incorporate the logic of all mathematical and scientific discourse in a single unified system," in *Sec. 2.5.1*; the presence of action for which something is fitted, as shown in the belief in the free market ideology in the older days, which served the function of supporting laissez-faire economics at the time, in *Sec. 2.5.2*; the presence of action for which something or someone is fitted, as shown in the use of the Bible in medieval Europe, which served the function of understanding reality along the hermeneutic line, in *Sec. 3.2.1*; the presence of action for which something or someone is fitted, as shown in the use of "a paper template and the sample of statements or other stimuli printed on individual cards" for doing Q sorts in the older days, which served the function of a relatively less technologically advanced society at the time, in *Sec. 3.2.2*; the presence of action for which something or someone is fitted, as shown in the use of "arbitrary standards set up by regional or local authorities" for measurement in antiquity, which served the function of the pre-scientific lifeworld at the time, in *Sec. 3.3.1*; the presence of action for which something or someone is fitted, as shown in the influence of classical mechanics in the older days, when it served the function of holding a deterministic worldview at that time, in *Sec. 3.3.2*; the presence of action for which something or someone is fitted, as shown in the popularity of the idea of the "survival of the fittest" for the pursuit of "private interests" in strategic action in the 19th century, which served the function of perpetuting the ideology of "laissez-faire capitalism" at the time, in *Sec. 3.4.1*; the presence of action for which something or someone is fitted, as shown in the use of mathematical folklore among those "aboriginal or ancient peoples," which served the function of maintaining the pre-scientific lifeworld at that time, in *Sec. 3.4.2*; the presence of action for which something or someone is fitted, as shown in the silence of Heidegger on the influence of Okakura Kakuzo's concept of *das-in-der-Welt-sein* [being in the world] expressed in *The Book of Tea* to describe Zhuangzi's philosophy, which Imamichi's teacher had offered to Heidegger in 1919, after having studied with him the year before"—which served the function of Eurocentrism in the Western world at the time, without giving much credit to Eastern philosophy, in *Sec. 3.5.1*; and the presence of action for which something or someone is fitted, as shown in the belief in "a unitary, autonomous Subject" in the older days of "social contract" theory, which served the function of perpetuating the "liberal" tradition at the time, in *Sec. 3.5.2*).

On the other hand, there is nonfunctionality in outcome (e.g., the relative lack of action for which something is fitted, as shown in the use of mental models nowadays, such that the relatively more deterministic, logical worldview in Aristotle's logic is no longer as much functional in our times as it was before, when the fast-paced, flexible nature of mental models is attractive to some people who wants to simplify the world as a way to make sense of it, in *Sec. 2.2.1*; the relative lack of action for which something is fitted, as shown in the increasing critique of the Weberian account of 4 types of rationality nowadays, such that they are no longer as much functional as they used to be, due to different feminist expectations and campaigns for a different voice on rationality with a feminine focus on equality, affect, and so on in our times, in *Sec. 2.2.2*; the relative lack of action for which something is fitted, as shown in the use of modern health care and education in the U.S. nowadays, such that the form of divination is not as functional there as it could in Tibet, because of different expectations and standards based on a relatively advanced, hi-tech society in America, in *Sec. 2.3.1*; the relative lack of action for which something is fitted, as shown in the rise of complexity science nowadays, such that it is used to understand the complexity of things in the world without appealing to a divine design, that is, in terms of "emergence" without any "guiding hand," such that creation theory is no longer as much functional in our time as it was dominant in Medieval Europe, in *Sec. 2.3.2*; the relative lack of action for which something is fitted, as shown in the rise of laws and organizations for consumer protection nowadays, such that the selfish focus on cost in the older days is no longer much functional in our times, because of different standards and expectations for safety, etc., in *Sec. 2.4.1*; the relative lack of action for which something is fitted, as shown in the rise of bounded rationality in the 1960s and 1970s, such that scientific optimism is no longer functional to the extent that it was in earlier times, due to different expectations and beliefs in our time, with a different interest in the cultural and social contexts of organizational behavior, in *Sec. 2.4.2*; the relative lack of action for which something is fitted, as shown in "Gödel's incompleteness theorems," which later challenged Hilbert's impossible task to find "a complete and consistent set of axioms for all mathematics," such that this futile quest for "the logic of all mathematical and scientific discourse in a single unified system" is no longer much functional in our time, because of a more critical rejection of any attempt to find "the foundations of mathematics" nowadays, in *Sec. 2.5.1*; the relative lack of action for which something is fitted, as shown in the rise of Keynsian economics in modern times, such that the ideology of free market is no longer dominant or functional nowadays, due to different expectations and beliefs in "the inherent unpredictability of economic systems," especially since the Great Depression and government intervention in American economic life ever since, which is the remaining baston of world capitalism, in *Sec. 2.5.2*; the relative lack of action for which something or someone is fitted, as shown in the use of the objectivist version of Bayesian methods nowadays, such that the biblical hermenuetics is no longer functional "in the process of science and especially in statistical methods" in our time, since "to meet the needs of science…, Bayesian statisticians have developed 'objective' methods for specifying prior probabilities," in *Sec. 3.2.1*; the relative lack of action for which something or someone is fitted, as shown in the use of sophisticated computer software applications for doing Q sorts nowadays, such that the use of "a paper template and the sample of statements or other stimuli printed on individual cards" in the older days is no longer much functional in our computer age, in *Sec. 3.2.2*; the relative lack of action for which something or someone is fitted, as shown in the use of "modern standards" for measurement set up since the scientific revolution, such that the older "arbitrary standards set up by regional or local authorities" for measurement in antiquity are no longer much functional in our relatively more scientific age, in *Sec. 3.3.1*; the relative lack of action for which something or someone is fitted, as shown in the rise of quantum mechanics in the 20th century, such that the older deterministic worldview of classical mechanics is no longer function in our relatively more scientific age, in *Sec. 3.3.2*; the relative lack of action for which something or someone is fitted, as shown in the popularity, among some people nowadays, of the idea of "consensus" and "mutual understanding," even if to be understood as an ideal type only, such that the idea of the "survival of the fittest" for the pursuit of "private interests" in strategic action is no longer much influential in our democratic age nowadays, when capitalism is becoming more and more "regulated" for the broader interests of society, in *Sec. 3.4.1*; the relative lack of action for which something or someone is fitted, as shown in the use of formal mathematics nowadays, such that the use of mathematical folklore is no longer as functional nowadays in our relatively more scientifically advanced society, in *Sec. 3.4.2*; the relative lack of action for which something or someone is fitted, as shown by "some researchers in phenomenology" nowadays, who "see possibilities of establishing dialogues with traditions of thought outside of the so-called Western philosophy, particularly with respect to East-Asian thinking," such that the old-fashioned Eurocentric prejudice in the old days is no longer as influential in our age of globalization, in *Sec. 3.5.1*; and the relative lack of action for which something or someone is fitted, as shown in the rise of postmodernism in our time, with its "deconstruction of the subject as a core-concept of metaphysics," such that the old-fashioned belief in "a unitary, autonomous Subject" in the older days of "social contract" theory is no longer much functional in our contemporary postmodernist era, in *Sec. 3.5.2*).

And the reverse direction also holds true.

Again, it should be stressed, however, that there are different shades of gray (or different degrees of truth) in the two opposites, and the classification is not necessarily mutually exclusive either. In fact, even when some combinations of the two occur, they only end up sharing the same dialectic relationship, but in a different degree.

Also, there are some other relationships (as a kind of family resemblance) which have something in common with the principle, but they are not exactly the same but only more or less comparable, which varies from case to case.

Good examples of family resemblance in relation to the principle are usefulness-nonusefulness, valueness-nonvalueness, usefulness-uselessness, and the like.

29th Thesis: The Intentionality-Nonintentionality Principle

The twenty-ninth thesis is called the intentionality-nonintentionality princple (on the planning for outcomes) in the category of "outcome" in existential dialectics (which is first worked out in *FPHW*), in that, if there is the planning or design of something for a certain outcome, there is likewise the relative absence of planning or design of something for a certain outcome. There is no intentionality without nonintentionallity—and vice versa.

An intention for an outcome does not have to be certain, so the outcome can even be counter-intentional in being the opposite of what was originally planned. In this sense, counter-intentionality is a special case of non-intentionality; in other words, there are unintented consequences which can be just the opposite of what was originally planned.

An intention for an outcome is mostly relevant to intelligent life, but there are non-intelligent life and non-living matter in the universe too, so nonintentionality is also important.

In addition, an intention does not have to be obvious or straightforward, as William Shakspeare once famously said: "I must be cruel only to be kind."

In any event, on the one hand, there is intentionality in outcome (e.g., the planning or design of something for a certain outcome, as shown in the creation of formal rules of inference in Aristotle's logic in the older days, for an understanding of the world in a relatively more deterministic, logical worldview, in *Sec. 2.2.1*; the planning or design of something for a certain outcome, as shown in the Weberian account of 4 types of rationality for a better understanding of social action, in *Sec. 2.2.2*; the planning or design of something for a certain outcome, as shown in the work on the "classical" interpretation of probability by James Bernoulli and Pierre-Simon Laplace in the older days, in *Sec. 2.3.1*; the planning or design of something for a certain outcome, as shown in the use of complexity study for the understanding of linkages among different parts in a system, in *Sec. 2.3.2*; the planning or design of something for a certain outcome, as shown by those in reliability engineering who work on the exact measurement of the performance of "a device and its intended function during a specified period of time under stated conditions," in *Sec. 2.4.1*; the planning or design of something for a certain outcome, as shown in the creation of the "rationality as optimization" in the early part of the 20th century in organizational behavior for "increased productivity," in *Sec. 2.4.2*; the planning or design of something for a certain outcome, as shown in "Hilbert's program to find a complete and consistent set of axioms for all mathematics," in *Sec. 2.5.1*; the planning or design of something for a certain outcome, as shown in the pursuit of each individual's interests in the free market, in *Sec. 2.5.2*; the planning or design of something for a certain outcome, as shown in the work of de Finetti on the "Dutch book approach" as a way to justify Bayesian probability, in *Sec. 3.2.1*; the planning or design of something for a certain outcome, as shown in the construction of factor analysis like R analysis by Charles Spearman and others in the older days, in *Sec. 3.2.2*; the planning or design of something for a certain outcome, as shown in the "arbitrary standards set up by regional or local authorities" in antiquity, for measurement as a way to facilitate local commerce, in *Sec. 3.3.1*; the planning or design of something for a certain outcome, as shown in the original intention of Heisenberg to use the uncertainty principle to shed light on the "tradeoff between the precisions" of the position and momentum of a sub-atomic particle, in *Sec. 3.3.2*; the planning or design of something for a certain outcome, as shown in the original intention of Habermas to address the social issue of power by providing a power-free communication like communicative action as a solution, in *Sec. 3.4.1*; the planning or design of something for a certain outcome, as shown in the original intention or hope of Euclid to use his "system of proving geometry theorems" as a foundation of mathematics, in *Sec. 3.4.2*; the planning or design of something for a certain outcome, as shown in the original intention of Husserl to use phenomenology for the study of "the structures of subjective experience and consciousness," in *Sec. 3.5.1*; and the planning or design of something for a certain outcome, as shown in the original intention of those in the "liberal theory of the social contract" to use the idea of "a unitary, autonomous Subject" as its "foundation," in *Sec. 3.5.2*).

On the other hand, there is nonintentionality in outcome (e.g., the relative absence of planning or design of something for a certain outcome, such that the outcome can even be counter-intentional in being the opposite of what was originally planned, as shown in the rise of mental models nowadays, for the simplification of reality as a way to make sense of it, contrary to the relatively more deterministic, logical worldview in Aristotle's logic, in *Sec. 2.2.1*; the relative absence of planning or design of something for a certain outcome, such that the outcome can even be counter-intentional in being the opposite of what was originally planned, as shown in the critique of the Weberian account of 4 types of rationality by [A.] Etzioni [1988], who reframes thought on decision-making to argue for a reversal of the position put forward by Weber," in that "purposive/instrumental reasoning is subordinated by normative considerations [ideas on how people 'ought' to behave] and affective considerations [as a support system for the development of human relationships]," contrary to the original Weberian construction, in *Sec. 2.2.2*; the relative absence of planning or design of something for a certain outcome, such that the outcome can even be counter-intentional in being the opposite of what was originally planned, as shown in the attempt to revive the "classical" interpretation by some people nowadays, by relating it to "the general interest in Bayesian probability," contrary to the original intention of James Bernoulli and Pierre-Simon Laplace in the older days, in *Sec. 2.3.1*; the relative absence of planning or design of something for a certain outcome, such that the outcome can even be counter-intentional in being the opposite of what was originally planned, as shown by R. Tulip [2012], who used complexity study for "a rational foundation for systematic Biblical theology"—contrary to the original mission of complexity study in science, in *Sec. 2.3.2*; the relative absence of planning or design of something for a certain outcome, such that the outcome can even be counter-intentional in being the opposite of what was originally planned, as shown in the distortion of perception by consumers, when "one bad experience can be magnified in the mind of the customer, inflating the perceived unreliability of the product,...regardless of actual reliability data"—contrary to the original mission to measure actual reliability, in *Sec. 2.4.1*; the relative absence of planning or design of something for a certain outcome, such that the outcome can even be counter-intentional in being the opposite of what was originally planned, as shown in the rise of "bounded rationality" in our time, which proposes the opposite view that "rationality of individuals is limited by the information they have, the cognitive limitations of their minds, and the finite amount of time they have to make a decision," contrary to the original mission of "rationality as optimization," in *Sec. 2.4.2*; the relative absence of planning or design of something for a certain outcome, such that the outcome can even be counter-intentional in being the opposite of what was originally planned, as shown in "Gödel's incompleteness theorems...that establish inherent limitations of all but the most trivial axiomatic systems capable of doing arithmetic" and show that "Hilbert's program to find a complete and consistent set of axioms for all mathematics is impossible, giving a negative answer to Hilbert's second problem," contrary to the original mission of Hilbert, in *Sec. 2.5.1*; the relative absence of planning or design of something for a certain outcome, such that the outcome can even be counter-intentional in being the opposite of what was originally planned, as shown in the failure of the free market to reach "the equilibrium of full employment" during the Great Depression, when each individual pursued his own self-interests in the free market, contrary to the original design of "the Invisible Hand" in the free-market ideology, in *Sec. 2.5.2*; the relative absence of planning or design of something for a certain outcome, such that the outcome can even be counter-intentional in being the opposite of what was originally planned, as shown in the work of Ian Hacking, who showed that "traditional Dutch book arguments did not specify Bayesian updating: they left open the possibility that non-Bayesian updating rules could avoid Dutch books," and "neither the Dutch book argument, nor any other in the personalist arsenal of proofs of the probability axioms, entails the dynamic assumption. Not one entails Bayesianism"—contrary to the original intention of de Finetti to justify Bayesian probability with the "Dutch book approach," in *Sec. 3.2.1*; the relative absence of planning or design of something for a certain outcome, such that the outcome can even be counter-intentional in being the opposite of what was originally planned, as shown in the construction of Q analysis by William Stephenson and others later on, such that its focus is on "correlations between subjects across a sample of variables"—contrary to the original intention of Spearman for factor analysis, which is more on "correlations between variables [say, height and age] across a sample of subjects" instead, in *Sec. 3.2.2*; the relative absence of planning or design of something for a certain outcome, such that the outcome can even be counter-intentional in being the opposite of what was originally planned, as shown in the "modern standards" for measurement set up since the scientific revolution, along the more scientific line, contrary to the original pre-scientific "arbitrary standards" in antiquity, in *Sec. 3.3.1*; the relative absence of planning or design of something for a certain outcome, such that the outcome can even be counter-intentional in being the opposite of what was originally planned, as shown in the use of the uncertainty principle by those people nowadays for the understanding of free will in philosophy, contrary to the original intention of Heisenberg, in *Sec. 3.3.2*; the relative absence of planning or design of something for a certain outcome, such that the outcome can even be counter-intentional in being the opposite of what was originally planned, as shown in the critique of the historian Ian McNeeley [2003], who argued "Jürgen Habermas subscribes to an unrealistic ideal of power-free communication," and "the discursive fiction of consensus achieved through rational argumentation might be used as a legitimating prop for social action to the detriment of marginalized members—this is the basis of much feminist critique of Habermas' notions," contrary to the original intention of Habermas to resolve the issue of power in social action, in *Sec. 3.4.1*; the relative absence of planning or design of something for a certain outcome, such that the outcome can even be counter-intentional in being the opposite of what was originally planned, as shown in the work of Putnam, which showed that "Euclid's system of proving geometry theorems was unique to the classical Greeks and did not evolve similarly in other mathematical cultures in China, India, and Arabia," so "a truly international culture of mathematics would...necessarily be at least 'quasi'-empirical," contrary to the original intention or hope of Euclid, in *Sec. 3.4.2*; the relative absence of planning or design of something for a certain outcome, such that the outcome can even be counter-intentional in being the opposite of what was originally planned, as shown by Garfinkel who spoke of "phenomenological texts and findings as being, 'appropriated' and intentionally 'misread,' for the purposes of exploring topics in the study of social order"—contrary to the original intention of Husserl for phenomenology, in *Sec. 3.5.1*; and the relative absence of planning or design of something for a certain outcome, such that the outcome can even be counter-intentional in being the opposite of what was originally planned, as shown in the works of Marx and Freud, who "provided a point of departure for questioning the notion of a unitary, autonomous Subject" and "opened up the way for the deconstruction of the subject as a core-concept of metaphysics"—contrary to the original intention of those in the "liberal theory of the social contract" to use the idea of "a unitary, autonomous Subject" as its "foundation," in *Sec. 3.5.2*).

And the reverse direction also holds true.

Again, it should be stressed, however, that there are different shades of gray (or different degrees of truth) in the two opposites, and the classification is not necessarily mutually exclusive either. In fact, even when some combinations of the two occur, they only end up sharing the same dialectic relationship, but in a different degree.

Also, there are some other relationships (as a kind of family resemblance) which have something in common with the principle, but they are not exactly the same but only more or less comparable, which varies from case to case.

Good examples of family resemblance in relation to the principle are purposiveness-nonpurposiveness, intention-accidentalness, design-randomness, etc.

30th Thesis: The Survivability-Nonsurvivability Principle

The thirtieth thesis is called the survivability-nonsurvivability princple (on the survival of things) in the category of "outcome" in existential dialectics (which is first worked out in *FPHPROB*), in that, if there are some entities which survive at certain times, there are likewise other entities which do not survive. There is no survivability without nonsurvivability—and vice versa.

In nature, for illustration, there are some animals which survive in hard times, whereas others do not—and this is true, not just among different species but also among different members of the same species (and the phenomenon of "selection" in evolutionary biology has much to say on this, as already addressed in *BNN*).

Of course, the principle is not restricted solely to the animal world, since it can be used to other cases concerning non-animals too (e.g., trees, lakes, etc. in the natural world—or empires, companies, etc. in the human world).

But one should not confuse this principle with the expansion-contraction principle, since those entities which survive, say, can still expand or contract over time (e.g., those animals which survive can be more capable in some activities but not in others, for instance).

Nor should this principle be confused with the stability-reaction principle, in that the condition of survivability or non-survivability may lead to further changes later on (as shown in the emergence of new species after the extinction of some old ones, as already explained in *BNR*).

In any event, on the one hand, there is survivability in outcome (e.g., the survival of those conclusions which pass the tests of counterexamples in mental models, in *Sec. 2.2.1*; the survival of those notions of rationality when they pass the test of what Richard Brandt proposed as "a form of cognitive-psychotherapy," in *Sec. 2.2.2*; the relative survival of the "modern" interpretation of probability nowadays, in *Sec. 2.3.1*; the relative survival of complexity study in modern science since its beginnng in the 1940s for the understanding of linkages among different parts in a system, without the help of any "guiding hand" or God, in *Sec. 2.3.2*; the survival of those devices which pass reliability testing, in *Sec. 2.4.1*; he survival of "bounded rationality" since 1960s unto now, in *Sec. 2.4.2*; the survival of propositional logic in modern times, in *Sec. 2.5.1*; the survival of those components in an electronic system which perform as designed, in *Sec. 2.5.2*; the survival of the "axiomatic approach," insofar as "other axiomatizations have been suggested by various authors to make the theory more rigorous" nowadays, in *Sec. 3.2.1*; the relative survival of using sophisticated computer software applications for doing Q sorts nowadays, in *Sec. 3.2.2*; the survival, in our time, of many "modern standards" for measurement set up since the scientific revolution, like mathematical modeling aforecited, in *Sec. 3.3.1*; the relative survival of quantum mechanics in the 21st century, after extensive debate about its validity in much of the 20th century, in *Sec. 3.3.2*; the survival of those who win in the battle of "influence" and "power" in strategic action, such that they become dominant, in *Sec. 3.4.1*; the survival of the sociology of mathematical practice in our time, in *Sec. 3.4.2*; the relative survival of Husserl's earlier phenomenology known as "realist phenomenology," as many "members of the Munich group," for example, "preferred the earlier realist phenomenology of the first edition of the *Logical Investigations*," in *Sec. 3.5.1*; and the relative survival of "the issue of subject" as "one of the major intractable problems in philosophy of mind" in our time, especially by those in contemporary analytic philosophy, in *Sec. 3.5.2*).

On the other hand, there is non-survivability in outcome (e.g., the non-survival of those conclusions which do not pass the tests of counterexamples in mental models, in *Sec. 2.2.1*; the non-survival of those notions of rationality when they fail the test of what Richard Brandt proposed as "a form of cognitive-psychotherapy," in *Sec. 2.2.2*; the relative non-survival of the "classical" interpretation of probaiblity nowadays, in *Sec. 2.3.1*; the relative non-survival of creation theory in modern science, when it tried to understand the linkages among different parts in a system with the help of the "guiding hand" of a supernatural being, although it remains relevant in Western theology, in *Sec. 2.3.2*; the non-survival of those devices which do not pass reliability testing, in *Sec. 2.4.1*; the non-survival of "the rational planning model," which "has fallen out of mass use as of the last decade," in *Sec. 2.4.2*; the non-survival of traditional syllogistic logic in modern times, as "propositional logic" is now "seen as an advancement from the earlier syllogistic logic," in *Sec. 2.5.1*; the non-survival of those components in an electronic system which fail to the point that it has to be discarded, in *Sec. 2.5.2*; the non-survival of the "axiomatic approach," insofar as "Cox's 1961 development [mainly copied by Jaynes] is non-rigorous, and in fact a counterexample has been found by Halpern," in *Sec. 3.2.1*; the relative non-survival of the use of "a paper template and the sample of statements or other stimuli printed on individual cards" for doing Q sorts nowadays, in *Sec. 3.2.2*; the non-survival, in our time, of many "arbitrary standards set up by regional or local authorities" in antiquity, in *Sec. 3.3.1*; the relative non-survival of classical mechanics in the 21st century, after it was replaced by quantum mechanics and the theory of relativity in the 20th century, although it retains some "limited validity" as an approximation to special relativity and quantum mechanics, in *Sec. 3.3.2*; the non-survival of those who lose in the battle of "influence" and "power" in strategic action, such that they become marginalized or even eliminated, in *Sec. 3.4.1*; the non-survival, in our time, of the old-fashioned "standard account of the development of geometry in ancient Egypt, followed by Greek mathematics and the emergence of deductive logic," since mathematicians now realized that "several ancient societies built impressive mathematical systems and carried out complex calculations based on proofless heuristics and practical approaches," different from the "standard" account of the development of geometry," in *Sec. 3.4.2*; the relative non-survival of Husserl's later phenomenology known as "transcendental phenomenology," since "many phenomenologists took a critical stance towards his new theories," in *Sec. 3.5.1*; and the relative non-survival of "the notion of a unitary, autonomous Subject" in our time, especially among those in postmodernism—although it was "the foundation of the liberal theory of the social contract" in the older days, in *Sec. 3.5.2*).

And the reverse direction also holds true.

Again, it should be stressed, however, that there are different shades of gray (or different degrees of truth) in the two opposites, and the classification is not necessarily mutually exclusive either. In fact, even when some combinations of the two occur, they only end up sharing the same dialectic relationship, but in a different degree.

Also, there are some other relationships (as a kind of family resemblance) which have something in common with the principle, but they are not exactly the same but only more or less comparable, which varies from case to case.

Good examples of family resemblance in relation to the principle are survival-extinction, existence-nonexistence, life-death, etc.

31st Thesis: The Post-Human Interpretiveness

And the thirtieth thesis is about the role of "post-humans," which I originally proposed in my previous books, starting with the first book titled *The Future of Human Civilization* in 2000 and all others afterwards.

As already pointed out in *Sec. 1.8*, I need to emphasize, as this is something that I used to repeat (*almost verbatim*) from my previous books, two clarifications here about the term "post-human" as a neologism in my works.

Firstly, the word "post-human" here should *not* be confused with another term which looks similar but has a totally different meaning in the literature of postmodernism, namely, "post-humanism"—which constitutes a critique of "humanism" as tradionally understood (especially, though not exclusively, in relation to the idea of progress in science and reason in the Enlightenment project). (WK 2008)

My works reject the project of "postmodernism" and propose the future world of what I orignally called "after-postmodernity" in *FHC* and *FCD*, for instance.

And secondly, the word "post-human" here should also *not* be confused with a similar term which is used to champion the ideology of technology for the future co-existence between humans and cyborgs in "trans-humanism." (WK 2008a)

Instead, my term "post-human" in relation to "posthuman-ism" also rejects "transhumanism" (especially, though not exclusively, in relation to the promises of technology) and refers to something else altogether, that is, the future extinction of humans and its post-human successors in deep space and beyond unto multiverses.

My critique of "transhumanism" was more extensively elabroated in *Sec. 2.4.1* of *BEPE*—just as my critique of different futuristic scenarios of fictional artificial intelligence in science fiction was given in *Sec. 3.5.2* of *FPHCOM*.

With these two clarifications in mind (as summarized in *Table 1.31*)—the post-human interpretiveness in the study of probability can be addressed in relation to five main directions, based on my *interpretivist theory of probability*, as analysed in this book (together with my visions as already worked out in my previous books).

Firstly, the post-human interpretiveness in the study of probability will learn from my interpretivist theory of probability, in relation to the dialectic context of no analysis of objectivity without that of subjectivity (and vice versa), in the context of probability.

Secondly, the post-human interpretiveness in the study of probability will learn from my interpretivist theory of probability in relation to the four perspectives of the mind, nature, society, and culture.

Thirdly, the post-human interpretiveness in the study of probability will learn from my interpretivist theory of probability in relation to the non-privilege of any specific theory over others in the literature and the non-integration of them all (as they are not necessarily compatible with each other).

Fourthly, the post-human interpretiveness in the study of probability will have to confront the new challenges as posed by the long-term civilizational development of intelligent life in the distant future, both here on this planet Earth and elsewhere in deep space until multiverses.

As this is something that I regularly pointed out in my previous books for background information (and summarize here, *almost verbatim*, with some updates), I already worked out, in my numerous books, what these new challenges will be and provide different original visions to meet them—especially in regard to the following future evolutions:

- of the mind (e.g., "supersession computing," "metamorphic humor," "contrarian personality," "hyper-martial body," "hyper-sexual body," "floating consciousness," "hyper-spatial consciousness," "thinking machines," "unfolding unconsciousness," "thinking robots," "genetically altered superior beings," "cyborgs," and others)
- of nature (e.g., "resilient natural resources," "transdisciplinary architecture," "creational chemistry," "multilateral sound," "resettlement geology," "post-cosmology," "the alteration of space-time," "the creation of new matter-energy," "selective geometry," and the like)
- of society (e.g., "post-capitalism," "post-democracy," "authoritarian liberal democracy," "transfigurative waste," "panoramic transportation," "cyclical-progressive migration," "reflective criminology," "interventive-reshaping geography," "detached gambling," "heterodox education," "multifaceted war and peace," "virtual organizations," the movement of "cyclical progression" at both structural and systemic levels, "ambivalent technology," and so forth)
- of culture (e.g., "modificative sports," "inquisitive culinary art," "interactive semantics," "transdisciplinary performing arts," "comparative-impartial literature," "post-human mind games," "comparative-substitutive religion," "post-ethics," "post-civilization," "transformative aesthetic experience," "contrastive mathematical-logic," and the whatnot)
- in history (e.g., "multifold history," the age of "after-postmodernity," etc.)
- in methodology (e.g., "sophisticated methodologial holism," "critical-dialectic formal science," "complex data analysis," etc.)
- in ontology (e.g., "existential dialectics," "contrastive rationality," etc.)

Of course, the examples (as listed above) are not exhaustive but illustrative, since my numerous books have worked out many other visions (as already roughly summarized in *Sec.1.8* and *Sec. 1.9*). Many, though not all, of my visions on the mind, nature, society, culture, history, methodology, and ontology (as cited above) are summarized in the tables as shown in Chapter One and Chapter Four (especially in *Table 4.43* about my original theories on numerous topics).

And finally, or fifthly, the post-human interpretiveness in the study of probability will learn from my interpretivist theory of probability in regard to the need to go beyond objectivity and subjectivity (as further elaborated in the next section).

Towards the Post-Human Interpretiveness

This "post-human" interpretiveness in the study of probability to transcend both objectivity and subjectivity can be understood in terms of nine great future transformations of probability, in relation to the mind, nature, society, and culture—as explained below (and summarized in *Table 1.2*).

Nine Great Future Transformations of Probability

The nine great future transformations of probability (in relation to the mind, nature, society, and culture) can be explained hereafter, with the caveat, however, that the classification in terms of nine is solely illustrative, as there is no objective basis that it must be classified in term of nine (as this is something that I often stressed and repeated in my previous books, albeit in the context of different topics).

The First Great Future Transformation of Probability

The first great future transformation of probability in the post-human era concerns what I originally call *new goals,* in relation to the mind.

This search for "new goals" will give future advanced humans and later their successors, namely, the post-humans, the capacity to be motivated to understand probability in a way that current humans cannot.

There are already some indications that the future of probability is moving toward this direction.

Consider, for illustration, the different "goals" as already seen in the history of probability, such as the exploration of imagination (*Sec. 2.2.1*), the goal of a rational way of life (*Sec. 2.2.2*), the quest for prediction (*Sec.*

2.3.1), the quest for mastering complexity (*Sec. 2.3.2*), the search for reliability in products and services (*Sec. 2.4.1*), the quest for optimality (*Sec. 2.4.2*), the search for formal systems (*Sec. 2.5.1*), the search for historical knowledge (*Sec. 2.5.2*), the search for the understanding of personal belief (*Sec. 3.2.1*), the quest for the understanding of association in thinking (*Sec. 3.2.2*), the quest for the knowledge of uncertainty (*Sec. 3.3.1*), the quest for the knowledge of quantum uncertainty (*Sec. 3.3.2*), the quest for power-free communication (*Sec. 3.4.1*), the quest for the understanding of mathematical practice (*Sec. 3.4.2*), the quest for the knowledge of subjective experience (*Sec. 3.5.1*), and the quest for the knowledge of the "subject" (*Sec. 3.5.2*), for instance.

More importantly, this future quest for new goals should also be understood in relation to my previous works on the future "post-human" supersession of the human mind from the combined perspectives of the mind, nature, society, and culture—with good examples like "supersession computing," "metamorphic humor," "contrarian personality," "hyper-martial body," "hyper-sexual body," "floating consciousness," "hyper-spatial consciousness," "unfolding unconsciousness," "thinking robots," "thinking machines," "genetically altered superior beings," "cyborgs," and others.

In any event, all these are just some illustrative (relatively primitive) examples in our time, of course, but the important point to remember here is that the search for different new goals will further transform probability at the more advanced level in the post-human future that the human world has never known.

The Second Great Future Transformation of Probability

The second great future transformation of probability in the post-human era concerns what I originally call *new means,* in relation to the mind.

This quest for "new means" will give future advanced humans and later their successors, namely, the post-humans, the capacity to find novel options to achieve the goals in regard to probability in a way that current humans cannot.

There are already some indications that the future of probability is moving toward this direction.

Consider, for illustration, the different "means" as already seen in the history of probability, such as mental models for learning (*Sec. 2.2.1*), the control of emotions for a good rationale (*Sec. 2.2.2*), dice games for the understanding of discrete probability (*Sec. 2.3.1*), computation for the understanding of complexity (*Sec. 2.3.2*), systems engineering logic for reliability (*Sec. 2.4.1*), heuristics for satisficing decision making (*Sec. 2.4.2*), the use of axioms and inference rules for the formulation of formal systems (*Sec. 2.5.1*), the idea of special cause for the unpredictiability of certain historical events (*Sec. 2.5.2*), the axiomatic justification of Bayesian probability (*Sec. 3.2.1*), data collection for Q analysis (*Sec. 3.2.2*), systematic error as a way to understand measurement uncertainty (*Sec. 3.3.1*), quantum mechanics as a way to understand uncertainty (*Sec. 3.3.2*), the study of speech acts for communicative competence (*Sec. 3.4.1*), proof by authority as a way to do quasi-mathematics (*Sec. 3.4.2*), bracketing for pure consciousness (*Sec. 3.5.1*), and German idealism as a way to study the "subject" (*Sec. 3.5.2*), for instance.

All these are just some illustrative (relatively primitive) examples in our time, of course, but the important point to remember here is that the search for different new means will further transform probability at the more advanced level in the post-human future that the human world has never known.

The Third Great Future Transformation of Probability

The third great future transformation of probability in the post-human era concerns what I originally call *new comparisons*, in relation to nature.

This search for "new comparisons" will give future advanced humans and later their successors, namely, the post-humans, the ability to learn from animals, plants, and the universe, through comparison and contrast, about probability in a way that current humans cannot.

There are already some indications that the future of probability is moving toward this direction.

Consider, for illustration, the different forms of comparison and contrast, such as as already seen in the history of probability, such as the imitation of polygons from natural objects (*Sec. 2.2.1*), discreteness in the physical world for probability study (*Sec. 2.3.1*), the comparison of gas in a container with disorganized complexity (*Sec. 2.3.2*), the contrast between Mars and Earth in making the Mars Rover in reliability engineering (*Sec. 2.4.1*), the comparison between the universe and human society through divination for non-optimal decision making (*Sec. 2.4.2*), the analogy of "tree" from nature for "truth-trees" in propositinal logic (*Sec. 2.5.1*), the analogy of "noise" from nature to understand "a fault in a power supply" as "noise" in an electronic system (*Sec. 2.5.2*), measurement uncertainty in relation to both the skill of the operator and the environment (*Sec. 3.3.1*), the analogy of quantum mechanics to the study of information in society (quantum information science) (*Sec. 3.3.2*), mathematical folklore for understanding both the universe and society

(*Sec. 3.4.2*), and the comparison and contrast between the "subject" and the "object" in different philosophical traditions (*Sec. 3.5.2*), for instance.

All these are just some illustrative (relatively primitive) examples in our time, of course, but the important point to remember here is that the search for different new comparisons will further transform probability at the more advanced level in the post-human future that the human world has never known.

The Fourth Great Future Transformation of Probability

The fourth great future transformation of probability in the post-human era concerns what I originally call *new integrations*, in relation to nature.

This search for "new integrations" will give future advanced humans and later their successors, namely, the post-humans, the ability to live and interact with nature for probability in a way that current humans cannot.

There are already some indications that the future of probability is moving toward this direction.

Consider, for illustration, the different integrations for sports as already seen in the history of probability, such as the revelation of the universe in the game of cleromancy (*Sec. 2.3.1*), the consideration of time in solving complex problems (*Sec. 2.3.2*), a Mars Rover to adjust to different natural environments (*Sec. 2.4.1*), the integration of "astrology" with everyday decision making (*Sec. 2.4.2*), the adjustment of an electronic system to prevent "ionizing radiation" or "electromagnetic interference" from "causing failures in multiple subsystems" (*Sec. 2.5.2*), the adjustment of a model of weighing by accounting for air buoyancy (*Sec. 3.3.1*), the relationship between the experimenter and measurement uncertainty in quantum mechanics (*Sec. 3.3.2*), the integration of mathematics about natural objects with social constructionism in society (*Sec. 3.4.2*), the relationship between an "intuited object" and the intentionality of consciousness (*Sec. 3.5.1*), and the harmony between the subject and the Tao in Eastern thought (*Sec. 3.5.2*), for instance.

More importantly, this future quest for new integrations should also be understood in relation to my previous works on the future "post-human" adventures into different dimensions of space-time unto multiverses from the combined perspectives of the mind, nature, society, and culture—with good examples like "the alteration of space-time" (*Table 1.3* and *Table 1.4*), "multiverses" (*Table 1.7* and *Table 1.8*), "hyperspace" (*Table 1.9* and *Table 1.25*), "time travel" (*Table 1.10* and *Table 1.11*), and others.

In addition, in my previous book titled *The Future of Post-Human Geology* (2010), I already envisioned the three great future transformations of geology, namely, "solar geology," "galactic geology," and "cosmic geology," in the context of my "resettlement theory of geology," in that humans and later post-humans will explore and migrate to other planets, solar systems, galaxies, and cosmos in the long haul.

Then, in another book of mine titled *The Future of Post-Human Geography* (2011), I linked the two books together, since geography has much to learn from geology (as in "physical geography," which, as already introduced in *Sec. 1.3* of *FPHGEOG*, relies heavily on the earth sciences) and discussed the search for "new explorations" in the context of the three forms of environments outside the planet Earth, namely, what I originally called "solar geography," "galactic geography," and "cosmic geography"—as a further elaboration of the three great future transformations of geology (in *FPHGEOL*).

Later, in still another book of mine titled *The Future of Post-Human Migration* (2012), I envisioned the future integrations in different forms, namely, what I originally called "solar migration," "galactic migration," "clustery migration," "cosmic migration," and "multiversal migration."

And then, in an additional book of mine titled *The Future of Post-Human Transportation* (2012), I envisioned the future integrations in different forms, namely, what I originally called "solar transportation," "galactic transportation," "clustery transportation," "cosmic transportation," and "multiversal transportation."

This extension is consistent with my other theory, namely, "the theory of the cyclical progression of system integration and fragmentation" in *FCD* (2002)—at the social-systemic level (as already clarified in *Sec. 1.8* and shown below):

Local → Regional → Global →Solar →
Galactic → Clustery…→ Multiversal

All these are just some illustrative (relatively primitive) examples in our time, of course, but the important point to remember here is that the search for different new integrations will further transform probability at the more advanced level in the post-human future that the human world has never known.

The Fifth Great Future Transformation of Probability

The fifth great future transformation of probability in the post-human era concerns what I originally call *new praxes*, in relation to society.

This search for "new praxes" will give future advanced humans and later their successors, namely, the post-humans, the ability to engage in practice (like activism, politics, personal experience, etc.) for the future of probability in a way that current humans cannot.

There are already some indications that the future of probability is moving toward this direction.

Consider, for illustration, the different praxes as already seen in history of probability like the application of mental models to the field of physics (*Sec. 2.2.1*), the practical level of rationality (*Sec. 2.2.2*), the application of discretness to the game of cleromancy (*Sec. 2.3.1*), the existence of "intractable" problems in practice (*Sec. 2.3.2*), the application of reliability theory to engineering problems (*Sec. 2.4.1*), the application of satisficing/maximizing decision-making strategies to the field of "classical twin studies" (*Sec. 2.4.2*), the application of inference rules to "domain-specific" contexts (*Sec. 2.5.1*), the application of probability to the study of mortality (*Sec. 2.5.2*), the application of probability to gambling (*Sec. 3.2.1*), the application of Q analysis to different fields (*Sec. 3.2.2*), the application of measurement uncertainty to industrial or scientific fields (*Sec. 3.3.1*), the application of probability to quantum mechanics (*Sec. 3.3.2*), the application of communicative action to the field of reducting social conflict in sociology (*Sec. 3.4.1*), the application of informal mathematics to the field of development psychology (*Sec. 3.4.2*), the application of Husserl's phenomenology to the field of psychology (*Sec. 3.5.1*), and the application of Freud's psychoanalysis to the study of split subject by Lacan (*Sec. 3.5.2*), for instance.

All these are just some illustrative (relatively primitive) examples in our time, of course, but the important point to remember here is that the search for different new praxes will further transform probability at the more advanced level in the post-human future that the human world has never known.

The Sixth Great Future Transformation of Probability

The sixth great future transformation of probability in the post-human era concerns what I originally call *new technologies*, in relation to society.

This search for "new technologies" will give future advanced humans and later their successors, namely, the post-humans, the ability to come up with different techniques, materials, and experiments for probability in a way that current humans cannot.

There are already some indications that the future of probability is moving toward this direction.

Consider, for illustration, the use of different technologies as already seen in the history of probability, such as human-computer interaction for mental models (*Sec. 2.2.1*), the technique of determining optimality for rational behavior (*Sec. 2.2.2*), the use of computer to calculate discreteness (*Sec. 2.3.1*), the use of computation to solve complex problems (*Sec. 2.3.2*), the technique of reliability modeling (*Sec. 2.4.1*), the mathematical method of formulating optimization problems (*Sec. 2.4.2*), the technique of formulating propositional calculus (*Sec. 2.5.1*), the technology to build "safety-critical systems using multiple redundant channels" in an "electronic system" (*Sec. 2.5.2*), the techniques of Bayesian statistics (*Sec. 3.2.1*), the technique of "Q sorts" (*Sec. 3.2.2*), the techniques of Type A and Type B evaluations (*Sec. 3.3.1*), the mathematical technique of formulating the uncertainty principle (*Sec. 3.3.2*), the technique of communicative competence (*Sec. 3.4.1*), the mathematical technique of formul proof (*Sec. 3.4.2*), the technique of transcendental reduction (*Sec. 3.5.1*), and the technique of Freudian psychoanalysis (*Sec. 3.5.2*), for instance.

All these are just some illustrative (relatively primitive) examples in our time, of course, but the important point to remember here is that the search for new technologies will further transform probability at the more advanced level in the post-human future that the human world has never known.

The Seventh Great Future Transformation of Probability

The seventh great future transformation of probability in the post-human era concerns what I originally call *new stratifications*, in relation to society.

This emergence of "new stratifications" will lead future advanced humans and later their successors, namely, the post-humans, to face new inequalities in the context of probability in a way that current humans cannot.

There are already some indications that the future of probability is moving toward this direction.

Consider, for illustration, the different inequalities as already seen in the history of probability, such as competing explanations of human reasoning (*Sec. 2.2.1*), the maintenance of male power in Weber's rationality at the expense of female one (*Sec. 2.2.2*), winners and losers in dice games (*Sec. 2.3.1*), the asymmetry between tractable and intractable problems (*Sec. 2.3.2*), the rivalry between those for the physics of failure approach and those for the parts stress modeling approach (*Sec. 2.4.1*), maximizers vs. satisficers (*Sec. 2.4.2*), higher-order logics vs. propositional calculus (*Sec. 2.5.1*), the disagreement between those for common causes and those for special causes (*Sec. 2.5.2*), the dispute between the objectivists and the subjectivists (*Sec. 3.2.1*), the debate between R analysis and Q analysis (*Sec. 3.2.2*), the competition between the "interval" way of measurement uncertainty and the "uniform probability distribution" way of measurement uncertainty (*Sec. 3.3.1*), the dispute between Einstein and Heisenberg (*Sec. 3.3.2*), the debate between Habermas and Gadamer (*Sec. 3.4.1*), the divide between formal mathematics in the Western world and folk mathematics in the underdeveloped world (*Sec. 3.4.2*), the disagreement between Husserl and Heidegger (*Sec. 3.5.1*), and the debate between the German idealists and the postmodernists (*Sec. 3.5.2*), for instance.

All these are just some illustrative (relatively primitive) examples in our time, of course, but the important point to remember here is that the emergence of new stratifications will further transform probability at the more advanced level in the post-human future that the human world has never known.

The Eighth Great Future Transformation of Probability

The eighth great future transformation of probability in the post-human era concerns what I originally call *new sub-fields*, in relation to culture.

This search for "new sub-fields" will give future advanced humans and later their successors, namely, the post-humans, the ability to explore new specialties for probability in a way that current humans cannot.

There are already some indications that the future of probability is moving toward this direction.

Consider, for illustration, the different sub-fields as already seen in the history of probability, such as the field of mental modeling (*Sec. 2.2.1*), the field of social action (*Sec. 2.2.2*), the field of discrete probability (*Sec. 2.3.1*), the field of complexity science (*Sec. 2.3.2*), the field of reliability engineering (*Sec. 2.4.1*), the field of organizational behavior (*Sec. 2.4.2*), the field of propositional calculus (*Sec. 2.5.1*), the field of common cause (*Sec. 2.5.2*), the field of Bayesian probability (*Sec. 3.2.1*), the field of Q methodology (*Sec. 3.2.2*), the field of metrology (*Sec. 3.3.1*), the field of quantum mechanics (*Sec. 3.3.2*), the field of pragmatics (*Sec. 3.4.1*), the field of the sociology of mathematical knowledge (*Sec.*

3.4.2), the field of phenomenology (*Sec. 3.5.1*), and the field of postmodernism (*Sec. 3.5.2*), for instance.

All these are just some illustrative (relatively primitive) examples in our time, of course, but the important point to remember here is that the search for new sub-fields will further transform probability at the more advanced level in the post-human future that the human world has never known.

The Ninth Great Future Transformation of Probability

And the ninth great future transformation of probability in the post-human era concerns what I originally call *new interpretations*, in relation to culture.

This search for "new interpretations" will give future advanced humans and later their successors, namely, the post-humans, the ability to come up with different evaluations and views about probability in a way that current humans cannot.

There are already some indications that the future of probability is moving toward this direction.

Consider, for illustration, the different interpretations as already seen in the history of probability, such as the argument for mental models (*Sec. 2.2.1*), the Weberian interpretation of rationality (*Sec. 2.2.2*), the argument for normal distribution in dice-rolling (*Sec. 2.3.1*), Bremermann's interrpetation of the fundamental limit on computation (*Sec. 2.3.2*), the physics of failure approach to reliability modeling (*Sec. 2.4.1*), the argument for decision making optimality (*Sec. 2.4.2*), Gödel's incompleteness theorems (*Sec. 2.5.1*), the argument for common-cause variation (*Sec. 2.5.2*), the Dutch book approach to probability (*Sec. 3.2.1*), the interpretation of R anaylsis (*Sec. 3.2.2*), the JCGM's "subjectivist" interpretation of measurement uncertainty (*Sec. 3.3.1*), the Copenhagen's interpretation of quantum mechanics (*Sec. 3.3.2*), the argument for communicative action (*Sec. 3.4.1*), the argument for folk mathematics (*Sec. 3.4.2*), Husserl's argument for pure consciousness (*Sec. 3.5.1*), and Nagel's view on the scientific inquiry about the "subject" (*Sec. 3.5.2*), for instance.

All these are just some illustrative (relatively primitive) examples in our time, of course, but the important point to remember here is that the search for new interpretations will further transform probability at the more advanced level in the post-human future that the human world has never known.

Enormous Implications

These nine great future transformations of probability have enormous implications for some soul-searching questions which have challenged some of the finest minds in the history of probability (in the context of objectivity and subjectivity).

For illustration, consider the following fundamental questions in the history of probability:

- Ex: is probability really so subjective as "the logic of partial belief"?
- Ex: are probabilities to be "objective, logical relations between propositions"?
- Ex: is the study of probability adequate for the understanding of human reasoning?
- Ex: to what extent can probability be applied to different fields?
- Ex: to what extent is mathematics (including probability study) socially constructed?

Questions and disagreements like this will receive new answers in the future history of probability in the advanced human and later post-human eras, at least in two fundamental ways.

The first fundamental way is to examine the relationships between objectivity and subjectivity—such that their possibility (or impossibility) and desirability (or undesirability) are not to the extent that the respective defenders would like us to believe.

The different views about probability, as already addressed in *Sec. 1.1*, *Sec. 1.5*, and the rest of the book, are just different ways to understand probability and its future, with each from its own vantage point.

Each view has its own merits and defects and is preferred over others by different proponents with their own different interests, but none of them is possible (or impossible) and desirable (or undesirable) to the extent that its respective advocates would like us to believe.

And the second fundamental way is to inquire about, and then transcend, in the long haul, the debate about objectivity and subjectivity, such that it will be shifted to different directions, as shown in the nine great future transformations that I have envisioned and outlined above. To transcend them all is to move into these different directions, in conjunction with the various visions of my previous books (as already summarized in *Sec. 1.8* and also earlier in this chapter, with the further help of the tables at this end of Chapter One and this chapter).

Five good examples can be summarized below:

- to seek alternative forms of induction (like new kinds of heuristics in uncertain situations—e.g., in *Sec. 2.4.2*)
- to use different forms of deduction (like different rules of inference, both general and context-specific—e.g., in *Sec. 2.5.1*)
- to explore the world of imagination (like different mental models—e.g., in *Sec. 2.2.1*)
- to explore the power of unconsciousness (be they cognitive, emotional, and behavioral—e.g., in *FPHU*)
- to create new forms of rationality for goals and means (e.g., in *Sec. 3.4.1*—and also in *FPHML*)
- to make good use of qualitative analysis on the "subject," "practice," and "interpretation" (e.g., in Chapter 3—and also in *FPHDA*), not just quantitative analysis

Of course, these examples are illustrative, not exhaustive.

Lest any misunderstanding occurs, one must remember that these nine great future transformations of probability will bring neither utopia nor dystopia, regardless of some optimistic expectations about using computation and mathematical modeling in our time, just to cite two obvious examples—as this is something that I have warned time and again in all of my previous books, almost verbatim.

In any event, as I so often stressed in my previous books, it is not up to us current humans to decide whether or not, or to what extent, the post-human future of probability will be for better and for worse, because it will be judged by the values and beliefs of future post-humans on earth and in deep space unto multiverses—in a way that the entire history of the human world has never known.

Thus is the future of probability—and of its post-human fate.

Table 4.1. Sophisticated Methodological Holism (Part I)

• "My methodological holism implies the partiality-totality principle in the ontology of existential dialectics (see the table on the partiality-totality principle for summary), which is against the varieties of (a) reductionism and (b) reverse-reductionism, in relation to (i) concept, (ii) theory, (iii) methodology, and (iv) ontology." (*FC*)

• "[M]y methodological holism here is not opposed to methodological individualism but includes it (and, for that matter, other methodologies too)…." (*FPHC*) For this reason (and others too, as summarized hereafter), my version of methodological holism is sophisticated—not vulgar as sometimes used by inapt scholars using the same term. (*FC*)

• "[M]y methodological holism does not democratically presume that all levels are equally valid, as all levels are not created equal. In other words, in relation to issue X, level A may be more relevant than level B, but in relation to Y, level B can be more relevant than level A instead." (*FPHC*) One excellent example of this vulgar democratic presumption is what I called in *BNN* "the compromise fallacy." (*FC*)

• My methodological holism does not presume that a lower level of analysis is more important than a higher level, solely because the former serves as the foundation for the latter—and vice versa, for that matter. One excellent example of this reductionistic presumption is what I called in *FPHST* "the foundation fallacy." (*FPHST*, *FC*)

(continued on next page)

Table 4.1. Sophisticated Methodological Holism (Part II)

• "[M]y methodological holism does not make any a-priori postulation that there must be a definite (and, for that matter, indefinite) number of levels" in any analysis. (*FPHC*) Nor does it dogmatically require that there must be a certain combination of levels of analysis in a given inquiry. (*FC*)

• "[M]ethodological holism, in my usage, does not assume that all levels…can necessarily be integrated, since methodological holism is not aimed to search for the holy grail of 'an integral theory…' (as is the case for Wilber). In other words, it allows that sometimes some levels may experience irreducible gaps between them, to be understood, at best, as empirical correlations, not as causal relations…." (*FPHC*)

• "[D]ifferent levels may overlap and even interact with each other in a given context (but sometimes may not), and the fact that I even proposed different ways of re-classifying the levels (whenever needed) in *FDC* reinforces this point….The dual danger here is either forcefully making different levels interact when they are just different (or, metaphorically speaking, apples and oranges) or inappropriately ignoring their interactions when some situations instead require them." (*FPHC*)

• "[T]o understand different levels from their own (unique) perspectives (as required by my methodological holism) is not the same as trying to reduce them to a preferred level in the process of learning from other levels. This second kind of multidisciplinary work is not genuine and does no justice to the unique complexities and merits inherent at each level." (*FPHC*)

(continued on next page)

Table 4.1. Sophisticated Methodological Holism (Part III)

• "[My] methodological holism walks a fine line between the artificial classification (separation) of levels and the simultaneous incorporation of them, if only for the sake of human scholarly endeavor. It should be reminded that nature does not impose upon itself the academic classification of the levels of analysis as humans have. The enterprise of classification is therefore anthropocentric." (*FPHC*)

• "[M]y methodological holism advocates neither epistemic subjectivism nor epistemic non-subjectivism (e.g., realism, idealism, and historicism), neither epistemic relativism (e.g., subjectivism, historicism) nor epistemic absolutism (e.g., realism, positivism), neither epistemic reductionism nor epistemic emergencism, and neither epistemic objectivism (e.g., realism, idealism) nor epistemic historicism....Neither does methodological holism, in my usage, accept the false meta-conceptual dichotomy between nominalism and realism....These false dichotomies...are to be transcended. In other words, methodological holism does not fully accept epistemic realism, positivism (a form of epistemic idealism), historicism, subjectivism, and reductionism in epistemology and philosophy of science but learns from the strengths and weaknesses of all of the opposing approaches without siding with any of them...." (*FPHC*)

• "Sophisticated methodological holism is subject to the constraints as imposed by the syntax of existential dialectics (e.g., the partiality-totality principle and the predictability-unpredictability principle). Even in predictability, outcomes are subject to uncertainty, the degree of which varies from case to case." (*FC*)

(continued on next page)

Table 4.1. Sophisticated Methodological Holism (Part IV)

• "Sophisticated methodological holism—when applied, especially though not exclusively, as illustrated in my numerous works—can enrich the understanding of reality in some distinctive ways. Here are three examples (as revealed in each of my books)." (*FPHCT*; *FPHL*)

—"Firstly, it provides a comprehensive analysis of a subject matter, from which much can be learned about reality," "both in relation to the perspectives of the mind, nature, society, and culture and also in relation to a new classification of the subject matter."

—"Secondly, it suggests some visions of the future in relation to the subject matter in question."

—"And thirdly, it proposes some insights on meta-theory (e.g., methodology and ontology) in general—with the clear understanding, however, of the dilemma of specific vs. general ontology (as shown in the table on the syntax of existential dialectics in the context of the dilemma of ontology). For this very reason, all of these ways are important, without reducing one into the analysis of another."

• "In the end, my meta-theory (both sophisticated methodological holism and existential dialectics) serves as a foundation to unify all domains of knowledge into *an unified theory of everything* (by way of some ontological principles and the comprehensive perspectives of the mind, nature, society, and culture). This is so, without commiting the sins of reductionism and reverse-reductionism (as often seen in many holistic approaches, with the fad of sysems approach—be it about systems theory, chaos theory, complexity theory, or else—as a most recent notorious example, which I debunked in *The Future of Complexity* and also in *The Future of Post-Human Formal Science*). My distinctive approach makes good use of different schools of thought without favoring any of them nor trying to integrate them (as they are not necessarily compatible with each other), so as to adjust for subjectivity, diversity, conflict, and complexity, for example. In this sense, the word "unified" does not have to mean integrative approach, in a narrow sense, nor systems approach, in a broad one, both of which I reject while learning from them." (*FPHFS*; *FPHS*)

Sources: A summary of *Sec.1.2* in *FPHC*—and also from *BNN*, *FPHST*, *ALD*, *FC*, *FPHCT*, and the rest of all other books of mine. See the books for more detail.

Table 4.2. On Reductionism and Reverse-Reductionism (Part I)

- **The Partiality-Totality Principle**
  —The partiality-totality principle in the ontology of existential dialectics targets against the varieties of reductionism and reverse-reductionism (as already worked out in my previous books).

- **Against the Varieties of Reductionism**
  —*Conceptual Reductionism*
  - Some illustrative instances involve myriad dualities like mind vs. body, self vs. world, democracy vs. non-democracy, and the like (as already addressed in *FHC*, *FPHC*, and *BDPD*, for instance).

  —*Theoretical Reductionism*
  - A fascinating case study concerns what I originally called "the foundation fallacy" in *FPHST*, in any attempt to naively understand space-time from the physical perspective as the foundation and, consequently, to dangerously dismiss other perspectives.
  - In *FAE*, I elaborated further these versions of reductionism in the literature on aesthetics (e.g., form vs. content, representation vs. expression, critics vs. artists, and externalism vs. internalism).
  - In *FIA*, I revealed other forms of reductionism in the literature on information architecture (e.g., the constructivist argument).
  - In *FPHU*, I showed the persistent legacy of reductionism, this time, in the literature on anomalous experience (e.g., the obsession with physics, chemistry, and biology for explaining anomalous experience).
  - In *FPHE*, I examined another case of reductionism in action, in the context of engineering (e.g., technical constraints vs. normative constraints).
  - In *FPHMM*, I elaborated one more version of reductionism, in relation to the three domains of communication (e.g., the competing views on sending, connecting, and receiving).
  - In *FPHCT*, I explored another version of reductionism, in relation to invention and innovation (e.g., the bio-psychological argument vs. the socio-cultural arguments).

(continued on next page)

Table 4.2. On Reductionism and Reverse-Reductionism (Part II)

- **Against the Varieties of Reductionism *(cont'd)***
  - *—Theoretical Reductionism (cont'd)*
    - In *FPHG*, I identify another version of reductionism, in relation to infinity, symmetry, and dimensionality (e.g., the Euclidean argument vs. the non-Euclidean arguments).
    - In *FPHUP*, I analyzed another version of reductionism, in relation to density and void (e.g., the engineering argument and the ecology argument).
    - In *FPHL*, I examined another version of reductionism, in relation to structure and context (e.g., the structuralist argument and the contextualist argument).
    - In *PFHO*, I explored another version of reductionism, in relation to communcation, decision-making, and leadership (e.g., the rational-system argument vs. the natural-system argument).
    - In *PFHMA*, I revealed another version of reductionism, in relation to the martial body and spirit (e.g., the spiritual argument vs. the materialist argument vs. the defensive argument).
    - In *PFHS*, I analyzed another version of reductionism, in relation to the sexual body and spirit (e.g., the naturalist argument vs. the constructivist argument).
    - In *PFHLAW*, I examined another version of reductionism, in relation to law (e.g., the necessity argument vs. the contengency argument).
    - In *FPHWP*, I showed another version of reductionism, in relation to war and peace (e.g., the aggressivist argument vs. the pacifist argument).
    - In *BEPE*, I analyzed another version of reductionism, in relation to morality and immorality (e.g., the objectivist argument vs. the non-objectivist argument vs. the skeptical argument).
    - In *BCOS*, I examined another version of reductionism, in relation to the contested beginnings and speculative ends of the universe (e.g., the scientific argument vs. the religious argument vs. the esoteric argument vs. the metaphysical argument).
    - In *FPHP*, I show another version of reductionism, in relation to normality and abnormality (e.g., the natural argument vs. the social argument vs. the cultural argument vs. the mental argument).

(continued on next page)

Table 4.2. On Reductionism and Reverse-Reductionism (Part III)

- **Against the Varieties of Reductionism** ***(cont'd)***
  —*Theoretical Reductionism (cont'd)*
    - In *FPHGEOL*, I scrutinized another version of reductionism, in relation to statics and dynamics (e.g., the catastrophe argument vs. the uniformity argument vs. the revision argument).
    - In *FPHCHESS*, I showed another version of reductionism, in relation to tactics and strategy (e.g., the natural argument vs. the social argument vs. the cultural argument vs. the mental argument).
    - In *FPHR*, I revealed another version of reductionism, in relation to secularness and sacredness (e.g., the critical argument vs. the skeptical argument vs. the theist argument).
    - In *FPHEDU*, I examined another version of reductionism, in relation to teaching and learning (e.g., the teacher-centered argument vs. the student-centered argument vs. the balanced argument).
    - In *FPHH*, I explored another version of reductionism, in relation to joking and laughing (e.g., the natural argument vs. the social argument vs. the cultural argument vs. the mental argument).
    - In *FPHA*, I proposed another version of reductionism, in relation to sound and silence (e.g., the natural argument vs. the social argument vs. the cultural argument vs. the mental argument).
    - In *FPHGAM*, I proposed another version of reductionism, in relation to risk and caution (e.g., the natural argument vs. the social argument vs. the cultural argument vs. the mental argument).
    - In *FPHCOM*, I worked out another version of reductionism, in relation to hardware and software (e.g., the natural argument vs. the social argument vs. the cultural argument vs. the mental argument).
    - In *FPHLIT*, I proposed another version of reductionism, in relation to fiction and non-fiction (e.g., the natural argument vs. the social argument vs. the cultural argument vs. the mental argument).
    - In *FPHCHEM*, I proposed another version of reductionism, in relation to substances and their changes (e.g., the natural argument vs. the social argument vs. the cultural argument vs. the mental argument).

(continued on next page)

Table 4.2. On Reductionism and Reverse-Reductionism (Part IV)

- **Against the Varieties of Reductionism** ***(cont'd)***
  —*Theoretical Reductionism (cont'd)*
    - In *FPHDA*, I proposed another version of reductionism, in relation to quantitative and qualitative research (e.g., the natural argument vs. the social argument vs. the cultural argument vs. the mental argument).
    - In *FPHGEOG*, I proposed another version of reductionism, in relation to environments and their interactions (e.g., the natural argument vs. the social argument vs. the cultural argument vs. the mental argument).
    - In *FPHPA*, I proposed another version of reductionism, in relation to the body and its presence (e.g., the natural argument vs. the social argument vs. the cultural argument vs. the mental argument).
    - In *FPHSEM*, I proposed another version of reductionism, in relation to internality and externality (e.g., the natural argument vs. the social argument vs. the cultural argument vs. the mental argument).
    - In *FPHARCH*, I proposed another version of reductionism, in relation to form and function (e.g., the natural argument vs. the social argument vs. the cultural argument vs. the mental argument).
    - In *FPHCRIM*, I proposed another version of reductionism, in relation to heroes and villains (e.g., the natural argument vs. the social argument vs. the cultural argument vs. the mental argument).
    - In *FPHHIST*, I proposed another version of reductionism, in relation to universality and relativity (e.g., the natural argument vs. the social argument vs. the cultural argument vs. the mental argument).
    - In *FPHMIG*, I proposed another version of reductionism, in relation to Sameness, Otherness, and identity (e.g., the natural argument vs. the social argument vs. the cultural argument vs. the mental argument).

(continued on next page)

Table 4.2. On Reductionism and Reverse-Reductionism (Part V)

- **Against the Varieties of Reductionism *(cont'd)***
  —*Theoretical Reductionism (cont'd)*
    - In *FPHTRAN*, I proposed another version of reductionism, in relation to networks and operations (e.g., the natural argument vs. the social argument vs. the cultural argument vs. the mental argument).
    - In *FPHCA*, I proposed another version of reductionism, in relation to ingredients and techniques (e.g., the natural argument vs. the social argument vs. the cultural argument vs. the mental argument).
    - In *FPHW*, I proposed another version of reductionism, in relation to uselessness and usefulness (e.g., the natural argument vs. the social argument vs. the cultural argument vs. the mental argument).
    - In *BNR*, I proposed another version of reductionism, in relation to diversity and discontinuity (e.g., the natural argument vs. the social argument vs. the cultural argument vs. the mental argument).
    - In *FPHSPORT*, I proposed another version of reductionism, in relation to training and winning (e.g., the natural argument vs. the social argument vs. the cultural argument vs. the mental argument).
    - And in *FPHPROB*, I propose another version of reductionism, in relation to objectivity and subjectivity (e.g., the natural argument vs. the social argument vs. the cultural argument vs. the mental argument).

  —*Methodological Reductionism*
    - A good instance concerns the debate between different versions of qualitative and quantitative methods (as already analyzed in *FC* and also *FHC*). In *FPHML*, I examined similar reductionism, this time, in the literature on mathematical logic (e.g., the obsession with consistency, soundness, and completeness). And in *FPHFS*, I also explored the problems of reductionism in the context of formal science (e.g., the analytical argument).

(continued on next page)

Table 4.2. On Reductionism and Reverse-Reductionism (Part VI)

- **Against the Varieties of Reductionism *(cont'd)***
  —*Ontological Reductionism*
    - An excellent example is the debate between emergentism and reductionism in complexity theory and also in psychology (as elaborated in *FPHC*, in the context of Being and Becoming).

- **Against the Varieties of Reverse-Reductionism**
  —*Conceptual Reverse-Reductionism*
    - Any concept of "art" (e.g., fine art, cave art, outsider art, junk art) is deemed acceptable in postmodernism (as already addressed in Ch.4 of *FHC*).

  —*Theoretical Reverse-Reductionism*
    - There are numerous art and literary theories co-exist. Take the case of literary studies, as there are now Literary Structuralism, Marxist Literary Criticism, New Criticism, Phenomenology, Hermeneutics, Language-Game Literary Criticism, Feminist Literary Criticism, Reception Theory, Reader Response Criticism, Poststructuralism, Semiotics, Pyschoanalytic Literary Criticism, just to cite some well-known ones, with no one being said to be better than any others (as detailedly analyzed in Ch.4 of *FHC*). (S. Raman 1997) In *BNN*, I even introduced "the compromise fallacy" as another good illustration of theoretical reverse-reductionism, in misleadingly treating both genetic and environmental approaches as equally valid.

  —*Methodological Reverse-Reductionism*
    - There is the "anything-goes" mentality in postmodernism (e.g., doing art without praxis, doing art with praxis, and doing art by sublation), as analyzed in Ch.4 of *FHC*. And in *FPHFS*, I also exposed the problems of reverse-reductionism in the context of formal science in relation to systems theory.

(continued on next page)

Table 4.2. On Reductionism and Reverse-Reductionism (Part VII)

- **Against the Varieties of Reverse-Reductionism** ***(cont'd)***
  —*Ontological Reverse-Reductionism*
  - There are likewise no privileged ontology, and the door is open for anything in postmodernism (e.g., the equal status of the ontology of Being vs. that of Becoming, as already addressed in Ch.4 of *FHC*—and also in *FPHC*). In *FAE*, I also introduced another version of reverse-reductionism, that is, "the pluralist fallacy," in the context of understanding aesthetic experience, for instance—although this fallacy has been committed not exclusively in relation to the ontological level (but also at the conceptual, theoretical, and methodological ones).

Sources: From my previous books.

Table 4.3. The Concepton of Existential Dialectics (Part I)

- **Sets and Elements**
  - —Sets
    - Ex: the Same
    - Ex: the Others
  - —Elements
    - Ex: whites in 20th century America (in the set of "the Same")
    - Ex: Iraq during the U.S. invasion in 2003 (in the set of "the Others")

- **Relations, Operations, Functions**
  - —Relations (e.g., "belongs," "equals to," "is greater than")
    - Ex: symmetric interactions within the Same (or the Others)
    - Ex: asymmetric interactions between the Same and the Others
  - —Operations (e.g., "and," "or," "not," "if…then")
    - Ex: if the Same oppresses the Others, it will also oppress itself.
    - Ex: the Same is not the Others.
  - —Functions (e.g., goals)
    - Ex: the Same is hegemonic in relation to the Others.

- **Truth Values**
  - —"1" if True (in Symbolic Logic)
    - Ex: the proposition that imperial Japan was hegemonic to China during WWII
  - —"0" if False (in Symbolic Logic)
    - Ex: the proposition that Grenada invaded France in 2003
  - —"1" & "0" if Both True and False (in Dialectic Logic)
    - Ex: the proposition that the rabbit-duck picture refers to a duck
  - —"~1" & "~0" if Neither True Nor False (or N/A, in Dialectic Logic)
    - Ex: the proposition that God really exists

(continued on next page)

Table 4.3. The Conception of Existential Dialectics (Part II)

- **Axioms and Postulates**
  - —Axioms
    - Ex: the reflexive axiom—"any quantity is equal to itself"
  - —Postulates
    - Ex: the SSS postulate—"if the three sides of a triangle are congruent to their corresponding parts, then the triangles are congruent"

- **Theorems and Principles**
  - —Theorems (and Principles) in Existential Dialectics
    - *In Relation to Method*
      - –#30: The formalness-informalness principle
      - –#29: The absoluteness-relativeness principle
      - –#28: The partiality-totality principle
      - –#27: The predictability-unpredictability principle
      - –#26: The explicability-inexplicability principle
      - –#25: The fiction-reality principle
      - –#24: The cognitiveness-noncognitiveness principle
    - *In Relation to Structure*
      - –#23: The finiteness-transfiniteness principle
      - –#22: The preciseness-vagueness principle
      - –#21: The simpleness-complicatedness principle
      - –#20: The openness-hiddenness principle
      - –#19: The denseness-emptiness principle
    - *In Relation to Process*
      - –#18: The change-constancy principle
      - –#17: The order-chaos principle
      - –#16: The slowness-quickness principle
      - –#15: The expansion-contraction principle

(continued on next page)

Table 4.3. The Conception of Existential Dialectics (Part III)

- **Theorems and Principles (*cont'd*)**
  —Theorems (and Principles) in Existential Dialectics
  - *In Relation to Agency*
    –#14: The theory-praxis principle
    –#13: The convention-novelty principle
    –#12: The evolution-transformation principle
    –#11: The symmetry-asymmetry principle
    –#10: The softness-hardness principle
    –#9: The seriousness-playfulness principle
    –#8: The activeness-inactiveness principle
    –#7: The selfness-otherness principle
  - *In Relation to Outcome*
    –#6: The regression-progression principle
    –#5: The same-difference principle
    –#4: The stability-reaction principle
    –#3: The functionality-nonfunctionality principle
    –#2: The intentionality-nonintentionality principle
    –#1: The survivability-nonsurvivability principle

Notes: The categories and examples in each are solely illustrative (not exhaustive). The comparison is also relative (not absolute), nor are they mutually exclusive. As generalities, they allow exceptions.
Sources: Starting from Ch.6 of *BCPC* and also from other books of mine

Table 4.4. The Syntax of Existential Dialectics I: The Principles (Part I)

- **In Relation to Method**

  —*The Formalness-Informalness Principle*

  (On the Formal Requirements of Logical Systems)

  - The formal requirements of a logical system (e.g., consistency, soundness, and completeness) have both usefulness and non-usefulness, to the extent that, if there are formal systems requiring them, there are alternative ones which do not. Thus, it does *not* exclude classical logics but simply goes beyond both classical and non-classical logics, while learning something from each. There is no formalness without informalness—and vice versa.
  - For instance, existential dialectics can make use of both classical logics under certain conditions (e.g., especially, though not exclusively, when they are clear-cut, etc.) and non-classical logics under alternative conditions (especially, though not exclusively, when they are "unknown," "irrelevant," "ambiguous," "possible," with "different degrees of truth," empirically inconsistent in a desirable way, etc.). (WK 2008u)
  - Family resemblance: e.g., logicalness-nonlogicalness, rationality-nonrationality, etc.
  - Sources: From *FPHML*, *FPHH*, *FPHGAM*, *FPHSEM*, and my later books.

(continued on next page)

Table 4.4. The Syntax of Existential Dialectics I: The Principles (Part II)

- **In Relation to Method *(cont'd)***

—*The Absoluteness-Relativeness Principle*
(On the Multiplicity of Things)

- There is the multiplicity of things in reality, be they about entities, qualities (or properties), and relationships, such that what is acceptable from one standpoint may not be so from another. For instance, if there is something absolute, there is likewise something relative. There is no absoluteness without relativeness—and vice versa.
- Both absoluteness and relativeness here are also relevant to different modalities often cited in the literature on ontology, such as possibility (e.g., something "can" happen) and its opposite (e.g., impossibility) in alethic modalities, probability (e.g., something "will" happen) and its opposite (e.g., improbability) in temporal modalities, and necessity (e.g., something "should" happen) and its opposite (e.g., contingency) in alethic modalities.
- Family resemblance: e.g., independence-dependence, isolatedness-interconnectiveness, universalness-relativeness, immanence-transcendence, autonomy-interdependency, internalness-externalness, etc.
- Sources: From *FPHK*. See also my later books.

(continued on next page)

Table 4.4. The Syntax of Existential Dialectics I: The Principles (Part III)

• **In Relation to Method *(cont'd)***

—*The Partiality-Totality Principle*

(On the Relationships between Whole and Parts)

- The whole is not the sum of the parts. There is no partiality without totality—and vice versa.
- Any inquiry about a phenomenon in the work is to guard against the varieties of (a) reductionism and (b) reverse-reductionism.
- Reductionism and reverse-reductionism can be (i) conceptual, (ii) theoretical, (iii) methodological, and (iv) ontological.
- Works on emergence (as in *FC*), gestalt psychology (as in *FPHGEOL*), the hermeneutic circle (as in *FPHDA*), and mental language (as in *FPHSEM*), to name just a few of them, have contributed to the understanding of whole and parts.
- Family resemblance: e.g., individualisticness-holisticness, partness-wholeness, analysis-synthesis, etc.
- Sources: Especially from *FC* & *FPHGEOL*. See also my later books.

(continued on next page)

Table 4.4. The Syntax of Existential Dialectics I: The Principles (Part IV)

- **In Relation to Method *(cont'd)***
  —*The Predictability-Unpredictability Principle*
  (On the Occurrence of Events)
  - Both predictability and unpredictability have a major role to play in the occurrence of things, so that neither determinism nor indeterminism wins the centuries-old fight. There is no predictability without unpredictability—and vice versa.
  - There are events which are predictable, just as there are those which are not. Or what is regarded as unpredictable at one point in time may turn out to be predictable later, and, conversely, what is deemed as predictable may turn out to not be so predictable. Even in predictability, outcomes are subject to uncertainty, the degree of which varies from case to case.
  - Works on complexity theory (as in *FC*), quantum mechanics (as in *FPHST* and *FPHG*), frequency probability, subjective (or Bayesian) probability (as in *FPHPROB*), "special causes" (as in *FPHGAM*), and measurement uncertainty (as in *FPHPROB*), just to name a few of them, have contributed to the understanding of uncertainity.
  - The predictability-unpredictability principle is not to be confused with the absoluteness-relativeness principle; although both can address the issue of probability, the former is concerned with the extent of certainty in prediction, whereas the latter focuses on the relativity of viewpoints.
  - Family resemblance: e.g., sureness-arbitrariness, certainty-uncertainty, deterministicness-indeterministicness, sureness-doubtfulness, likeliness-riskiness, deterministicness-randomness, reliability-unreliability, etc.
  - Sources: Especially from *FC* & *FPHGAM.* See also my later books.

(continued on next page)

**Table 4.4. The Syntax of Existential Dialectics I:
The Principles
(Part V)**

- **In Relation to Method *(cont'd)***

—The Explicability-Inexplicability Principle
(On the Underlying Mechanisms of Things)

- Both explicability and inexplicability are part of the understanding of things. There is no explicability without inexplicability—and vice versa.
- This principle tells us the duality of the research dilemma, in that, if reality can be explained in some ways, it also has its other ways which are not quite explainable, at a given point in time.
- Family resemblance: e.g., underlyingness-regularness, causation-regularness, causation-correlation, explanation-description, explanation-narration, explanation-interpretation, etc.
- Sources: Especially from *FPHU*—and also *FPHC*. See also my later books.

—The Fiction-Reality Principle
(On the different faces of Reality)

- If there is reality, there is also fiction, especially for any agent with some kind of intelligence life, be it about humans, animals, or, later, post-humans. There is no reality without fiction—and vice versa.
- The distinction between reality and fiction is not absolute, since there can be a mixture of both. Fiction can have non-fictional elements, just as non-fiction can have fictional ones. So, the distinction is in degree, not in kind.
- Family resemblance: e.g., fiction-nonfiction, fiction-fact, imagination-fact, subjectivity-objectivity, imagination-nonfiction, inaccuracy-accuracy, etc.
- Sources: First from *FPHLIT*.

(continued on next page)

Table 4.4. The Syntax of Existential Dialectics I: The Principles (Part VI)

- **In Relation to Method *(cont'd)***

  —*The Cognitiveness-Noncognitiveness Principle*
  (On the Understanding of Meaning)

  - If a view of things is based on reason and evidences, it is also affected by non-cognitive factors (such that it is neither true nor false in the latter case)—and vice versa. In other words, there is no cognitiveness without noncognitiveness—and vice versa.
  - Noncognitive factors can come in all shapes and sizes, and good examples include envy, jealousy, power, nationality, race, gender, age, class, greed, lust, status, faith, anger, sadness, joy, fear, wish, etc.
  - Communication, even in its ideal type, is seldom solely based reason and evidences, as falsely assumed during the Age of the Enlightenment and its continued supporters nowadays. What is understood about something is often twisted on the basis of something else too. In these cases, what is at stake is often neither true nor false (which is something that classical logic has sidelined and ignored).
  - Family resemblance: e.g., proposition-attitude, cognition-emotion, cognition-prescription, justification-nonjustification, theory-extratheory, etc.
  - Sources: First from *BNR*.

(continued on next page)

Table 4.4. The Syntax of Existential Dialectics I: The Principles (Part VII)

- **In Relation to Structure**

—*The Finiteness-Transfiniteness Principle*

(On the Nature of Numbers)

- If there are finite things, there are likewise transfinite ones. There is no finiteness without transfiniteness—and vice versa.
- To avoid confusion, my usage of the word "transfinite" here differs radically from the one used by Cantor (and other mathematicians) for "relative" infinity—and is more limited, in light of the problems confronting any attempt to understand the idea of infinity, be it by intuition, imagination, and conception (as detailedly analyzed in *Sec. 2.2.3* of *FPHG*).
- Instead, by "transfinity," I allow numbers which can be many times larger—or smaller, for that matter—than the finite things that we encounter in daily life, but they do not have to be related to the idea of infinity at all (which may not exist).
- Of course, there may be some *borderline* cases, in which it is not clear whether the number in question is transfinite (in my usage) or simply a mathematical convenience. A good example of a borderline case is the Planck unit of length for "the smallest space possibly measured in nature," which is "less than billionths of trillionths of trillionths of an inch" (or something like 1.6×10^{-35} meters).
- That said—my usage of "transfinity" can also resolve (or better, dissolve) an age-old problem in philosophy known as "Zeno's paradoxes" (as already explained in *Sec. 2.2.3* of *FPHG*).
- Family resemblance: e.g., boundedness-quasiunboundedness, smallness-largeness, boundedness-unboundedness, etc.
- Source: From *FPHG*. See also my later books.

(continued on next page)

**Table 4.4. The Syntax of Existential Dialectics I:
The Principles
(Part VIII)**

- **In Relation to Structure *(cont'd)***
  —*The Preciseness-Vagueness Principle*
  (On the Refinement of Things)
  - Both preciseness and vagueness are important, not that one is better than the other, but that both are used, in different degrees of preference, in accordance to the contextual application from the perspectives of nature, the mind, culture, and society. Even when both are used in a combination, the dilemma is shifted instead to one of combinational concern. There is no preciseness without vagueness—and vice versa.
  - In relation to taxonomy, preciseness has its taxonomic clarity, just as vagueness has its classificatory flexibility, for instance. And in relation to network, vagueness has its explorative liberty, just as preciseness has its conceptual definitiveness, for instance.
  - Family resemblance: e.g., clarity-ambiguity, directness-indirectness, quantitativeness-qualitativeness, describability-nondescribability, sharpness-roughness, specificity-generality, concreteness-abstractness, specificity-obscurity, translatability-untranslatability, clearcutness-transitionalness, etc.
  - Sources: Especially from *FIA*. See also my later books.

(continued on next page)

Table 4.4. The Syntax of Existential Dialectics I: The Principles (Part IX)

- **In Relation to Structure** ***(cont'd)***

  —*The Simpleness-Complicatedness Principle*

  (On the Interconnection among Things)

  - Both simpleness and complicatedness are vital, without favoring one over the other, in that each is utilized, depending on the basis of the perspectives of nature, the mind, culture, and society. And even when a combination of them is preferred, the dilemma is only shifted to a combinational degree of concern. There is no simpleness without complicatedness—and vice versa.
  - For illustration, in relation to taxonomy, simpleness has its heuristic usefulness, just as complicatedness has its realistic representation, for instance. In relation to network, simpleness has its economical attractiveness, just as complicatedness has its practical reliability, for instance. And in relation to logic, there is the simple zeroth-order logic, just as there is the complicated higher-order logic.
  - Family resemblance: e.g., inflexibility-flexibility, standardization-specialization, imperfectness-perfectness, superficiality-depth, shallowness-deepness, economicalness-elaboratedness, plainness-circumspection, onesidedness-multisidedness, succinctness-duplicatedness, simpleness-sophisticatedness, crudeness-subtleness, etc.
  - Sources: Especially from *FIA* and *FPHPROB*. See also my later books.

(continued on next page)

Table 4.4. The Syntax of Existential Dialectics I: The Principles (Part X)

- **In Relation to Structure *(cont'd)***
  —*The Openness-Hiddenness Principle*
  (On the Detection of Things)
  - Reality has its hidden face, just as it is open to outside view in some other ways. There is no openness without hiddenness—and vice versa.
  - For instance, in the context of anomalous experience, certain aspects of reality can be open for examination (e.g., the experiment with SPECT images, the transmission of telepathemic bit, and the use of metonymies and metaphors). Other aspects, however, remain hidden, and examples include the elusive deeper nature of spiritual reality in the intangible realm, the undetectability of different branched universes, the underlying mechanisms of signals traveling between different folds of physical space-time, and the dependence on language and the bias for science and logic).
  - Family resemblance: e.g., overtness-covertness, publicness-privateness, openness-closedness, transparency-secrecy, openness-biasedness, participation-obstruction, exposition-illusion, surface-depth, etc.
  - Sources: Especially from *FPHU*. See also my later books.

(continued on next page)

Table 4.4. The Syntax of Existential Dialectics I: The Principles (Part XI)

- **In Relation to Structure *(cont'd)***
  —*The Denseness-Emptiness Principle*
  (On the Distribution of Entities in Space)
  - Both density and void are needed, in relation to the mind, nature, culture, and society, albeit in different ways. There is no denseness without emptiness—and vice versa.
  - For clarity, the term "void" is used here only as an approximation of emptiness (depending on the degree of the lack of density), since, in physics, it is well known that "empty" space is not really empty all the way, because it can be full of energy (e.g., random quantum fluctuations at the sub-atomic level, and, for that matter, dark energy in the universe) and matter (e.g., different versions of sub-atomic particles, and, for that matter, dark matter in the universe).
  - Family resemblance: e.g., fullness-voidness, nearness-farness, concentration-dispersion, denseness-nothingness, presence-absence, focus-unfocus, solidness-void, something-nothing, heaviness-lightness, composition-emptiness, emphasis-disemphasis, etc.
  - Sources: First worked out in *FPHUP*. See also my later books.

- **In Relation to Process**
  —*The Change-Constancy Principle*
  (On the Alteration of Things)
  - Change occurs over time, although constancy is also allowed. There is no change without constancy—and vice versa.
  - Asymmetry undergoes changes over time, so does symmetry.
  - Old players fade away, and new ones emerges, with ever new causes and ever new forms.
  - Family resemblance: e.g., dynamicness-staticness, instability-stability, etc.
  - Sources: First named in *BCPC*. Especially from *FHC*, *FCD*, and *FPHC*. See also other books of mine (from that point on).

(continued on next page)

Table 4.4. The Syntax of Existential Dialectics I:
The Principles
(Part XII)

- **In Relation to Process *(cont'd)***
  —*The Order-Chaos Principle*
  (On the Pattern of Things)
  - Both order and chaos are vital in the process of change in the world. The preference for order is biased, since it does not give sufficient attention to the vital role of chaos in the transformation of the world (without somehow reducing it for the understanding of order). There is no order without chaos—and vice versa.
  - The scientific search for order in the world is often a hidden bias in its ontological obsession with order, since chaos is often treated as the "bad" guy, with order as the "good" guy (for the end goal of science).
  - Neither order nor chaos is the final end of the world, and one is not to be treated as the means for the other in the transformation of things. Both are fundamental in their recurrent dialectical interactions with each other over time, without reducing one for the other.
  - Family resemblance: e.g., lawfulness-disorder, order-disorder, linearity-nonlinearity, linearity-irreversalness, pattern-chaos, unity-disunity, etc.
  - Sources: Especially from *FC*. See also my later books.

(continued on next page)

Table 4.4. The Syntax of Existential Dialectics I: The Principles (Part XIII)

- **In Relation to Process *(cont'd)***
  —*The Slowness-Quickness Principle*
  (On the Speed of Change)
  - Both slowness and quickness co-exist, with their own internal tension, to the extent that each fights for its own relevance with the other, in accordance to the perspectives of nature, the mind, culture, and society, without one being the victor and the other being the vanquished in the long haul. Even when both are chosen in other cases, this dilemma is only transferred into something else with a combinational character. There is no slowness without quickness—and vice versa.
  - In relation to taxonomy, quickness has its efficient usability, just as slowness has its aesthetic appeal, for instance. And in relation to network, quickness has its adventurous readiness, just as slowness has its risk-adverse convenience, for instance.
  - Family resemblance: e.g., inconvenience-convenience, passiveness-activenessness, gradualness-abruptness, deceleration-acceleration, difficulty-ease, nonresilience-resilience, patience-impatience, etc.
  - Sources: Especially from *FIA*. See also my later books.

(continued on next page)

Table 4.4. The Syntax of Existential Dialectics I: The Principles (Part XIV)

- **In Relation to Process *(cont'd)***
  —*The Expansion-Contraction Principle*
  (On the Growth of Things)
  - Entities in the world can both expand in some ways and contract in other ones, as part of their nature. There is no expansion without contraction—and vice versa.
  - For instance, in the context of anomalous experience, one of its most fascinating expressions is none other than the analysis of different causes of the slow but steady expansion (and, furthermore, if put in a different way, unfolding) of unconsciousness.
  - Family resemblance: e.g., conquest-autarky, rise-fall, spread-shrink, extendingness-shorteningness, widening-narrowing, expansion-destruction, expansion-decline, construction-demolition, creation-destruction, construction-deconstruction, expansion-contraction, expansion-constriction, etc.
  - Sources: Especially from *FPHU*. See also my later books.

(continued on next page)

Table 4.4. The Syntax of Existential Dialectics I: The Principles (Part XV)

- **In Relation to Agency**
  —*The Theory-Praxis Principle*
  (On the Duality of Knowledge))
  - This has to do with the duality of knowledge in existential dialectics, in that, if there is theoretical construction, there is likewise its practical application, both technical and normative. There is no theory without praxis—and vice versa.
  - Family resemblance: e.g., discovery-application, knowledge-action, invention-innovation, knowledge-motivation, theory-intentionality, universality-particularity, discourse-action, knowledge-interest, potentiality-actuality, theory-practice, etc.
  - Source: Especially from *FPHE*. See also my later books.

(continued on next page)

Table 4.4. The Syntax of Existential Dialectics I: The Principles (Part XVI)

- **In Relation to Agency *(cont'd)***
  - —*The Convention-Novelty Principle*
    (On the Nature of Creative Thinking)
    - If there is conventional wisdom, there is likewise novel challenge, to the extent that both convergent and divergent thinking are part of life. There is no convention without novelty—and vice versa.
    - In addition, in this context of the convention-novelty principle, there are, in the absence of better words, what I want to call (a) *creative techniques* and (b) *creative traits*, which, when satisfied—in relation to the larger context of the mind, nature, society, and culture—can be used to enhance creative works. The list of creative techniques and traits (as summarized in *Table 4.1* of *FPHCT*) are not exhaustive, of course, but the examples here constitute a great beginning of understanding the structure of creative thinking.
    - Yet, lest the reader gets carried away by the euphoria about creative thinking in our time, it should be reminded that creative thinking has its own possibilities and limits (as shown in *Table 2.1* of *FPHCT* on invention), just as it has its own promises and pitfalls (as shown in *Table 3.1* of *FPHCT* on innovation). In the end, it should also be stressed that creative thinking has its own desirability and dark sides (as shown in *Table 4.2* of *FPHCT*, *BNN*, and *FPHMIG*).
    - This has important daily implications for an average individual, and the word "average" here allows different degrees of variation from one individual to another in a population, at any given point in history.
    - Family resemblance: e.g., normalness-nonnormalness, conformity-nonconformity, convergence-divergence, familiarity-freshness, inertia-newness, equilibrium-disturbance, etc.
    - Sources: First from *FPHCT*. See also my later books.

(*continued on next page*)

Table 4.4. The Syntax of Existential Dialectics I: The Principles (Part XVII)

- **In Relation to Agency *(cont'd)***
  —*The Evolution-Transformation Principle*
  (On the Multiple Kinds of Agency)
  - This principle (and the symmetry-asymmetry principle) are both about *the agency of change*. The word "agency," in a formal definition, refers to "a person or thing through which power is exerted or an end is achieved." (MWD 2007b) It therefore does not have to necessarily involve an intelligent lifeform.
  - Because of this dual meaning in agency, the evolution-transformation principle is more concerned with *the multiple kinds of agency*, that is, both about the *evolution* in the state of nature (e.g., an object of natural beauty) and the *transformation* in the world of intelligent lifeforms (e.g., a work of art, an air-conditioner, etc.). There is no evolution without transformation—and vice versa.
  - And the transformative part of the principle precisely refers to the other dimension in the dual meaning of agency, in giving technology a major role to play in the change of the world, which is something that I extensively analyzed in *FHC* in the context of the technophilic lifeworld, espeically though not exclusively since modern times. The same theme is also elaborated in other books of mine.
  - Family resemblance: e.g., nonwillingness-willingness, naturalness-technologicalness, evolution-creation, naturalness-nonnaturalness, inorganicness-volition, naturalness-artificialness, inorganicness-motivation, etc.
  - Sources: Especially from *FAE*. See also other books of mine.

(*continued on next page*)

Table 4.4. The Syntax of Existential Dialectics I: The Principles (Part XVIII)

- **In Relation to Agency** ***(cont'd)***
  —*The Symmetry-Asymmetry Principle*
  (On the Relationships among Existents)
  - There is no symmetry without asymmetry—and vice versa.
  - For instance, the Same can be symmetric and asymmetric towards the Others. But in case of asymmetry, oppression and self-oppression can occur. So, when the Same is asymmetric towards the Others, the Same can also be relatively asymmetric towards itself in self-oppression, just as the Others can be likewise towards themselves.
  - The subsequent oppressiveness is dualistic, as much by the Same against the Others and itself, as by the Others against the Same and themselves.
  - Both oppression and self-oppression can be achieved by way of downgrading differences between the Same and the Others and of accentuating them.
  - This is true, even though not all forms of asymmetry have to be about oppression and self-oppression.
  - In addition, from Chapter Three of *FPHG*, symmetry is not perfect, to be understood in an approximate sense under many life circumstances.
  - Family resemblance: e.g., balance-extremity, reflexiveness-unreflexiveness, balance-unbalance, equality-inequality, center-periphery, compatibility-incompatibility, proportion-disproportion, harmoniousness-unharrmoniousness, etc.
  - Sources: From all my books, starting with *FHC*. First named in *BCPC*. See, for instance, *FCD* for more details on the Same and the Others.

(*continued on next page*)

Table 4.4. The Syntax of Existential Dialectics I: The Principles (Part XIX)

- **In Relation to Agency *(cont'd)***
  —*The Softness-Hardness Principle*
  (On the Force of Change)
  - This has to do with *the force of change* under the category about agency in existential dialectics, in that any change by an agent, be it organic (like humans) or non-organic (like natural objects), can occur in a forceful (aggressive) or gentle (pacific) way, which can come in all shapes and sizes, of course. There is no softness without hardness—and vice versa.
  - Family resemblance: e.g., peacefulness-violence, cooperation-competition, acceptance-discrimination, love-hatred, peace-war, pacificity-aggression, dovishness-hawkishness, reward-punishment, kindness-hostility, admiration-belittlement, inclusion-exclusion, friendliness-unfriendliness, like-dislike, etc.
  - Sources: Especially from *ALD*. See also my later books.

(continued on next page)

Table 4.4. The Syntax of Existential Dialectics I: The Principles (Part XX)

- **In Relation to Agency *(cont'd)***
  —*The Seriousness-Playfulness Principle*
  (On the Extent of Seriousness)
  - If there is seriousness, there is also playfulness, especially for any agent with some kind of intelligence life, be it about humans, animals, or, later, post-humans. There is no seriousness without playfulness—and vice versa.
  - Lest any misunderstanding occurs, the word "playfulness" here should not be confused with other terms like "laughter," "smile," "joke," "tickleness," and "laugh-like vocalization" (in some animals), for instance.
  - Being playful can come in all shapes and sizes and therefore is not necessarily tied up with any of these terms, alhough it can result from any of them, or a combination of them, or something else altogether.
  - This principle, together with some other principles (especially though not exclusively, like the formalness-informalness principle), have important implications for what constitutes rationality, be it about "rationality of application" at the practical level or "rationality of knowledge" at the meta-theoretical level (as summarized in *Sec. 4.20* of *FPHH*).
  - Family resemblance: e.g., seriousness-jokingness, seriousness-humorousness, seriousness-wittiness, seriousness-nonseriousness, seriousness-relaxedness, seriousness-restingness, Apollo-Dionysus, etc.
  - Sources: First from *FPHH & FPHGAM* See also my later books.

(*continued on next page*)

Table 4.4. The Syntax of Existential Dialectics I: The Principles (Part XXI)

- **In Relation to Agency *(cont'd)***
  - —*The Activeness-Inactiveness Principle*
    (On the Extent of Activeness)
    - If there is activeness, there is also inactiveness, especially (though not exclusively) for any agent with some kind of intelligence life, be it about humans, animals, or, later, post-humans. There is no activeness without inactiveness—and vice versa.
    - Being active can come in all shapes and sizes, as a person can be inactive because he is tired, old, sick, sleepy, uninterested, etc., just as an object can be inactive because it is at rest, out of work, etc.
    - This principle, together with some other principles (especially though not exclusively, like the formalness-informalness principle and seriousness-playfulness principle), have important implications for what constitutes rationality, be it about "rationality of application" at the practical level or "rationality of knowledge" at the meta-theoretical level (as summarized in *Sec. 4.20* of *FPHH*).
      - Family resemblance: e.g., alertness-unalertness, activeness-decrepidness, alertness-sleepiness, awakenness-hibernatedness, activeness-tiredness, alertness-uninterestedness, alertness-unwareness, soberness-unawareness, soberness-inactiveness, alertness-inactiveness, etc.
    - Sources: First from *FPHTRAN*.

(*continued on next page*)

Table 4.4. The Syntax of Existential Dialectics I: The Principles (Part XXII)

- **In Relation to Agency *(cont'd)***
  —*The Selfness-Otherness Principle*
  (On the Magnitude of Self-Centeredness)
  - If there is the tendency to be mostly centered on one's self or group, there is likewise the tendency to be mostly centered on others, especially (though not exclusively) for any agent with some kind of intelligence life, be it about humans, animals, or, later, post-humans. There is no selfness without otherness—and vice versa.
  - Being self-centered and being other-centered do not have to be mutually exclusive, as the two can come in all shapes and sizes, or in different degrees. And other-centeredness does not entail desirability, any more than self-centeredness implies non-desirability, as both can go either way.
  - In addition, whether self-centeredness constitutes a moral alternative or, conversely, an alternative to morality is of course debatable (as addressed in my other book titled *Beyond Ethics to Post-Ethics*).
  - Good examples of family resemblance in relation to the principle are egoisticness-altruisticness, selfishness-unselfishness, selfness-selflessness, egocentricness-othercentricness, and so on.
  - Source: First from *FPHCA*

(*continued on next page*)

Table 4.4. The Syntax of Existential Dialectics I: The Principles (Part XXIII)

- **In Relation to Outcome**

—*The Regression-Progression Principle*
(On the Direction of History)

- Neither the cyclical nor the linear views are adequate for explaining many phenomena at all levels. There is regression without progression—and vice versa.
- History progresses to more advanced forms, but with a regressive touch. Examples include no freedom without unfreedom, no equality without inequality, and no civilization without barbarity.
- This is not an inevitable law, but merely a highly likely empirical trend.
- Family resemblance: e.g., cost-benefit, undesirability-desirability, badness-goodness, risk-opportunity, dysfunctionality-functionality, cyclicity-linearity, etc.
- Sources: From all my books, starting with *FHC*. First named in *BCPC*.

—*The Same-Difference Principle*
(On the Metamorphosis of Change)

- An entity, as it evolves over time, can be both different from and similar to its opposing alternatives and does not have to be solely more different from them over time. There is no similarity without difference—and vice versa.
- Opposites are not absolute in a black-or-white fashion; so, an entity can become relatively more similar to (or more different from) its opposite over time.
- Family resemblance: e.g., homogeneity-heterogeneity, we-they, similarity-contrast, etc.
- Sources: Especially from ALD. See also my later books

(*continued on next page*)

**Table 4.4. The Syntax of Existential Dialectics I:
The Principles
(Part XXIV)**

- **In Relation to Outcome *(cont'd)***
  —*The Stability-Reaction Principle*
  (On the Feedback Mechanism for Further Changes)
  - Once an outcome becomes stabilized over time (which varies from case to case), it will undergo further changes in its multifaceted interactions with the outside world (e.g., with other entities, energy, space, and so on).
  - There is no stability without reaction for further changes—and vice versa.
  - Family resemblance: e.g., stability-forwardness, balance-reaction, balance-forwardness, rest-motion, normalcy-feedback, stability-breakup, consolidation-modification, etc.
  - Sources: First from *FPHCHEM.*

(*continued on next page*)

Table 4.4. The Syntax of Existential Dialectics I: The Principles (Part XXV)

- **In Relation to Outcome *(cont'd)***

  —*The Functionality-Nonfunctionality Principle*
  (On the Presence of Function)

  - If there is the presence of action for which something or someone is fitted, there is likewise the relative lack of action for which something or someone is fitted. There is no functionality without nonfunctionallity—and vice versa.
  - A *function* should not be confused with an *alternative function*, which is just another kind of function; just as *nonfunction* should not be confused with *dysfunction* (which has more to do with malfunction and is thus related more to the regression-progression principle).
  - In addition, being functional (or useful) and being nonfunctional (or useless) do not have to be mutually exclusive, as the two can come in all shapes and sizes, or in different degrees. And what is functional at one time may become nonfunctional at another time, and vice-versa.
  - Good examples of family resemblance in relation to the principle are usefulness-nonusefulness, usefulness-uselessness, valueness-nonvalueness, and the like.
  - Source: First from *FPHCA*

(*continued on next page*)

Table 4.4. The Syntax of Existential Dialectics I: The Principles (Part XXVI)

- **In Relation to Outcome *(cont'd)***
  —*The Intentionality-Nonintentionality Principle*
  (On the Planning for Outcomes)
  - If there is the planning or design of something for a certain outcome, there is likewise the relative absence of planning or design of something for a certain outcome. There is no intentionality without nonintentionallity—and vice versa.
  - An intention for an outcome does not have to be certain, so the outcome can even be counter-intentional in being the opposite of what was originally planned. In this sense, counter-intentionality is a special case of non-intentionality; in other words, there are unintented consequences which can be just the opposite of what was originally planned.
  - An intention for an outcome is mostly relevant to intelligent life, but there are non-intelligent life and non-living matter in the universe too, so nonintentionality is also important.
  - In addition, an intention does not have to be obvious or straightforward, as William Shakspeare once famously said: "I must be cruel only to be kind."
  - Good examples of family resemblance in relation to the principle are purposiveness-nonpurposiveness, intention-accidentalness, design-randomness, etc.
  - Source: First from *FPHW*

(*continued on next page*)

Table 4.4. The Syntax of Existential Dialectics I: The Principles (Part XXVII)

- **In Relation to Outcome *(cont'd)***
  - *—The Survivability-Nonsurvivability Principle (On the Survival of things)*
    - If there are some entities which survive at certain times, there are likewise other entities which do not survive. There is no survivability without nonsurvivability—and vice versa. In nature, for illustration, there are some animals which survive in hard times, whereas others do not—and this is true, not just among different species but also among different members of the same species (and the phenomenon of "selection" in evolutionary biology has much to say on this, as already addressed in *BNN*).
    - Of course, the principle is not restricted solely to the animal world, since it can be used to other cases concerning non-animals too (e.g., trees, lakes, etc. in the natural world—or empires, companies, etc. in the human world).
    - But one should not confuse this principle with the expansion-contraction principle, since those entities which survive, say, can still expand or contract over time (e.g., those animals which survive can be more capable in some activities but not in others, for instance).
    - Nor should this principle be confused with the stability-reaction principle, in that the condition of survivability or non-survivability may lead to further changes later on (as shown in the emergence of new species after the extinction of some old ones, as already explained in *BNR*).
    - Good examples of family resemblance in relation to the principle are survival-extinction, life-death, existence-nonexistence, and the like.
    - Source: First from *FPHPROB*

Notes: The examples here are solely illustrative (not exhaustive). The comparison is also relative (not absolute), nor are they mutually exclusive. As generalities, they allow exceptions.

Table 4.5. The Syntax of Existential Dialectics II: The Principles as Short Cuts (Part I)

• The principles should be treated with caution, lest misunderstanding occurs, since they do not constitute rigid dualities (or dichotomies).

• The reason is that each pair in an ontological principle consists of two opposites, which are, however, merely short cuts both for multiple variations and degrees, as well as for different interactions with multiple other entities unlike them. In this light, each pair can end up having hundred (or even thousand, if not more) different versions, which interact with hundred (or even thousand, if not more) other entities. There are two clarifications here.

• Firstly, this conception of shortcuts is not mutually exclusive nor absolute, in that the opposites can come in all shapes and sizes, with different degrees. For instance, by analogy, just as there are different degrees of the two colors "white" and "black"—there are likewise different degrees of the opposites in each ontological pair, to the extent that there can be multiple entities (not only two) interacting with each other in each pair.

• And secondly, this conception of shortcuts do not ignore other possible entities in interacting with the two opposites (with their different versions) in each pair. For instance, by analogy, there are not only the two colors "white" and "black" as opposites, since there are other colors too besides them like "yellow," "green," "purple," or else. The same logic can be applied to each ontological pair, in that they also interact with other entities, not with only two of them (with their different versions).

• Therefore, with these two clarifications in mind—each pair in an ontological principle serves only as an abbreviation for something more complicated and, therefore, although it contains two opposites, it should not be confused as a duality (dualism).

(continued on next page)

Table 4.5. The Syntax of Existential Dialectics II: The Principles as Short Cuts (Part II)

• Existential dialectics rejects any dualism (or dichotomy) as too rigid and instead allows the multiplicity of entities, to the extent that between the two opposites in each pair exist many other alternatives to choose from. They are named in that short form for aesthetic elegance, instead of listing all possible entities between the two opposites in the title.

• To be dialectic is to go beyond any rigid dichotomy and transcend into something different altogether in the long haul. One may be tempted to call the dialectic logic here with a different name like existential "multilectics" (instead of "dialectics"), but this naming is incorrect (or even misleading), for the two reasons aforestated.

• After all, each ontological pair come in all shapes and sizes (with different degrees) and do not exist by themselves but also interact with other entities unlike them. The virtue of revealing an ontological pair is to show how they relate within themselves (in multiple versions) and also interact with others unlike them (also in multiple versions).

• This will be clear in the table on "the dialectic constraints imposed by the principles."

Source: From Ch.1 of *FPHK*

Table 4.6. The Syntax of Existential Dialectics III: The Principles as Family Resemblances

• Each ontological principle is generic, with some other comparable ontological pairs to be put in the same family (like a *family resemblance*). Ludwig Wittgenstein (1953) once suggested the idea of "family resemblance" in explaining different games classified under the same family called *games*.

• Why should, for instance, playing football and chess as playing "games," when it is well understood that football is not the same as chess? The answer is that, although each game is different and has different rules, many of them (though not all) share, more or less, some commonalites (e.g., scoring as necessary for winning).

• And this is so, even though some games share more than some others in any given selection of critieria, and no two games are exactly identical. So, his point here is that there is no essential core which is common to all games, and the best that one can look for is some characteristics which are common to many (but not all) games. (A. Biletzki 2006)

• By the same logic—in the previous section on selection criteria, the flexibility-inflexibility pair can be put in the family resemblance of the simpleness-complicatedness principle, although the two pairs are not exactly identical. Likewise, the directness-indirectness pair can be put in the family resemblance of the preciseness-vagueness principle, although, again, the two pairs are not exactly identical.

• In this sense, which specific pair in a family should be used to designate the name of the family can be at times a bit arbitrary, but with good reason.

• For illustration, in the context of method, the partiality-totality principle can take the different form like individualisticness-holisticness, just as the explicability-inexplicability principle can take the different form like underlyingness-regularness—although each two pairs are not exactly identical and have slightly different meanings and usages. A more comprehensive listing of this family resemblance for all other principles is shown in the table on the syntax of existential dialectics.

Source: From Ch.1 of *FPHK*

Table 4.7. The Syntax of Existential Dialectics IV: The Dialectic Constraints Imposed by the Principles (Part I)

- **Co-Existent and Asymmetric**

—The principles, as they constitute the syntax of existential dialectics, are dialectic in character, such that, when they are applied, they impose dialectic constraints on how reality is to be understood. Consider, say, the symmetry-asymmetry principle as an illustration here, in order to summarize two main characters of the dialectic constraints in question.

—Firstly, to be dialectic here is to go beyond the narrow dichotomies (and, for that matter, any rigid multi-dimensional classificatory scheme), be they about "self" vs. "world," "freedom" vs. "unfreedom," "barbarity" vs. "civilization," "individuality" vs. "communality," and so on.

—One way to do so (to go beyond) is to consider them all in terms of co-existence (without favoring one over the rest). For instance, my theory of "post-civilization" (to be summarized later in the section on the pragmatics of existential dialectics) is to go beyond barbarity and civilization in terms of understanding barbarity and civilization as being co-existent. And the same logic can be said in relation to my theories of "post-democracy," "post-capitalsm," and others (also to be introduced later in the section on the pragmatics of existential dialectics), in regard to freedom vs. unfreedom, equality vs. inequality, communality vs. individuality, spirituality vs. materiality, and so on.

—But to consider them all (in the dichotomies—and, for that matter, in any rigid multi-dimensional classifactory schme) as co-existent is not the same as to imply that the opposites in any classificatory scheme are all equal, since, in accordance to the symmetry-asymmetry principle (as an illustration here), if they are equal in terms of being considered as co-existent, they are asymmetric in terms of being unequal in dialectic interaction (e.g., X can be more dominant than Y in case A, or Y is more dominant than X in case B).

(continued on next page)

Table 4.7. The Syntax of Existential Dialectics IV: The Dialectic Constraints Imposed by the Principles (Part II)

—For this reason, there are different versions of "post-democracy" and "post-capitalism" in my theories. As an illustration, in version I of the theory of post-democracy, freedom is more dominant than equality, whereas in version II of the theory of post-democracy, equality is more so than freedom.

—But this "X more than Y" has to be understood in the context of dialectic logic (not in conventional logic), in that both "X" and "Y" are important in post-democracy (in the context of dialectic logic), but in an asymmetry way. By contrast, in conventional logic, it often favors one over the other—be it in regard to privileging freedom over equality in Fascism, favoring freedom relatively more than equality in Liberal Democracy, or favoring equality relatively more than freedom in Socialist Democracy. In the latter two cases (about Liberal Democracy and Socialist Democracy), the difference between dialectic logic and conventional logic can be one in degree, not in kind—in this sense, albeit not in other senses.

—The same logic can be said about the relatonships between individuality and communality, between spirituality and materiality, and between formal legalism and informal legalism in the different versions of my theory of post-capitalism.

- **Transcendent**

—Secondly, to be dialectic is to go beyond the narrow dichotomies (and, for that matter, any rigid multi-dimensional classificatory scheme) in another way, this time, in a transcendent way, that is, in exploring other possibilities or even other issues not considered within the narrow confines of narrow dichotomies (and, for that matter, any rigid multi-dimensional classificatory scheme).

(continued on next page)

Table 4.7. The Syntax of Existential Dialectics IV: The Dialectic Constraints Imposed by the Principles (Part III)

—As an analogy, to go beyond the narrow color dichotomy of "black" and "white" is not just to choose both "black" and "white" (as in the first meaning) but also to explore other color options (e.g., "green," "purple," "blue," etc.—and, alternatively, "shade," "line," "curve," etc.). By the same logic, to go beyond "democracy" is to transcend democracy (as in version III of the theory of "post-democracy") and to explore other possibilties of lifeforms (e.g., floating consciousness, hyper-spatial consciousness, etc., to live beyond the narrow obsession with freedom and equality).

—This dialectic character of the principles in existential dialectics has important implications for the pragmatics of existential dialectics (as will be clear shortly, in the section on the pragmatics of existential dialectics).

Notes: The examples here are solely illustrative (not exhaustive). The comparison is also relative (not absolute), nor are they mutually exclusive. As generalities, they allow exceptions.
Sources: First explicitly stated in *ALD*. Also from all other books of mine.

Table 4.8. The Syntax of Existential Dialectics V: Further Clarifications (Part I)

• **Pioneering**

—Firstly, the total number of ontological principles is unknown, to be discovered later, as our knowledge of the world becomes more advanced.

—Consequently, the principles as introduced in my books are not exhaustive, with new ones being added, whenever more of them are discovered in later research.

—At least, future generations can pick up where I leave off and continue the discovery.

—In this sense, my work should be treated as a pioneering effort for the development of a systematic, comprehensive analysis of a new general ontology for the future of knowledge.

• **Flexible**

—Secondly, the principles are not rigidly classified, as they can be reclassified in a different way.

—For instance, the preciseness-vagueness principle is classified under the category of "structure" but can be reclassified under the category of "method," although in so doing, it has a different meaning in the context of method.

—The same logic applies to the same-difference principle under the category of "outcome," which can be reclassified under the category of "structure," although in so doing, once more, it has a different meaning in the context of structure.

• **Mutually Constraining**

—Thirdly, the principles are to be understood together, not that each prinicple is to be analyzed independently of others. After all, the principles are mutually constraining, in that they work together as a whole.

(continued on next page)

Table 4.8. The Syntax of Existential Dialectics V: Further Clarifications (Part II)

- **Selectively Useful**
  —Fourthly, the principles are relevant to all subject matters, but some principles are more useful to some subject matters than others—as implied in the symmetry-asymmetry principle. This is true, even if different studies of the same kind can yield different views about the degree of relevance for each ontological principle, depending on the specific nature of a research in question, needless to say.

- **Anti-Reductionistic**
  —And fifthly, the principles constitute only two levels of analysis, this time, at the ontological and methodological levels—while other levels of analysis (from the perspectives of the mind, nature, society, and culture) are also needed, in order to understand reality in its totality (as explained in the section on "sophisticated methodological holism," so as to avoid the dual dangers of reductionism and reverse-reductionism).

Notes: The examples here are solely illustrative (not exhaustive). The comparison is also relative (not absolute), nor are they mutually exclusive. As generalities, they allow exceptions.
Sources: From my previous books like *FHC*, *FCD*, *FPHC*, *BDPD*, *BCPC*, *BNN*, *FC*, *FAE*, and so on.

Table 4.9. The Syntax of Existential Dialectics VI: The Dilemma of Specific vs. General Ontology (Part I)

• Ontology is often more complicated and imposes some seemingly insurmountable difficulties concerning what constitutes a kind of ontology which can be valid enough to be accepted by the wider intellectual community in question. An excellent example concerns what I want to call *the dilemma of ontology* in relation to the relationship between "specific ontology" and "general ontology." (M. Bunge 1999; R. Corazzon 2007)

• On the one hand, "general ontology," as Mario Bunge (1999) put it, "studies all existents." But, on the other hand, "special ontology studies one genus of thing or process—physical, chemical, biological, social, etc." (M. Bunge 1999; R. Corazzon 2007)

• Consequently, the relationship between the former and the latter is that, as an illustration, "whereas general ontology studies the concepts of space, time, and event, the ontology of the social investigates such general sociological concepts as those of social system, social structure, and social change." (M. Bunge 1999; R. Corazzon 2007)

• With this formal definition in mind, the relationship between the two forms of ontology only poses *the dilemma of ontology*, in that general ontology is possible to the extent that its general contructs must be broad enough to serve as the "lowest common denominator" for all disciplines, but this character of the lowest common denominator is not only hard to find but also its usefulness is limited. (WK 2008b)

• But specific ontology fares no better either, in that its specific constructs primarily serve a specific discipline in question, although other disciplines can appropriate them for their own purposes, but in a different context (especially, though not exclusively, when used as a metaphor from one field to another). For instance, the Kantian category of understanding under the heading of "quantity" constitutes a type of specific ontology, which is specific to the discipline of mathematics, although other disciplines can make use of the categories in their own context.

(continued on next page)

Table 4.9. The Syntax of Existential Dialectics VI: The Dilemma of Specific vs. General Ontology (Part II)

• A solution lies in my proposal of "sophisticated methodological holism," which requires all levels of analysis in any subject matter. See the tables on sophisticated methodological holism, and on reductionism and reverse-reductionism, for more details.

Source: From *Sec. 1.4* of *FPHK*

Table 4.10. The Syntax of Existential Dialectics VII: Types of Inappropriate Family Resemblances (Part I)

- **Inadequate Family Resemblances**
  —Ex: consistency-inconsistency: the formalness-informalenss principle
  —Ex: cognition-noncognition: the formalness-informalness principle

- **Compound Family Resemblances**
  —Ex: reality-nonreality: the formalness-informalness principles, the absoluteness-relativeness principle
  —Ex: normality-abnormality: the symmetry-asymmetry principle, the density-emptiness principle, the convention-novelty principle, the same-difference principle
  —Ex: structure-context: the simpleness-complicatedness principle, the theory-praxis principle
  —Ex: freedom-unfreedom: the expansion-contraction principle
  —Ex: morality-immorality: the denseness-emptiness principle, the symmetry-asymmetry, the same-difference principle
  —Ex: particularness-universalness: the preciseness-vagueness principle, the simpleness-complicatedness principle
  —Ex: pureness-mixedness: the absoluteness-relativeness principle, the simpleness-complicatedness principle, the same-difference principle
  —Ex: shortsightedness-foresightedness: the partiality-totality principle, the simpleness-complicatedness principle
  —Ex: shorttermness-longtermness: the partiality-totality principle, the simpleness-complicatedness principle, the slowness-quickness principle

- **One-Sided Family Resemblances**
  —Ex: sound-silence: asymmetry (in the symmetry-asymmetry principle)
  —Ex: vision-blindness: asymmetry (in the symmetry-asymmetry principle)

(continued on next page)

Table 4.10. The Syntax of Existential Dialectics VII: Types of Inappropriate Family Resemblances (Part II)

- **Vague Family Resemblances**
  —Ex: shortness-longness, narrowness-broadness, smallness-largeness, smallness-greatness, microness-macroness, and littleness-muchness: three different possibilities, namely, (a) finiteness (in the finiteness-transfiniteness principle), if both opposites are defined on a finite scale, (b) transfiniteness (in the finiteness-transfiniteness principle) if both opposites are defined on a transfinite scale, and (c) the finiteness-transfiniteness principle if one opposite is defined on a finite scale and the other opposite is defined on a transfinite scale.

- **One-Sided and Compound Family Resemblances**
  —Ex: extroversion-introversion: one-sided, in being part of the side (category) of "normality" in the normality-abnormality pair. And compounded, because the normality-abnormality pair is in turn related to the symmetry-asymmetry principle, the density-emptiness principle, the convention-novelty principle, the same-difference principle
  —Ex: organizedness-spontaneousness: one-sided, in being part of the side (category) of "normality" in the normality-abnormality pair. And compounded, because the normality-abnormality pair is in turn related to the symmetry-asymmetry principle, the density-emptiness principle, the convention-novelty principle, the same-difference principle
  —Ex: risk-caution: one-sided, in being part of the side (category) of "normality" in the normality-abnormality pair. And compounded, because the normality-abnormality pair is in turn related to the symmetry-asymmetry principle, the density-emptiness principle, the convention-novelty principle, the same-difference principle

Notes: The examples here are solely illustrative (not exhaustive). The comparison is also relative (not absolute), nor are they mutually exclusive. As generalities, they allow exceptions.
Source: First summarized in *BEPE* (and updated in later books)

Table 4.11. The Semantics of Existential Dialectics (Part I)

- **Abstract Structure vs. Specific Meanings**
  —The syntax of existential dialectics so understood in terms of ontological principles only gives us the structure of ontology in the world, in an abstract (general) sense. These principles by themselves do not tell us the specific meanings in a given context.
  —In order to grasp the specific meanings of the principles in a given context, it is necessary to study the semantics of existential dialectics. And the analysis of the ontological principles as family resemblances in the section on syntax is only a starting point (and thus overlaps a bit with the semantics of existential dialectics here).
  —With this caveat in mind, the reason that I have often gone in great lengths in my previous books on different subjects is to explain the specific meanings of the principles when applied in different contexts.

- **Specific Meanings in Specific Fields**
  —For instance, in *FPHST*, I used the first three principles (i.e., the change-constancy principle, the regression-progression principle, and the symmetry-asymmetry principle) to propose "the perspectival theory of space-time," for a better way to understand space and time—especially, though not exclusively, in relation to future post-human history (as summarized in *Table 3.6*, *Table 3.7*, *Table 3.8*, and *Table 3.9*). In so doing, I had to introduce concepts and theories specific to the field of physics and other related fields (e.g., "absolute space" and "absolute time" in "classical mechanics" and "relative space-time" in "the theory of relativity").
  —In *BNN*, I also exploited the three principles to propose the "transcendent" approach to the study of genes and memes as a new way to understand the interaction between nature and nurture. In so doing, I had to explore concepts and theories in the world of evolutionary theory (e.g., "mutation," "variation," "adaptation," "selection," and "inheritance" in Darwinian evolutionary theory) and neural biology (e.g., "chromosome," "gene," "DNA," "RNA," "protein," "neuron," "neural network," and "behavior").

(continued on next page)

Table 4.11. The Semantics of Existential Dialectics (Part II)

- **Two Types of Translation (Interpretation)**
  —In the technical jargons of semantics, these exercises in my books perform two essential functions of translation (or interpretation), namely, (a) "intensive" interpretation (about the properties or relations of the principles in existential dialectics), and (b) "extensive" interpretation (about the classes or cases to which the properties and relations of the principles in existential dialectics are applied). (WK 2011oo)

- **Limits of Translation (Interpretation)**
  —But there are limits of translation (or interpretation), and an excellent example is that "there is a problem with the 'objectivity' of translation, as it is relative to different cultural (or sub-cultural) contexts. For instance, one of the still unresolved problems in linguistics is: 'Is there an objective gauge for the quality of translation?'" (*FPHCOM* & WK 2011pp)
  —This is one serious problem about the logic of translation and interpretation that I already addressed in *Sec. 3.5.1* of *FPHCOM (*and also, for that matter, in *FHC*, in the context of hermeneutics).

Notes: The categories and examples in each are solely illustrative (not exhaustive). The comparison is also relative (not absolute), nor are they mutually exclusive. As generalities, they allow exceptions.
Sources: First explicitly stated in *ALD* (and later revised in *FPHCOM*). Also from all other books of mine.

Table 4.12. The Pragmatics of Existential Dialectics (Part I)

- **The Two-Way Street Connecting Theory and Meta-Theory**
  —The pragmatics of existential dialectics is not a one-way street (that is, using the ontological principles for theoretical insights in praxis) but a two-way one, that is, (a) from meta-theory to theory, and (b) from theory to meta-theory.
  —(a) On one side of the street, the ontological principles can inspire some theoretical insights in praxis, that is, in relation to some specific fields.
  —(b) On the other (opposing) side of the street, however, the study of a subject matter in the specific fields in turn reveals some more hitherto unknown ontological principles to be discovered and identified. For this reason, three new principles were added in *FC*, one in *FAE*, and two in *ALD*, on top of the original three in *BCPC*—after some research on the specific subject matters.

- **Direct and Indirect Applications**
  —*Direct*
  - The logic of existential dialectics can shed some theoretical insights on diverse phenomena in the world, and good instances are the pertinent use of the principles of existential dialectics for the theoretical insights on the freedom/unfreedom dialectics, the equality/inequality dialectics, and the wealth/poverty dialectics in my previous works.
  - My latest books like *FPHST* and *BNN* also use the principles to reveal some theoretical insights on the perspectives of space and time (as in *FPHST*) and of nature and nurture (as in *BNN*).

  —*Indirect*
  - The theoretical insights can further be used to reveal other phenomena directly from them (viz., the theoretical insights) and therefore indirectly from the principles themselves. A good instance is the use of the theoretical insights on the freedom/unfreedom and equality/inequality dialectics for the understanding of the civilization/barbarity dialectics.

(continued on next page)

Table 4.12. The Pragmatics of Existential Dialectics (Part II)

- **Direct and Indirect Applications (*cont'd*)**
  - Even in indirect applications, however, a phenomenon under study can still be directly related back to the principles themselves. In the example as cited above, the civilization/barbarity dialectics can be directly related to the principles of existential dialectics without the intermediate role of the freedom/unfreedom and equality/inequality dialectics.

- **Multiple Levels of Application**

  —The theoretical insights can be applied to different levels of analysis, even though in a given example, it may refer to one level only. For instance, in the example concerning the freedom/unfreedom dialectics, it can be used at the structural level (e.g., in relation to the theory of cyclical progression of hegemony), but it can be exploited as well for other levels (e.g., the theory of post-capitalism at the institutional level).

Notes: The categories and examples in each are solely illustrative (not exhaustive). The comparison is also relative (not absolute), nor are they mutually exclusive. As generalities, they allow exceptions.
Sources: From Ch.6 of *BCPC*. See also other books of mine.

Table 4.13. The Freedom-Unfreedom Dialectics (Part I)

- **On Having**
  - —*In Relation to the Technological*
    - (1) if freer from submission to Nature, then less free from ecological degradation (Deep and Social Ecology), even if in a hi-tech form
    - (2) if freer from technological inconvenience / backwardness, then less free from technological control and the loss of privacy
    - (3) if freer from technological (material) backwardness, then less free from the abusive (barbaric) maltreatment of the primitive Others
  - —*In Relation to the Everyday*
    - (1) if freer from abject poverty, then less free from artificial needs/discontents (Frankfurt School)
    - (2) if freer from sensual suppression, then less free from violent sublimation (Freud)
    - (3) if freer from the snobbishness of high culture, then less free from the shabbiness (leveling-off effect) of mass culture (Tocqueville)
    - (4) if freer from the inefficiency of traditional "compassionate economy," then less free from the bondage of a "ruthless [competitive] economy" (Keynes)
    - (5) if freer from anarchy in the state of nature (system fragmentation), then less free from government regulations and controls in system integration

(continued on next page)

Table 4.13. The Freedom-Unfreedom Dialectics (Part II)

• **On Belonging**

—*In Relation to the Good and the Just*

(1) if freer from disciplinary society, then less free from society of control (Foucault)

(2) if freer from the tyranny of one or a few, then less free from the tyranny of the majority (or sometimes, minority veto)

(3) if freer from elitist decision making, then less free from political gridlock/cleavage

(4) if freer from arbitrary (discretionary) administration, then less free from bureaucratic irrationality (Weber) and legal trickery (loopholes)

• **On Being**

—*In Relation to the True*

(1) if freer from unscientific dogmas, then less free from instrumental abyss (nihilism). Or conversely, if freer from meaninglessness, then less free from dogmas.

(2) if freer from the bondage of partiality/partisanship (e.g., prejudice, discrimination), then less free from the danger of impartiality and neutrality (e.g., opportunism, unrealisticness, lack of compassion, inaction)

(3) if freer from making generalizations, then less free from being unable to understand much of anything

—*In Relation to the Holy*

(1) if freer from collective conscience, then less free from social loneliness

(2) if freer from religious absoluteness, then less free from spiritual emptiness

—*In Relation to the Beautiful/Sublime*

(1) if freer from artistic non-autonomy, then less free from aesthetic disillusion (deconstruction)

(continued on next page)

Table 4.13. The Freedom-Unfreedom Dialectics (Part III)

Notes: The examples in each category are solely illustrative (not exhaustive), and the comparison is relative (not absolute), nor are they necessarily mutually exclusive. And some can be easily re-classified elsewhere. As generalities, they allow exceptions.
Sources: A reconstruction from Ch.10 of *FCD*, based on *FHC*

Table 4.14. The Equality-Inequality Dialectics (Part I)

- **On Having**
  - *—In Relation to the Technological*
    - (1) if more equal in treating Nature with spiritual unity, then less equal in suppressing the dominant drive to transcend it altogether
  - *—In Relation to the Everyday*
    - (1) if more equal in building social plurality, then less equal in leveling-off effects (e.g., the subsequent relative intolerance of high/intellectual ethos in mass culture industry)
    - (2) if more equal in socioeconomic distribution beyond a certain point, then less equal in efficiency (e.g. resentment, the erosion of work ethics)
    - (3) if more equal in urging an affirmative action program, then less equal in creating victim mentality (in oneself), stigma (from others), reverse discrimination (against the once privileged), and mediocracy (against the more able)

- **On Belonging**
  - *—In Relation to the Good and the Just*
    - (1) if more equal in banning monarchic/oligarchic exclusion, then less equal in producing "the tyranny of the majority" or of "minority veto"
    - (2) if more equal in encouraging participatory decision making, then less equal in inducing political divisiveness (gridlock/cleavage in power blocs) and organizational oligarchy
    - (3) if more equal in institutionalizing a decentralized bureaucracy, then less equal in falling into more territorial/turf politics (intrigues)

(continued on next page)

Table 4.14. The Equality-Inequality Dialectics (Part II)

- **On Being**
  - —In Relation to the Beautiful / Sublime
    - (1) if more equal in accepting diverse styles ("anything goes" mentality), then less equal in artistic good quality (in leveling-off effects against the best)
  - —In Relation to the True
    - (1) if more equal in tolerating multiple viewpoints (no matter how extreme), then less equal in epistemic standards
  - —In Relation to the Holy
    - (1) if more equal in celebrating any cults and sects (no matter how questionable), then less equal in spiritual depth and authenticity

Notes: The examples in each category are solely illustrative (not exhaustive), and the comparison is relative (not absolute), nor are they mutually exclusive. And some can be easily reclassified elsewhere. As generalities, they allow exceptions.

Sources: A reconstruction from Ch.10 of *FCD*, based on *FHC*

Table 4.15. The Duality of Oppression in Existential Dialectics: Oppression and Self-Oppression (Part I)

- **From the Same to the Others and Itself**
  —The Oppression by the Same against the Others
  - *By way of downgrading differences*
    –Ex: on judiciary caprice for corporate crimes (*Sec.2.2.1.2.1*)
    –Ex: on the deceptive politics of liberation (*Sec.3.5*)
    –Ex: on the humanitarian mystique (*Sec.4.4*)
    –Ex: on the fad of emotional intelligence (*Sec.5.3*)
  - *By way of accentuating differences*
    –Ex: on the legal sophistry of self-defense (*Sec.2.3*)
    –Ex: on the legal semantics of proportionality (*Sec.2.4*)
    –Ex: on the tricky politics of external threat (*Sec.3.4*)
    –Ex: on the appeal of the Far Right for democracy (*Sec.5.4*)
    –Ex: on the democratic axis of evil (*Sec.5.5*)
    –Ex: on the democratic way of brutality and revenge (*Sec.5.6*)
    –Ex: on democratic autocracy (*Sec.6.4*)

  —The Oppression by the Same against Itself
  - *By way of downgrading differences*
    –Ex: on the politics of fear (*Sec.2.2*)
    –Ex: on the trickery of compassionate conservatism (*Sec.3.2*)
    –Ex: on the deceptive politics of patriotism (*Sec.3.3*)
  - *By way of accentuating differences*
    –Ex: on the caprice of due process on domestic suspects (*Sec.2.2*)
    –Ex: on the false security/freedom dilemma (*Sec.6.5.2*)

(continued on next page)

Table 4.15. The Duality of Oppression in Existential Dialectics: Oppression and Self-Oppression (Part II)

- **From the Others to the Same and Themselves**
  - —The Oppression by the Others against the Same
    - *By way of downgrading differences*
      - –Ex: on judiciary caprice in the reverse direction (*Sec.2.2.1.2.2*)
      - –Ex: on equal pay (*Sec.6.2.1.1*)
      - –Ex: on equal representation (*Sec.6.2.1.2*)
      - –Ex: on affirmative action program (*Sec.6.3.1.1*)
      - –Ex: on same-sex marriage (*Sec.6.3.1.2*)
    - *By way of accentuating differences*
      - –Ex: on sexual harassment (*Sec.6.2.2.1*)
      - –Ex: on physical violence (*Sec.6.2.2.2*)
      - –Ex: on sexual exploitation (*Sec.6.2.2.3*)
  - —The Oppression by the Others against Themselves
    - *By way of downgrading differences*
      - –Ex: on the reverse-class mystique (*Sec.4.2*)
      - –Ex: on the reverse-black mystique (*Sec.4.3*)
      - –Ex: on self-discrimination by downgrading (*Sec.6.3.2.2*)
    - *By way of accentuating differences*
      - –Ex: on self-discrimination by accentuating (*Sec.6.3.2.1*)

Notes: The examples are solely illustrative (not exhaustive), nor are they mutually exclusive. As generalities, they allow exceptions. Also, both forms of oppression co-exist in all of the examples, so the listing of them are only meant in a relative, not absolute, sense.
Source: A summary of the sections (as cited) in Chs.2-6 of *BDPD*. See text for more info and references.

Table 4.16. The Structure of Existential Dialectics I: The Freedom/Unfreedom and Equality/Inequality Dialectics

• Each freedom and equality produces its own unfreedom and inequality, regardless of whether the pair occurs in political society (with the nation-state), in civil society (with some autonomy from the state), or elsewhere (e.g., in the private sphere of individual homes)—and regardless of whether freedom and equality are understood as "negative" or "positive."

• Oppression is dualistic, as much by the Same against the Others and itself, as by the Others against the Same and themselves.

• Both forms of oppression and self-oppression can be achieved by way of downgrading differences (between the Same and the Others) and of accentuating them.

• The relationships are relatively asymmetric between the Same and the Others and relatively symmetric within them. This is true, even when the Same can be relatively asymmetric towards itself in self-oppression, just as the Others can be likewise towards themselves.

• Symmetry and asymmetry change over time, with ever new players, new causes, and new forms.

Notes: The examples in each category are solely illustrative (not exhaustive) nor necessarily mutually exclusive, and the comparison is relative (not absolute). As generalities, they allow exceptions. "Negative" freedom is freedom "from" (e.g., freedom from poverty), whereas "positive" freedom is freedom "to" (e.g., freedom to the state of enlightenment). "Negative" equality is "procedural" equality (e.g., equality of opportunity), while "positive" equality is "substantive" equality (e.g., equality of outcome). Existential dialectics impose constraints on freedom and equality in democracy, non-democracy, and post-democracy. There is no utopia, in the end; even should there be one, dystopia would exist within it.
Sources: From *Table 1.5* of *BDPD*—and also from *FHC*, *FCD*, and *FPHC*

Table 4.17. The Structure of Existential Dialectics II: The Wealth/Poverty Dialectics

• There is no wealth without poverty, just as there is no poverty without wealth.

• The wealth/poverty dialectics occurs in the realms of having, belonging, and being, in relation to the material, relational, and spiritual.

• The wealth/poverty dialectics also expresses itself at the multiple levels of analysis in accordance to methodological holism, be they about the micro-physical, the chemical, the biological, the psychological, the organizational, the institutional, the structural, the systemic, the cultural, and the cosmological.

• The wealth/poverty dialectics is a different manifestation of existential dialectics in general, subject to the principles in its logic of ontology—just as the freedom/unfreedom and equality/inequality dialectics are likewise.

• There is no economic utopia, in the end; even should there be one, dystopia would exist within it.

Notes: The main points here are solely illustrative (not exhaustive) nor necessarily mutually exclusive, and the comparison is relative (not absolute). As generalities, they allow exceptions.
Sources: From *BCPC*. See also *FCD* and *FHC*.

Table 4.18. The Structure of Existential Dialectics III: The Civilization/Barbarity Dialectics

• There is no civilization without barbarity.

• The civilization/barbarity dialectics applies in the four civilizing processes (e.g., the rationalizing process, the pacifying process, the stewardizing process, and the subliming process).

• The civilization/barbarity dialectics is another (different) manifestation of existential dialectics in general, subject to the principles in its logic of ontology—just as the freedom/unfreedom and equality/inequality dialectics and the wealth/poverty dialectics are likewise.

• There is no utopia, in the end; even should there be one, dystopia would exist within it.

Notes: The main points here are solely illustrative (not exhaustive) nor necessarily mutually exclusive, and the comparison is relative (not absolute). As generalities, they allow exceptions.
Sources: From *BCIV*. See also *FCD*, *FHC*, and *BDPD*.

Table 4.19. The Double Sides of Virtual Organizations (Part I)

- **Psychological**
  —Ex: virtual psychosis
  —Ex: impersonality and loneliness in quaternary social relations

- **Organizational**
  —Ex: the race for power and interests
  —Ex: the world of unequal successes
  —Ex: the bureaucratic life of its own
  —Ex: lesser accountability and transparency

- **Economic**
  —Ex: the economic divides
  —Ex: the erosive impact of commercialization

- **Political**
  —Ex: the anti-authoritarian myth
  —Ex: different power struggles among groups (e.g., Conservative, Reformist, Radical)

- **Structural**
  —Ex: different localities (e.g., climates, scenery)
  —Ex: divided domains (e.g., core, peripheral, sub-peripheral)
  —Ex: substitution effect of social capital

- **Systemic**
  —Ex: uneven advances in transportation and communications, and geographical migration

- **Cultural**
  —Ex: conflicting civilizational fabrics (e.g., Confucian, Islamic, Western)
  —Ex: the bias of the Liberal Democratic agenda
  —Ex: the addiction to moral fanaticism

(continued on next page)

Table 4.19. The Double Sides of Virtual Organizations (Part II)

Notes: These categories and examples are solely illustrative (not exhaustive), and some of the items can be reclassified somewhere else. Nor are they mutually exclusive. Since they are generalities, exceptions are expected.
Source: From Ch.7 of *FCD*

Table 4.20. Beyond the World of Titans, and the Remaking of the World Order

- **Hyper-Empires**
  —Ex: The Chinese Union
  —Ex: The Indian Union

- **Meso-Empires**
  —Ex: The European Union
  —Ex: The North American Union

- **Micro-Empires**
  —Ex: The Latin American Union
  —Ex: The Middle Eastern Union

- **The Rest of the World**
  —Odd Powers
    - Ex: Japan
    - Ex: Russia

  —The Poor Club
    - Ex: The African Union

  —Ambivalent Regions
    - Ex: Southeast Asia
    - Ex: Oceania
    - Ex: South Asia
    - Ex: Central Asia
    - Ex: Southern/Eastern Europe
    - Ex: North Africa
    - Ex: Central America
    - Ex: Others (e.g., the Korean peninsula)

Source: A summary of Chs.2-5 (of *BWT*)

Table 4.21. The Origins of Authoritarian Liberal Democracy

- **The Geopower of Nature (Ch.4 of *ALD*)**
  - —Ex: Power Character and Geographical Strategy
  - —Ex: Living Space and Territorial Expansion
  - —Ex: Strategic Heartland and Containment

- **The Biopsychology of the Mind (Ch.5 of *ALD*)**
  - —Ex: The Bell Curve and Mass Intelligence
  - —Ex: Group Analysis and Mass Knowledge
  - —Ex: Groupthink and Elite (Mis)calculation

- **The Disciplinary Control of Society (Ch.6 of *ALD*)**
  - —Ex: Social Organizations of Ruthlessness
  - —Ex: Social Institutions of Greed
  - —Ex: Social Structure of Exclusion
  - —Ex: Social Systems of Violence

- **The Molding Force of Culture (Ch.7 of *ALD*)**
  - —Ex: The Tradition of Conquest
  - —Ex: The Rationalization of Unreason

Notes: The examples in each category are solely illustrative (not exhaustive), and they are also relative (not absolute), nor are they mutually exclusive. As generalities, they allow exceptions.
Source: A summary of Chs.4-7 of *ALD*

**Table 4.22. The Theory of Post-Democracy I:
The Priority of Freedom over Equality
(Part I)**

- **Differences**
  - —*For the aggressive Lions (the strong Elitists)*
    - •Setting up rank distinctions among unequals (e.g., between inferior humans and superior post-humans, or later among inferior post-humans and superior ones, relatively speaking)
    - •Yearning for being not only distinguished from unequals, but also the first among equals (the best of the very best)
    - •Soul-searching for a high spiritual culture (not the trashy one for the masses). Mass culture is a dirty joke for them.
  - —*For the manipulative Foxes (the weak Counter-Elitists)*
    - •Seeking a gentle hegemony by way of more communitarian concerns (for inferior humans and, later, inferior post-humans)
    - •Being more sympathetic to less formal-legalistic institutions and values

- **Similarities**
  - —*For both Lions and Foxes*
    - •Exploring different spheres of non-human consciousness in the cosmos (something vastly superior than the human one)
    - •Recognizing the democratic illusions (e.g., no freedom without unfreedom, no equality without inequality, or simply no justice without injustice, and vice versa)

(continued on next page)

Table 4.22. The Theory of Post-Democracy I: The Priority of Freedom over Equality (Part II)

Notes: The two callings and examples in each category are solely illustrative (not exhaustive), since there will be many different post-human value ideals in the distant future of post-human civilization. The comparison is also relative (not absolute) towards post-democracy, so this is not just a version of free-market democracy (nor Fascism/Nazism, as shown in the table later on democracy, non-democracy, and post-democracy). Nor are they mutually exclusive. As generalities, they allow exceptions. And the specific forms of post-human post-democratic ideals need to be further developed in future after-postmodern history, as they will be different from the ones we now know. The point here is to solely give an extremely rough picture of a small part of the world to come that we have never known.
Source: From Ch.10 of *FCD*. Refer to text for more info and references.

Table 4.23. The Theory of Post-Democracy II: The Priority of Equality over Freedom

- **Hybrid Versions of**
  —Ex: the Trans-Feminine Calling
  —Ex: the Trans-Sinitic Calling
  —Ex: the Trans-Islamic Calling
  —Ex: the Trans-Outerspace Calling

- **Qualifications**
  —These four versions of post-capitalist value ideals need not automatically be post-democratic, just as capitalism does not necessarily mean democracy. They are two different entities—though closely related.
  —But up to a certain threshold of elevating equality at the farther expense of freedom, the democratic ideals will be overcome and cease to exist.
  —The overcome will not be socialist or communist, but post-democratic with no freedom without unfreedom and no equality without inequality, subject to the constraints of existential dialectics.

Notes: The callings are solely illustrative (not exhaustive), since there will be many different post-human value ideals in the distant future of post-human lifeforms. The comparison is also relative (not absolute), nor are they mutually exclusive. As generalities, they allow exceptions. And the specific forms of post-human post-democratic ideals need to be further developed in future after-postmodern history, as they will be different from the ones we now know. The point here is to solely give an extremely rough picture of a small part of the world to come that we have never known.
Source: From Ch.10 of *FCD*. Refer to text for more info and references.

Table 4.24. The Theory of Post-Democracy III: The Transcendence of Freedom and Equality (Part I)

- **Transcending Freedom in Floating Existence**
  —*Freedom*: seeking an ultimate elimination of the body. Being without the body. The aim is to transcend freedom in the end into a metaphysical state (i.e., beyond the physique).
  —*Unfreedom*: yet facing difficult trade-offs. The sacrifice of bodily existence and its joyfulness. An eternal boredom in floating existence in dark deep space, though with alternative pleasures. There is no free lunch even in the state of transcending freedom.

- **Transcending Equality in the Rivalry of Cosmic Hegemony**
  —*Inequality*: competing to outlast other lifeforms in floating existence, or just marginalizing them for one's hegemonic expansiveness in the rest of the cosmos (and even beyond). Universalism is only for the mediocre.
  —*Equality*: accepting only those of one's rank as equal partners in the vast spacetime for cosmic supremacy. Even here, the aim is to transcend equality into a metaphysical state.

(continued on next page)

Table 4.24. The Theory of Post-Democracy III: The Transcendence of Freedom and Equality (Part II)

Notes: Do not confuse this transcendence of freedom and equality (as one version of post-democracy) with the naïve temptation to transcend the freedom/unfreedom and equality/inequality dialectics. Existential dialectics hold true for freedom and equality in all cultures and societies—past, present, or future (i.e., democracy, non-democracy, and post-democracy), regardless of whether freedom and equality are conventionally understood as "negative" or "positive."
Also, the two features and examples in each are solely illustrative (not exhaustive), since there will be many different post-human value ideals in the distant future of post-human lifeforms. The comparison is also relative (not absolute), nor are they mutually exclusive. As generalities, they allow exceptions. And the specific forms of post-human ideals even for these radically alien floating lifeforms (and others unknown to us) need to be further developed in future after-postmodern history, as they will likely be different from the ones herein illustrated. The point here is to solely give a very rough picture of a small part of the extremely alien world to come that we have never known.
Source: From Ch.10 of *FCD*. Refer to text for more info and references.

Table 4.25. Democracy, Non-Democracy, and Post-Democracy (Part I)

- **Democracy**
  - —*Theoretical Constructs*
    - •The pursuit of freedom and equality (in various degrees), regardless of whether freedom and equality can be understood as "negative" or "positive"
      - (1) more equality than freedom: The relative priority of the good over the right
      - (2) more freedom than equality: The relative priority of the right over the good
  - —*Types*
    - •Only (1): Different versions of communitarian moral universalism
    - •Only (2): Different versions of liberal moral universalism
    - •(1) or (2): Different versions of anarchic (non-nation-state) moral universalism
    - •(1) or (2): Different versions of postmodern moral localism

- **Non-Democracy**
  - —*Theoretical Constructs*
    - •The focus on (1') equality or (2') freedom, but not both, regardless of whether freedom and equality can be understood as "negative" or "positive"
  - —*Types*
    - •Only (1'): Different versions on the Far Left (e.g., Stalinism, Robespierrianism)
    - •Only (2'): Different versions on the Far Right (e.g., Nazism, absolute monarchism)

(continued on next page)

Table 4.25. Democracy, Non-Democracy, and Post-Democracy (Part II)

- **Post-Democracy**

—*Theoretical Constructs*

•The priority of (1'') equality over freedom, or (2'') freedom over equality, or (3'') the transcendence of freedom and equality, regardless of whether freedom and equality are "negative" or "positive." In degree, (1'') or (2'') is less than (1') or (2') but more than (1) or (2)—respectively.

•Like democracy and non-democracy, post-democracy is also subject to the freedom/unfreedom and equality/inequality dialectics (or existential dialectics in general). Unlike them, post-democracy acknowledges the constraints of existential dialectics and no longer value freedom and equality as sacred virtues. There is no utopia, in the end; even were there one, dystopia would exist within it.

—*Types*

•(1''): Different versions of trans-Sinitic value ideals
•(1''): Different versions of trans-feminine value ideals
•(1''): Different versions of trans-Islamic value ideals
•(1''): Different versions of trans-outerspace value ideals
•(2''): Different versions of post-human elitist value ideals
•(3''): Different versions of the value ideals of floating consciousness (etc.)

(continued on next page)

Table 4.25. Democracy, Non-Democracy, and Post-Democracy (Part III)

Notes: The examples are solely illustrative (not exhaustive), nor are they mutually exclusive. As generalities, they allow exceptions. "Negative" freedom is freedom "from" (e.g., freedom from poverty), whereas "positive" freedom is freedom "to" (e.g., freedom to the state of enlightenment). "Negative" equality is "procedural" equality (e.g., equality of opportunity), while "positive" equality is "substantive" equality (e.g., equality of outcome). Existential dialectics impose constraints on freedom and equality in democracy, non-democracy, and post-democracy, regardless of whether freedom and equality can be understood as "negative" or "positive" in conventional discourse. Therefore, do not confuse the transcendence of freedom and equality in (3'') with the naïve temptation to transcend existential dialectics. There is no utopia, in the end; even should there be one, it would not exist without dystopia embedded within it.
Sources: A summary, based on my previous works, especially Ch.5 of *FHC*, Chs.5-10 of *FCD*, Chs.2-4 of *FPHC*, and Chs.1 & 7 of *BDPD*. The reader should consult the books for more analysis, as this is only a summary here.

Table 4.26. Multiple Causes of the Emergence of Post-Democracy (Part I)

- **At the Micro-Physical Level**
  —Ex: intelligent life without the human physical-chemical system
  —Sources: Ch.7 of *FHC*; Chs.9-10 of *FCD*; Ch.1 of *FPHC*

- **At the Chemical Level**
  —Ex: space radiation and toxins
  —Sources: Ch.7 of *FHC*; Chs.9-10 of *FCD*

- **At the Bio-Psychological Level**
  —Ex: exo-biological evolution in deep space
  —Ex: genetic engineering of new beings
  —Ex: limits of cognitive partiality
  —Ex: illusions of emotional neutrality
  —Ex: human biological inequality
  —Ex: the rise of unfolding unconsciousness
  —Sources: Ch.2 & Chs.9-10 of *FCD*; Ch.7 of *FHC*; Ch.4 of *BCPC*; *FPHU*

- **At the Institutional Level**
  —Ex: the flawed logic of equality
  —Ex: the conflicting nature of governance
  —Sources: Ch. 5 of *FHC*; Chs. 6 & 10 of *FCD*; Ch. 3 of *FPHC*; Chs.2-5 of *BDPD*

- **At the Organizational Level**
  —Ex: e-civic alienation
  —Ex: the dark sides of formal-legalistic routines
  —Sources: Ch.3 of *FHC*; Ch.7 of *FCD*; Ch.3 of *FPHC*

- **At the Structural Level**
  —Ex: ever new forms of inequities, at home and abroad
  —Ex: the emergence of China, women, and Islam as major actors
  —Sources: Chs.5-6 of *FHC*; Chs.7, 9 & 10 of *FCD*; Chs.4-5 of *BDPD*

(continued on next page)

Table 4.26. Multiple Causes of the Emergence of Post-Democracy (Part II)

- **At the Cultural Level**
  —Ex: freedom/unfreedom dialectics
  —Ex: equality/inequality dialectics
  —Ex: system fragmentation and integration
  —Sources: Ch.5 of *FHC*; Chs. 3, 9 & 10 of *FCD*; Ch.4 of *FPHC*; Ch.1 of *BDPD*; Ch.4 of *BCPC*

- **At the Systemic Level**
  —Ex: space habitats (in zero-gravity) and colonization
  —Ex: ultra advanced future info systems
  —Ex: qualitative demography
  —Sources: Ch.7 of *FHC*; Chs.9 &10 of *FCD*

- **At the Cosmological Level**
  —Ex: the colonization of multiverses
  —Ex: the alteration of space-time and the creation of new matter-energy
  —Ex: the expansion of floating consciousness
  —Ex: the spread of hyper-spatial consciousness
  —Sources: Ch.7 of *FHC*; Chs.9 &10 of *FCD*; Ch.4 of *FPHC*; *FPHST*

Notes: The examples in each category are solely illustrative (not exhaustive), and some of the items can be reclassified somewhere else. Nor are they always mutually exclusive. Since they are generalities, exceptions are expected.
Sources: Especially from *FHC*, *FCD*, *FPHC*, *BCPC*, *BDPD*, *FPHST*, and *FPHU*. See also other books and my perspectives on civilizational holism.

Table 4.27. Some Clarifications about Post-Capitalism and Post-Democracy (Part I)

• The prefix "trans-" in the first category of post-capitalism (with its four versions) refers to something "going beyond" (not "uniting" or "combining"). Ex: *Sec.10.3.3* of *FCD*; *Sec.2.4* & *Sec.4.4* of *FPHC*; *Sec.7.2* of *BCPC*

• Such terms like "post-democracy," "post-capitalism," "post-human elitist," "trans-feminine calling," and the like as used in my works are more for our current intellectual convenience than to the liking of future humans and post-humans, who will surely invent more tasteful neologisms to call their own eras, entities, and everything else, for that matter. But the didactic point here is to use the terms to foretell what the future might be like, not that its eras and entities must be called so exactly and permanently.
Ex: *Sec.11.1* of *FCD*; *Sec.7.2* of *BCPC*

• The four versions in the first category of post-capitalist value ideals need not automatically be post-democratic, just as capitalism does not necessarily mean democracy. They are two different entities—though closely related. But up to a certain threshold of elevating equality at the farther expense of freedom, the democratic ideals will be overcome and cease to exist. The same is true for the post-human elitist calling in the second category of post-capitalism in relation to post-democracy, depending on the extent to which freedom is elevated at the expense of equality.
Ex: *Sec.10.4.3.3* of *FCD*; *Table 3.9* of *FPHC*; *Table 7.6* of *BDPD*

• The comparison in each of the three realms of existence in all forms of post-capitalism is not absolute, but relative. Examples include "communal" vs. "individualistic," and the like.
Ex: Notes in *Table 10.8*, *Table 10.9*, *Table 10.10*, & *Table 10.11* of *FCD*; Chs.2-4 of *FPHC*; *Sec.7.2* of *BCPC*

(continued on next page)

Table 4.27. Some Clarifications about Post-Capitalism and Post-Democracy (Part II)

• The emergence of post-capitalism (and post-democracy, for that matter) has multiple causes (to not be reduced to one or only a few).
Ex: Ch.10 of *FCD*, Chs.2-4 of *FPHC*; *Sec.1.3* & *Sec.7.2* of *BCPC* (or *Table 1.8* & *Table 7.11*)

• The specific forms of post-capitalism (and post-democracy, for that matter) need to be further developed in future after-postmodern history, as they will be different from the ones we now know. The point here is to solely give an extremely rough sketch of a world to come that we have never known.
Ex: *Sec.10.3.3* & *Sec.10.4.3.3* of *FCD*; *Table 10.14* & *Table 10.15* of *FCD*; *Sec.7.2* of *BCPC*

• All forms of post-capitalism are not part of a "teleological law," but of "historical trends" only. The same is also true for all forms of post-democracy.
Ex: *Sec.7.1* of *FHC*; *Sec.9.5.3.2* & *Sec.10.3.4.2* of *FCD*; *Sec.7.2* of *BCPC*

• Post-capitalism is not better than capitalism in an "absolute" sense but only fits in better, on the basis of the historical contingency of culture, society, nature, and the mind in some future eras. The same is true for post-democracy in relation to democracy. The term "better" is historically relative.
Ex: *Sec.10.3.3* of *FCD*; *Sec.1.7* of *BDPD*; *Sec.1.5* of *BCPC*

(continued on next page)

Table 4.27. Some Clarifications about Post-Capitalism and Post-Democracy (Part III)

• All forms of post-capitalism and post-democracy are subject to the constraints of existential dialectics. In the process, the dialectic direction is to go beyond the conventional "either-or" dichotomies (e.g., freedom vs. unfreedom, equality vs. inequality, freedom vs. equality, individuality vs. communality, spirituality vs. materiality, formal legalism vs. informal legalism, etc.). As is true in post-civilization, to go beyond the dichotomies is to acknowledge the co-existence of both in each dichotomy, although the degree of scaling one over the over varies from case to case (e.g., the theory of post-capitalism I, the theory of post-capitalism II, the theory of post-democracy I, the theory of post-democracy II, etc.)—but is not to be extreme in largely favoring one over the other, *on average* (all things considered). There is no utopia to be had in the end; even should there be one, dystopia would exist within it.
Ex: Ch.5 of *FHC*; *Sec.10.4.4.2* of *FCD*; *Sec.1.5* of *BDPD*; *Sec.1.3* of *BCPC*; *BCIV*

• All forms of post-capitalism, however different from each other though they are, share one common feature, in that they all inspire for a higher spiritual culture. The same is also true for post-democracy.
Ex: *Sec.10.3*, *Sec.10.4* & *Sec.10.5* of *FCD*; *Chs.2-4* of *FPHC*; *Sec.7.2* of *BCPC*

• All forms of post-capitalism try to avoid the excess in capitalist consumerism by favoring more basic than artificial needs in having, but the quality and quantity of these "basic" needs will be measured by future standards, not by our current ones. Standards are historically relative.
Ex: *Sec.10.3*, *Sec.10.4* & *Sec.10.5* of *FCD*; Ch.2 of *FPHC*; *Sec.7.2* of *BCPC*

(continued on next page)

Table 4.27. Some Clarifications about Post-Capitalism and Post-Democracy (Part IV)

• All forms of post-capitalism make use of a different degree of political authority with advanced info systems in future history and strives for higher spiritual cultures (especially in the post-human age), while acknowledging the constraints of existential dialectics and no longer valuing free market (as in capitalism) and economic control (as in non-capitalism) as sacred virtues.
Ex: *Sec.10.3.4.2*, *Sec.10.3*, *Sec.10.4* & *Sec.10.5* of *FCD*; Chs.2-4 of *FPHC*; *Sec.1.5* of *BDPD*; *Sec.7.2* of *BCPC*

Notes:: The main points here are solely illustrative (not exhaustive) nor necessarily mutually exclusive, and the comparison is relative (not absolute). As generalities, they allow exceptions. The sections as cited are only illustrative (not exhaustive).
Sources: From *FHC*, *FCD*, *FPHC*, and *BDPD*

Table 4.28. The Theory of Post-Capitalism I.1: By Group— Ex: Spiritual/Communal in the Trans-Feminine Calling

- **More Communal Than Individual**
  —*Sharing*: learning from others, as different ideas mutually enrich
  —*Cooperative*: encouraging a sense of shared leadership and teamwork

- **More Informal-Legalistic Than Formal-Legalistic**
  —*Specific*: listening more from the heart than from the head, to know a person as a concrete, not as an abstract, unit
  —*Affective*: thinking and acting with others on a more affective tone. Business can mix with an emotional touch.
  —*Ascriptive*: hiring (or firing) can be done on the basis of merit (or lack of it), but deep solidarity (sisterhood) is important too.
  —*Particularistic*: making decisions on the basis of cost-benefit analysis, but a given group relationship is vital

- **More Spiritual Than Secular**
  —*Long-Term Looking*: sharing for a long-term relationship (e.g., love, friendship), not just for a short-term gain
  —*Loving/Caring*: showing compassion for the sufferings of others, without quickly blaming and pre-judging
  —*Respectful*: showing acceptance about others' feelings (and thoughts)

Notes: The categories and examples are solely illustrative, since there can be different versions, and the comparison is relative (not absolute), nor are they mutually exclusive. As generalities, they allow exceptions. The specific forms of the trans-feminine version need to be further developed in future after-postmodern history, as they will be different from the ones we now know, since the prefix "trans-" here means going beyond or deconstructing the feminine values, while using them as the inspirational point at the beginning.
Source: From Ch.10 of *FCD*. Refer to text for more info and references.

Table 4.29. The Theory of Post-Capitalism I.2: By Nation-State— Ex: Spiritual/Communal in the Trans-Sinitic Calling

- **More Communal Than Individualistic**
  —*Centralized*: being more top-down in management
  —*Collective*: encouraging more group cooperation
  —*Social*: investing in trust and connection

- **More Informal-Legalistic Than Formal-Legalistic**
  —*Specific*: knowing more of those related or connected
  —*Affective*: behaving in a paternalistic, hierarchical way
  —*Ascriptive*: favoring family members and those related
  —*Particularistic*: building connection (guanxi) as imperative

- **More Spiritual Than Secular**
  —*Expansionist*: diffusing civilizational values (e.g., the superiority complex of civilizationalism)
  —*Holistic*: synthesizing things into a panoramic horizon
  —*Historical*: learning from the lessons of the ancient past
  —*Respectful*: deferential to elders and superiors

Notes: The categories and examples are solely illustrative, since there can be different versions, and the comparison is relative (not absolute), nor are they mutually exclusive. As generalities, they allow exceptions. The specific forms of the trans-Sinitic version need to be further developed in future after-postmodern history, as they will be different from the ones we now know, since the prefix "trans-" here means going beyond or deconstructing the Sinitic values, while using them as the inspirational point at the beginning.
Source: From Ch.10 of *FCD*. Refer to text for more info and references.

Table 4.30. The Theory of Post-Capitalism I.3: By Region— Ex: Spiritual/Communal in the Trans-Islamic Calling

- **More Communal Than Individualistic**
  —*Collective*: building the webs of relationships to bind individuals
  —*Sharing*: cultivating the established "wisdom" through common experience
  —*Cooperative*: stressing harmony, solidarity, and commonality

- **More Informal-Legalistic Than Formal-Legalistic**
  —*Specific*: making efforts to know well the participants (family and larger community) in matters of common concern
  —*Affective*: mixing work with language and ritual on explicit religious (Islamic) ideals, texts, stories, and examples
  —*Ascriptive*: privileging local history and custom on relationships among kinship groups
  —*Particularistic*: preferring an unbiased insider with ongoing connections to all parties

- **More Spiritual Than Secular**
  —*Historical*: learning from the lessons of the past as a source of stability and guidance
  —*Deferential*: showing respect for age, experience, status, and leadership in communal affairs
  —*Honorable*: emphasizing face, dignity, prestige, and fairness
  —*Compassionate*: giving mercy and charity ("Zahah") to others

Notes: The categories and examples are solely illustrative (not exhaustive), and the comparison is relative (not absolute), nor are they mutually exclusive. As generalities, they allow exceptions. The specific forms of the trans-Islamic version need to be further developed in future after-postmodern history, as they will be different from the ones we now know, since the prefix "trans-" here means going beyond or deconstructing the Islamic values, while using them as the inspirational point at the beginning.
Sources: From Ch.10 of *FCD*. Refer to text for more info and references, especially from the works by George Irani (2000) and C. Murphy (September 19, 2001).

Table 4.31. The Theory of Post-Capitalism I.4: By Universe— Ex: Spiritual/Communal in the Trans-Outerspace Calling

- **More Communal Than Individual**
  —*Cooperative*: requiring teamwork in small space habitats
  —*Sharing*: learning from, and enjoying being with, each other in a small group in outer space

- **More Informal-Legalistic Than Formal-Legalistic**
  —*Specific*: knowing more about each other to facilitate living and working together in space, both as fellow astronauts and space-mates
  —*Affective*: being friendly and social to each other as vital to working and living in small space quarters
  —*Ascriptive*: nurturing comaraderie among fellow astronauts as if they are family members over time
  —*Particularistic*: building work relationship with enduring memory in a space mission

- **More Spiritual Than Secular**
  —*Long-Term*: looking beyond selfish materialistic concerns in a precarious space environment with potential life or death
  —*Loving/Caring*: cultivating deep bondage for the success of a long term space mission
  —*Transcendent*: searching for life meaning in outer space

Notes: The calling and examples in each category are solely illustrative (not exhaustive), since there will be many different outer-space value ideals in the distant future of space colonization. The comparison is also relative (not absolute), nor are they mutually exclusive. As generalities, they allow exceptions. And the specific forms of trans-outer-space calling need to be further developed in future after-postmodern history, as they will be different from the ones we now know, since the prefix "trans-" here means going beyond or deconstructing the current outer-space values, while using them as the inspirational point at the beginning. The point here is to solely give an extremely rough picture of a small part of the world to come that we still do not know much about.
Source: From Ch.10 of *FCD*. Refer to text for more info and references.

Table 4.32. The Theory of Post-Capitalism II: Spiritual/Individualistic in the Post-Human Elitist Calling (Part I)

- **More Individualistic Than Communal**
  —Setting up rank distinctions among unequals (e.g., between inferior humans and superior post-humans, or later among inferior post-humans and superior ones, relatively speaking)
  —Yearning for being not only distinguished from unequals, but also the first among equals (the best of the very best)
  —Recognizing the constraints of equality/inequality dialectics (or existential dialectics in general)

- **More Spiritual Than Secular**
  —Soul-searching for a high spiritual culture (not the trashy one for the masses). Mass culture is a dirty joke for them.
  —Exploring different spheres of non-human consciousness in the cosmos (something vastly superior than the human one)
  —Recognizing the constraints of freedom/unfreedom dialectics (or existential dialectics in general)

- **Qualifications**
  —Although post-human elitist post-democracy is comparable to post-human elitist post-capitalism in some respects, the former does not necessarily imply the latter (post-human elitist post-capitalism), just as democracy does not have to entail capitalism. They are two different (though related) entities.
  —But up to a certain threshold of incorporating government intervention with advanced info systems in future civilizations for higher spiritual concerns at the expense of the free market and materialist pursuit, the capitalist ideal will be overcome.
  —The overcome will not be Fascist or feudalistic, but post-capitalist, subject to the constraints of existential dialectics.

(continued on next page)

Table 4.32. The Theory of Post-Capitalism II: Spiritual/Individualistic in the Post-Human Elitist Calling (Part II)

Notes: The calling and examples in each category are solely illustrative (not exhaustive). The comparison is also relative (not absolute), nor are they mutually exclusive. As generalities, they allow exceptions. And the specific forms of post-human elitist post-capitalism need to be further developed in future after-postmodern history, as they will be different from the ones we now know, while using them as the inspirational point at the beginning. The point here is to solely give an extremely rough picture of a small part of the world to come that we still do not know much about.
Sources: From Ch.10 of *FCD* (and also *FPHC*, *BDPD*, and *BCPC*). Refer to the text for more info and references.

Table 4.33. Capitalism, Non-Capitalism, and Post-Capitalism (Part I)

- **Capitalism**
  - —*Theoretical Constructs*
    - •Allocation of scarce resources among alternative wants largely by free market for competition (whose characteristics in its ideal form include, for instance, no barrier to entry or exit, homogeneity, perfect information, a large number of buyers/sellers, and perfect factor mobility)
    - •More formal-legalistic than informal-legalistic, more individualistic than communal, and more material (secular) than spiritual
    - •Either (1) minimal government or (2) relatively active government
  - —*Types*
    - •Only (1): Different versions of market capitalism (e.g., the U.S.)
    - •Only (2): Different versions of welfare capitalism (e.g., Sweden)

- **Non-Capitalism**
  - —*Theoretical Constructs*
    - •Allocation of scarce resources among alternative wants mainly by political authority for policies (which can be regulative, redistributive, symbolic, and participatory)
    - •More informal-legalistic than formal-legalistic
    - •Either (1') more individualistic (for the elites), often (though not always) for material (secular) concerns, or (2') more communal (for the masses), often (though not always) for spiritual concerns
  - —*Types*
    - •Only (1'): Different versions on the Right (e.g., Fascist corporate-state economy for the glory of the new Rome, medieval lord-vassal-serf economy for the power of the feudalistic order)
    - •Only (2'): Different versions on the Left (e.g., Soviet command economy for the creation of the New Socialist Man)

(continued on next page)

Table 4.33. Capitalism, Non-Capitalism, and Post-Capitalism (Part II)

- **Post-Capitalism**

—*Theoretical Constructs*

- •Allocation of scarce resources among alternative wants largely by political authority with advanced info systems in future civilizations, subject to existential dialectics. In degree of allocating by authority, post-capitalism is more than capitalism but less than non-capitalism.
- •More spiritual than secular (material)
- •Either (1'') more individualistic or (2") more communal
- •Like capitalism and non-capitalism, post-capitalism is also subject to the freedom/unfreedom and equality/inequality dialectics (or existential dialectics in general). There is no utopia, in the end; even were there one, dystopia would exist within it.
- •Unlike capitalism and non-capitalism, post-capitalism makes use of a different degree of political authority with advanced info systems in future civilizations and strives for higher-spiritual cultures (especially in the post-human age), while acknowledging the constraints of existential dialectics and no longer valuing free market (as in capitalism) and economic control (as in non-capitalism) as sacred virtues.

—*Types*

- •Only (1"): Different versions of post-human elitist value ideals
- •Only (2''): Different versions of trans-Sinitic value ideals
- •Only (2''): Different versions of trans-feminine value ideals
- •Only (2''): Different versions of trans-Islamic value ideals
- •Only (2''): Different versions of trans-outerspace value ideals

(continued on next page)

Table 4.33. Capitalism, Non-Capitalism, and Post-Capitalism (Part III)

Notes: The calling and examples in each category are solely illustrative (not exhaustive). The comparison is also relative (not absolute), nor are they mutually exclusive. As generalities, they allow exceptions. And the specific forms of each calling need to be further developed in future after-postmodern history, as they will be different from the ones we now know, while using them as the inspirational point at the beginning. The point here is to solely give an extremely rough picture of a small part of the world to come that we still do not know much about.
Source: From Ch.10 of *FCD*. Refer to the text for more info and references.

Table 4.34. Multiple Causes of the Emergence of Post-Capitalism (Part I)

- **At the Micro-Physical Level**
  —Ex: intelligent life without the human physical-chemical system
  —Ex: mastering of quantum mechanics, electromagnetism, and other fields for the understanding of a broad range of anomalous experiences and the application for artificial intelligence for spiritual quest
  —Sources: Ch.7 of *FHC*; Chs.9-10 of *FCD*; Ch.1 of *FPHC*

- **At the Chemical Level**
  —Ex: space radiation and toxins
  —Sources: Ch.7 of *FHC*; Chs.9-10 of *FCD*

- **At the Bio-Psychological Level**
  —Ex: exo-biological evolution in deep space
  —Ex: genetic engineering of new beings
  —Ex: limits of human cognition
  —Ex: the rise of unfolding unconsciousness
  —Sources: Ch.2 & Chs.9-10 of *FCD*; Ch.7 of *FHC*; *FPHU*

- **At the Institutional Level**
  —Ex: the flawed logic of the free market
  —Ex: the need of a post-autistic economics
  —Sources: Ch.10 of *FCD*

- **At the Organizational Level**
  —Ex: the dark sides of formal-legalistic routines
  —Sources: Ch.3 of *FHC*; Ch.7 of *FCD*; Ch.3 of *FPHC*

- **At the Structural Level**
  —Ex: ever new forms of inequities, at home and abroad
  —Ex: the emergence of China, women, and Islam as major actors
  —Sources: Chs.5-6 of *FHC*; Chs.7, 9 & 10 of *FCD*; Chs.4-5 of *BDPD*

(continued on next page)

Table 4.34. Multiple Causes of the Emergence of Post-Capitalism (Part II)

- **At the Cultural Level**
  —Ex: freedom/unfreedom dialectics
  —Ex: equality/inequality dialectics
  —Sources: Ch.5 of *FHC*; Chs.3 & 10 of *FCD*; Ch.4 of *FPHC*; Ch.1 of *BDPD*

- **At the Systemic Level**
  —Ex: space habitats (in zero-gravity) and colonization
  —Ex: ultra advanced future info systems
  —Ex: qualitative demography
  —Sources: Ch.7 of *FHC*; Chs. 9 & 10 of *FCD*

- **At the Cosmological Level**
  —Ex: the colonization of multiverses
  —Ex: the alteration of space-time and the creation of new matter-energy
  —Ex: the expansion of floating consciousness
  —Ex: the spread of hyper-spatial consciousness
  —Sources: Ch.7 of *FHC*; Chs. 9 & 10 of *FCD*; Ch.4 of *FPHC*; *FPHST*

Notes: The examples in each category are solely illustrative (not exhaustive), and some of the items can be reclassified somewhere else. Nor are they always mutually exclusive. Since they are generalities, exceptions are expected.
Sources: Especially from *FHC*, *FCD*, *FPHC*, *BCPC*, *BDPD*, *FPHST*, and *FPHU*. See also other books and my perspectives on civilizational holism.

Table 4.35. The Theoretical Debate on Civilization

- **The Progressive Theory of Civilization**
  —*Thesis*: The "civilizing" process is "good," as opposed to the "barbarizing" process as something "bad," relatively speaking.
  —*Discourse*: Especially, though not exclusively, in the Enlightenment era and a bit before. Example: Thomas Hobbes—in that the tribes in primitive societies were "savages."

- **The Romantic Theory of Barbarity**
  —*Thesis*: The "civilizing" process is "bad," as opposed to the "barbarizing" process as something "good," relatively speaking.
  —*Discourse*: Especially, though not exclusively, in the Counter-Enlightenment circle. Example: Jean-Jacques Rousseau—in that civilization "corrupts" men, and the "savages" are in fact "noble."

- **The Moderate Theory of Civilization**
  —*Thesis*: The "civilizing" process is "good," but there is a price to pay, especially in systematic (compulsive) self-control.
  —*Discourse*: Especially, though not exclusively, in some late modern and postmodern circles. Example: Norbert Elias—in that social manners become more refined in the civilizing process, but self-control also becomes more systematic.

- **The Theory of Post-Civilization**
  —*Thesis*: The civilizing process is as evil and good as barbarity, and each cannot exist without the other, to be eventually superseded by post-civilization unto the post-human age.
  —*Discourse*: Proposed by Peter Baofu. See the rest of *BCIV* for more analysis.

Source: From *BCIV* on the theoretical debate

Table 4.36. No Freedom Without Unfreedom in the Civilizing Processes (Part I)

- **The Rationalizing Process (at the Level of Culture)**
  —if freer from the dominance of unreason (as in barbarism) in the civilizing process, then less free from the rationalizing process (be it in the form of the principle of either transcendence or immanence)
  —if freer from the principle of immanence in the rationalizing process, then less free from the inclination to commit terror in the name of reason and the relative underdevelopment of non-reason (e.g., in relation to yoga and meditation)
  —if freer from the principle of transcendence in the rationalizing process, then less free from the relative underdevelopment of reason (e.g., in relation to systematic methodology) and the occurrence of oppression in the name of non-reason

- **The Pacifying Process (at the Level of Society)**
  —if freer from the dominance of pillage (as in savagery) in the civilizing process, then less free from the pacifying process (be it in the form of external control or self-control)
  —if freer from self-control in the pacifying process, then less free from the temptation of expansionist oppression and rebellious mindset in external control
  —if freer from external control in the pacifying process, then less free from the gruesome psychological self-torture and conformism in self-control

- **The Stewardizing Process (at the Level of Nature)**
  —if freer from the dominance of nature (as in the state of nature) in the civilizing process, then less free from the stewardizing process (be it in the form of the stewardship of creation or the covenant with nature)
  —if freer from the stewardship of creation in the stewardizing process, then less free from material underdevelopment, relatively speaking, and spiritual exclusion in the covenant with nature
  —if freer from the covenant with nature in the stewardizing process, then less free from ecological degradation and spiritual disconnection from nature in the stewardship of creation

(continued on next page)

Table 4.36. No Freedom Without Unfreedom in the Civilizing Processes (Part II)

- **The Subliming Process (at the Level of the Mind)**
  - —if freer from the dominance of spontaneity (as in the wild state of the mind) in the civilizing process, then less free from the subliming process, be it in the form of (cyclical-centric) self-refinement or (linear-centric) self-discipline
  - —if freer from (cyclical-centric) self-refinement in the subliming process, then less free from the (linear-centric) self-regimen (as a form of neurosis)
  - —if freer from (linear-centric) self-discipline in the subliming process, then less free from the (cyclical-centric) self-torture (equally as a form of neurosis)

Notes: The examples in each category are solely illustrative (not exhaustive), and the comparison is relative (not absolute), nor are they necessarily mutually exclusive. And some can be easily re-classified elsewhere. As generalities, they allow exceptions.
Sources: From *BCIV*. See also *FHC*, *FCD*, *FPHC*, *BDPD*, and *BCPC*.

Table 4.37. No Equality Without Inequality in the Civilizing Processes (Part I)

- **The Rationalizing Process (at the Level of Culture)**
  —if more equal for the role of rationalization in the rationalizing process (of civilizational making), then less equal for that of mythicization (as in barbarism)
  —if more equal for the principle of transcendence in (linear-centric) rationalizing process, then less equal for the principle of immanence
  —if more equal for the principle of immanence in (cyclical-centric) rationalizing process, then less equal for the principle of transcendence

- **The Pacifying Process (at the Level of Society)**
  —if more equal for pacification in civilizational making, then less equal for the institution of pillaging and others (as in savagery)
  —if more equal for external control, relatively speaking, in pacifying process, then less equal for self-control
  —if more equal for self-control, relatively speaking, in pacifying process, then less equal for external-control

- **The Stewardizing Process (at the Level of Nature)**
  —if more equal for stewardship in the stewardizing process (of civilizational making), then less equal for reverent (submissive) existence (as in barbarism)
  —if more equal for the stewardship of creation in (linear- centric) stewardizing process, then less equal for the (cyclical-centric) covenant with nature for harmonious co-existence
  —if more equal for the (cyclical-centric) covenant with nature in the stewardizing process, then less equal for the (linear-centric) stewardship of nature for domination

(continued on next page)

Table 4.37. No Equality Without Inequality in the Civilizing Processes (Part II)

- **The Subliming Process (at the Level of the Mind)**
  —if more equal for the role of reason in the subliming process, then less equal for that of unreason (as in the natural state of wildness)
  —if more equal for the primacy of reason in (linear-centric) subliming process, then less equal for other faculties (e.g., intuition, existential feelings, and analogous thinking) in cyclical-centric one
  —if more equal for the exercise of other faculties (e.g., intuition, existential feelings, and analogous thinking) in cyclical-centric subliming process, then less equal for the role of reason in linear-centric counterpart

Notes: The examples in each category are solely illustrative (not exhaustive), and the comparison is relative (not absolute), nor are they mutually exclusive. And some can be easily reclassified else-where. As generalities, they allow exceptions.
Sources: From *BCIV*. See also *FHC*, *FCD*, *FPHC*, *BDPD*, and *BCPC*.

Table 4.38. Five Theses on Post-Civilization

• Post-civilization no longer treats civilization as good and barbarity as evil (relatively speaking), nor does it nostalgically regard barbarity as good and civilization as evil (relatively speaking again). Civilization is as evil and good as barbarity.

• Post-civilization also no longer accepts the dichotomy between civilization and barbarity. Civilization cannot exist without barbarity. It is no longer necessary to preserve civilization, any more than it is imperative to destroy barbarity. To go beyond civilization and barbarity is to acknowledge the co-existence of both, although the degree of scaling one over the over varies from case to case—but is not to be extreme in largely favoring one over the other, *on average* (subject to the constraints of existential dialectics).

• Post-civilization is thus subject to the constraints of existential dialectics. There is no freedom without unfreedom, and no equality without inequality, for instance. There will be no utopia; even should there be one, there would be dystopia embedded within it.

• Post-civilization will eventually replace civilization (as a form of life settlement), to be dominated by post-capitalist and post-democratic lifeforms here on earth and in deep space (besides other alien lifeforms that we have never known), unto the post-human age in multiverses. Those few post-humans who keep civilization will live in a "post-human civilization," while the rest (the majority), who choose post-civilization, will evolve towards the state of "post-human post-civilization." One therefore should not confuse "post-human civilization" with "post-human post-civilization," as the two are not the same.

• Post-civilization will confront psychosis as a primary problem in the culture of virtuality unto the post-human age, just as civilization has neurosis as a primary one of its own (although both neurosis and psychosis are major problems in both).

Notes: The comparison in each category is relative (not absolute), nor are they necessarily mutually exclusive. And some can be easily re-classified elsewhere. As generalities, they allow exceptions.
Sources: From *BCIV*. See also *FHC*, *FCD*, *FPHC*, *BDPD*, and *BCPC*.

Table 4.39. Barbarity, Civilization, and Post-Civilization

- **The Rationalizing Process (at the Level of Culture)**
  —*Barbarity*
  - More mythicizing than rationalizing, relatively speaking

  —*Civilization*
  - More rationalizing than mythicizing, relatively speaking

  —*Post-Civilization*
  - Beyond the dichotomy, subject to existential dialectics

- **The Pacifying Process (at the Level of Society)**
  —*Barbarity*
  - More pillaging than pacifying, relatively speaking

  —*Civilization*
  - More pacifying than pillaging, relatively speaking

  —*Post-Civilization*
  - Beyond the dichotomy, subject to existential dialectics

- **The Stewardizing Process (at the Level of Nature)**
  —*Barbarity*
  - More revering than stewardizing, relatively speaking

  —*Civilization*
  - More stewardizing than revering, relatively speaking

  —*Post-Civilization*
  - Beyond the dichotomy, subject to existential dialectics

- **The Subliming Process (at the Level of the Mind)**
  —*Barbarity*
  - More impulsing than subliming, relatively speaking

  —*Civilization*
  - More subliming than impulsing, relatively speaking

  —*Post-Civilization*
  - Beyond the dichotomy, subject to existential dialectics

Notes: The comparison in each category is relative (not absolute), nor are they necessarily mutually exclusive. And some can be easily re-classified elsewhere. As generalities, they allow exceptions.
Sources: From *BCIV*. See also *FHC*, *FCD*, *FPHC*, *BDPD*, and *BCPC*.

Table 4.40. Types of Super Civilization in the Cosmos (Part I)

- **Type I**
  —a civilization which gains control of and uses the total energy output "falling on its planet from its sun for interstellar communication" (or, in general, space colonization). For N. Kardashev, who proposed the first three types, human civilization is currently Type Zero (Type O), which is below even Type I, since its present energy consumption for all purposes, let alone for interstellar communication, is still 10,000 times less.

- **Type II**
  —a civilization which gains control of and uses directly the total energy output of its sun for interstellar communication (or, in general, space colonization).

- **Type III**
  —a civilization which gains control of and uses the total energy output of its galaxy for interstellar communication (or, in general, space colonization).

- **Type IV**
  —a civilization which gains control of and uses the total energy output of its cluster of galaxies for interstellar communication (or, in general, space colonization).

- **Type V**
  —a civilization which gains control of and uses the total energy output of its supercluster of galaxies for interstellar communica-tion (or, in general, space colonization).

(continued on next page)

Table 4.40. Types of Super Civilization in the Cosmos (Part II)

- **Type…n**
  —So continues the series in what I call the cyclical progression of hegemony in the cosmos and beyond.

Notes: The Russian astrophysicist Nikolai Kardashev proposed the first three types of super civilization in terms of total energy out-put for interstellar communication. (CSM 1979) I extend his argument further to propose Type IV, Type V, Type VI, and Type…n, in the context of my claim about the cyclical progression of he-gemony in the cosmos and beyond.
Sources: From *Table 9.4* of *FCD*. See *FHC*, *FCD*, and *FPHC* for more info.

Table 4.41. The Civilizational Project from Pre-Modernity to After-Postmodernity (Part I)

	Pre-Modern	*Modern*	*Postmodern*	*After-Postmodern*
Main narratives	•Sacralness •Courtliness •Vitalism •Animism	•Freedom •Equality •Fraternity	•Multiplicity •Hybridization	•Naked contingency •Cyclical progression of hegemony
Main institutions	•Monarchy •Aristocracy •Feudalism •Holy order •Primitivism	•Capitalism •Liberalism •Socialism •Nazism •Fascism	•Capitalism •Liberalism •Postmodern politics of difference	•Post-Capitalism •Post-Democracy •Others
Main technological and economic revolutions	•Agricultural	•Service •Industrial	•Informational	•Biological •Material •Energy •Space •Others

(continued on next page)

Table 4.41. The Civilizational Project from Pre-Modernity to After-Postmodernity (Part II)

	Pre-Modern	*Modern*	*Postmodern*	*After-Postmodern*
Main agents	•Males •Upper strata •Mini-states	•Males •Upper strata •Whites •Empires	•Males •Upper strata •Whites •Others •Supra-states •IO's	•Post-humans •Humans •Others
Main impacts	•Local	•Inter-national	•Global	•Outer-space •Multiverse
Main outcomes	•Towards moderntiy •Rise of linear- & cyclical-centric civiliza-tions	•Towards post-modernity •Dominance of linear-centric civilization	•Towards after-post-moderntiy •Linear-centric civiliza-tion in crisis	•Towards human (& maybe post-human) extinction •Rise of post-civiliza-tion, especially in post-human forms of space-time

Notes: The examples in each category are solely illustrative (not exhaustive) nor necessarily mutually exclusive, and the comparison is relative (not absolute). As generalities, they allow exceptions.

Sources: From *Table 10.16* of *FCD*—and also from *BCIV* on post-civilization (and *FPHST*)

Table 4.42. Civilizational Holism
(Part I)

• **At the Micro-Physical Theoretical Level**
—Ex: Mastering of quantum mechanics, electromagnetism, and other fields for the understanding of a broad range of anomalous experiences and theapplication for artificial intelligence (*Sec.1.4.1* of *FPHC*)

• **At the Chemical Theoretical Level**
—Ex: Unprecedented expansion of (and violence to) the mind through ever new forms of drugs (and virtual technologies, for that matter) (Ch.9 of *FCD*)

• **At the Biological Theoretical Level**
—Ex: Humans are not biologically equal, on the basis of race, gender, ethnicity, age, and whatnot. (*Sec.2.6* & Ch.10 of *FCD*; *BNN*) And post-humans will experience the same fate, in an even more amazing way.

• **At the Psychological Theoretical Level**
—Ex: Human cognitive impartiality and emotional neutrality are quite limited. (*Secs.2.4-2.5* of *FCD*)
—Ex: Rise of Floating Consciousness (Ch.10 of *FCD*; Chs.1 & 4 of *FPHC*) and Unfolding Unconsciousness (*FPHU*)

• **At the Organizational Theoretical Level**
—Ex: Administrative colonization of deep space, with less legal-formalism in some corners. (Chs.9-10 of *FCD*)

• **At the Institutional Theoretical Level**
—Ex: Both capitalism and democracy will not last, to be superseded by different versions of post-capitalism and post-democracy in after-postmodernity. (Ch.10 of *FCD*)

(continued on next page)

Table 4.42. Civilizational Holism (Part II)

- **At the Structural Theoretical Level**
  —Ex: Social stratification reappears in ever new forms, also with new causes and new players in the cyclical progression of hegemony. (Chs.8-10 of *FCD*)
  —Ex: The world of hyper-empires, and the union of the unions (*BWT*)

- **At the Systemic Theoretical Level**
  —Ex: Outerspace expansion: local → regional → global → solar → galactic → clustery → multiversal (Ch.9 of *FCD*)
  —Ex: Demographic transition: human extinction, and the rise of post-humans (e.g., cyborgs, thinking machines, thinking robots, genetically altered superior beings, floating consciousness, hyper-spatial consciousness) (Ch.4 of *FPHC*; Ch.10 of *FCD*; & Ch.7 of *FHC*)
  —Ex: New technological forces in material sciences, electronic and communication sciences, energy sciences, biosciences, manufacturing and engineering sciences, and space sciences (Ch.10 of *FCD* & Ch.7 of *FHC*)
  —Ex: Systematic dominance towards nature for space colonization (Chs.9-10 of *FCD*; Chs.2 & 7 of *FHC*)

- **At the Cultural Theoretical Level**
  —Ex: The post-human transcendence of freedom and equality (Ch.10 of *FCD*)
  —Ex: Methodological Holism (Ch.1 of *FCD*; Ch.1 of *FPHC*; *Sec.2.1* & *Sec.2.5* of *BCPC*)
  —Ex: The Evolution from Barbarity to Post-Civilization (*BCIV*)

- **At the Cosmological Theoretical Level**
  —Ex: Mastering of dark matter and dark energy, and the exploration of multiverses (Ch.4 of *FPHC*; Ch.10 of *FCD*; & Ch.7 of *FHC*)
  —Ex: Alternation of space-time (*FPHST*)
  —Ex: The emergence of hyper-spatial consciousness (*FPHC*)

(continued on next page)

Table 4.42. Civilizational Holism (Part III)

- **At Other Levels**
  —Ex: Historical: pre-modernity → modernity → postmodernity → after-postmodernity (human distinction, and the rise of post-humans, including floating consciousness, hyper-spatial consciousness, and unfolding unconsciousness) (Ch.7 of *FHC*; Ch.10 of *FCD*; *FPHC*; *FPHU*)

Notes: These examples are solely illustrative (not exhaustive), and some of the items can be reclassified somewhere else. Nor are they always mutually exclusive. Since they are generalities, exceptions are expected. And the comparison is relative, not absolute.
Sources: From *Table 5.1* of *FPHC*—with details from *FHC*, *FCD*, and the rest of my books. Not every aspect in each category as discussed in all my books are presented here, since there are too many issues. For more info, also consult the table on theories on civilizational holism and, of course, the books themselves.

Table 4.43. Theories on Civilizational Holism (Part I)

I. Theories in Relation to Nature

—*At the Chemical Theoretical Level*

• 65. Creational Theory of Chemistry (Peter Baofu) (*FPHCHEM*)

(• 60). Resilient Theory of Natural Resources (Peter Baofu) (*BNR*)

—*At the Macro-Physical (Cosmological) Theoretical Level*

• 64. Resettlement Theory of Geology (Peter Baofu) (*FPHGEOL*)

• 63. Theory of Post-Cosmology (Peter Baofu) (*BCOS*)

• 62. Theory of Hyper-Spatial Consciousness (Peter Baofu) (Ch.4 of *FPHC*; *FPHG*)

(• 60). Resilient Theory of Natural Resources (Peter Baofu) (*BNR*)

(• 59). Panoramic Theory of Transportation (Peter Baofu) (*FPHTRAN*)

(• 57). Multilateral Theory of Acoustics (Peter Baofu) (*FPHA*)

(• 56). Selective Theory of Geometry (Peter Baofu) (*FPHG*)

(• 55). Perspectival Theory of Space-Time (Peter Baofu) (*FPHST*)

(• 54). Dialectic Theory of Complexity (Peter Baofu) (*FC*)

(• 31). Theory of Floating Consciousness (Peter Baofu) (Ch.10 of *FCD*; Chs.1 & 4 of *FPHC*)

• 61. Theory of the Geopower of Nature (Ch.4 of *ALD*)—refined as the Interventive-Reshaping Theory of Geography (Peter Baofu) (*FPHGEOG*)

(continued on next page)

Table 4.43. Theories on Civilizational Holism (Part II)

I. Theories in Relation to Nature (*cont'd*)

—*At the Micro-Physical Theoretical Level*

- 60. Resilient Theory of Natural Resources (Peter Baofu) (*BNR*)
- 59. Panoramic Theory of Transportation (Peter Baofu) (*FPHTRAN*)
- 58. Transcendent Theory of Architecture (Peter Baofu) (*FPHARCH*)
- 57. Multilateral Theory of Acoustics (Peter Baofu) (*FPHA*)
- 56. Selective Theory of Geometry (Peter Baofu) (*FPHG*)
- 55. Perspectival Theory of Space-Time (Peter Baofu) (*FPHST*)
- 54. Dialectic Theory of Complexity (Peter Baofu) (*FC*)

II. Theories in Relation to Culture

—*At the Cultural Theoretical Level*

- 53. Modificative Theory of Sports (Peter Baofu) (*FPHSPORT*)
- 52. Inquisitive Theory of Culinary Art (Peter Baofu) (*FPHCA*)
- 51. Interactive Theory of Semantics (Peter Baofu) (*FPHSEM*)
- 50. Transdicsiplinary Theory of Performing Arts (Peter Baofu) (*FPHPA*)
- 49. Comparative-Impartial Theory of Literature (Peter Baofu) (*FPHLIT*)

(continued on next page)

Table 4.43. Theories on Civilizational Holism (Part III)

II. Theories in Relation to Culture (*cont'd*)

—*At the Cultural Theoretical Level* (*cont'd*)

- • 48. Mediative-Variative Theory of Chess (Peter Baofu) (*FPHCHESS*)
- • 47. Theory of Post-Ethics (Peter Baofu) (*BEPE*)
- • 46. Dualistic Theory of Mass Culture (Peter Baofu) (Ch.2 of *FHC*)
- • 45. Comparative Theory of Religion—also known as the Comparative-Substitutive Theory of Religion (Peter Baofu) (Ch.3 of *FHC*; Ch.9 of *FCD*; Ch.1 of *FPHK*; *FPHR*)
- • 44. Theory of Post-Civilization (Peter Baofu) (*BCIV*)
- • 43. Theory of the Trinity of Modernity to Its After-Postmodern Counterpart (Peter Baofu) (*FHC*; Ch.10 of *FCD*)
- • 42. Transformative Theory of Aesthetic Experience (Peter Baofu) (*FAE*)
- (• 21). Theory of Post-Capitalism (Peter Baofu) (Ch.10 of *FCD*; Chs.2 & 4 of *FPHC*; *BCPC*)
- (• 20). Theory of Post-Democracy (Peter Baofu) (Ch.10 of *FCD*; Chs.3 & 4 of *FPHC*; *BDPD*)
- (• 5). Theory of Methodological Holism (Peter Baofu) (Ch.1 of *FCD*; Ch.1of *FPHC*; *Sec.2.1* & *Sec.2.5* of *BCPC*; *FC*; *FPHK*; etc.)

(continued on next page)

Table 4.43. Theories on Civilizational Holism (Part IV)

II. Theories in Relation to Culture (*cont'd*)

—*At the Cultural Theoretical Level* (*cont'd*)

(• 4). Theory of Existential Dialectics,
or the Holistic Theory of Knowledge (Peter Baofu)
(*FHC*; *FCD*; *FPHC*; *BDPD*; *FC*; *FAE*; *ALD; FIA; FPHK*; etc,)

(• 3). Contrastive Theory of Rationality (Peter Baofu)
(*FPHML*)

III. Theories in Relation to the Mind

—*At the Biological Theoretical Level*

• 41. Theory of Contrastive Advantages (Peter Baofu)
(*Sec.2.6* & Ch.10 of *FCD; BNN*)

(• 31).Theory of Floating Consciousness (Peter Baofu)
(Ch.10 of *FCD*; Chs.1 & 4 of *FPHC*)

—*At the Psychological Theoretical Level*

• 40. Complex Theory of Data Analysis (Peter Baofu)
(*FPHDA*)

• 39. Supersession Theory of Computing (Peter Baofu)
(*FPHCOM*)

• 38. Metamorphic Theory of Humor (Peter Baofu)
(*FPHH*)

• 37. Contrarian Theory of Personality (Peter Baofu)
(*FPHP*)

• 36. Theory of Virtual Sexuality (Peter Baofu)
(*FPHS*)

• 35. Expansive-Contractive Theory of Martial Arts (Peter Baofu)
(*FPHMA*)

• 34. Multilogical Theory of Learning (Peter Baofu)
(*FPHL*)

• 33. Comprehensive Theory of Creative Thinking (Peter Baofu)
(*FPHCT*)

(continued on next page)

Table 4.43. Theories on Civilizational Holism (Part V)

III. Theories in Relation to the Mind (*cont'd*)

—At the Psychological Theoretical Level (*cont'd*)

- 32. Theory of Unfolding Unconsciousness—also known as the Unfolding Theory of Anomalous Experience (Peter Baofu) (*FPHU*)
- 31. Theory of Floating Consciousness (Peter Baofu) (Ch.10 of *FCD*; Chs.1 & 4 of *FPHC*)
- 30. Theory of Cognitive Partiality (Peter Baofu) (*Sec.2.4* of *FCD*; *Sec.4.5.1.1* of *BCPC*)
- 29. Theory of Emotional Non-Neutrality (Peter Baofu) (*Sec.2.5* of *FCD*; *Sec.4.5.2* of of *BCPC*)
- 28. Theory of Behavioral Alteration (Peter Baofu) (*Sec.4.5.3* of *BCPC*)

IV. Theories in Relation to Society

—At the Organizational Theoretical Level

- 27. Theory of E-Civic Alienation (Peter Baofu) (Ch.7 of *FCD*)
- 26. Combinational Theory of Organization (Peter Baofu) (*FPHO*; Ch.6 of *ALD*)

—At the Institutional Theoretical Level

- 25. Reflective Theory of Criminality (Peter Baofu) (*FPHCRIM*)
- 24. Detached Theory of Gambling (Peter Baofu) (*FPHGAM*)
- 23. Heterodox Theory of Education (Peter Baofu) (*FPHEDU*)
- 22. Reconstruction Theory of Law (Peter Baofu) (*FPHLAW*)
- 21. Theory of Post-Capitalism (Peter Baofu) (Ch.10 of *FCD*; Chs.2 & 4 of *FPHC*; *BCPC*)
- 20. Theory of Post-Democracy (Peter Baofu) (Ch.10 of *FCD*; Chs.3 & 4 of *FPHC*; *BDPD*)

(continued on next page)

Table 4.43. Theories on Civilizational Holism (Part VI)

IV. Theories in Relation to Society (*cont'd*)

—*At the Institutional Theoretical Level (cont'd)*

- 19. Dynamic Theory of Comparative Political Systems (Peter Baofu) (*ALD*)
- (• 9.) Theory of the Cyclical Progression of Empire-Building (Peter Baofu) (*BWT*)

—*At the Systemic Theoretical Level*

- 18. Transfigurative Theory of Waste (Peter Baofu) (*FPHW*)
- 17. Theory of the Cyclical Progression of Migration (Peter Baofu) (*FPHMIG*)
- (• 64). Resettlement Theory of Geology (Peter Baofu) (*FPHGEOL*)
- 16. Contingent Theory of Urban Planning (Peter Baofu) (*FPHUP*)
- 15. Totalistic Theory of Communication (Peter Baofu) (*FPHMM*; *FCD*; *FHC*)
- 14. Ambivalent Theory of Technology (Peter Baofu) (*FPHE*; *FCD*; *FHC*)
- 13. Multifaceted Theory of War and Peace (Peter Baofu) (Ch.9 of *FCD*; Ch.1 of *FPHK*)
- 12. Theory of Post-Humanity (Peter Baofu) (Ch.7 of *FHC*; Chs.3, & 10 of *FCD*; Chs.1, 3 & 4 of *FPHC*; and other books of mine)
- 11. Theory of the Cyclical Progression of System Integration and Fragmentation (Peter Baofu) (Chs.9-10 of *FCD*)
- 10. Synthetic Theory of Information Architecture (Peter Baofu) (*FIA*)

—*At the Structural Theoretical Level*

- 9. Theory of the Cyclical Progression of Hegemony (Peter Baofu) (Chs.9-10 of *FCD*; Chs.1, 3 & 4 of *FPHC*; *BDPD*)
- 8. Theory of the Cyclical Progression of Empire-Building (Peter Baofu) (*BWT*)

(continued on next page)

Table 4.43. Theories on Civilizational Holism (Part VII)

V. Meta-Theories (in Relation to Theories)

—*At the Methodological Meta-Theoretical Level*

(• 51).Interactive Theory of Semantics (Peter Baofu) (*FPHSEM*)

(• 40). Complex Theory of Data Analysis (Peter Baofu) (*FPHDA*)

• 7. Interpretivist Theory of Probability (Peter Baofu) (*FPHROB*)

• 6. Critical-Dialectic Theory of Formal Science (Peter Baofu) (*FPHFS*)

• 5. Theory of Methodological Holism (Peter Baofu) (Ch.1 of *FCD*; Ch.1of *FPHC*; *Sec.2.1* & *Sec.2.5* of *BCPC*; *FC*; *FPHK*; etc.)

—*At the Ontological Meta-Theoretical Level*

• 4. Theory of Existential Dialectics, or the Holistic Theory of Knowledge (Peter Baofu) (*FHC*; *FCD*; *FPHC*; *BDPD*; *FC*; *FAE*; *ALD; FIA; FPHK*; etc,)

• 3. Contrastive Theory of Rationality (Peter Baofu) (*FPHML*)

VI. Theories in Relation to the Rest

—*At the Meta-Historical Level*

• 2. Multifold Theory of History (Peter Baofu) (*FPHHIST*)

—*At the Empirical-Historical Level*

• 1. Theory of the Evolution from Pre-Modernity to After-Postmodernity (Peter Baofu) (*FHC*; Ch.9-10 of *FCD*; *FPHC*)

Notes: All these theories are my constructions, as some of the main contributions of my grant project on civilization and its future. These examples are solely illustrative (not exhaustive), and some of the items can be reclassified somewhere else. Nor are they always mutually exclusive. Since they are generalities, exceptions are expected.

Sources: From my previous books.

Table 4.44. Three Great Future Transformations of Mind Games

- **Virtual Games**
  —Ex: online chess
  —Ex: virtual experience

- **Novel Games**
  —Ex: new chess variants
  —Ex: new chess engines

- **Post-Human Mind Games**
  —Ex: the quest for broader/deeper mental benefits of chess playing
  —Ex: games designed for the evolution of the mind into different body-less forms (e.g., "floating consciousness," "hyper-spatial consciousness," "unfolding unconsciousness")

Notes: The examples in the categories are solely illustrative (not exhaustive), and the comparison is relative (not absolute), nor are they necessarily mutually exclusive. And some can be easily re-classified elsewhere. As generalities, they allow exceptions.
Sources: From *Sec. 4.16* of *FPHCHESS*. See text for more info. Also, consult *FCD* for strategy and tactics in warfare and *FPHO* for strategy and tactics in organization.

BIBLIOGRAPHY

Abrams, William. 2003. *A Brief History of Probability*, <http://www.secondmoment.org/articles/probability.php>.

Aschbacher, Michael. 2004. "The Status of the Classification of the Finite Simple Groups." Notices of the American Mathematical Society. <http://www.ams.org/notices/200407/fea-aschbacher.pdf>.

Banfield, Edward C. 1957. "The Decision-Making Schema." *Public Administration Review*, 17 (4): 278-285.

Baofu, Peter. 2013. *The Future of Post-Human Thanatology: Towards a New Theory of Death and After-Death*. Berlin, Germany: Logos Verlag Berlin.

_____.2013a. *The Future of Post-Human Health Care: Towards a New Theory of Mind and Body*. Hauppauge, NY: Nova Science Publishers, Inc.

_____.2013b. *The Future of Post-Human Morphology: Towards a New Theory of Typologies and Rules*. Hauppauge, NY: Nova Science Publishers, Inc.

_____.2013c. *The Future of Post-Human Probability: Towards a New Theory of Objectivity and Subjectivity*. Hauppauge, NY: Nova Science Publishers, Inc.

_____.2013d. *The Future of Post-Human Sports: Towards a New Theory of Training and Winning*. Cambridge, England: Cambridge Scholars Publishing.

_____.2013e. *Beyond Human Resources to Post-Human Resources: Towards a New Theory of Diversity and Discontinuity*. Cambridge, England: Cambridge Scholars Publishing.

_____.2012. *The Future of Post-Human Waste: Towards a New Theory of Uselessness and Usefulness*. Cambridge, England: Cambridge Scholars Publishing.

_____.2012a. *The Future of Post-Human Culinary Art: Towards a New Theory of Ingredients and Techniques*. Cambridge, England: Cambridge Scholars Publishing.

_____.2012b. *The Future of Post-Human Transportation: A Preface to a New Theory of Networks and Operations*. Cambridge, England: Cambridge Scholars Publishing.

_____.2012c. *The Future of Post-Human Migration: A Preface to a New Theory of Sameness, Otherness, and Identity*. Cambridge, England: Cambridge Scholars Publishing.

_____.2012d. *The Future of Post-Human History: A Preface to a New Theory of Universality and Relativity*. Cambridge, England: Cambridge Scholars Publishing.

_____.2012e. *The Future of Post-Human Criminality: A Preface to a New Theory of Heroes and Villains*. Cambridge, England: Cambridge International Science Publishing, Ltd.

_____.2012f. *The Future of Post-Human Architecture: A Preface to a New Theory of Form and Function*. Cambridge, England: Cambridge International Science Publishing, Ltd.

_____.2012g. *The Future of Post-Human Semantics: A Preface to a New Theory of Internality and Externality*. Cambridge, England: Cambridge Scholars Publishing.

_____.2012h. *The Future of Post-Human Performing Arts: A Preface to a New Theory of the Body and its Presence*. Cambridge, England: Cambridge Scholars Publishing.

_____.2011. *The Future of Post-Human Geography: A Preface to a New Theory of Environments and their Interactions*. Cambridge, England: Cambridge International Science Publishing, Ltd.

_____.2011a. *The Future of Post-Human Data Analysis: A Preface to a New Theory of Quantitative and Qualitative Research*. Cambridge, England: Cambridge International Science Publishing, Ltd.

_____.2011b. *The Future of Post-Human Chemistry: A Preface to a New Theory of Substances and their Changes*. Cambridge, England: Cambridge Scholars Publishing.

_____.2011c. *The Future of Post-Human Literature: A Preface to a New Theory of Fiction and Non-Fiction*. Cambridge, England: Cambridge International Science Publishing, Ltd.

_____.2011d. *The Future of Post-Human Computing: A Preface to a New Theory of Hardware, Software, and the Mind*. Cambridge, England: Cambridge International Science Publishing, Ltd.

_____.2011e. *The Future of Post-Human Gambling: A Preface to a New Theory of Risk and Caution*. Cambridge, England: Cambridge International Science Publishing, Ltd.

_____.2011f. *The Future of Post-Human Acoustics: A Preface to a New Theory of Sound and Silence*. Cambridge, England: Cambridge International Science Publishing, Ltd.

_____.2011g. *The Future of Post-Human Humor: A Preface to a New Theory of Joking and Laughing*. Cambridge, England: Cambridge International Science Publishing, Ltd.

_____.2011h. *Beyond Ethics to Post-Ethics: A Preface to a New Theory of Morality and Immorality*. Charlotte, NC: Information Age Publishing.

_____.2011i. *The Future of Post-Human Education: A Preface to a New Theory of Teaching and Learning*. Cambridge, England: Cambridge International Science Publishing, Ltd.

_____.2010. *The Future of Post-Human Religion: A Preface to a New Theory of Spirituality*. Cambridge, England: Cambridge International Science Publishing, Ltd.

_____.2010a. *The Future of Post-Human Chess: A Preface to a New Theory of Tactics and Strategy*. Cambridge, England: Cambridge International Science Publishing, Ltd.

_____.2010b. *The Future of Post-Human Geology: A Preface to a New Theory of Statics and Dynamics*. Cambridge, England: Cambridge International Science Publishing, Ltd.

_____.2010c. *Beyond Cosmology to Post-Cosmology: A Preface to a New Theory of Different Worlds*. Cambridge, England: Cambridge International Science Publishing, Ltd.

_____.2010d. *The Future of Post-Human Personality: A Preface to a New Theory of Normality and Abnormality*. Cambridge, England: Cambridge International Science Publishing, Ltd.

_____.2010e. *The Future of Post-Human War and Peace: A Preface to a New Theory of Aggression and Pacificity*. Cambridge, England: Cambridge Scholars Publishing.

_____.2010f. *The Future of Post-Human Formal Science: A Preface to a New Theory of Abstraction and Application*. Cambridge, England: Cambridge Scholars Publishing.

_____.2010g. *The Future of Post-Human Law: A Preface to a New Theory of Necessity, Contingency, and Justice*. Cambridge, England: Cambridge Scholars Publishing.

_____.2010h. *The Future of Post-Human Sexuality: A Preface to a New Theory of the Body and Spirit of Love Makers*. Cambridge, England: Cambridge Scholars Publishing.

_____,2009. *The Future of Post-Human Martial Arts: A Preface to a New Theory of the Body and Spirit of Warriors*. Cambridge, England: Cambridge Scholars Publishing.

_____.2009a. *The Future of Post-Human Organization: A Preface to a New Theory of Communication, Decision-Making, and Leadership*. Cambridge, England: Cambridge Scholars Publishing.

_____.2009b. *The Future of Post-Human Language: A Preface to a New Theory of Structure, Context, and Learning*. Cambridge, England: Cambridge Scholars Publishing.

_____.2009c. *The Future of Post-Human Urban Planning: A Preface to a New Theory of Density, Void, and Sustainability*. Cambridge, England: Cambridge

Scholars Publishing.

______.2009d. *The Future of Post-Human Geometry: A Preface to a New Theory of Infinity, Symmetry, and Dimensionality*. Cambridge, England: Cambridge Scholars Publishing.

______.2009e. *The Future of Post-Human Creative Thinking: A Preface to a New Theory of Invention and Innovation*. Cambridge, England: Cambridge Scholars Publishing.

______.2009f. *The Future of Post-Human Mass Media: A Preface to a New Theory of Communication*. Cambridge, England: Cambridge Scholars Publishing.

______.2009g. *The Future of Post-Human Engineering: A Preface to a New Theory of Technology*. Cambridge, England: Cambridge Scholars Publishing.

______.2008. *The Future of Post-Human Mathematical Logic: A Preface to a New Theory of Rationality*. Cambridge, England: Cambridge Scholars Publishing.

______.2008a. *The Future of Post-Human Knowledge: A Preface to a New Theory of Methodology and Ontology*. Oxford, United Kingdom: Chandos Publishing (Oxford) Ltd.

______.2008b. *The Future of Post-Human Unconsciousness: A Preface to a New Theory of Anomalous Experience*. Cambridge, England: Cambridge Scholars Publishing.

______.2008c. *The Future of Information Architecture: Conceiving a New Way to Understand Taxonomy, Network, and Intelligence*. Oxford, United Kingdom: Chandos Publishing (Oxford) Ltd.

______.2007. *The Rise of Authoritarian Liberal Democracy: A Preface to a New Theory of Comparative Political Systems*. Cambridge, England: Cambridge Scholars Publishing.

______.2007a. *The Future of Aesthetic Experience: Conceiving a Better Way to Beauty, Ugliness, and the Rest*. Cambridge, England: Cambridge Scholars Publishing.

______.2007b. *The Future of Complexity: Conceiving a Better Way to Understand Order and Chaos*. London, United Kingdom: World Scientific Publishing Co.

______.2007c. *Beyond the World of Titans, and the Remaking of World Order: A Preface to a New Logic of Empire-Building*. Cambridge, England: Cambridge Scholars Publishing.

______.2006. *Beyond Nature and Nurture: Conceiving a Better Way to Understand Genes and Memes*. Cambridge, England: Cambridge Scholars Publishing.

______.2006a. *The Future of Post-Human Space-Time: Conceiving a Better Way to Understand Space and Time*. New York: Peter Lang Publishing, Inc.

______.2006b. *Beyond Civilization to Post-Civilization: Conceiving a Better Model of Life Settlement to Supersede Civilization*. New York: Peter Lang

Publishing, Inc.

_____.2005. *Beyond Capitalism to Post-Capitalism: Conceiving a Better Model of Wealth Acquisition to Supersede Capitalism.* New York: The Edwin Mellen Press.

_____.2004. *The Future of Post-Human Consciousness.* New York: The Edwin Mellen Press.

_____.2004a. *Beyond Democracy to Post-Democracy: Conceiving a Better Model of Governance to Supersede Democracy.* Volume 1. New York: The Edwin Mellen Press.

_____.2004b. *Beyond Democracy to Post-Democracy: Conceiving a Better Model of Governance to Supersede Democracy.* Volume 2. New York: The Edwin Mellen Press.

_____.2002. *The Future of Capitalism and Democracy.* Maryland: The University Press of America.

_____.2000. *The Future of Human Civilization.* Volume 1. New York: The Edwin Mellen Press.

_____.2000a. *The Future of Human Civilization.* Volume 2. New York: The Edwin Mellen Press.

Barrouillet, P. et al. 2000. "Conditional Reasoning by Mental Models: Chronometric and Developmental Evidence." *Cognit.*, 75: 237-266.

Barrow, John D. 1998. *Impossibility: The Limits of Science and the Science of Limits.* Oxford: Oxford University Press.

Batley, Richard; & Andrew Daly. 2006. "On the Equivalence between Elimination-by-Aspects and Generalised Extreme Value Models of Choice Behaviour." *Journal of Mathematical Psychology*, 50 (5): 456–467.

Bell Jr., Whitfield J., ed. 1956. *Mr. Franklin: A Selection from His Personal Letters.* New Haven, CT: Yale University Press.

Ben-David, Joseph; & Teresa A. Sullivan. 1975. "Sociology of Science." *Annual Review of Sociology*, 1 (1): 203–222.

Benveniste, Guy. 1994. *Mastering the Politics of Planning.* San Francisco, CA: Jossey-Bass.

Bernardo, José M.; & Adrian Smith. 1994. *Bayesian Theory.* NY: John Wiley.

Biletzki, Anat; & Anat Matar. 2006. "Ludwig Wittgenstein" (Nov. 17). *Stanford Encyclopedia of Philosophy.* <http://plato.stanford.edu/entries/wittgenstein/>.

Bishop, C. M. 2007. *Pattern Recognition and Machine Learning.* Springer.

Boghossian, Paul. 2006. *Fear of Knowledge: Against Relativism and Conmstructivism.* England: Oxford University Press.

Bohr, Niels. 1958. *Atomic Physics and Human Knowledge.* New York: Wiley.

Bracken, Stacey Storch; & Janet E. Fischel. 2006. "Assessment of Preschool

Classroom Practices: Application of Q-Sort Methodology." *Early Childhood Research Quarterly* (Elsevier) 21 (4): 417–430. <http://www.sciencedirect.com/science/article/pii/S0885200606000652>.

Brandt, Richard B. 1959. Ethical Theory: The Problems of Normative and Critical Ethics. NY: Prentice-Hall.

Bremermann, H. J. 1977. "Transcomputability and Complexity." *The Encyclopaedia of Ignorance*. M. Smith & R. Duncan, eds. London: Routledge and Keagan Paul.

Buckley, Maureen E.; Daniel Klein; C. Emily Durbin; Elizabeth Hayden; & Kirstin Moerk. 2002. "Development and Validation of a Q-Sort Procedure to Assess Temperament and Behavior in Preschool-Age Children." *Journal of Clinical Child & Adolescent Psychology*, 31 (4): 525–539.

Bunge, Mario. 1999. *Dictionary of philosophy*. Amherst, MA: Prometheus Books.

Byrne, R. M. J.; & P. Johnson-Laird. 2009. "'If' and the Problems of Conditional Reasoning." *Trends in Cognitive Sciences*, 13: 282–28.

_____.2005. *The Rational Imagination: How People Create Counterfactual Alternatives to Reality*. Cambridge, MA: MIT Press.

Cassidy, David. 2012. "Quantum Mechanics: 1925-1927: The Uncertainty Principle" (December 06). *American Physical Society*. <http://www.aip.org/history/heisenberg/p08.htm>.

_____.2009. *Beyond Uncertainty: Heisenberg, Quantum Physics, and the Bomb*. New York: Bellevue Literary Press.

Chase, Scott. 1993. "Tachyons" (March). *Physics FAQ*. <http://math.ucr.edu/home/baez/physics/ParticleAndNuclear/tachyons.html>.

Cheng, P. C. & K. Holyoak. 2008. "Pragmatic Reasoning Schemas." *Reasoning: Studies of Human Inference and its Foundations*. J. Adler and L. Rips, eds. England: Cambridge University Press.

Cohen, Michael D. 2007. "Administrative Behavior: Laying the Foundations for Cyert and March." *Organization Science*, 18 (3): 503–506.

Collins, H. M.; & S. Yearley. 1992. "Epistemological Chicken." *Science as Practice and Culture*. A. Pickering, ed. Chicago, IL: Chicago University Press.

Cooke, M. 1994. *Language and Reason: A Study in Habermas's Pragmatics*. Cambridge, MA: MIT Press.

Corazzon, Raul. 2007. "Definitions of Ontology: From Nicolai Hartmann to the Present Time." <http://www.formalontology.it/section_4_two.htm>.

Corbett, Dan. 2008. "Introduction to String Theory" (November 23). *ThinkQuest*. <http://library.thinkquest.org/27930/stringtheory1.htm>.

Cosmides, L. et al. 2005. "Detecting Cheaters." *Trends in Cognitive Sciences*, 9: 505–506.

Cox, D. R. 2006. *Principles of Statistical Inference*. England: Cambridge

University Press.

Cox, Richard T. 2001. *Algebra of Probable Inference*. Baltimore, MD: The Johns Hopkins University Press.

Daston, Lorraine; & Peter Galison. 2007. *Objectivity*. New York: Zone Books.

Davies, P. C. W. 2004. "Does Quantum Mechanics Play a Non-Trivial Role in Life?" *BioSystems*, 78: 69–79.

de Finetti, B. 1974. *Theory of Probability*. New York: J. Wiley & Sons, Inc.

DeGroot, Morris. 1970. *Optimal Statistical Decisions*. New York: McGraw-Hill.

Dupré, Maurice J.; & Frank Tipler. 2009. "New Axioms For Bayesian Probability," *Bayesian Analysis*, 3: 599-606.

Eagleton, M., ed. 2003. *A Concise Companion to Feminist Theory*. Oxford, England: Blackwell Publishing.

Elster, C. 2007. "Calculation of Uncertainty in the Presence of Prior Knowledge." *Metrologia*, 44: 111-116.

Etzioni, A. 1988. "Normative-Affective Factors: Towards a New Decision-Making Model." *Journal of Economic Psychology*, 9: 125–150.

Everett, Hugh. 1957. "'Relative State' Formulation of Quantum Mechanics" (July). *Reviews of Modern Physics*, vol. 29, no.3. <http://www.univer.omsk.su/omsk/Sci/Everett/paper1957.html>.

Famous-Quotes.com (FQ). 2012. "Stephen Chow" (December 10). <http://www.1-famous-quotes.com/quotes/category/Chinese+Actor>.

Feller, W. 1968. *An Introduction to Probability Theory and its Applications*. Volume 1. New York: Wiley.

Finetti, Bruno de. 1970. "Logical Foundations and Measurement of Subjective Probability." *Acta Psychologica*, 34: 129–145.

Firehammer, Reginald. 2005. "Basic Principles of Ontology" (January 05). <http://usabig.com/autonomist/philosophy/ontology2.html>.

Ford, David N.; & John Sterman. 2009. "Expert Knowledge Elicitation to Improve Mental and Formal Models". Cambridge, MA: Massachusetts Institute of Technology. <http://web.mit.edu/jsterman/www/ford_sterman_elicit_1.pdf>.

Freund, John. 1993. *Introduction to Probability*. New York: Dover.

Garfinkel, Harold. 2002. *Ethnomethodology's Program*. New York, NY: Rowman and Littlefield.

Golembiewski, Robert T. 1988. "Perspectives on Simon's Administrative Behavior: Stock-Taking on the Fortieth Anniversary—Part I." *Public Administration Quarterly*, 12 (3): 259-382.

Gott, J. Richard. 2002. *Time Travel in Einstein's Universe*. Boston : Houghton Mifflin.

Gow, James Iain. 2003. "Decision Man: Herbert Simon in Search of Rationality." *Canadian Public Administration/Administration Publique du Canada*, 46 (1): 120–126.

Greiner, Walter; & Berndt Müller. 1994. *Quantum Mechanics Symmetries*. Berlin, Germany: Springer-Verlag.

Guth, Alan. 1997. *The Inflationary Universe: The Quest for a New Theory Of Cosmic Origins*. New York: Helix Books.

Guthrie, Kenneth. 1988. *The Pythagorean Sourcebook and Library: An Anthology of Ancient Writings which Relate to Pythagoras and Pythagorean Philosophy*. Grand Rapids, MI: Phanes Press.

Habermas, J. 1987. *The Theory of Communicative Action. Vol 2: Lifeworld and System: A Critique of Functionalist Reason*. T. McCarthy, tr. Boston, MA: Beacon Press.

_____.1984. *The Theory of Communicative Action. Volume 1, Reason and the Rationalization of Society*. Cambridge, England: Polity Press.

_____.1979. *Communication and the Evolution of Society*. Toronto, Canada: Beacon Press.

Hacking, I. 2006. *The Emergence of Probability: A Philosophical Study of Early Ideas about Probability, Induction and Statistical Inference*. England: Cambridge University Press.

_____.1967. "Slightly More Realistic Personal Probability." *Philosophy of Science*, 34 (4): 311–325.

_____.1965. *The Logic of Statistical Inference*. England: Cambridge University Press.

Halpern, J. 1999. "A Counterexample to Theorems of Cox and Fine." *Journal of Artificial Intelligence Research*, 10: 67-85.

Hamming, R. W. 1980, "The Unreasonable Effectiveness of Mathematics." *The American Mathematical Monthly*, 87 (2).

Heidegger, Martin. 1971. "A Dialogue on Language between a Japanese and an Inquirer." *On the Way to Language*. New York: Harper & Row.

Hicks, Stephen. 2004. *Understanding Postmodernism: Skepticism and Socialism from Rousseau to Foucault*. Tempe, AZ: Scholargy.

Hinntikka, Jaakko. 1962. *Knowledge and Belief*. New York: Cornell University Press.

Hofstadter, Douglas. 1979. *Gödel, Escher, Bach*. New York: Basic Books.

Hogg, Robert V.; Allen Craig; & Joseph McKean. 2004. *Introduction to Mathematical Statistics*. Upper Saddle River: Pearson.

Hunter, Geoffrey. 1971. *Metalogic: An Introduction to the Metatheory of Standard First-Order Logic*. CA: University of California Press.

Hurley, Patrick. 2007. *A Concise Introduction to Logic*. Stamford, CT: Wadsworth Publishing.

Husserl, Edmund. 1970. *The Crisis of the European Sciences and Transcendental Phenomenology*. Evanston, IL: Northwestern University Press.

Imamichi, Tomonubu. 2006. *In Search of Wisdom. One Philosopher's Journey*. Tokyo, Japan: International House of Japan.

Ivancevic, Vladimir G.; & Tijana T. Ivancevic. 2008. *Quantum Leap: From Dirac and Feynman, across the Universe, to Human Body and Mind*. Hackensack, NJ: World Scientific.

Jacobs, Jane. 1961. *The Death and Life of Great American Cities*. New York: Random House.

Jarvie, Ian Charles; Karl Milford; & David Miller. 2006. *Karl Popper: A Centenary Assessment*. Burlington, VT: Ashgate Publishing.

Jay, Martin. 1982. "Should Intellectual History Take a Linguistic Turn? Reflections on the Habermas-Gadamer Debate." *Modern European Intellectual History: Reappraisals and New Perspectives*. Steven Kaplan & Dominick LaCapra, eds. Ithaca, NY: Cornell University Press.

Jaynes, E.T. 2003. *Probability Theory: The Logic of Science*. England: Cambridge University Press.

_____.1986. "Bayesian Methods: General Background." *Maximum-Entropy and Bayesian Methods in Applied Statistics*. J. H. Justice, ed. Cambridge, England: Cambridge Univ. Press.

Johnson, Neil F. 2007. *Two's Company, Three is Complexity: A Simple Guide to the Science of all Sciences*. Oxford: Oneworld.

Johnson-Laird, P. N. 2006. *How We Reason*. England: Oxford University Press.

_____.& R. Byrne. 2002. "Conditionals: A Theory of Meaning, Inference, and Pragmatics." *Psychol. Rev.*, 109: 646–678.

Joyce, James. 2003. "Bayes' Theorem" (June 28). *Stanford Encyclopedia of* Philosophy. <http://plato.stanford.edu/entries/bayes-theorem/>.

Kaku, Michio. 1994. *Hyperspace: A Scientific Odyssey Through Parallel Universes, Time Warps, and The Tenth Dimension*. Illustrations by Robert O'Keefe. Oxford: Oxford University Press.

Kassian, Mary A. 2005. *The Feminist Mistake: The Radical Impact of Feminism on Church and Culture*. England: Crossway.

Kompridis, Nikolas. 2010. "So We Need Something Else for Reason to Mean." *International Journal of Philosophical Studies*, 8: 3, 271-295.

Krantz, Steven George. 2005. *Mathematical Apocrypha Redux: More Stories and Anecdotes of Mathematicians and the Mathematical*. Washington, DC: Mathematical Association of America.

LaCapra, Dominick. 1977. "Habermas and the Grounding of Critical Theory." *History and Theory*.

Larmore, Charles. 1987. Patterns of Moral Complexity. Cambridge, MA: Cambridge University Press.

Leon-Garcia, Alberto. 2008. *Probability, Statistics and Random Processes for*

Electrical Engineering. Upper Saddle River, NJ: Pearson.

Lukas, Carrie L. 2006. *The Politically Incorrect Guide to Women, Sex, and Feminism*. Washington, DC: Regnery Publishing.

McCarthy, Thomas. 1978. *The Critical Theory of Jurgen Habermas*. Cambridge, MA: The MIT Press.

Marenbon, John. 2007. Medieval Philosophy: An Historical and Philosophical Introduction. UK: Routledge.

McAleer, Neil. 1987. *The Mind-Boggling Universe*. New York: Doubleday & Co.

McCarthy, Thomas. 1973. "A Theory of Communicative Competence." *Philosophy of the Social Sciences: An International Journal* (June).

McGrayne, Sharon Bertsch. 2011. *The Theory That Would Not Die*. New Haven, CT: Yale University Press.

McNeeley, Ian. 2003. *The Emancipation of Writing*. Berkeley, CA: Univ. of California Press.

Merriam-Webster's Collegiate Dictionary Online (MWD). 2012. "Objectivity" (October 28). <http://www.m-w.com/dictionary/objectivity>.

_____2012a. "Subjectivity" (October 28). <http://www.m-w.com/dictionary/Subjectivity>.

_____2007. "Agency" (March 20). <http://www.m-w.com/dictionary/agency>.

Mosterín, Jesús. 2008. *Lo Mejor Posible: Racionalidad y Acción Humana*. Madrid: Alianza Editorial.

_____.2002. "Acceptance Without Belief." *Manuscrito*, XXV: 313–335.

Natanson, M. 1973. *Edmund Husserl: Philosopher of Infinite Tasks*. Evanston, IL: Northwestern University Press.

Nova Online (NO). 2000. "Time Can Vary." <http://www.pbs.org/wgbh/nova/time/thinktime.html>.

O'Brien, D. 2009. "Human Reasoning Requires a Mental Logic." *Behav. Brain Sci.*, 32: 96–97.

Oaksford, M.; & N. Chater. 2007. *Bayesian Rationality*. England: Oxford University Press.

Oberauer K. 2006. "Reasoning with Conditionals: A Test of Formal Models of Four Theories." *Cognit. Psychol.*, 53: 238–283.

Parkes, Graham, ed. 1987. *Heidegger and Asian Thought*. Honolulu, Hawaii: University of Hawaii Press.

Patai, Daphne; & Noretta Koertge. 2003. *Professing Feminism: Education and Indoctrination in Women's Studies*. Lanham, MD: Lexington Books.

Paulos, John Allen. 2011. "The Mathematics of Changing Your Mind" (August 05). *The New York Times*

Popper, Karl. 1959. *The Logic of Scientific Discovery*. London: Hutchinson & Co.

Postmes, T; Russell Spears; & Sezgin Cihangir. 2001. "Quality of Decision Making and Group Norms." *Journal of Personality and Social Psychology*:

80 (6): 918–930.

Previte, Josephine; Barbara Pini; & Fiona Haslam-McKenzie. 2007. "Q Methodology and Rural Research." *Sociologia Ruralis* (European Society for Rural Sociology) 47 (2): 135–147.

Prior, Arthur N. 1967. *Past, Present and Future*. Oxford: Clarendon.

_____.1955. *Formal Logic*. Oxford: Clarendon Press.

Putnam, Hilary. 1985. *Mathematics, Matter and Method. Philosophical Papers*, vol. 1. Cambridge, England: Cambridge University Press.

_____& Paul Benacerraf, eds. 1983. *Philosophy of Mathematics, Selected Readings*. Cambridge, UK: Cambridge University Press.

_____.1975. *Mind, Language, and Reality. Philosophical Papers*, Volume 2. Cambridge, England: Cambridge University Press.

Raadgever, G. T.; E. Mostert; & N. van de Giesen. 2008. "Identification of Stakeholder Perspectives on Future Flood Management in the Rhine Basin Using Q Methodology." *Hydrology and Earth System Sciences*, 12 (4): 1097-1109. <http://adsabs.harvard.edu/abs/2008HESS...12.1097R>.

Raman, Shankar. 1997. Lectures on Literary Theory. Cambridge, MA: MIT (spring).

Reason, James. 1990. *Human Error*. UK: Ashgate.

Rescher, Nicholas. 1971. *Temporal Logic*. New York, Springer-Verlag.

_____.1968. *Topics in Philosophical Logic*. New York: Humanities Press.

Rollinger, Robin. 1999. *Husserl's Position in the School of Brentano*.London: Kluwer.

Safranski, R. 1998. *Martin Heidegger: Between good and evil*. Cambridge, MA: Harvard University Press.

Schlafly, Phyllis. 1977. *The Power of the Positive Woman*. New York: Arlington House Publishers.

Shlain, Leonard. 1991. *Art and Physics: Parallel Visions in Space,Time, and Light*. New York: William Morrow and Co., Inc.

Schroyens, W. et al. 2003. "In Search of Counterexamples: Deductive Rationality in Human Reasoning." *Quart. J. Exp. Psychol.*, 56(A): 1129–1145.

Shewhart, Walter A. 2012. "Statistical Inference." *Surfstat.australia: An Online Text in Introductory Statistics*. <http://web.archive.org/web/20061007055524/http://www.anu.edu.au/nceph/surfstat/surfstat-home/5-1-2.html>.

_____.1931. *Economic Control of Quality of Manufactured Product*. New York: D. Van Nostrand Company, Inc.

Simonson, I.; & A. Sela. 2011. "On the Heritability of Consumer Decision Making: An Exploratory Approach for Studying Genetic Effects on Judgment and Choice." *Journal of Consumer Research*, 37 (6): 951-966.

Smith, David Woodruff. 2007. *Husserl*. London, England: Routledge.

Sokal, Alan. 1996. "A Physicist Experiments With Cultural Studies." *Lingua Franca*. <http://www.physics.nyu.edu/faculty/sokal/lingua_franca_v4/lingua_franca_v4.html>.

_____.1996a. "Transgressing the Boundaries: Towards a Transformative Hermeneutics of Quantum Gravity." *Social Text*, 46/47 (spring/summer). <http://www.physics.nyu.edu/faculty/sokal/transgress_v2/transgress_v2_singlefile.html>.

Sokolowski, Robert. 2000. *Introduction to Phenomenology*. Cambridge, England: Cambridge University Press.

Solomon, Robert C. 2005. "Subjectivity." *Oxford Companion to Philosophy*. Ted Honderich. England: Oxford University Press.

Sommers, Christina Hoff. 1995. *Who Stole Feminism?: How Women Have Betrayed Women*. New York: Simon & Schuster.

Stanford Encyclopedia of Philosophy (SEP). 2012. "Ancient Logic" (November 05). <http://plato.stanford.edu/entries/logic-ancient/>.

Stigler, Stephen M. 1986. *The History of Statistics*. Cambridge, MA: Harvard University press.

_____.1978. "Mathematical Statistics in the Early States." *Annals of Statistics*, 6: 239–265.

Suchman, Lucy. 2007. *Human-Machine Reconfigurations: Plans and Situation Action*. Cambridge, England: Cambridge University Press.

Teixeira, Joao. 1998. "Computational Complexity and Philosophical Dualism." *Philosophy and Cognitive Science*. Paper originally presented at the Twentieth World Congress of Philosophy in Boston. <http://www.bu.edu/wcp/Papers/Cogn/CognTeix.htm>.

ThinkExist.com (TE). 2012. "Ken Auletta" (October 30). <http://en.thinkexist.com/search/searchquotation.asp?search=objectivity&page=2>.

_____.2012a. "Berenice Abbott" (December 03). <http://en.thinkexist.com/search/searchquotation.asp?search=subjective>.

Tulip, R. J. 2012. "Complexity and Christian Faith: A Fractal Theology" (November 02). <http://rtulip.net/yahoo_site_admin/assets/docs/Complexity_Faith.16004252.htm>.

Tversky, Amos; & Shmuel Sattath. 1979. "Preference Trees." *Psychological Review*, 86 (6): 542–573.

_____. 1972. "Elimination by Aspects: A Theory of Choice." *Psychological Review*, 79 (4): 281–299.

Ulanowicz, Robert. 1997. *Ecology, the Ascendant Perspective*. New York: Columbia University Press.

van Frassen, B. 1989. *Laws and Symmetry*. England: Oxford University Press.

Verschueren, N. et al. 2005. “Everyday Conditional Reasoning: A Working Memory-Dependent Tradeoff between Counterexample and Likelihood Use.” *Mem. Cognit.*, 33: 107-119.

Wigner, Eugene. 1960. “The Unreasonable Effectiveness of Mathematics in the Natural Sciences.” *Communications on Pure and Applied Mathematics*, 13.

Wald, Abraham. 1950. *Statistical Decision Functions*. NY: Wiley.

Weaver, Warren. 1948. “Science and Complexity.” *American* Scientist, 36 (4): 536–44.

Weber, Max. 1930. *The Protestant Ethic and the Spirit of Capitalism*. Talcott Parsons, tr. New York: Scribner.

Weise, K.; & W. Wöger. 1992. “A Bayesian Theory of Measurement Uncertainty.” *Meas. Sci. Technol.*, 3: 1-11, 4.8.

Western Electric Company (WEC). 1956. *Introduction to Statistical Quality Control handbook*. Indianapolis, Indiana: Western Electric Co..

Wikipedia (WK). 2012. “Probability Interpretations” (October 26). <http://en.wikipedia.org/wiki/Probability_interpretations>.

_____.2012a. “Frank P. Ramsey” (October 26). <http://en.wikipedia.org/wiki/Frank_P._Ramsey>.

_____.2012b. “Probability” (October 28). <http://en.wikipedia.org/wiki/Probability>.

_____.2012c. “Probability Theory” (October 28). <http://en.wikipedia.org/wiki/Probability_theory>.

_____.2012d. “Random Variable” (October 28). <http://en.wikipedia.org/wiki/Random_variable>.

_____.2012e. “Stochastic Process” (October 28). <http://en.wikipedia.org/wiki/Stochastic_process>.

_____.2012f. “Randomness” (October 28). <http://en.wikipedia.org/wiki/Randomness>.

_____.2012g. “Event (Probability Theory)” (October 28). <http://en.wikipedia.org/wiki/Event_(Probability_theory)>.

_____.2012h. “Probability Distribution” (October 28). <http://en.wikipedia.org/wiki/Probability_distribution>.

_____.2012i. “Objectivity (Science)” (October 28). <http://en.wikipedia.org/wiki/Objectivity_(science)>.

_____.2012j. “Objectivity (Philosophy)” (October 28). <http://en.wikipedia.org/wiki/Objectivity_(philosophy)>.

_____.2012k. “Subjectivity” (October 28). <http://en.wikipedia.org/wiki/Subjectivity>.

_____.2012l. “Subjectivity (Philosophy)” (October 28). <http://en.wikipedia.org/

wiki/Subjectivity_(philosophy)>.

_____.2012m. “Subjective” (October 28). <http://en.wikipedia.org/wiki/Subjective>.

_____.2012n. “Bayesian Probability” (October 28). <http://en.wikipedia.org/wiki/Bayesian_probability>.

_____.2012o. “Reasoning” (October 28). <http://en.wikipedia.org/wiki/Reasoning>.

_____.2012p. “Mental Model” (October 29). <http://en.wikipedia.org/wiki/Mental_model>.

_____.2012q. “Dice” (October 29). <http://en.wikipedia.org/wiki/Dice>.

_____.2012r. “Uncertainty Principle” (October 29). <http://en.wikipedia.org/wiki/Uncertainty_principle>.

_____.2012s. “Quantum Mechanics” (October 29). <http://en.wikipedia.org/wiki/Quantum_mechanics>.

_____.2012t. “Optimal Decision” (October 29). <http://en.wikipedia.org/wiki/Optimal_decision>.

_____.2012u. “Satisfying” (October 29). <http://en.wikipedia.org/wiki/Satisfying>.

_____.2012v. “Rationality” (October 29). <http://en.wikipedia.org/wiki/Rationality>.

_____.2012w. “Feminism” (October 30). <http://en.wikipedia.org/wiki/Feminism>.

_____.2012x. “Common Cause and Special Cause” (October 30). <http://en.wikipedia.org/wiki/Common_cause_and_special_cause>.

_____.2012y. “Cultural Consensus Theory” (October 30). <http://en.wikipedia.org/wiki/Cultural_consensus_theory>.

_____.2012z. “Discrete” (November 01). <http://en.wikipedia.org/wiki/Discrete>.

_____.2012aa. “Discrete Group” (November 01). <http://en.wikipedia.org/wiki/Discrete_group>.

_____.2012bb. “Dice” (November 01). <http://en.wikipedia.org/wiki/Dice>.

_____.2012cc. “Normal Distribution” (November 01). <http://en.wikipedia.org/wiki/Normal_distribution>.

_____.2012dd. “Central Limit Theorem” (November 01). <http://en.wikipedia.org/wiki/Central_limit_theorem>.

_____.2012ee. “Discrete Logarithm” (November 01). <http://en.wikipedia.org/wiki/Discrete_logarithm>.

_____.2012ff. “Talk: Probability Theory” (November 01). <http://en.wikipedia.org/wiki/Talk:_probability_theory>.

_____.2012gg. “Classical Definition of Probability” (November 02).

<http://en.wikipedia.org/wiki/Classical_definition_of_probability>.

_____.2012hh. “Mo Divination” (November 02). <http://en.wikipedia.org/wiki/Mo_divination>.

_____.2012ii. “Complexity” (November 02). <http://en.wikipedia.org/wiki/Complexity>.

_____.2012jj. “Cybernetics” (November 02). <http://en.wikipedia.org/wiki/Cybernetics>.

_____.2012kk. “Reliability Theory” (November 03). <http://en.wikipedia.org/wiki/Reliability_theory>.

_____.2012ll. “Reliability Engineering” (November 03). <http://en.wikipedia.org/wiki/Reliability_engineering>.

_____.2012mm. “Consumer Protection” (November 03). <http://en.wikipedia.org/wiki/Consumer_protection>.

_____.2012nn. “Organizational Behavior” (November 04). <http://en.wikipedia.org/wiki/Organizational_behavior>.

_____.2012oo. “Decision Making” (November 04). <http://en.wikipedia.org/wiki/Decision_making>.

_____.2012pp. “Rational Planning Model” (November 04). <http://en.wikipedia.org/wiki/Rational_planning_model>.

_____.2012qq. “Heuristic” (November 04). <http://en.wikipedia.org/wiki/Heuristic>.

_____.2012rr. “Administrative Behavior” (November 04). <http://en.wikipedia.org/wiki/Administrative_behavior>.

_____.2012ss. “Bounded Rationality” (November 04). <http://en.wikipedia.org/wiki/Bounded_rationality>.

_____.2012tt. “Propositonal Calculus” (November 05). <http://en.wikipedia.org/wiki/Propositional_calculus>.

_____.2012uu. “Incompleteness Theorem” (November 05). <http://en.wikipedia.org/wiki/Incompleteness_theorem>.

_____.2012vv. “Psychology of Reasoning” (November 05). <http://en.wikipedia.org/wiki/Psychology_of_reasoning>.

_____.2012ww. “History of Logic” (November 05). <http://en.wikipedia.org/wiki/History_of_logic>.

_____.2012xx. “Q Methodology” (December 04). <http://en.wikipedia.org/wiki/Q_methodology>.

_____.2012yy. “William Stephenson (Psychologist)” (December 04). <http://en.wikipedia. org/wiki/William_Stephenson_(Psychologist)>.

_____.2012zz. “Factor Analysis” (December 04). <http://en.wikipedia. org/wiki/Factor_analysis>.

_____.2012aaa. "Measurement Uncertainty" (December 05). <http://en.wikipedia. org/wiki/Measurement_uncertainty>.

_____.2012bbb. "Metrology" (December 05). <http://en.wikipedia. org/wiki/ Metrology>.

_____.2012ccc. "Uncertainty" (December 06). <http://en.wikipedia. org/wiki/ Uncertainty>.

_____.2012ddd. "Many-Worlds Interpretation" (December 06). <http://en.wikipedia. org/wiki/Many-worlds_interpretation>.

_____.2012eee. "Classical Mechanics" (December 06). <http://en.wikipedia. org/wiki/Classical_mechanics>.

_____.2012fff. "Universal Pragmatics" (December 07). <http://en.wikipedia. org/wiki/Universal_pragmatics>.

_____.2012ggg. "Communicative Action" (December 07). <http://en.wikipedia. org/wiki/Communicative_action>.

_____.2012hhh. "Sociology of Scientific Knowledge" (December 08). <http://en.wikipedia.org/wikiSociology_of_scientific_knowledge/ >.

_____.2012iii. "Mathematical Practice" (December 08). <http://en.wikipedia.org/ wiki/Mathematical_practice >.

_____.2012jjj. "Quasi-Empiricism in Mathematics" (December 08). <http://en.wikipedia.org/wiki/Quasi-empiricism_in_mathematics>.

_____.2012kkk. "Folk Mathematics" (December 08). <http://en.wikipedia.org/ wiki/Folk_mathematics>.

_____.2012lll. "Mathematical Folklore" (December 08). <http://en.wikipedia.org/ wiki/Mathematical_folklore>.

_____.2012mmm. "Informal Mathematics" (December 08). <http://en.wikipedia. org/wiki/Informal_mathematics>.

_____.2012nnn. "Philosophy of Mathematics" (December 08). <http://en.wikipedia.org/wiki/Philosophy_of_mathematics>.

_____.2012ooo. "Social Constructionism" (December 08). <http://en.wikipedia.org/ wiki/Social_constructionism>.

_____.2012ppp. "Phenomenology (Philosophy" (December 09). <http://en.wikipedia.org/ wiki/Phenomenology_(philosophy)>.

_____.2012qqq. "Ethnomethodology (Philosophy" (December 09). <http://en.wikipedia.org/ wiki/Ethnomethodology>.

_____.2012rrr. "Subject (Philosophy)" (December 10). <http://en.wikipedia.org/ wiki/Subject_(philosophy)>.

_____.2012sss. "Double Bind" (December 10). <http://en.wikipedia.org/ wiki/Double_bind>.

_____.2011nn. "Design Elements and Principles" (September 30).

<http://en.wikipedia.org/wiki/Design_elements_and_principles>.

_____.2011oo. "Interpretation (Logic)" (April 13). <http://en.wikipedia.org/wiki/Interpretation_(Logic)>.

_____.2011pp. "Unsolved Problems in Linguistics" (February 20). <http://en.wikipedia.org/wiki/Unsolved_problems_in_linguistics>.

_____.2008. "Posthumanism" (October 20). <http://en.wikipedia.org/wiki/Posthumanism>.

_____.2008a. "Transhumanism" (October 20). <http://en.wikipedia.org/wiki/Transhumanism>.

_____.2008b. "Lowest Common Denominator" (March 26). <http://en.wikipedia.org/ wiki/Least_common_denominator>.

_____.2008c. "Infinity" (November 11). <http://en.wikipedia.org/wiki/Infinity>.

_____.2008d. "Transfinite Number" (November 10). <http://en.wikipedia.org/wiki/Transfinite_ number>.

_____.2008e. "Quantum Mechanics" (November 26). <http://en.wikipedia.org/wiki/Quantum_mechanics>.

_____.2008f. "Copenhagen Interpretation" (November 26). <http://en.wikipedia.org/wiki/Copenhagen_interpretation >.

_____.2008g. "Many Worlds Interpretation" (November 26). <http://en.wikipedia.org/wiki/Many_Worlds_Interpretation>.

_____.2008h. "Hyperspace (Science Fiction)" (November 23). <http://en.wikipedia.org/wiki/Hyperspace_(science_fiction)>.

_____.2008i. "Time Travel" (November 26). <http://en.wikipedia.org/wiki/Time_travel#Time_dilation>.

_____.2008j. "Wormhole" (November 25). <http://en.wikipedia.org/wiki/Wormhole>.

_____.2007. "Linguistics" (March 04). <http://en.wikipedia.org/wiki/Linguistics>.

_____.2006. "Self-Organization." <http://en.wikipedia.org/wiki/Self_organization>.

Wikiversity (WKV). 2008. "Why 10 Dimensions" (November 23). <ttp://en.wikiversity.org/wiki/Why_10_dimensions>.

Wittgenstein, Ludwig.1953. *Philosophical Investigations*. G. E. M. Anscombe and R. Rhees, eds. G.E.M. Anscombe, tr. Oxford: Blackwell.

Wolpert, R. L. 2004. "A Conversation with James O. Berger." *Statistical Science*, 9: 205–218.

Yablo, S. 1998. "A Paradox of Existence" (November 08). <http://www.mit.edu/~yablo/apex.html#fn1>.

Zahavi, Dan. 2003. *Husserl's Phenomenology*, Stanford, CA: Stanford University Press.

Zayed, J. M.; N. Nouvel; U. Rauwald; & O. A. Scherman. 2010. "Chemical Complexity—Supramolecular Self-assembly of Synthetic and Biological Building Blocks in Water." *Chemical Society Reviews*, 39: 2806–2816. <http://pubs.rsc.org/en/Content/ArticleLanding/2010/CS/b922348g>.

INDEX

•A•

•B•

•C•

•D•

•E•

•F•

•G•

•H•

•I•

•J•

•K•

•N•

•O•

•P•

•Q•

•R•

•S•

•T•

•U•

•V•

•W•

•X•

•Y•

•Z•